The Landscapes of Dieter Kienast

Anette Freytag

The Landscapes of Dieter Kienast

With photographs
by Georg Aerni
and Christian Vogt

gta Verlag

7 Preamble
8 Foreword
10 More Than Just Gray

I
17 The Nature of the City: Traces of Kassel and the Beauty of Weeds

22 A New Type of Green Urban Planner
26 A New Theory of Planning Open Spaces: The Kassel School
30 Lucius Burckhardt's Critique of Planning
38 Kienast and the Kassel School
56 The Third Stocktaking of the World
62 The Natural and the Artificial: The Dry Grassland Biotope in Basel
78 Sociopolitical Allegories
85 Vegetation and Use: The École cantonale de langue française in Bern
112 New Images of Urban Nature
120 Processes of Reinterpretation: The Front Yard and Rear Courtyard for Ernst Basler + Partner in Zurich
133 Conditions for Designing Urban Open Space: The Garden Courtyard for the Schweizerische Rückversicherungs-Gesellschaft in Zurich

II
157 Forms of Use: Aesthetic Experience and Coping with Everyday Life

161 "Using Open Space Means Coping with Everyday Life": Kienast's First Design for the Grounds of a Housing Development
182 Form as the Antithesis of the Natural Garden: The Brühlwiese Municipal Park in Wettingen
219 Transparency and Collage: Making City and Countryside Legible
245 One's Own Garden as a Field for Experimentation
265 Mourning and the Experience of Nature: Consolation through Beauty as a Way of Coping with Everyday Life

III
293 Drawing and Perceiving: Media of Representation

298 From Zoning to Designing
306 "This Thing We Call Art": Dieter Kienast and the Phenomenology of Postmodernism—Art, Architecture, Literature
320 A Vocabulary for the Landscape I: Shaping the Terrain
341 A Vocabulary for the Landscape II: Drawing and Representation as Processes for Understanding Topography, Form, and Material
367 Mechanisms of Representation: Texts, Photographs, Exhibitions, and Video

403 Ten Theses on Landscape Architecture
407 Biography
408 List of Works
415 Bibliography
427 Illustration Credits
428 Index
430 Acknowledgements

*Qui cache son fou,
meurt sans voix.*

Henri Michaux,
L'Espace du dedans

Preamble

This book is the first comprehensive monograph on Dieter Kienast, the pioneering Swiss landscape architect of the late twentieth century.

Dieter Kienast was the most influential landscape architect of his generation in Switzerland, in terms of design practice, landscape innovation, and education. He studied at the Gesamthochschule Kassel during the 1970s, in a period that was paramount to the rise of the environmental movement, which rapidly became the dominant ideology in landscape architecture circles across German-speaking Europe. In her book, Anette Freytag pinpoints specifically how this shift in ideology affected Dieter Kienast's own personal views and later work. After completing his doctoral dissertation on phytosociology at the Gesamthochschule Kassel, faithful to the profoundly iconoclastic spirit of the times, he elevated the study of common weeds to an art form. In 1980, Kienast created one of the most admired "natural" landscapes in Switzerland by artificial means: the Dry Grassland Biotope at the Grün 80 horticultural exhibition in the Botanical Garden of Basel-Brüglingen, which remains to this day a true masterpiece of "designed ecology."

Behind the ecologist, however, stood the author, artist, and teacher Dieter Kienast. By laying out his entire professional and academic career, Anette Freytag shows the complexity and contradictions of the period with regard to landscape aesthetics—where his attempt to reconcile the form and history of landscape with ecological processes was a constant concern. Kienast experimented with collage and drawing early on in his work, for Kassel in the 1970s was not only the seat of radical reform in the education of landscape architects; it was also the crucible of documenta 6 in 1977, probably one of the most significant artistic manifestations of that decade, where countless artists, filmmakers, and photographers—from Joseph Beuys to Michael Heizer—exhibited their works. The show influenced Dieter Kienast profoundly, and the juxtaposition of artistic freedom and ecological rigor became the credo of his entire oeuvre as a designer and of later projects by his firms.

From the early 1980s to the mid-1990s, Dieter Kienast consolidated his design practice through two successive partnerships, working together with the most talented young Swiss architects of his time on some of the most significant projects of that decade. Moreover, he invested considerable energy in landscape education reform during that same period at three successive schools, of which the ETH Zurich—Eidgenössische Technische Hochschule (Swiss Federal Institute of Technology) would be the final venue of his academic career before his tragic death in 1998.

Dieter Kienast's work escapes any simple definition and cannot be easily categorized stylistically. He came from a family of gardeners in Zurich, completed first an apprenticeship in gardening, and preserved his direct and modest attitude toward cultivated land throughout his entire life. Even the way he worked at the drafting table, holding a smoldering pipe in one hand and carefully applying graphite to his drawings with the other, recalls the unrelenting gestures that a gardener repeats day after day on his or her own patch of land. Anette Freytag has produced a true milestone in that her research has reinstated the figure of Dieter Kienast in the history of contemporary European landscape architecture, addressing many questions that remain open still today. She describes his life as one of personal development that is exemplary for the era, in which creative talent, hope, and scholarly issues come together to explore the possible role of nature in our cities. Undoubtedly, this book is bound to fill a significant gap in our understanding of landscape architecture, and in the related teaching activity, for generations to come.

Christophe Girot

Foreword

It is no easy undertaking to write a critical study on Dieter Kienast and his work, especially since he was an unorthodox, critical person and a designer of great influence. He was also an artist who promoted himself as a brand through his plans, texts, exhibitions, photographs, and videos. Moreover, the untimely death of this extraordinary human being also created its own myths. Remembering him thus poses a series of obstacles that can obscure an unclouded and honest perspective.

His oeuvre can only be studied by examining in detail the meaning of garden and landscape architecture as a discipline. How may we develop an understanding that ventures beyond dualistic constructs and brings to light the multilayered, polyvalent complexities dictated by life itself? Garden and landscape architecture is a true and vital cultural discipline. Thus, a study of the work of Dieter Kienast must fuse his personal and professional approach with themes not only of garden and landscape design but also of culture and nature at large. It means shedding light on the value of his work and its contribution to emancipating this profession, and on how professional and social conditions influenced Kienast and his decisions. In this context, it is extremely important to investigate the relationship between art and science. Kienast's reflections on nature and landscape, space and form, materials and craft, typology and social objectives are just as crucial. What exactly did he see as the role of thought and conceptualization, and their mediation through drawing, representation, and writing?

Kienast's life spanned the eventful second half of the twentieth century, representing a period of profound transformation in the profession of design, whose history and significance have been studied more with an eye to urban planning and architecture than in the context of garden design and landscape. Revealing the premises of the timeliness of garden and landscape architecture as part of national and international developments thus offers numerous insights into the founding and evolution of this profession. The science of phytosociology—the subject of Kienast's dissertation in Kassel—has, for example, almost never been regarded as an integral part of the inherent identity of modern garden and landscape architecture. For Kienast and others, however, it was an essential factor in bridging design and science, and in developing new approaches to nature inside and outside of urban contexts.

A study such as the present one demands great skill, originality, and also methodology in order to develop an approach that is at once critical and engaging. Talent is a prerequisite for doing research that will assemble all the necessary documentation for analysis, while at the same time it must attempt a fundamental interpretation and cut deep to the essence of a personality, an oeuvre, a discipline, and a culture of thinking, understanding, reflecting, and acting. In this respect, Anette Freytag has developed a thematically productive and informative approach that not only creates a monument to Dieter Kienast but also recapitulates the history of developments in the garden and landscape architecture of his time—and, as the reader will discover, well beyond. Her approach is firmly rooted in the study of archival and published sources, thus ensuring that the related findings are new and substantiated. History may be of value and interest as a past era, but even beyond studying the past there is a realm in which a life lived in a visionary way, an accomplished oeuvre, will remain a crucial touchstone for the profession and for culture today. History is not closed off, because it continues to have an effect, often implicitly, through our deeds and actions; it lives on beyond the stereotypes that often shape our understanding.

Anette Freytag's study is being published at the right moment, because today we have to think about a profession that has been experiencing a second wave of emancipation since the 1980s and 1990s. It is manifested at a time when society is confronted with great challenges in what nature, landscape, and city must mean for human culture in a deeply changing world. The Anthropocene exhorts us to contribute to the poignant themes and problems that arise from our human needs and interventions concerning the environment. Landscape architecture as a cultural discipline is not about naively "beautifying" our lifeworld while evading discussions of identity and memory, places and sites, or on ecology and biodiversity, nature and culture. Its perspectives demand public awareness, a designing and writing of public history, and a public critique so that we become clear about the reflection, necessity, and aesthetics meant to be expressed by garden, landscape, and urban space.

The text of this book may thus serve as a theoretical and cultural beacon that can provide orientation to the profession and to all those interested in the role of designing with nature as a whole—with an eye to reflection, evaluation, and self-exploration. It ties together the significance of design, writing, and representation in the context of Dieter Kienast's private and professional life and opens up a window to the artistic, intellectual, scholarly, and social discipline of garden and landscape architecture. The author explains the significance of this discipline on its own accord, but always in relation to other fields and to culture at large. The life of plants, the composition of materials, and ultimately our sensory experiences—at their core reflecting an essence of true aesthetics—are all at the heart of Kienast's work. He translated them into the space as place. These themes are still relevant to our current environment and are closely linked, in the end, to our individual selves. We have to discover anew that we cannot speak about identity and the meaning of life without developing a true relationship with the city, landscape, and nature in connection with our own culture. The relevance of Anette Freytag's study on Dieter Kienast thus transcends both time and place.

Erik A. de Jong

More Than Just Gray

"Die mit den Förmchen spielen" (Playing with Little Forms) was the title given by the landscape architect Reto Mehli to a text in 1996, two years before Dieter Kienast's death, signifying a kind of settling of accounts with the "prodigal son" of the Kassel School. It was a contribution to a Festschrift for the phytosociologist and landscape planner Karl Heinrich Hülbusch, the spiritual father of the Kassel School and Dieter Kienast's doctoral advisor. In it, Mehli pointed to a "transformation of the 'formal idiom'" in garden architecture and lamented the "development of the 'stage model' as a new muddle/model."[1] He argued that the practical aspect of open spaces, which was the primary concern of the program in landscape planning at the Gesamthochschule Kassel, no longer played a role.[2] Garden architects, he wrote, had begun to see themselves as "set designers," whose works were "a space for experience and a world of backdrops" at the very same time: "By agreement with their clients, garden architects design open space images intended to be artistic in the sense of 'symbolic capital,' as a luxury good for upper-income groups."[3] Kienast, too, was said to be this sort of corrupt "player with forms," who "to judge from his success in competitions [is] probably the best-known garden architect in Switzerland, and can also point to international successes."[4]

Mehli accused Kienast of placing himself above other human beings in his park designs, arguing that he ignored their socioeconomic conditions in favor of a purely abstract design. He wrote that the result was semiotically hyper-coded spaces that were neither intelligible to nor practical for their users: "Or who carries shopping bags home on a zigzag course?"[5] Mehli denied the open spaces designed by Kienast any social relevance—and their users the competence to derive any aesthetic value from them: "Signs of use become signs of aesthetics that can no longer be decoded via a learned and familiar code."[6] Mehli discredited Kienast's formal idiom as useless by declaring it to be an abstruse puzzle. He ignored the phenomenological dynamic inherent in the viewer's relationship to open space. Mehli saw Kienast's gardens and parks as mere semiotic games—not theatrical stages for spiritual experience.

The narrow-mindedness of this point of view may have contributed to the Kassel School largely being ignored in Kienast's posthumous reception. That already began in the obituaries of 1999: Udo Weilacher did not mention Kassel; Peter Wullschleger mentioned Kienast's studies and dissertation there without further comment; Lisa Diedrich at least regarded Kienast's phytosociological dissertation as a propaedeutic for his later works.[7] Peter Paul Stöckli, Kienast's first employer and firm partner for many years, emphasized the latter's studies and dissertation on the spontaneous vegetation of the city of Kassel, which made Kienast one of the first landscape architects in Switzerland with a doctorate, only to describe his turn away from a scholarly career in favor of the practice of planning.[8] Later, summary assessments, such as that of Marc Treib,[9] by contrast, entirely ignore Kienast's beginnings. Individual studies concentrate on the work of the late 1980s and especially the 1990s. Kienast's contribution to the landscape design of housing developments in the 1970s is ignored entirely, as is the genesis of the municipal park in Wettingen or the relationship of use, form, and spontaneous vegetation.

Just as the Kassel School's vituperation made it blind to Kienast's aesthetic, his supporters demonstratively rejected Kassel in the reception of his work. This mutual ignoring can be explained, first, by the contrary cultural contexts of Kienast's oeuvre: whereas the early works were created in the context of the socially committed and sociocritical student movement,[10] his later clients were distinguished private parties and wealthy companies. Over the years, the leftist rebel—who seemed "to embody absolutely the unwieldy quality of his 'pugnacious planning culture,' a giant with a constantly tousled shock of hair, in his long pullover, unimaginable without his heavy tobacco pipe, seemingly always coming directly from the front lines of the garden"[11]—evolved into a *homme de lettres* of the Swiss architectural scene,[12] whose astuteness overshadowed what remained his basic political stance. Second, a new generation of landscape architects had elevated Dieter Kienast to its "role model":[13] They believed that his turn to form-oriented concepts represented a break with the teachings of the Kassel School and above all with the *Naturgarten* (natural garden) movement, which was a powerful influence on landscape architecture in German-speaking Europe. Kienast provided the discipline with a new self-confidence, since he was a recognized partner with renowned Swiss architectural firms and won prestigious competitions in Europe.

Kienast himself also contributed, consciously or unconsciously, to his connection to the Kassel School being forgotten. The presentation of his works in black-and-white photographs by Christian Vogt, in panorama and square photographic formats, which Dieter Kienast and Günther Vogt initiated, became canonical as a result of the four monographs that appeared between 1997 and 2004.[14] With their obligatory black borders, they turn Kienast's gardens into abstract works of art.[15] Although leaving the edge of the negative is a reminiscence of photojournalism, it also expresses the fact that a work of art was being created in parallel. The cool gray shades transported the sites portrayed out of the gray of everyday reality. The socially committed élan of Kienast's early days seemed to give way to pure elegance: the sites seemed "inaccessible."[16] But the impression was deceiving: Kienast is more than gray. His works preserve continuity with the ideas of the Kassel School.[17]

Rather than approaching Dieter Kienast's oeuvre in the classic structure for a monograph, this book will examine his work through three prisms: "The Nature of the City," "Forms of Use," and "Drawing and Perceiving." These are the leitmotifs of the three chapters of the book. Each spans an arc covering his entire creative period, and they are interwoven by the essential themes for Kienast. The reflection grows more detailed from chapter to chapter. Kienast's eight years in Kassel are the missing link between the unknown early work and the works from the 1980s onward that made him internationally famous. In the vacant lots of Kassel left behind by the war, Kienast discovered the "beauty of weeds" and explored the conditions of vegetation in the city—special knowledge that later distinguished him from the Swiss Naturgarten movement. With his doctorate in phytosociology, he was guided by the effect that designed open spaces have on their users, and his relationship to the naturalness and artificiality of these sites was correspondingly pragmatic: they were intended not as idylls but rather as places of social and aesthetic experience and practice.

Conversely, the ambiguity of aesthetics and action was omnipresent in Kassel, the city of documenta, during the 1970s and 1980s, when noted artists—above all Joseph Beuys—combined political statements with productive creativity. At the same time, the boundaries between documentation and artifact were blurred: maps, graphs, and searching for clues became the focus of artistic work. Dieter Kienast's phytosociological maps were not aesthetically motivated, but rather guided by the same interest in nature and the historical present. Like contemporaneous artists, as a natural scientist and designer he chose an inductive approach to reality. Through his work on the methods of phytosociology and spontaneous urban vegetation, he developed a new aesthetic for designing with nature in the city. He moved from learning to read urban vegetation to making his works legible by using plants as signs and symbols.

The context of Kassel makes it clear that labeling Kienast one-sidedly as "postmodern" is an oversimplification, even if several of his methods can certainly be called that: the integration of local traces, of classical models of garden architecture, or of written quotations in a collage-like manner; designing open spaces in fragmentary conditions according to the method of "transparent organization of space"; and the demonstration and documentation of his works using presentation plans based on the techniques of montage and collage. By contrast, Kienast's striving for aesthetic perfection, which began at the same time, is indebted to the spirit of modernism as is his ambition to emancipate people through the design of open spaces. Whereas in his early work this was conceived in terms of the users' unconditional ability to appropriate and alter the grounds, in Kienast's later spaces there is no longer any relativity. Here, too, individuality should be expressed, though in the yearning for perfection: the concise design that reflects consciousness of form. From now on, every subsequent intervention constricts and destroys the space of possibility, whether in a private garden, an urban park, or a housing development. This absoluteness was often criticized, above all by public clients, who attacked Kienast's open spaces for lacking utility. This conflict over use reveals Dieter Kienast to have been a figure on the threshold between the politicization of the 1970s and the aesthetics of postmodernism. His work embodied the tension between those two parallel developments, which were not as contrary as the divided reception of Kienast's work would suggest. The earth pyramids in the municipal park in Wettingen exemplify this: the geometric elements were conceived as a playground for children and conform both to Kienast's rigorous formal idiom and to his utopia of experiencing architectural design in a new way. The functional aspect and the cover with a calcareous meadow follow the same principles as the child-appropriate hill landscape in Niederhasli, which Kienast designed while still in Kassel.

Today the ideals of the Kassel School have faded, just as the social spaces in Niederhasli have fallen prey to weathering. The reasons lie in changes in the habits of use in a society that has become more individualized and digitalized and in the transformation of the concept of nature. The approaches and methods of the Kassel School continued to have an effect in Kienast's gardens in a correspondingly sublimated way. The Kassel School accused him of escapism. But Kienast did not so much turn away from the school as subject its ideas to several processes of reinterpretation: he responded to the call for a new way of living with spaces intended to make possible a new way of seeing. He focused wellbeing and mental acuity within such spaces on cognition, and he integrated aesthetic experience into his gardens and grounds as a specific way of coming to terms with the quotidian in a positive sense, which in his view every design of an open space should offer. From that time forward, his landscape architecture addressed the haptic and mental senses, guided by the question: "How can someone best experience it?"[18]

Kienast's interest in processes of perception is also reflected in the changing ways of presenting his work. In addition to his own thoroughly crafted plans and the black-and-white photographs of Christian Vogt, the videos of Marc Schwarz influenced the reception of Kienast's work. The exhibitions Kienast conceived demonstrated that he preferred the experiment to the museum: on almost choreographed paths through the exhibition, the public is offered ever-new planes of contemplation, clearly in an effort to establish an analogy to the effect that garden spaces have on their visitors. In his basic, plural understanding of the creative process, the landscape architect turns out to be a seismograph and osmotically creative person who always allows current theoretical discourses and media developments to flow into his activity. Thanks to this attitude, Kienast found his way out of the "gray aesthetic" that was threatening to obscure his work. This constant participation in the progressive impetus of a wide variety of disciplines goes back to Kienast's time in Kassel.

This book will not follow a teleological approach in the sense of "Kienast's path from Kassel to Zurich." His studies were not a point of departure, but rather a point of reference for this research that resulted, as a reaction to the aforementioned blind spot, in discussions of Kienast thus far. The book brings together the various strands of his reception and opens up a new perspective on his work.

Dieter Kienast did not leave any "late work" behind: he died at the age of fifty-three. The unfinished aspect of his oeuvre makes it possible to approach his work along different paths. The chapters "The Nature of the City," "Forms of Use," and "Drawing and Perceiving" focus on three aspects in Kienast's work that constantly interact:[19] Kienast, the drawing designer, cannot be separated from the gardener and the phytosociologist. For that reason, the three chapters remain open in a sense, or rather create space "for to end yet again."[20]

On the one hand, the selection of works was based on gaps in the scholarship, with the following projects discussed in detail for the first time: Kienast's first large project, designing the grounds of the housing development in Niederhasli (1972–75); the Brühlwiese Municipal Park in Wettingen (1979–84); and the spa gardens in Bad Zurzach (1983–85). On the other hand, I wanted to fan out as broad of a spectrum of his work as possible from all his creative phases in private, semipublic, and public spaces.

Engaging with Kienast's plans means steeping oneself in their graphic design, for the modification of design principles in the terrain went hand in hand with a transformation in the methods and means for representing them on paper. In short, Kienast switched from the chart to the felt-tip pen; from the felt-tip pen to ink, pencil, and crayon; and from there to montage, collage, and software-based techniques. As he concentrated on aesthetic experience, the verve of the felt-tip pen gave way to exact strokes, delicate hatching, and filigree patterns. Looking at a plan provides information about the formal vocabulary that Kienast developed over time for representing a landscape in drawings. Especially in the representation plans produced after the project, he made an effort to record not only the forms but also the material qualities of his sites and to penetrate into the "essence" of things in the process.

Everyone who knew Dieter Kienast describes him as charismatic. Never having "experienced" him personally was both an advantage and a disadvantage during the many years of research, which entailed preparing, categorizing, and entering into a database the roughly 330 objects in his archives. This monograph also includes a geographically arranged list of works and a bibliography. Exchange with people from his personal life, representatives of the Kassel School, and long-standing colleagues was also indispensable to approaching and understanding his personality and work. The polyphony of these voices echoes in the book, and, even though there were often overtones of the controversies over his legacy, all sides nevertheless unconditionally supported the effort to follow Dieter Kienast's tracks in order to find my own path to a critical assessment.

Kienast himself would not have wanted a "tribute." One feature that repeatedly stands out when studying his thinking is his ironic distance from anything solemn and pompous—including himself—and a certain casualness with which he brushed off anything fraught with meaning. Despite all my efforts to preserve the necessary critical distance, Kienast's personality indirectly supported this book: his stubbornness when solving problems or climbing steep mountains, his constantly "seeking mind," and his nonchalance in activating everything and everyone that helped him to mobilize his work all served as encouragement over the years to clear the many hurdles—in keeping with Dieter Kienast's principle, as told by Peter Paul Stöckli: "Get started, create, and then learn more in the process."

1. Reto Mehli, "Die mit den Förmchen spielen: Über die 'Bühnenbildnerei' in der Gartenarchitektur," in *Freiraum und Vegetation: Festschrift zum 60. Geburtstag von Karl Heinrich Hülbusch am 21. Mai 1996,* ed. Helmut Böse-Vetter, Notizbuch der Kasseler Schule 40, no. 2 (Kassel: Arbeitsgemeinschaft Freiraum und Vegetation, 1996), 77–88, here 77: "neues Leit(d)bild" (a pun on *Leitbild,* or "role model," and *Leid,* or "suffering, harm." Trans.).

2. Mehli justified this transformation in the classical Marxist way: "As a result of the recession from 1973 to 1975, a completely different form of capital accumulation developed: the so-called 'regime of flexible capital accumulation.' This led to a redistribution of labor markets, rapidly changing production processes, and increasing competition between cities. . . . In order to identify them as financial sites worthy of investment, municipal authorities aesthetically pepped up their inner cities, squares, open spaces, and peripheries with promenades, glass shopping palaces, and large urban parks." Ibid., 77, with reference to David Harvey, *The Condition of Postmodernity: An Enquiry into the Origins of Cultural Change* (Cambridge, MA: Blackwell, 1990).

3. Mehli, "Die mit den Förmchen spielen," 78.

4. Ibid., 77.

5. Ibid., 84. Mehli was referring to Kienast's competition entry for the Günthersburgpark in Frankfurt am Main (1991–92).

6. Reto Mehli, "Das Lei(d)tbild 'Landschaft': Zur Kritik ästhetischer Leitbilder in der Gartenarchitektur," in *Reise oder Tour?,* Notizbuch der Kasseler Schule 26 (Kassel: Arbeitsgemeinschaft Freiraum und Vegetation, 1992), 128–56, here 142.

7. See Udo Weilacher, "Die Sinnlichkeit architektonischer Strenge: Zum Tod des Landschaftsarchitekten Dieter Kienast," *archithese* 29, no. 1 (1999): 74–75; Peter Wullschleger, "Ein letzter Garten: Zum Tod des Landschaftsarchitekten Dieter Kienast," *Der Gartenbau* 120, no. 2 (1999): 13, reprinted in *Werk, Bauen + Wohnen* 86, no. 4 (1999): 62–63; Lisa Diedrich, "Dieter Kienast, 1945–1998," *Bauwelt* 90, no. 4 (1999): 150.

8. Peter Stöckli, "Am Ende der Strasse: Ein Nachruf auf Dieter Kienast," *anthos* 38, no. 1 (1999): 58–59.

9. Marc Treib, "The Hedge and the Void: The Landscapes of Dieter Kienast and an Overview of His Career," *Landscape Architecture* 93, no. 1 (2003): 79–107. See also Thomas Proksch, "Dieter Kienast und seine Gärten," in *Zauber der Gärten: Ideen zum Nachgestalten* (Leopoldsdorf: Agrarverlag, 2000), 15–17.

10. See Stöckli, "Am Ende der Strasse," 58.

11. Arthur Rüegg, speech at the memorial for Dieter Kienast at ETH Zurich, January 23, 1999, typescript, Zurich, p. 2, Archiv Arthur Rüegg.

12. Weilacher, "Die Sinnlichkeit architektonischer Strenge," 74.

13. For a detailed discussion, see Udo Weilacher, "Scanning Tracks: Effects of the Principal Creative Work of Dieter Kienast," trans. Felicity Gloth, in *Dieter Kienast,* photographs by Christian Vogt (Basel: Birkhäuser, 2004), 284–95.

14 *Kienast: Gärten / Gardens,* photographs by Christian Vogt (Basel: Birkhäuser 1997); *Kienast Vogt: Aussenräume / Open Spaces*, photographs by Christian Vogt (Basel: Birkhäuser, 2000); *Kienast Vogt: Parks und Friedhöfe / Parks and Cemeteries,* photographs by Christian Vogt (Basel: Birkhäuser, 2002); *Dieter Kienast,* photographs by Christian Vogt (Basel: Birkhäuser, 2004).

15 The Kienast critic Mehli registered this presentation with disdain: "It is in fact already revealing enough when garden architects present their designs like paintings in a museum to an illustrious audience." See Mehli, "Die mit den Förmchen spielen," 78.

16 See Ilse Helbich, "Von unbetretbaren Gärten," in Anette Freytag and Wolfgang Kos, "Neue Parkideen in Europa: Zwischen Arkadien und Restfläche," radio broadcast, *Diagonal: Radio für Zeitgenossen,* Österreichischer Rundfunk (ORF), Österreich 1 (Ö1), first broadcast on October 10, 1998.

17 First references to this continuity were made in Lucia Grosse-Bächle, *Eine Pflanze ist kein Stein: Strategien für die Gestaltung mit der Dynamik von Pflanzen; Untersuchung an Beispielen zeitgenössischer Landschaftsarchitektur,* PhD diss., Universität Hannover, published as volume 71 of the *Beiträge zur räumlichen Planung* (Hannover: Institut für Freiraumentwicklung und Planungsbezogene Soziologie, 2003), 151–87. Udo Weilacher integrated the results of this research into his summary assessment published in 2004 (Weilacher, "Scanning Tracks"). On the occasion of the symposium "Natur entwerfen: Zur Aktualität des Werks von Dieter Kienast (1945–1998)" at the Schaulager Basel on December 5, 2008, various aspects of Kienast's connection to Kassel were discussed: in the video documentation *D. K.: Eine Spurensuche* by Marc Schwarz and Annemarie Bucher (DVD, Zurich: schwarzpictures, 2008), in a lecture by the present author, and in a podium discussion in which Annemarie Burckhardt, who died in 2012, participated.

18 Dieter Kienast on the occasion of the planning of the Masoala Rain Forest Hall at the Zoo Zürich; quoted in Schwarz and Bucher, *D. K.,* as a paraphrase by the zoo's director, Alex Rübel.

19 Other important themes in Kienast's work, such as his working with the boundary as a connection to and separation from the surrounding landscape, or his relationship to his role models in garden architecture, could only be touched on in this framework, as is also true of Kienast's teaching at various universities.

20 Samuel Beckett, *For to End Yet Again and Other Fizzles* (London: John Calder, 1976), 9.

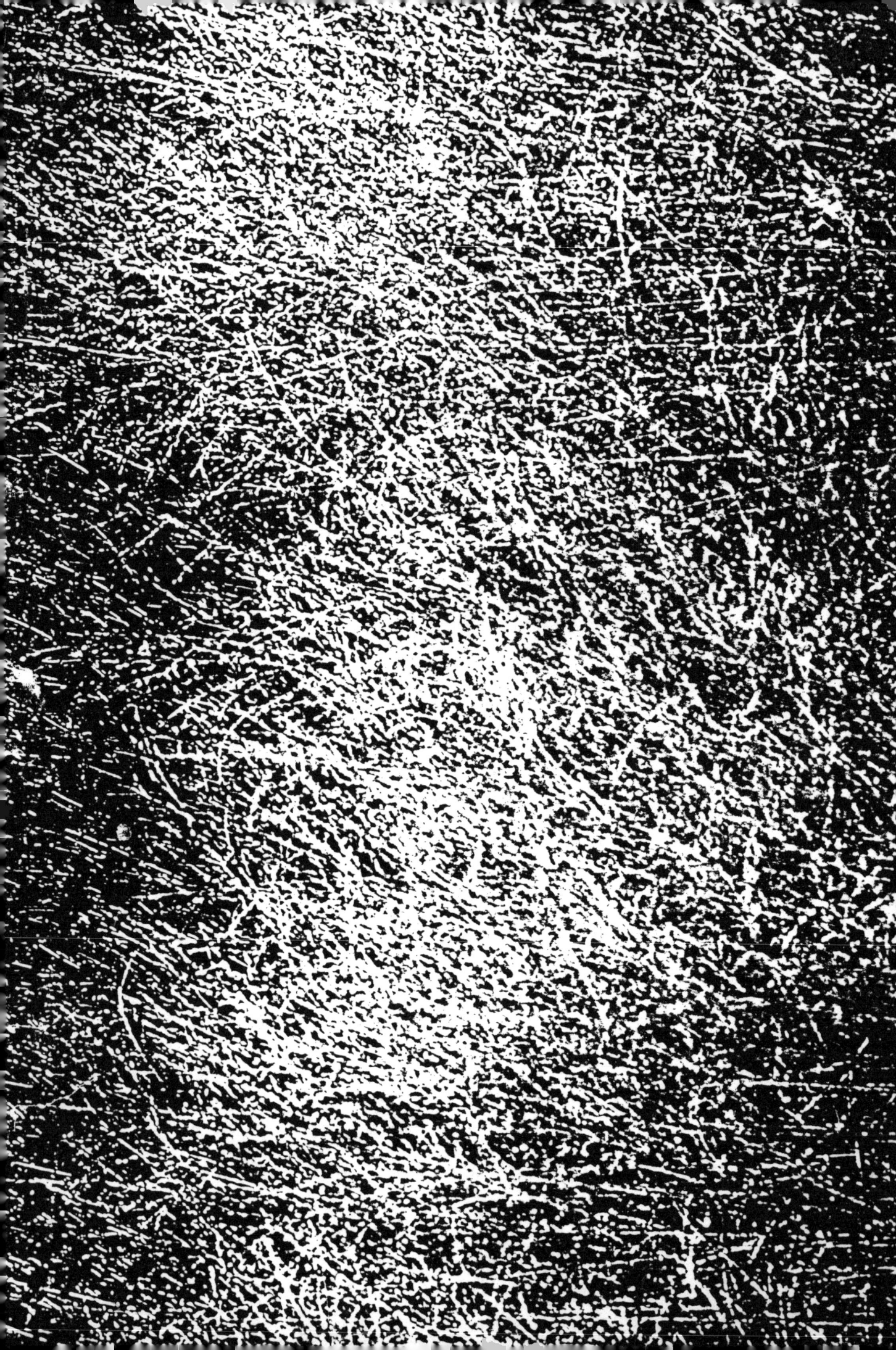

I
The Nature
of the City

Traces of Kassel and the Beauty of Weeds

Two souls dwelled in his breast—that of the gardener and that of the phytosociologist—as Dieter Kienast confessed in 1979:
I don't like to do without celandine, *Chelidonium majus,* flowering on the edge of the path, or without *Draceana* in a wooden tub decorating the seating area. The wild thyme, *Thymus serpyllum,* slowly spreading on the gravel surface is just as dear to me as the old Antarctic beech, *Nothofagus antartica,* whose leaves—I don't know why—remind me of distant lands.[1]

This ode to the diversity of plants—elsewhere transformed into a sociopolitical allegory with the words "Imagine: peaceful coexistence, at least among plants!"—is found in a text in which Kienast used his knowledge of phytosociology and gardening to polemicize against nature conservation and "natural gardens" (1). "The nature of the city"[2] took on a new meaning for Kienast against the backdrop of his scientific training in Kassel. Inspired by his phytosociological studies, Kienast arrived at a new aesthetic for designing with nature in the city (2). His interest in spontaneous vegetation liberated him both from the imitation of rural models that had been practiced until that time—and that is still practiced today—and from a lush gardening of the city (3). In parallel with his long-nurtured courage to pursue reduction in design and in the selection of plants, the sensuousness of experiencing nature became the focus of his interest in design. In his "Zehn Thesen zur Landschaftsarchitektur" (Ten Theses on Landscape Architecture), he wrote in 1992:
The essential thing is to rediscover the plant as an urban element, and not to consider it simply as an ecological or dendrological factor, as an architectural filling element. We should learn that there are different shades of green, that plants rustle differently in the wind, that not just flowers but also fallen leaves have a fragrance. We should include shade, take account of the impression of the bare branches in winter, express the symbolism of plants, and feel their sensuousness.[3]

Kienast's work is an argument for the power of aesthetic experience to alter consciousness. Apart from the odor perceived by the nose and the noises heard by the ears, the iconicity and symbolism of plants—perceived by the eyes and linked with the intellect—played an essential role for him. His goal was to make every environment that he designed "legible" by means of the symbolism of plants and other materials he used. The garden, too, should once again convey meaning, sharpen awareness, and arouse the senses.[4]

In order to "read" a given image, viewers rely on their everyday knowledge and on collective memory. At Fürstenwald Cemetery in Chur, Switzerland, Kienast made use of a square arrangement of trees as a quotation of paradise. Situated in front of the viewing hall and the chapel, it puts the mourners in a frame of mind to bid farewell to the deceased and symbolically promises that they will enter paradise (4). This reading is at least suggested to passersby, since the square is a minimal variation on the garden of paradise and is an old and often used topos. By decoding a hidden message, they are drawn into the surroundings and also themselves give meaning to the situation. The aesthetic experience is heightened by their own interpretation.

The formal idiom that Kienast developed in the 1980s and 1990s was intended to reveal the qualities and new use of a given site and especially to accentuate its topography. To this end, Kienast emphasized contrasts and clear forms. Intellectuals such as the formalist literary theorist Viktor Shklovsky (1893–1984) have shown how forms focus awareness and can intensify an experience—in Kienast's case, the perception of space and of materials in the widest sense. A sensuousness is evoked, and often meaning as well. The external environment appears before the viewer's inner eye as an awareness of its form. Shklovsky referred to the conditions for such an effect as "defamiliarization"

1
The front yard of a house in the city, photographed by Dieter Kienast, who regarded its blend of plants as typical of the mixed nature of a city. The caption Kienast wrote for it reads: "Atmospheric front yard with peaceful coexistence of exotic and native plants: fuchsia and oleander alongside a pavement-crack and seam community."

2
Swiss Re (now Bank Vontobel) Garden Courtyard on Gotthardstrasse in Zurich. The sloping ground of the garden courtyard points to the underground parking garage. The pattern of the ground responds to the façade of the office building; it looks like its symbolic cast shadow. In the foreground, the drainage channels planted with irises and the perforated boundary wall; in the background, the katsura trees and the water basin. Photo: Christian Vogt, ca. 1996

3
Dry Grassland Biotope for Grün 80 in Basel. In 1978, Kienast had proposed planting a phytosociological garden on the grounds of the Grün 80 horticulture show in the Brüglingen district of Basel in order to illustrate succession processes. The Dry Grassland Biotope that was ultimately realized is now a nature reserve. Photo: Georg Aerni, 2012

4
Fürstenwald Cemetery in Chur. The square arrangement of linden trees in front of the viewing hall is symbolic of the garden of paradise. It reflects Kienast's study of the history of gardens. Photo: Georg Aerni, 2008

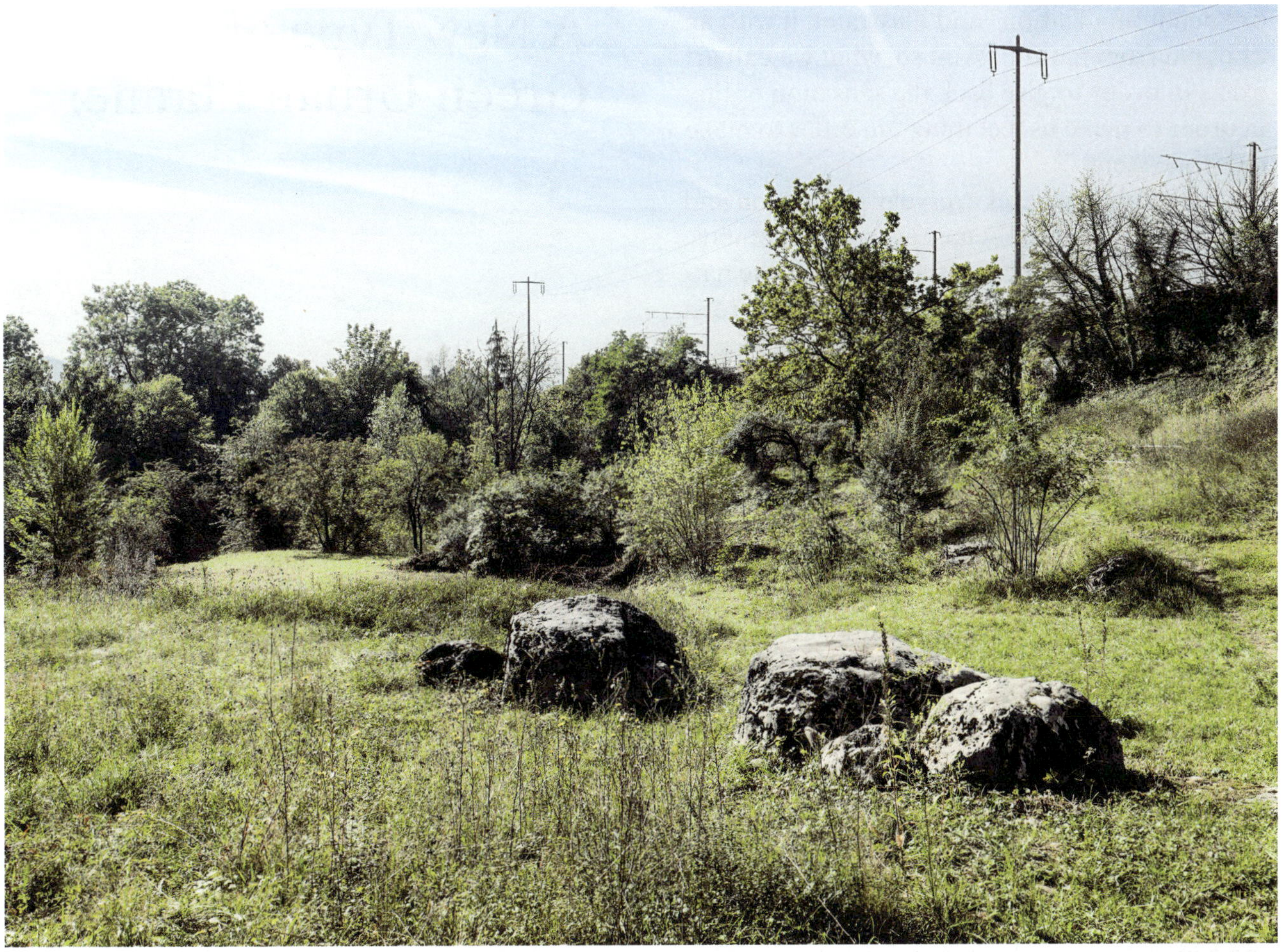

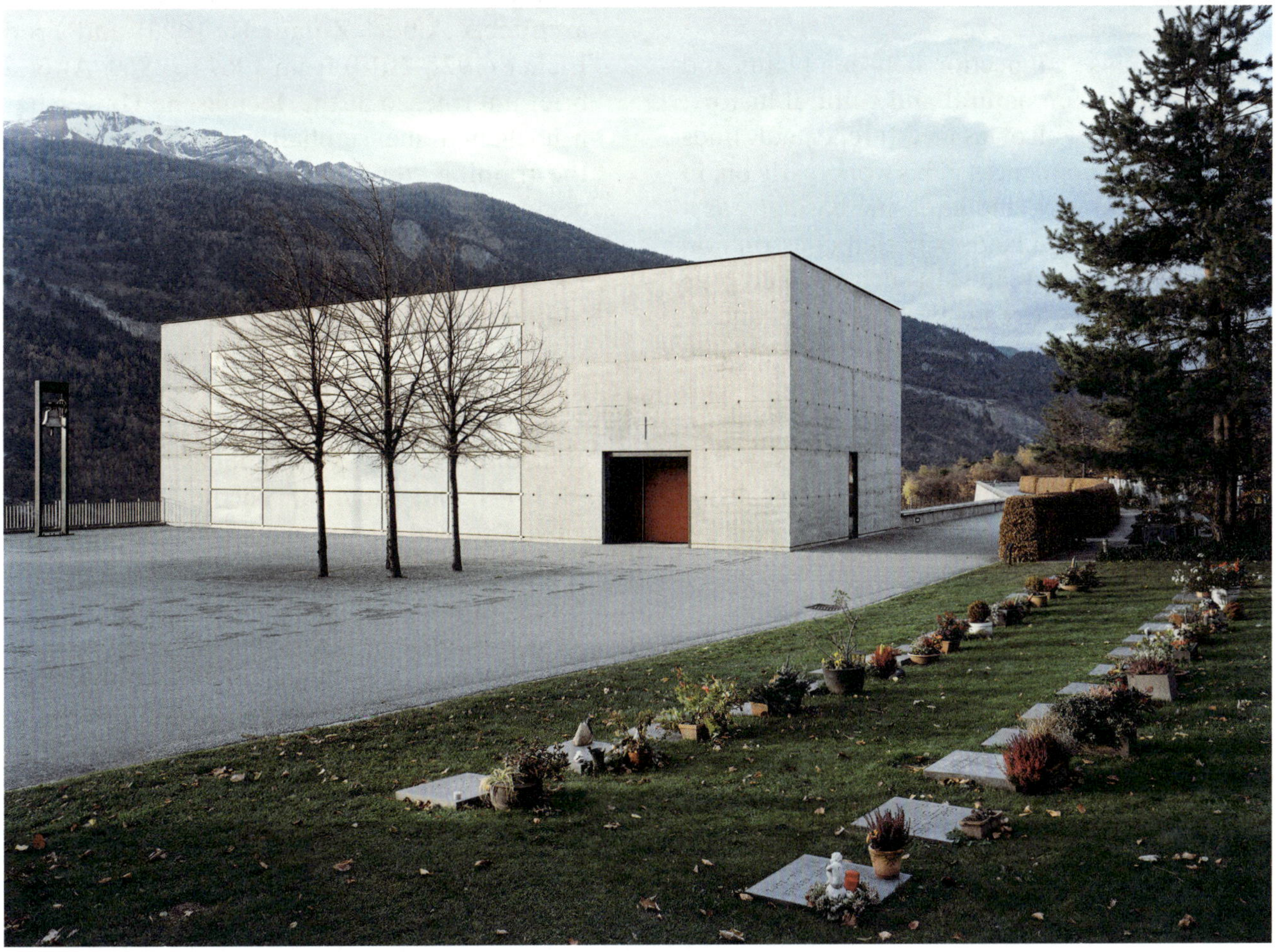

or "deautomatization" and illustrated it with an example from nature: "And so, what we call art exists in order to give back the sensation of life, in order to make us feel things, in order to make the stone stony."[5]

Abstraction vexes everyday perception and sharpens the senses. Kienast used such methods, placing the organic next to the inorganic, the natural next to the artificial; he combined his green with gray. This formal idiom inspired the viewer to discover symbols and increase the sensory experience. For example, in the courtyard garden of Swiss Re on Gotthardstrasse in Zurich, *Iris versicolor*—known in German as *Sumpfiris,* or "swamp iris"—was planted in the drainage channels for rainwater (2). The environment is deliberately designed artificially and reflects the urban space rather than simulating a rural idyll. The natural drainage makes it explicit that the built area is sealed off. The material qualities of water, plants, concrete, and asphalt are revealed in the dovetailing of naturalness and artificiality. Freely adapting Shklovsky, it can be said that *Sumpfiris* evokes—literally—an image of "it is damp here," which would have been only latent on the banks of a lake, because the differentiating contrasts are lacking in the natural environment of the iris.

The close connection between plants and humans, between natural and cultural history, inspired Kienast all of his life.[6] It repeatedly finds formal correspondences in his work: early on, in an allegorical way, and later, increasingly, as a vocabulary for the landscape that construction material literally creates. In such filigreed plan drawings, Kienast reenacted these semantics. Kienast's efforts to make open spaces legible was preceded by a process of learning to read the city, based on the profound knowledge of spontaneous vegetation that he had acquired as a budding landscape architect during his phytosociological studies in Kassel.

A New Type of Green Urban Planner

When Dieter Kienast began studying at the Lehrstuhl für Landschaftskultur (Chair of Landscape Culture) of Günther Grzimek (1915–1996) at the Hochschule für bildende Künste (University of the Fine Arts) in Kassel in 1970, he already had in his kit a basic education from the nursery of his parents, an apprenticeship in gardening with the Hottinger brothers, and further education in design in the form of an internship with the Swiss landscape architects Albert Zulauf (b. 1923) and Fred Eicher (1927–2010) from 1967 to 1969. After a brief intermezzo at the Technische Universität München-Weihenstephan, where Kienast found the training "too much like high school,"[7] he decided to go to Kassel. He had been advised to do so by the master of his apprenticeship, Eicher, who had studied there with Hermann Mattern (1902–1971), and by his future fellow student Thom Roelly (b. 1945). Roelly, who was the same age, already resided in Kassel when Kienast applied for admission there. Over the course of the year 1970, Kienast's wife, Erika (d. 2017), and daughter, Nicole (b. 1969), also moved to Kassel. In order to make a living, Kienast took small jobs redesigning gardens in Kassel and from 1972 onward regularly spent school breaks working in the Wettingen-based firm of landscape architect Peter Paul Stöckli (b. 1941), who was four years older and whom Kienast had met in 1967 during his internship with Albert Zulauf in Baden. Kienast was thus regularly active as a designer from the late 1960s and did many of the drawings by hand. He brought with him to Kassel the tools of the traditional crafts of gardening and landscape architecture, which were combined there with training as a landscape and open-space planner in a program that had been reoriented in the context of the sociopolitical-reform movements of the late 1960s.

The guidelines for that reformed training had been formulated in 1968 by Günther Grzimek in his "Leitbild für das Studium am Lehrstuhl für Landschaftskultur an der Hochschule der bildenden Künste in Kassel" (Misson Statement for Studies at the Chair of Landscape Culture at the University of Fine Arts in Kassel), with the goal of training "a new type of green urban planner":[8]

He practices cooperation and teamwork by discussing projects thoroughly; he learns to integrate the various artistic, technical, and biotechnical aspects of planning.... One crucial factor in the process of planning is precisely formulating the task and analyzing the problems. In this first phase in particular, the green urban planner has to turn a program determined by technology and politics into a functioning plan.... Apart from the formally and aesthetically successful solution, the technological function and the social purpose have to be understood as elements constitutive of the planning.[9]

The very language betrays the attitude of the time: although the Hochschule für bildende Künste in 1968 was still organized in so-called master classes, there was nothing left of the bourgeois idea of the creative genius who designs a park, for example, based on his intuition and powers of observation. According to this program, the "new type of green urban planner" is a political person who strives to do justice to the social purpose and technological function beyond formally and aesthetically successful solutions. It is about analyses, processes, discussions, about cooperation among specialists, and about developing a "planning idea" from "knowledge of the capabilities of the various greens."[10] Knowledge should not be an end in itself, but instead be acquired in order to apply it in practice. It is important for that reason that all of the students actually implement one of their projects. "The client takes the risk," it states drily.

The specific structure of studies and the teaching subjects, however, seemed far more traditional than the goals and methods formulated in this model might suggest. A basic course in which the students were to develop a "relationship to design and color" was followed by attendance of lectures and courses in philosophy, sociology, and psychology. Joint classes with the departments of architecture, painting, and sculpture were as much a part of this training as seminars, improvised tasks, competition entries, and excursions. In order to help students solve specific planning tasks, they received training in biology, ecology, technology, materials science, planting design, photography, drawing, urban planning, and the history of landscape architecture and rural development. In the seminar "Berufspraxis" (Professional Practice), in preparation for their future activity, they were taught about "the relationship of the green urban planner and the architect" and the "characteristics of other partners such as private clients, community or state clients, and companies."[11] From today's perspective, the formulation of this passage sounds odd, as if the point were to learn rules of etiquette for smooth interplay with authorities. But the goal was to train green urban planners with a view to their role as mediators between separate disciplines: to become coordinators focused on "cross section" and "use."[12]

An Interdisciplinary Education

In 1971, the Hochschule für bildende Künste in Kassel was integrated into the newly founded Integrierte Gesamthochschule Kassel (Integrated University of Kassel, since 2003 University of Kassel). The Lehrstuhl für Landschaftskultur was incorporated into the new Organisationseinheit Architektur, Stadt- und Landschaftsplanung (Organizational Unit of Architecture and Urban and Landscape Planning). The teaching at the Gesamthochschule was based on principles similar to those that Grzimek had developed previously at the Hochschule für bildende Künste, but it focused more on criticism of society and authorities. That was due not least to two of the personalities teaching there, and to the content of their teachings, both of whom influenced Kienast as well: the economist, sociologist, and architecture critic Lucius Burckhardt (1925–2003) and the phytosociologist and landscape planner Karl Heinrich Hülbusch (b. 1936). In 1971, Hülbusch gave a lecture course at the Lehrstuhl für Landschaftskultur—together with Jürgen von Reuß (b. 1937), who would also later become a professor in Kassel—and organized an excursion through the Ruhr region that same year.[13] In 1974, he was appointed Professor of Landscape Planning at the Gesamthochschule Kassel, where Lucius Burckhardt had held the Chair for the Social Economics of Urban Systems since 1973.[14] Previously, Burckhardt had taught at ETH Zurich—Eidgenössische Technische Hochschule (Swiss Federal Institute of Technology) from 1971 to 1973. In an educational experiment there, which has gone down in the annals of ETH Zurich as the "Lehrcanapé" (teaching sofa: i.e., a two-person chair), he taught students his understanding of planning and of a contemporary profile for the architect.[15] The critique of planning and the questioning of the self-image of the architect—which, for the president of ETH Zurich and the dean of the Division of Architecture (since 1999: Department of Architecture), had simply gone too far—was just right for the new syllabus in Kassel.

The Integrierte Gesamthochschule Kassel was officially founded on June 25, 1970, and the Organisationseinheit Architektur, Stadt- und Landschaftsplanung began offering courses in the winter semester of 1971–72.[16] The department had been formed as the merger of the architecture training of three schools: the Werkkunstschule (School of Applied Arts); the Hochschule für bildende Künste, which included Grzimek's Lehrstuhl für Landschaftskultur; and the Staatliche Ingenieurschule (State Engineering School).[17] The new concept of the *Gesamthochschule* combined three types of schools otherwise separated in Germany—*Universität* (university), *Fachhochschule* (technical college), and the *Kunsthochschule* (college of art)—and thus enabled the coexistence of scientific and artistic disciplines in one institution.[18] The former Klasse für Landschaftskultur (Landscape Culture Class) moved out of the buildings that had been constructed in 1962, following the tradition of Bauhaus ideas, on the former factory grounds of the Kassel company Henschel & Sohn, where locomotives and cannons had been produced. Kienast was thus operating in an environment for studies that was infused by interests in the natural and social sciences and the arts in equal measure. That would be reflected in his work all of his life.[19] His training as an engineer represented the fourth component. Thanks to the new Gesamthochschule, Kienast had the opportunity, with his degree from the Hochschule für bildende Künste, to participate in an Integrierte Abschlussphase (Integrated Degree Phase, IAP) of four semesters with concentrations in building production, building planning, urban planning, and landscape planning. He thus completed his studies with a degree in engineering in the winter semester of 1975–76.[20] This academic degree corresponded to that of a technical college. Kienast wrote a joint thesis with Thom Roelly titled "Freiraumplanung Kassel-Nordstadt" (Planning of Open Space in the Nordstadt District of Kassel); it focused in particular on the vegetation and how it could be used for analysis and planning. This thesis was then followed by a dissertation on the subject of spontaneous urban vegetation. In December 1978, Kienast received a doctorate in engineering for that work, which had been supervised by Karl Heinrich Hülbusch.

The course of study in architecture and urban and landscape planning at the newly founded Gesamthochschule was one of the many such courses of study at German universities that had been reformed in the 1970s. This represented a reaction to the student movement and the fervent demands of students to have more say in shaping their education. These changes had in common a critique of fine-arts-oriented training that seemed out of date and an effort to train planners with a sense of sociopolitical responsibility. There were echoes of this already in Günther Grzimek's mission statement at the Lehrstuhl für Landschaftskultur. The reorganization of the universities went a crucial step further with regard to grounding the intended reforms in a curriculum and the subject matter of the teaching itself: the frequent claim that students had to be responsible for themselves and the need for interdisciplinary training were reflected both in the curriculum and in the study bylaws.

The focus of the training in Kassel was—like that of the reformed curricula of other universities—the so-called *Projektstudium* (project-based studies). The students could largely choose the subjects to be addressed, and they could form mixed groups with architects and urban and landscape planners if so desired. Moreover, they were free to consult advisors from all departments. The one-semester projects were supplemented by shorter study projects of similar focus but with smaller tasks. In addition to interdisciplinary collaboration on the projects, the students were free to enroll in seminars in other departments. During the so-called orientation phase of two semesters and a required semester of professional practice, they were even required to do so. As a result of these regulations, architects could attend seminars on vegetation, and landscape planners could enroll in seminars on designing load-bearing structures. Only in the fourth semester did students choose a major that they would study in greater detail to receive their degree.

As signs of their self-organization and autonomy, no final exams were given in seminars and lectures, but instead papers were presented or summaries written. The way that Lucius Burckhardt used to greet students in Kassel summed up the expectations of the teachers when it came to students being responsible for and motivating themselves: "There's a party tonight, where you all receive your diploma. Then tomorrow we begin to study."[21] In the assessment of a retrospective seminar paper that compared the curricula for urban planning in Germany, the education in Kassel was the "most liberal," since the students could set their own project tasks and organize themselves; and, apart from presenting project and study works and certificates of attendance for seminars, there were no grades until the final degree exam.[22]

Dieter Kienast experienced this new joint training of architects and urban and landscape planners during a transitional phase before it became mandatory in the winter semester of 1975–76. His interest in interdisciplinary collaboration,[23] especially with architects and urban planners, may have been inspired by the Kassel model—just like his own teaching later, in which he initially placed explicit emphasis on critical resistance and heated discussions[24] before setting urban-planning or philosophical/artistic tasks for the students.[25] During his own professional career, however, he was rarely commissioned for urban-planning projects or analyses.[26]

Kienast's exploration of art in the context of his work can also be traced back to Kassel, where since 1955 the exhibition documenta has shown the current trends in art at intervals of first four then five years. In 1972, the fifth exhibition was organized by the Swiss curator Harald Szeemann under the motto "Better Seeing through documenta 5," breaking with the previous concept of the "Museum der 100 Tage" (Museum of 100 Days). Rather than simply exhibiting works, that legendary documenta made image and reality its themes: *Befragung der Realität—Bildwelten heute* (Questioning Reality—Pictorial Worlds of Today)[27] included spheres of non-artistic visual production, such as advertising slogans, devotionals, material from everyday culture, and science fiction illustrations. In addition to training perception, however, it focused on artists who emphasized subjective experience and personal identity—what Szeemann called "individual mythologies"[28]—in the context of the great debates over improving society. For the Swiss curator, documenta 5 represented "life concentrated in the form of an exhibition." He regarded the "survey exhibition" as an attempt to create an order that could be experienced through the senses or understood intuitively, a walk-in arrangement. The exhibition areas on the different image worlds of artistic and non-artistic production and the "individual mythologies" both supplemented and criticized one another. The phrase had come to Szeemann years before documenta 5, but he used it as the title of a section of the exhibition he had originally intended to call—with an eye to the work of Joseph Beuys—"Shamanism and Mysticism." Szeemann asserted that:

> Based on their outward appearance, the "individual mythologies" are a phenomenon without a common denominator, but they are understandable as part of an art history of intensity oriented not solely around formal criteria but also around the palpable identity of intention and expression.[29]

Harald Szeemann saw the connection between questioning image worlds and mythologies in a "history of intensity made to be experienced by the senses."[30] The artists exhibited in Kassel in 1972 included some who were often declared to be crazy by those outside of the art world, such as Viennese Actionists like Hermann Nitsch, Günter Brus, and Rudolf Schwarzkogler. The intensity of their art interested Szeemann, as did their attempt to replace an aesthetic *l'art pour l'art* with a "lived" *l'art pour l'art*—a radical statement of existence. This sort of rigorous, subjective, artistic realization, which could be perceived and understood more directly by the viewer because of its closeness to life, corresponded to Beuys's dictum that everyone is an artist,[31] since each of these "mythologists" cultivated his or her own individual creative expression. In that sense, this seemingly self-contained—and for many incomprehensible—section of documenta 5 was in keeping with the ideas of democratization that shaped the art of the 1960s and 1970s. At the same time, Szeemann saw in this creative production "the individual's attempts to counter the great disorder with an order of his own."[32]

These goals of documenta 5[33] are revealing in connection with works by Dieter Kienast in several respects: the question of producing images by using a certain type of vegetation, a specific material, and of finding a form occupied Kienast in all of his projects. His "images," which were adjusted to an experience in the space, were intended to increase the intensity of the experience of a site. From the 1980s onward, Kienast was deliberately working with contrasts of bright and dark, warm and cold, shady and sunny, and so on. He created spaces for sensuous experience in which the sense of vision is certainly dominant and captures the first impression, but these spaces directly stimulate other senses and the intellect as well.[34] Last but not least, the idea of producing one's own order—and countering the great disorder of the world with it—runs through Kienast's work like a golden thread. This makes it clear how much his work was embedded in the production of a time—whether artistic or non-artistic in nature—and how his grappling with art influenced his creative work on levels that at first glance are not affected at all by artistic questions. The oft-evoked dovetailing of art and life, of sharpening the senses, and of becoming aware establish the counterpoints to Kienast's further development.

A New Theory of Planning Open Spaces: The Kassel School

In a survey titled "Die Entwicklung der Landschaftsarchitekten und ihre Ausbildung in Deutschland" (The Evolution of Landscape Architects and Their Training in Germany), published in 1992, Heinz Hallmann, a professor of landscape architecture in Berlin, observed with regard to the new course of study established in Kassel in the early 1970s that probably "no other educational institution grappled as intensely with the questions of the use and appropriation of open spaces as they did in Kassel."[35] Using the term "Kassel School," he emphasized especially the "components of social space" and "the questions of grassroots participation in the planning and construction process," and he predicted that these approaches would have much greater influence both on the theory and on the political program of the discipline in the future.

According to the research of one of its proponents, Karl Heinrich Hülbusch, Professor of Landscape Planning in Kassel, the term Kassel School was coined in 1981 by Eike Schmidt, the editor in chief of the journal *Garten + Landschaft* at the time. In issue 11 of the journal, he had published a lecture presented at the "Tag der Landschaft" (Day of Landscape) in Kassel in which he addressed the phenomenon of the *Naturgarten* (natural garden).[36] Schmidt mentions as the Kassel School an approach developed by a group around Karl Heinrich and Inge Meta Hülbusch, as well as Helmut Böse, in which all planning began on principle with the study of the history of the site's use and the vegetation located there. Planners should incorporate this vegetation and only intervene modestly. Schmidt summed it up by saying that the Kassel School reads public and private open space as "the result of their appropriation by man to which the vegetation becomes adapted in the course of time resulting in an independent biotope which should only be altered with the greatest care by the designer."[37] Because it took into account the local vegetation in every plan, Schmidt connected the Kassel School with the Naturgarten movement that arose in the late 1970s. Two years later, the landscape architect and journalist Jürgen Milchert used the term Kassel School in a review of Helmut Böse's *Die Aneignung von städtischen Freiräumen* (The Appropriation of Urban Open Spaces) of 1981.[38] In that work, Böse explained his view of "social planning of open space," which Milchert saw as a summary of the essential statements of the Kassel School and their presentation to a specialist audience.[39]

Social Planning of Open Space

Building on an analysis of the crucial developments in green urban planning during the twentieth century, Böse contrasts in his work two positions that were topical in the late 1970s: the scientific/ecological and the artistic. Böse concluded that neither position took into account the social needs of users, which is why, he claimed, green urban planning was still in a "presocial stage":

Garden art and subsequent green urban planning have in common that they do not develop their solutions from the everyday lives of people. The "problems" were and are "artistic" in character. They aim to come to terms with materials and means that are not intended for use but should essentially speak for themselves—and above all for the artist.[40]

Social planning of open space, as Böse understood it, should be developed from the everyday situation of human beings (5). It has to react to their needs and habits and at the same time help them succeed. Böse called on planners to make opens spaces available that would offer different opportunities for appropriation and hence could be used by different groups in line with their own wishes. According to his summary, an open space designed in this way is, first, not completed at some specific point in time but is rather "finished" over the course of its use,[41] because the residents constantly change, as they move in and out, and new people adapt the open space to their needs. Second, this kind of planning avoids the criterion of "originality," which until then had been so important for the "garden artist" and then the "green urban planner."[42] (6)

Böse was thus criticizing a bourgeois concept of art and the work that had been attacked and broken open in the realm of art since the 1960s. The bogeymen were the same, but the fight against them had occurred much earlier. During the 1970s, garden art and green urban planning carried connotations of qualities open to criticism, such as "originality," "autonomy," and "self-referentiality," resulting in negative associations. They were meant to be replaced by "social open-space planning" with new content. This new approach to planning and targeting open spaces was not intended to claim from the outset that it was artistically valuable—which it certainly still could be—but rather to be derived solely from the needs of its users. It was felt that open spaces should provide people with opportunities for appropriation and participation. The integration of existing traces left by users, the provocation of action and events—in short, the interaction between people and open spaces—should free the grounds shaped by the garden artist, whose purpose is focused on a contemplative enjoyment of nature through walking, spending time, and perhaps also boating.

This focus on two mutually exclusive models is, of course, only possible in theory. In practice, the two attitudes are mixed, as are the actions of the users. The discussion as such is interesting in that the Kassel School, presumably unconsciously, was paraphrasing Claude Lévi-Strauss's idea of the *bricoleur* and was distinguishing that type from the specialist or engineer.[43] The figure of the engineer would correspond to a garden artist who designs everything down precisely to the last detail, leaving the users hardly any room to operate or change. The *bricoleur*, by contrast, develops a design via detours, provisional solutions, creative insights, and always guided by the situation. The postmodern quality that can be found in the work of the Kassel School and that Kienast cultivated in his projects is different to the superficial postmodernism that simply cites from a historic reservoir of forms. Instead, it is about people, use, experience, and about integrating and learning from history.[44]

One central motto of the Kassel attitude is "Vorbilder statt Leitbilder" (Models, Not

5
Courtyard in Berlin divided into three parts by chain-link fencing. After studying the patterns of how it was used, Helmut Böse concluded that it was better not to intervene in the design.

6
Courtyard of a block of IBA buildings on Ritterstraße in Berlin: decorative beds in the shape of a human brain were arranged around a playground. Helmut Böse regarded it as a hardly usable space, which the designers hoped would make them stand out artistically.

Guidelines), and another is that there is nothing left to invent in the world.⁴⁵ Among the earlier thinkers mentioned as those on whose work the Kassel School was built are Leberecht Migge, Alexander Mitscherlich, Jane Jacobs, and Martha Muchow; all of them criticized the living conditions in modern big cities. Under the banner "Models, Not Guidelines," and with a view to the specific designing of open spaces, they called for planners to study carefully tried-and-tested examples from history and the present in order to determine why they function so well, which parts function especially well, and what does not function. By "functioning" they meant how actively and creatively these open spaces are utilized and how the vegetation used there stands up to it. What is learned from this analysis should be introduced into new sites. In this way, planners were meant to avoid an individual, unmistakable stamp, to analyze the situation for current and future use, and to focus their designs entirely on that.

Internal and
External Perspectives

There is disagreement about the question of what defines the Kassel School and who belongs to it. There is an external perspective and an internal perspective that were defined retrospectively. As perceived from the outside—that is, by the specialists who coined the term in order to name a philosophy they could understand—the Kassel School stands for planning open spaces to serve the users, with an analysis of vegetation in connection with usage by people representing the essential basis on which to build. The open spaces should have a "utility value" and should encourage appropriation by the users. In open-space design, the use of vegetation again plays a crucial role because the understanding obtained from the analysis of a place can be applied. Moreover, all of the planning and design processes have to build on the principles of participation.

Whereas the external perspective generally seeks to equate the Kassel School with all of the subject matter taught in the architecture and urban and landscape planning program in Kassel in the 1970s and 1980s—to which Lucius Burckhardt, Peter Latz, Jürgen von Reuss, and others also contributed—the circle around Hülbusch and Böse attaches great importance to seeing the Kassel School as separate from the department. They define the Kassel School as the result of all the activity of the people who from 1985 onward were organized in the registered association Arbeitsgemeinschaft Freiraum und Vegetation (Open Space and Vegetation Working Team), which from that year on also published the book series *Notizbuch der Kasseler Schule* (Notebook of the Kassel School).⁴⁶ With harsh words, they reclaimed for themselves the term Kassel School:

> The works of the Kassel School are largely results of the learning-teaching research of the Arbeitsgemeinschaft Freiraum und Vegetation in the landscape planning course of study at the GhK [Gesamthochschule Kassel]. The Department of Urban and Landscape Planning has given the impression that the Kassel School is identical with this department. This cooptation of the concept—as well as the arbitrary use of the terms "open-space planning" and "landscape planning"—is very popular and is intended to distract from or conceal conventional green urban planning and land conservation. They did not take part in the works of the Kassel School. The Arbeitsgemeinschaft Freiraum und Vegetation is active in an open working agreement of professionals, teachers, and students actively learning, teaching, and researching. Since 1985, it has published with a nonprofit association, most of whose members are not only outside of Kassel but also active outside of the university: the *Notizbücher der Kasseler Schule*. The Kassel School takes its name from the place where many of the participants happen to work. All the efforts to put the place-name in the foreground, as opposed to the content and results of the work, are understandable if someone wants to appropriate content and even things out: but they are wrong, because the Kassel School has its name from the work and not the place.⁴⁷

The term Kassel School was coined in 1981 to describe retrospectively the teaching of the university professors who had been hired in Kassel for the new course of study from 1973 onward. That was three years after Dieter Kienast had left Kassel. Kienast did leave his mark on the research of the Kassel School in the form of his thesis on open-space planning in the Nordstadt district of Kassel and his dissertation on the spontaneous vegetation of the city of Kassel[48]—both works were supervised by Karl Heinrich Hülbusch. Conversely, the teachings of Hülbusch and Burckhardt left traces in many texts that Kienast wrote between 1975 and 1992.

Although Kienast was imbued with the ideals advocated in Kassel and with the methods of analysis and planning taught there, from the early 1980s he took his own paths and gradually developed a way of working that in the eyes of those in Kassel represented precisely what they were battling against, and that disconcerted them greatly against the backdrop of Kienast's training.[49] Böse's critique of interest in artistic questions and in the employment of materials and design means not determined by use, and his argument for observation and reaction in lieu of inventive and original design, already gives a sense that Kienast's projects since the mid-1980s had been moving away from the design principles advocated in Kassel. Indeed, Kassel was not the only early stimulus on Kienast: the influence of Fred Eicher was also crucial for his later work. Kienast synthesized these superficially irreconcilable attitudes and subjected the Kassel theories to a series of processes of reinterpretation.

The break with Hülbusch and the alienation from Burckhardt were inevitable, because Kienast implemented the methods advocated in Kassel in modified ways. Since the 1980s, he had been focusing on designing with his own creativity and originality. The connection to Burckhardt was still strong in terms of content, especially because, since the mid-1970s, he had turned to the works of the French landscape architect Bernard Lassus and the Scottish artist Ian Hamilton Finlay, both of whom also inspired Kienast.[50] The contact to Hülbusch ended in the 1980s, and his students vehemently criticized all of Kienast's works built after 1985.[51]

Lucius Burckhardt's Critique of Planning

Some of what Günther Grzimek formulated in the program to reform the training of a "new type of green urban planner" at the Hochschule für bildende Künste in Kassel had also been analyzed and questioned by Lucius Burckhardt, a sociologist, economist, and architecture critic based in Basel, in his critique of planning and urbanism. Burckhardt explicitly addressed the architectural profession. But his analyses also help to understand changes in the occupational profiles of the green urban planner and of the landscape planner, and to clearly work out how interdisciplinary education in Kassel differed from that of other schools reformed in similar ways.

The End of Polytechnic Solvability

For Burckhardt, the call for a new type of planner that developed over the course of the 1960s was based on a series of urban planning problems but also social ones that called into question the self-image and tasks of several disciplines at once: the *tabula rasa* architecture that had been taught at architecture schools and practiced until that point was beginning to reveal its first negative consequences. At the same time, the economic conditions were marked by inflation and by labor and financial shortages for the first time since the Second World War.[52] Apart from that, it became clear to city and country dwellers that, as their private comfort increased, the world around them was deteriorating—the redesign of many villages and cities to accommodate automotive traffic is mentioned here as but one example. This development reached a provisional climax in the early 1970s, which was portrayed par excellence in the series *Alle Jahre wieder saust der Presslufthammer nieder* (The Jackhammer Crashes Down, Again and Again, Year for Year, officially translated as *The Changing Countryside*) of 1973 by the Swiss book illustrator Jörg Müller.[53] On seven plates, Müller illustrated in detail the fictitious but realistic change of a landscape over a period of two decades (7). The current state is recorded in a panorama image every three years. Each plate is labeled with a day of the week and a date. This turns the visual fiction into a factual report: for instance, between Wednesday, May 6, 1953, and Tuesday, October 3, 1972, the idyllic village of Güllen is transformed into an urban agglomeration with a shopping center and an autobahn on-ramp.

In the center of the foldout plates, at first there is a pink single-family home with a garden and a knotty old tree. The vegetable bed is replaced by a canopy swing in 1966. Ultimately, the house, garden, and tree give way to an autobahn that cuts through the landscape, demonstrating that the individual human scale and also the natural scale have decreased in value. On the day the wrecking ball is demolishing the house, and the tree has already been felled, Müller ironically illustrates a planning firm in the office building that has been on the right of the image since 1966, in which engineers positioned at tall, turned-up drafting tables continue to design for a future that will bring so much change. More than a hundred thousand copies of the series, which sketched in miniature a horrifying image of the wild overdevelopment of Switzerland, were printed, and it was present in many Swiss homes as a book, poster, or wall calendar.

In response to the gradual destruction of villages, cities, and landscapes, as captured by Müller in his naive, childlike style, the "vision of an urban planning architecture" developed, in Burckhardt's view, over the course of the 1960s.[54] Building was to be understood as something that exists in a complex social context, and the interdependent and processual character of all planning and building recognized. For Burckhardt, this vision explains architects' search for ways to cope with the coming tasks in collaboration with other disciplines, such as those of green, landscape, and traffic planners. Very much in keeping with the utopian character of the 1960s, planners pursued the goal of an "even more comprehensive engagement with and responsibility to the future, the environment, and social justice."[55] However, according to Burckhardt, the unshakable self-confidence of the planners leads them to destroy the given structures and features of the very landscape and urban fabric they are dealing with—they treat them as a blank slate and by doing so, destroy them. Jörg Müller captures the same idea of the misguided planner in his depiction of the changing landscape in the year 1969: despite being able to see the destruction of the village and the surrounding landscape from their office windows in the building on the right, they remain blithely industrious, inured by their professional self-assurance and faith in progress.

For Burckhardt, the architect signified the great hope to fill the role of the new type of planner. In the 1960s, the architect had the aura of a technically versed, broadly educated, creative personality who stands out for "drawing a conclusion from information and needs presented in a disordered, sometimes contradictory way, and to turn the solution found into a real form."[56] That was how Burckhardt summed up the then current view of the role of the architect in 1965. This marked the exact beginning of his critique of the education of architects, in which great importance was attached to reducing a problem to its essential qualities and then solving it: repeatedly working out solutions was supposed to lead to the experience that would make the architect a "master" at an advanced age. This "intuitively ordering behavior" of the architect contrasted, according to Burckhardt, with the very different logic of a "planning behavior."[57] He defined planning behavior as "enlightened academicism"[58] and criticized it just as much as the conservative idea of the architect.

Günther Grzimek's reform program for training green urban planners followed the logic of planning behavior. Grzimek postulated that the green urban planner has to take a technically and politically determined plan and turn it into "a functioning plan."[59] For Lucius Burckhardt, however, the point was precisely the opposite: it has to be understood and recognized that for certain tasks there cannot even be a functioning plan and accordingly no "clean" solutions; instead a responsible planner admits this fact and the consequences for the planning. The core of Burckhardt's critique of planning[60] is that politicians act as if a building or even a street, a park, and so on, could be the solution to a social problem or a response to a social need. While planners of "enlightened academicism"

7
Four of the seven panoramas in
Jörg Müller's *The Changing
Countryside,* English trans. 1977
(orig. pub. 1973)

Friday, November 20, 1959
Sunday, April 17, 1966
Monday, July 14, 1969
Tuesday, October 3, 1972

may analyze and discuss based on the latest scientific methods and allow others to participate, they do not question the task as such and will hence delegate that social responsibility to a developer. Burckhardt called for every planner to instead explain the consequences of a chosen solution for those affected by the planning and why it and not some other solution was selected.

The planner's decision, in Burckhardt's view, is still always an ideological one and usually conforms to the values of the planner—here one could substitute "of the architect" or "of the green urban planner"—or "of the clients." The decision is thus sociopolitically determined. In the sense of a new honesty, Burckhardt called for making these processes transparent and understandable. This in turn made it possible to discuss them and call them into question. Following the example of the mathematician, sociologist, and design theorist Horst W. J. Rittel (1930–1990), Burckhardt encouraged an assessment of both problems and solutions in terms of their "malignancy" and suggested that planning also be understood as "transferring suffering."[61] Rather than developing a solution "intuitively" from one's own creative talent (as traditionally imparted in the training of architects, including garden architects) or (as the first reformers of the design disciplines wanted) starting out from an analysis of a program or set of problems, the call is for awareness of the problem and social responsibility on the part of all designers, regardless of discipline. In the future, they should justify every proposed decision on the basis of a scientific and methodological approach and at the same time abandon the idea that they can improve the world.

Burckhardt formulated this fundamental change in the attitude toward planning and toward training architects, where previously great value had been placed on making "the insoluble soluble," among other places in his 1989 text "Das Ende der polytechnischen Lösbarkeit" (The End of Polytechnic Solvability). This text concludes by noting that, from the outset, the students in the newly founded courses of study in Kassel had been guided to deal with "wicked problems," and that herein lay the innovation and strength of this training in Kassel: Problems are insolvable—owing to the fact that they are irrevocably bound up with the allocation of hardship. There are no optimal or final solutions to problems. . . . Unsolvability does not mean, however, that we should do nothing at all, or that technicians are not in demand—on the contrary! Problems must be dealt with. No doctor who discovers that his patient has a fatal disease abandons him to his fate; indeed the doctor knows then that the question of care is more important than ever. By contrast, the polytechnic tradition paints the engineer as the great solver of tasks; if ever things become problematic and he fails to cut the Gordian knot, he thinks his expertise is not required. But it is required, here especially: deontic questions in particular, which pertain to risk, appropriateness, environmental degradation, and the allocation of hardship, are not necessarily dealt with more objectively when expertise is brought to bear, but certainly more effectively, and in more differentiated terms, and with greater leeway for decision-making.[62]

The "Lehrcanapé"

Before Burckhardt was appointed Professor of Socioeconomics of Urban Systems at the Gesamthochschule Kassel in autumn of 1973, he had tried out his teaching method for three years, starting in the winter semester of 1970–71 in the Division of Architecture of ETH Zurich. As part of the student movement in Zurich, the students had demanded that the sociologist be offered, together with the architect Rolf Gutmann (1926–2002), a guest chair of design.[63] It was renamed a "Lehrcanapé" (teaching sofa) because a *Lehrstuhl* (teaching chair) can only seat one person, whereas a sofa offers room for two. When the two teachers proposed a "classic" design task of the sort typical at ETH Zurich—for example, "design a youth center on Paradeplatz in Zurich"—they asked the students not just to accept the proposal but to address the question of whether the proposed task made any sense: What does it mean to build a youth center on this site, one of the most prominent squares in Zurich, where two big Swiss banks had their headquarters? Wouldn't such a building necessarily be exploited by politicians as a superficial solution to the "youth problem"? What was the basis for this "youth problem" and what strategies and offerings would serve the youth in Zurich?

In one of the assessments of the Lehrcanapé that were published regularly as *Canapé News,* its philosophy was summed up as a kind of Socratic approach:

The sociology we teach prohibits such seemingly direct aids to a solution: it does not simplify them, it problematizes them; it causes the person dealing with it not to think about the *object* but first about *himself* and his position within the social structure of decision-making authorities. And therein lies the transfer value of our teaching: that such self-reflection does not leave the person who has consciously done it, and that it always reintroduces him anew to the problems of new tasks. We believe that it will turn the architect not into a "narrow-gauge" sociologist but rather into an architecture specialist who is aware of sociology.[64]

The consequence of this problem-based theory, which above all questioned the self-image of the budding architects and how they would approach tasks in the future in order to open up new perspectives and methods, was that in the end no hypothetical tasks were proposed at all. Instead, the students would work on existing real problems that they selected themselves:
We at the Lehrcanapé do not design hospitals that cannot be paid for, nor theaters in which no one will perform, nor housing for which no land can be purchased, no vocational schools for which there are no more teachers, students, builders, curricula, testing regulations, concepts.[65]

Both Burckhardt's critique of the long-postulated autonomy of the intuitively designing architect and his general questioning of setting hypothetical tasks for the purpose of practicing design cut to the quick of the principles of the architectural education at ETH Zurich. The Division of Architecture dismissed the critical teacher in the summer of 1973, and Burckhardt accepted the Chair of Socioeconomics of Urban Systems at the Gesamthochschule Kassel that autumn and continued his teaching there.

Strollology

Building on the critique of planning he had begun in Switzerland,[66] Burckhardt developed in Kassel the theory of the "smallest possible intervention" (*l'intervention minimale*) in exchange with the landscape architect and artist Bernard Lassus (b. 1929).[67] And in the 1980s, in connection with his interest in perceiving the city and the country, he developed "strollology" or "promenadology." Strollology was based on the experience of perceiving an urban and rural structure. It is not a coherent, precisely identifiable form of cognition but is grounded in actionist strolls in the city and its periphery, and the insights gained in the process were sometimes summarized ironically as science (8, 9).[68] The purpose of the actions was to obtain a new perspective on the seemingly familiar, and also on something that had not yet been called into question, and to gain knowledge in the process. "Reise nach Tahiti" (Voyage to Tahiti) was Burckhardt's name for a 1987 walk across the Dönche, a nature reserve near Kassel that was in part a military training ground. The viewing of the terrain was accompanied by the reading of texts aloud from a book published in 1777, in which the naturalist and ethnologist Georg Forster described his discovery of the island of Tahiti alongside James Cook. The eighteen-year-old had accompanied Captain Cook on his second voyage around the world (1772–75), and his description launched a new literary form of scientifically grounded travelogue. In 1778, he was appointed Professor of Natural History at the Collegium Carolinum in Kassel.[69] A street on the grounds of the Gesamthochschule is named after Forster. Burckhardt's choice of Forster's text and the superposition of the descriptions of his voyage on the landscape of the abandoned training ground—a kind of terra incognita—were certainly no coincidence. The strollers, as citizens and scientists from Kassel, were following in the tradition of the great discoverer, world traveler, educator, and politician. The message of this action is that in the twentieth century there are still things waiting to be discovered just beyond the front door.

The ironic defamiliarization to which the former training ground was subjected by the South Sea travelogue being read aloud opened up for the strollers another perspective on a landscape that had an entirely different cultural

8
Lucius Burckhardt's "Reise nach Tahiti" (Voyage to Tahiti) action in the Dönche Nature Reserve in Kassel, 1987. Photo: Klaus Hoppe

9
Drawing by Lucius Burckhardt of the destruction of the strollological context, so that elements of the landscape are no longer legible

10
"Das Zebra streifen" (Striping the Zebra) action, 1993. This transportable "zebra stripe" (striped crosswalk) carpet was long enough to span a six-lane street. It was designed by the artist Gerhard Lang and was opened up wherever the strollers wanted to cross a street. Photo: Angela Siever

connotation for them. It was also about rediscovering a piece of Tahiti in the Dönche and realizing that one sees a little of what one hears.[70] The participants could become aware of their own cultural construction as they observed landscapes and also question their own connection to the world. Using this example of peeling away the cultural construction to which we are subject as viewers of a landscape, Burckhardt formulated his critique of the common approach to nature preservation: identifying typical landscapes and then protecting them as such. He saw the identification of the typical as a threat to the diversity of all that existed, because, once the typical had been determined, the diverse would be harmonized with it through interventions of nature preservation—by human beings and according to their ideas.

The action "Das Zebra streifen" (Striping the Zebra) was organized by former students in 1993 to honor Lucius Burckhardt on the occasion of his retirement from the Gesamthochschule Kassel: the artist Gerhard Lang, the architect Ruth Jureczek, and the typographer Helmut Aebischer (10). Lang developed a transportable carpet as artwork, 4 meters across and 30 meters long, with a zebra pattern that could be placed in an urban space wherever the strollers wanted to overcome an obstacle "set up" by urban planning that hindered their ability to walk freely. The urban stroll became a procession, accompanied by six hundred students from the departments of architecture and urban and landscape planning. Lucius Burckhardt was euphoric:
Once again moving through the city the way it was supposed to be possible: you cross the street wherever you want and wherever you can shorten your path rather than where traffic planners had a zebra crossing painted, because they thought you would disturb the auto traffic least there.... Anyone who does not want to commit a punishable offense by crossing the road without permission can bring his own zebra crossing.[71]

Opening up one's handmade oversized zebra carpet wherever the strollers wanted to, rather than in the places the traffic and urban planners had seen fit, revealed the conflicts over the use of public space. Learning in Kassel resulted from logically resolving the conflicts that such actions uncovered. Burckhardt once held a seminar in a parking lot, setting up tables and chairs for the attendees, which provoked passionate, aggressive reactions from drivers, even though they had paid the required parking fee. The result: a parking car triggers no aggression, but people spending time in parking lots do. The playful, uninhibited quality of such actions, which are always ambiguous in their supposed *naïveté*, recalls related attitudes of renowned Swiss artists, such as Fischli/Weiss, Roman Signer, and Markus Raetz. This shows, first, the reciprocal openness and interest that critics of urban planning displayed toward artistic actions between 1960 and 2000 and, conversely, artists' interest in the critique of urban and landscape planning. Second, the playful, ironic, seemingly childish and innocent feature is striving for a strong critical effect—a subversive attitude—that is often found in the Swiss-German art of every generation.

The Stroll as Narrative

Another dimension of strollology is the effort to recreate contextualization, a narrative, through walking—or indeed gained knowledge that this narrative is no longer producible. Within the organization of a city or a landscape, there are organically evolved structures which can be traced back to types of production that have changed over the course of history (agriculture, commerce, industry), and also back to trade, to cultural traditions, or to political or religious practices. The centers of cities and villages have one or more market squares; situated next to them are churches, the town hall, and other administrative buildings. Somewhat outside the center of the city or village lie the train station and its neighborhood; on the edge of the city or village begins the countryside. These familiar structures enable us to orient ourselves even in places we have never been. As a result of various factors—demographic changes resulting from falling birth rates or exodus due to declining industries, new needs for residential space, overdevelopment of urban peripheries into agglomerations with large shopping centers, and so on—these ordering systems are no longer valid, especially in the outer areas or cities, or in the countryside between them. Thus, it is no longer possible to orient oneself based on cultural knowledge.

Burckhardt recorded this development in a series of drawings. In one such drawing, the upper image shows the gable of a chapel in the middle of a hilly landscape; the lower image shows the changing landscape and new context where the chapel is now on the edge of a parking lot (9). In a speech balloon it describes itself: "Je suis une chapelle de campagne!" The rural context has been so totally destroyed that what was once simple to read—Burckhardt would have said: the "strollological" context of meaning—has been lost. Hence it requires explanation: "I am a country chapel!"

Strollology combines the knowledge of how to track down and read the historical structures of the city and the countryside with the perception and description of contemporary development.[72] On the one hand, it helps us to understand new forms of city and countryside in the experience of walking through them. On the other hand, it casts a critical eye on these new forms of rampant overdevelopment based exclusively on economic criteria and no longer aesthetic ones. Burckhardt worked out critically that the distinguishing perception of city and countryside had broken down because the antitheses were becoming increasingly blurred. Burckhardt wanted to counteract this elimination of contrasts by urbanizing the countryside and ruralizing the city with a type of planning that consciously recreated the lost tension:

Anyone who designs a park today must incorporate the user's experience of the shift in context, the shift from the city to the park. In the past, people really did leave the city to visit a park. In eighteenth-century Paris, one still walked through narrow stone streets to reach the Tuileries Gardens. Today, one arrives there from the spacious tree-lined sidewalks of the Champs-Élysées. Moreover, on entering the Tuileries Gardens one finds just as much asphalt and just as much green as on the Champs-Élysées. So, if we were to build the Champs-Élysées today, we would have to create this suspense, this shift symbolically, would have to draw on the art of gardening and landscape planning to conjure an impression of this shift "from the city to the park." This is a wholly new challenge and one to which few landscape gardeners have yet risen.[73]

One narrative that can be derived from a stroll is the synthesis of a chain of impressions, just as the stroll itself is "thus a chain, a string of pearls made up of more expressive and then less expressive passages that are ultimately synthesized in the mind's eye."[74] According to Burckhardt, designing in today's ruralized urban areas and urbanized countryside needs this series of impressions and antitheses, the more and less expressive, so that strollers can orient themselves within it and find it beautiful. That is why Burckhardt strictly rejected connected strips of green in a city, for example. By constantly breaking through the stone city, making it impossible to experience it in its full force, the experience of green in the city is diminished as well.[75] When asked by a mayor once how the municipal park could be made more popular among the residents again, because they had become dissatisfied with the park even though its substance had not been altered in any way, Burckhardt responded that the residents no longer noticed the beautiful municipal part after they had crossed "the tree-lined shopping street and the Town Hall's rubber plant collection."[76] He proposed, as an emergency measure, that the municipal park should be fenced off and a guard placed at the gate so that residents would become aware of the park again as a green contrast with the city.[77]

Kienast and the Kassel School

The effort to make an environment legible by designing it deliberately and taking into account human needs became central themes of Dieter Kienast's work. In that sense, he continued the teachings of Lucius Burckhardt and Karl Heinrich Hülbusch. There was also a lively exchange of ideas between the latter two men as well: Burckhardt and the circle around Hülbusch were repeatedly involved in joint actions, including resistance to the Bundesgartenschau (Federal Horticulture Show) held in Kassel in 1981. The preparatory work for the show caused massive damage to the Karlsaue in Kassel, which Burckhardt and Hülbusch documented and protested in their publication *Durch Pflege zerstört: Die Kasseler Karlsaue vor der Bundesgartenschau* (Destroyed by Maintenance: The Karlsaue in Kassel before the Federal Horticultural Show) (11).[78] The Stadtgartenamt (Municipal Horticulture Department) and its particular stance on historical garden art and on public green spaces became their common enemy.

Burckhardt, Hülbusch, and their students were also involved in student protests against the school administration, as was Joseph Beuys (1921–1986) (12), in the design of the university grounds of the Henschelei, and in Beuys's action *7000 Eichen* (7,000 Oaks, 1982–87). Under the slogan "Stadtverwaldung statt Stadtverwaltung" (City Foresting Rather Than City Administration), as part of an action begun at documenta 7 in 1982, the artist wanted to cover the City of Kassel with trees (13).[79] As a statement against increasing urbanization, this artistic, social, and ecological action was intended to alter the urban habitat. Beuys stacked a wedge-shaped pile of seven thousand basalt stelae in front of the Museum Fridericianum, at the top of which he planted the first of the seven thousand oaks. Purchasing a basalt stela for 500 deutschmarks enabled one to plant a tree next to the stone in the City of Kassel. A coordination office took care of the financing, planning, and implementation.

Beuys saw himself as a worker on the project of "social sculpture": "Sculpture only has a value if it works on developing human consciousness.... I mean to say that developing human consciousness is itself already a sculptural process."[80] Public involvement was just as much part of the invisible sculpture as the vanishing of the pile of stones and the trees scattered across the topography of Kassel. It also resulted in a tension between the organic and inorganic, the growing and the formed, and the green and the gray, in that Beuys had a basalt stela placed next to each tree. In his own words, his concern was "that each would be a monument, consisting of a living part, the live tree, changing all the time, and a crystalle mass, maintaining its shape, size, and weight."[81]

Despite the title of the action—which Beuys associated with the symbolism of oaks[82]—the trees that were planted in Kassel were chosen based on their location. Beuys was assisted by Hülbusch and his phytosociology students in choosing the sites and by Burckhardt in promoting his ideas. The linking of questions of ecology, social resistance, form, and material in this action is an outstanding example of the issues that occupied Dieter Kienast's generation in the 1980s and influenced their works.[83] Consciousness, symbolism, and the tension of materials were fundamental to Kienast's design vocabulary for landscape in the 1980s and 1990s.

There were, however, not just commonalities but also significant differences. Both Burckhardt and Hülbusch harbored a deep mistrust of "brilliant" formal solutions. Kienast, however, had been trying since the mid-1980s to achieve just such brilliance. In his design, he became the master of fine details. Burckhardt and Hülbusch vehemently rejected any "autonomous architecture" whose designs were based on the intuition of a "master." Kienast agreed with their arguments, but his critique of architecture and society grew more subtle. Despite these differences, he was largely in agreement with the ideas of the Kassel School. Burckhardt and Hülbusch were highly critical of hierarchies and administrations, and of the planning and design tasks they formulated. Both pressed for planning and design that take seriously the needs and desires of the "affected" citizens and incorporate

11
Documentation of the destruction of the Karlsaue in Kassel by the preparations for the Bundesgartenschau (Federal Horticultural Show) in 1981

12
Joseph Beuys planting the "Daxner linden" in 1980 during student protests against the ministry of culture in Hesse for refusing to confirm Michael Daxner as the first rector elected by the convention of the Gesamthochschule Kassel in accordance with the rules. Photo: Helmut Böse

13
Joseph Beuys, *7000 Eichen: Stadtverwaldung statt Stadtverwaltung* (7,000 Oaks: City Foresting Rather Than City Administration), Kassel, 1982–87

14
Dieter Kienast's playful "do-it-yourself" planning model for the "Erde" (Earth) sector of the Schweizerische Landesausstellung für Garten- und Landschaftsbau (Swiss National Exhibition of Horticulture and Landscape Architecture), Grün 80, in Basel

these individuals in the process—especially when building housing and planning the associated open spaces. They emphasized the social responsibility of planners. Kienast had a similar attitude and worked hard for processes of participation, which were then expressed in the design of numerous works of the 1970s and 1980s.[84]

The professionalizing of amateurs and the simultaneous amateurizing of professionals, with particular emphasis on the crafts, was a fundamental theme of both Burckhardt and Hülbusch. Kienast regarded the craft components of his work as essential. In his rhetoric, he placed the gardener, as an artisan in the best sense, far above the artist. The professionalizing of amateurs was a great concern for him as well. For the Grün 80 (Green 80), the Schweizerische Landesausstellung für Garten- und Landschaftsbau (Swiss National Exhibition of Horticulture and Landscape Architecture), in Basel in 1980, he developed an experimental model for planning in which visitors could put together their own urban open space (14). Above the model was a beam with the words "Gray or Green" in three of Switzerland's official languages. The question of the green or gray city runs through Kienast's entire oeuvre.[85]

With concentrated interest, Burckhardt, Hülbusch, and Kienast studied historical role models from the fields of gardening, urban planning, architecture, and landscape design. All three were also concerned with—albeit in different ways—the visual character of the city and the countryside, issues of their legibility, and the question how planning could support this legibility and take advantage of their visual character. For Hülbusch's circle and for Dieter Kienast, understanding production processes and studying the local vegetation played a crucial role in the interpretation and subsequent planning. Burckhardt was more interested in aesthetic questions, issues of cultural constructions, and developing a design response to the disappearance of the boundaries between city and landscape. That quality became increasingly important in Kienast's work from the mid-1980s onward. The argument that Burckhardt had been making since the early 1990s in the context of strollology in favor of a kind of design that took aim at the ill-considered and undifferentiated mixing of urban and rural elements was echoed by the third of Kienast's "Zehn Thesen zur Landschaftsarchitektur" (Ten Theses on Landscape Architecture):
The traditional city–countryside rivalry has disintegrated, the boundaries have blurred. We assume that it is impossible either to dismantle the city or to rejuvenate the countryside. Nevertheless, the legibility, the perceptibility of the world is rooted in the principle of dissimilarity. Considering this synchrony of city and countryside, the coming task is to stop the further erosion of the inner boundaries and splits. They have to become sensuously perceptible again. [86]

These lines date from a period already fifteen years after Kienast left Kassel, and they underscore yet again that he and the Kassel School had arrived at parallel views and methods. Burckhardt's principles also accorded in many respects with those of the circle that claimed the name Kassel School for itself, with its own definition, in an effort to distinguish itself from the other teachers in Kassel. In that sense, it is understandable that the uninitiated also considered Burckhardt to be part of the Kassel School. The substantive differences between Burckhardt and the Hülbusch circle were that the latter emphasized far more strongly the sociological background when analyzing the visual quality of the landscape and placed far more weight on appropriating open spaces. In the 1980s, both Burckhardt and Kienast distanced themselves from the Hülbusch circle, whose position they perceived as increasingly dogmatic.

An Introduction
to Vegetation Science
and Phytosociology

The new theory of planning open spaces developed in Kassel called into question several of the principles of the design of gardens and landscapes as taught at art academies and universities until the 1970s (and beyond). Integrating the disciplines of architecture and urban and landscape planning into one course of study was intended to provide institutional support for the idea that "open space" should now be understood to include urban space as well—a way of thinking that, like Lucius Burckhardt's vision of "urban planning architecture," was the result of the developments of the 1960s. As far as the discussions of participation and the use value of open spaces were concerned, the new beginning was based, first, on a parallel professionalizing of amateurs and the simultaneous amateurizing of professionals, and, second, on collecting objective data and evaluating it using scientific methods. These methods were usually borrowed from the social sciences, with the result that sociology could be introduced into architecture. In the special case of the Kassel School, it also incorporated vegetation science and phytosociology, which were devoted to studying plants and their communities. The existing vegetation is read and interpreted as a clue to milieu, maintenance, and use, hence becoming the basis for statements about the history of a landscape, its changing utilization, and the social origins of its use. A critique of planning based on the methods of studying vegetation was what definitively distinguished Kassel from other universities.

Karl Heinrich Hülbusch led students on excursions to study vegetation and observe the landscape. His goal was to break them of the habit of "the *idée fixe* of making" that had been characteristic of generations of planners.[87] The crucial consideration for him was the interpretation of the analyzed landscape and the measures derived from it, which could possibly lead to an intervention that would change the landscape. Hülbusch wanted his students to learn to understand vegetation as "indicators of the methods and relationships of production" and as a "material expression of the social economy and of social history."[88] On these walks, landscape should be understood not as nature but as work that had concretized in a "decoration of nature," as something that resulted from the economy. Hülbusch tried to counter the paradigm of the transformation of nature into landscape—influenced by philosophers such as Georg Simmel[89] and Joachim Ritter[90]—with a different way of seeing by asserting a "purely" aesthetic view. That was the decisive difference between Hülbusch's walks and Burckhardt's, even though the latter was also about images and perception. In Hülbusch's view, every culturally constructed idea of landscape should be combated as much as possible by the careful observation of landscape. Vegetation was observed and interpreted in connection with a knowledge of economics and social history that the strollers had to contribute themselves or acquire over the course of time.[91]

Hülbusch considered knowledge that comes from practical experience to be particularly valuable, such as that of farmers, for example, who distinguish between meadow and grassland and can connect these forms of vegetation with the number of calves born, or who can determine from the color of butter the type of landscape on which the milk cows had grazed.[92] Of interest here, not least, is how knowledge also uses descriptive terms that refer simultaneously to the observed object, to work, and to use, such as "pasture," "meadow," "woodland pasture," "low or coppice forest," "high or timber forest," "selection-cut forest," and so on. From the appearance of a landscape—Hülbusch used the phrase "image of vegetation"—one can read not only work and use but also the history of government, taxes, sanctions, energy crises, laws of inheritance, changes in the awarding of subsidies, and more.[93] By mapping and interpreting the vegetation in the context of knowledge from social and economic history, one can then write a kind of explanatory narrative of a landscape's history and the conditions of production there. If landscape planners acquire this knowledge and these insights, they will, according to Hülbusch, be in a position to plan "more rationally."

Hülbusch's second central method of analysis was phytosociology. That is a separate research method within vegetation science in which the existing plants are grouped into plant communities according to their qualities specific to their location. Phytosociology had been the dominant method in Europe for describing plants and their communities since the 1930s. Schools taught different methods for mapping vegetation.[94] In his work, Dieter Kienast referred to two methods: that of the Swiss German botanist and phytosociologist Josias Braun-Blanquet (1884–1980) and that of the German botanist and phytosociologist Reinhold Tüxen (1899–1980). Tüxen had been Hülbusch's doctoral advisor and a member of Kienast's dissertation committee. In phytosociological analyses, the existing plants in a self-selected area are recorded according to type, coverage, and sociability and then classified in a hierarchically organized system (15). Using this system, recurring combinations of plants are identified, which can then be integrated into the existing phytosociological system or described as a new plant community.[95]

As he did with vegetation science, Hülbusch linked phytosociology to socioeconomic questions, such as the social class of the residents of a given area and their habits of use, but also to the question of the costs of maintaining streets or parks. The plants provide, so to speak, incorruptible, objective data, which Hülbusch taught his students to read and interpret. It was an obvious

SAGINO-BRYETUM ARGENTEI DIEM., SISS. et WESTH. 1940

		1 2 3 4 5 6	7 8 9 10 11 12 13	14 15 16 17 18 19	20 21 22
	Laufende Nr.				
	Nr. der Aufnahme	339385545266625627	5755465425436526585444	685570567661566582	588238782727
	Jahr der Aufnahme	76 76 76 74 76 76	76 76 76 76 76 76 76	76 76 76 76 76 76	76 76 76
	Probeflächengrösse	1 1 1 2 5 2	1 6 1 2 2 1 2	1 1 1 4 1 1	6 5 6
	Vegetationsbedeckung insgesamt	80 80 60 40 75 90	95 90 90 95 60 90 90	70 90 80 95 80 95	95 20 40
	Vegetationsbedeckung Kryptogamen	70 60 30 v 60 90	70 40 80 80 40 85 60	30 50 30 90 30 50	70 2 30
	Artenzahl	5 7 9 7 9 9	13 6 7 9 12 10 11	10 13 8 13 10 11	12 5 5

Ch.	Bryum argenteum	45 45 33 22 12 21	44 33 54 22 11 11 11	33 11 11 11 33 11	11 11 22
	Sagina procumbens	. . + +6 33 22	11 +2 22 33 22 22 33	. 33 33 22 33 55	33 . .
	Ceratodon purpureus	. . . 12 22 45	11 . + 55 34 . 33	22 22 22 55 11 33	34 11 33
D¹	Spergularia rubra	+ 22 22 22 12 21	11	. . . . . .	. . .
D²	Herniaria glabra	. . . . . .	22 44 33 33 +2 22 23	. . . . . .	. . .
d2	Tripleurospermum inodorum	. . . . . +	. . . + 22 + +	11 + 12 11 . +2	+ . .
	Senecio viscosus	. . . . . .	. . . . 22 + +	+ . . 11° + +	+ . .
	Reseda luteola	. . . . . .	. . . . . . +	. + 11 . + +	+ . .
D³	Eragrostis minor	. . . . . .	. . r	. . . . . .	11 22 22
D⁴	Capsella bursa-pastoris	. . . . . .	. . . .	. . . . . .	. . .
	Matricaria discoidea	. . . + . .	. . . .	. . . . . .	. . .
D⁵	Lepidium ruderale	+ 22	. . . .	. . . . . .	. . .
D⁶	Stellaria media	. . . . . .	. . . .	. . . . . .	. . .
d	Plantago major	. . . . . .	. . . .	. . . . . .	. . .
VOK.	Polygonum arenastrum BOREAU	22 22 33 . 11 11	+ 11 11 + + + +	. 11 22 r 23 +	+ 11 24
	Poa annua	11 22 . 12 21 r	+ . . . 11 + .	. + . 22 11 +	. r .
	Polygonum aviculare agg.	. . . 12 . .	. . . .	. . . . . .	. . .
	Polygonum aviculare L.	. . . . . .	. . . .	. . . . . .	. . +2
B.	Conyza canadensis	. . + . + +	22 . . r + 11 +	+ + 11 11° + +	+2 . .
	Solidago canadensis	. . . . . .	11 r . + 11° . .	+ r	. . .
	Taraxacum officinale	. r	r	. . . r . .	r . .
	Sonchus oleraceus	. . . . . .	. . . . + . .	. . . . . .	11 . .
	Arenaria serpyllifolia	. . + . + +	. . . .	. . . + . .	. . .
	Agrostis tenuis	. . . . . .	. . . .	. . . 11 . .	. . .
	Bromus tectorum	. . + . . .	. . . . 11 . +	. . . . . .	. . .
	Epilobium tetragonum	. . . . . .	r°	. r . . . +°	+° . .
	Epilobium spec.	. . . . . .	. . . .	. . . . . .	. . .
	Poa pratensis	. . . . . .	. . . .	. . . +2 . .	. . .
	Poa compressa	. . . . . +	. . . .	+	. . .
	Cerastium font. holosteoides	. . . . . .	. . . .	. . . . . .	. . .
	Lolium perenne	. + ...	+°	. . . r . .	. . .
	Artemisia vulgaris	. . . . . .	. . . + . + .	. . . . . .	. . .
	Hypericum perforatum	. . . . . .	. . . .	33	. . .
	Poa prat. angustifolia	. . . . . .	. . . .	. . . . . .	. . .
	Veronica arvensis	. . . . . .	. . . .	. . . . . .	. . .
	Sisymbrium officinale	. . . r . .	. . . .	. . . . . .	. . .

Kryptogamen und Flechten
	Pohlia nutans	. . . . . .	. . . . 55 . .	. . 11 . 11 .	. . .
	Thallus spec.	. . . . . .	. . . .	. . . . . .	. . .
	Barbula fallex	. . . . . .	. . . .	. . . . . .	. . .
	Barbula convoluta	. . . . . .	. . . .	. . . . . .	. . .
	Barbula unguiculata	. . . . . .	. . . .	. . . . . .	. . .
	Bryum caespiticeum	. . . . . .	. . . .	. . . . . .	. . .
	Riccia spec.	. . . . . .	. . . .	. . . . . .	. . 4
	Barbula hornschuchiana	. . . . . .	. . . .	. . . . . .	. . .

Ausserdem je einmal: in 7 Bromus sterilis +; in 11 Epilobium angustifolium +; in 14 Erigeron acris +; in 15 Cornus sanguinea r°; in 20 Solidago gigantea 11°; in 25 Sonchus asper r; in 29 Betula pendula juv. r ;

    Aufnahmen   1 - 6   Typ. Subassoziation, Var. v. Spergularia rubra
        "   7 - 13   Subass. v. Herniaria glabra, Var. v. Herniaria glabra
               14 - 19   Subass. v. Herniaria glabra, Typ. Var.
               20 - 28   Subass. v. Eragrostis minor
               29 - 36   Subass. v. Capsella bursa-pastoris
               37 - 49   Subass. v. Lepidium ruderale, Var. v. Capsella bursa-pastoris
               50 - 54   Subass. v. Stellaria media

15
Chart of a phytosociological mapping of the *Sagino-Bryetum Argentei* (Diem., Siss., Westh., 1940) plant community and its subassociations as found in Kassel. Dieter Kienast, "Die spontane Vegetation der Stadt Kassel" (Spontaneous Vegetation in the City of Kassel), dissertation, 1978, table 7

Tabelle 7

24	25	26	27	28		29	30	31	32	33	34	35	36		37	38	39	40	41	42	43	44	45	46	47	48	49		50	51	52	53	54
2	1	2	6	3	3	5	8	7	2	3	7	2	3	6	1	9	5	2	8	1	2	2	7	2	9	9	1	1	9	6	2	6	0	1	1	0	3	2	1	6	7	3	2	2	2	4	5	8	5	6	2	2	6	6	3	8	4	5	2	6	5	1	3	5	5	0	4	2	8	4	9	8	5	4	9	5	2	7	4	7	1	7	3	3	7	6	6	1	2	0
76	76	76	76	76		75	76	75	76	75	76	75	75		76	76	75	76	75	76	76	76	76	76	76	76	76		76	76	76	76	75
4	2	4	3	2		9	1	1	1	1	2	2	1		1	2	4	3	2	4	4	4	1	2	3	4	3		2	3	1	3	1
35	70	90	75	75		80	95	70	100	80	40	45	75		60	75	90	75	90	90	90	98	70	70	60	95	80		80	80	60	60	90
35	40	90	60	70		v	30	v	80	v	v	v	v		10	70	v	50	20	v	80	90	40	20	v	80	40		80	40	5	40	v
7	12	7	8	7		11	7	8	7	7	7	11	9		9	12	9	11	9	9	16	12	9	9	7	10	10		9	15	10	9	6

21	23	33	44	21		45	33	11	44	22	12	.	11		11	33	45	33	22	45	22	22	33	22	33	34	33		11	32	11	22	11
11	11	+2	11	+		21	22	12	33	11	22	+2	21		44	+	33	22	11	33	22	33	.	+2	22	22	11		+	11	33	.	33
33	33	-	22	44		22	.	.	33	.	.	+2	45		11	33	.	22	.	.	33	44	.	.	12	33	.		33	33	+2	34	.
.	.	.	.	.		.	.	.	.	.	.	.	.		.	.	.	.	.	.	.	.	.	.	.	.	.		.	.	.	.	.
.	.	.	.	.		12	.	.	.	.	.	.	.		.	.	.	.	.	.	.	.	.	.	.	.	.		.	.	.	.	.
.	r	.	.	.		.	.	.	.	.	.	.	.		.	.	.	.	.	.	.	.	.	.	.	.	.		.	.	.	.	.
.	.	.	.	.		.	.	.	.	.	.	.	.		.	.	.	.	.	.	.	.	.	.	.	.	.		.	.	.	.	.
11	22	33	12	+		.	.	.	.	.	.	.	.		.	.	.	.	.	.	.	.	.	.	.	.	.		.	.	.	.	.
.	.	+	+	11		21	22	+	r°	+	+	22	+2		+	+	+	+	+°	(r)	+	.	.	.	.	.	.		.	.	.	.	.
.	.	.	.	.		+	.	.	.	11	+	22	+°		11	.	.	+	.	.	11	+	21	+	.	.	.		+	11	.	.	.
.	.	.	.	.		.	.	.	.	.	.	.	.		.	+	+	+	+2	+	11	+	11	22	11	21	+		.	.	.	.	.
.	.	.	.	.		.	.	.	.	.	.	.	.		.	.	.	.	.	.	.	.	.	.	.	.	+		+	+	r	r°	+
.	.	.	.	.		+2	11	+1	11	+2	+	+	+		r°	r	.	+	32	.	11	+	11	.	11	+	11		.	.	+°	r°	.
11	r	.	22	+		.	33	.	.	.	.	.	.		12	22	.	33	44	.	11	.	12	+	.	+	22		.	+2	11	22	.
.	.	+	.	11		11	+2	11	+	22	22	22	12		+	+	12	11	11	12	11	.	+	.	11	+	22		+	+	22	21	22
.	.	.	.	.		13	.	11	.	33	21	+	+		.	+	.	11	.	.	11	.	.	.	.	.	.		.	.	.	.	11
.	.	.	+	.		.	.	.	.	.	.	.	.		.	.	.	.	.	.	.	.	.	.	.	.	.		.	.	.	11	.
.	11	.	.	.		21	11	.	.	.	.	.	.		.	+	r	+	.	r	+2	+	.	.	.	11	.		.	33	+	.	.
.	r°	+	.	.		.	.	.	.	.	.	.	.		.	.	.	.	.	r	.	.	.	r	r	.	.		+	.	.	.	.
.	.	.	.	.		.	.	.	+	.	.	.	.		.	.	.	.	.	.	+	+	.	22	.	.	.		.	+	.	.	+
.	.	.	.	.		.	.	.	.	.	.	.	.		.	.	.	.	.	.	r	+	.	11	.	.	+		.	+	+2	.	.
.	11	.	.	.		.	.	.	.	.	.	.	.		.	.	.	.	.	.	.	+2	.	.	.	+	.		.	.	.	.	.
.	.	.	.	.		.	.	.	.	12	+2	.	.		.	.	.	.	.	.	.	.	.	.	.	.	.		.	.	.	.	+2
.	r	.	.	.		.	.	.	.	.	.	.	.		.	.	.	.	.	.	r	.	.	+	.	.	.		.	.	.	.	.
.	.	21°	.	.		.	.	.	.	.	.	.	.		.	.	.	.	.	.	.	.	.	.	.	+	.		.	.	.	.	.
.	.	.	.	.		.	.	.	.	.	.	.	.		.	.	.	.	+2	+	.	+	.	.	12	.	.		.	.	.	.	.
.	.	.	.	.		.	.	+	.	.	.	.	.		.	.	.	.	.	.	.	.	.	.	.	.	.		.	.	.	.	+
.	.	.	.	.		.	.	.	.	.	.	.	.		.	.	.	.	.	.	.	.	.	11	.	.	.		.	.	.	.	.
.	.	.	.	.		.	.	.	.	.	.	.	.		.	.	.	.	.	.	.	.	.	.	.	.	.		.	+2	.	.	.
.	.	.	.	.		.	.	.	.	.	.	.	.		.	.	.	r	.	.	r	.	.	.	.	.	.		.	.	.	.	.
.	.	.	.	.		.	.	.	.	.	.	.	.		.	.	.	.	.	.	+	.	.	.	+2	.	.		.	.	.	.	.
.	.	.	.	.		.	.	.	+	.	.	.	.		.	.	.	.	.	.	11	.	.	.	.	.	.		.	33	.	+	.
+	.	.	.	.		.	.	.	.	.	.	.	.		.	.	.	.	.	.	.	.	.	.	.	.	.		.	.	.	.	.
.	.	.	.	.		.	.	.	.	.	.	.	.		.	.	.	.	.	.	11	.	.	.	.	.	.		.	.	.	.	.
.	.	.	.	.		.	.	.	.	.	.	.	.		.	.	.	.	.	.	.	.	.	.	.	.	.		22	.	.	.	.
.	.	.	.	.		.	.	.	.	.	.	.	.		.	.	.	.	.	.	.	.	.	.	.	.	.		.	.	.	.	33
11	.	.	.	.		.	.	.	.	.	.	.	.		.	.	.	.	.	.	.	.	.	.	.	.	.		.	.	.	.	.
.	11	.	.	.		.	.	.	.	.	.	.	.		.	.	.	.	.	.	.	.	.	.	.	.	.		.	.	.	.	.

in 35 Poa trivialis +2, Calystegia sepium r, Chenopodium album r; in 38 Malva neglecta r, Herniaria hirsuta 12; in 41 Tanacetum vulgare r°; in 44 Galinsoga parviflora r; in 45 Dactylis glomerata +; in 51 Epilobium montanum +2, Viola odorata +2, Geranium molle +2, Urtica dioica +°.

and yet characteristic idea to make a return to plants—the very material of the profession of gardeners and landscape architects—the point of departure for a new theory. This reorganization of the content of teaching literally went "back to the roots" of the métier and provided solid ground to legitimize the new theory for planning open space.

The scientific method of phytosociology was linked with issues of the social sciences in its analysis. The existing plant communities can be interpreted as evidence of human production, use, and maintenance. Kienast remarked on this in 1978, in the introduction to his dissertation, "Die spontane Vegetation der Stadt Kassel in Abhängigkeit von bau- und stadtstrukturellen Quartierstypen" (Spontaneous Vegetation in the City of Kassel in Relation to Types of Neighborhoods in Terms of Architectural and Urban Structure):

The desires and demands of the residents for the open space can . . . indeed be made tangible by observing the behavior of residents in the open space. . . . The observations have to be made in every block and neighborhood, because each case to be studied represents a special case at first. The question of the relationship between space and behavior—which presumes that the two variables can be isolated—is meaningless in this process, because space always implies behavior as well. . . .

Because human beings leave clues behind when spending time in open spaces, . . . these clues can permit inferences about their behavior there. This is only true, however, if the areas in question are the sort on which vegetation can grow. Sealed asphalt or concrete areas are not suited to this kind of observation, but coverings associated with water, natural stone, and even concrete composite pavement will enable vegetation to develop, if only minimally.[96]

In his 1996 essay "Schwierigkeiten mit dem Spurenlesen" (Difficulties with Reading Clues), the university professor Gerhard Hard, who is highly educated in biology, geography, and German language studies, summed up what Kienast's dissertation presented in detail: the possibilities of reading clues by evaluating the existing vegetation and the advantages of this method over other methods of study, such as a sociological survey based on questionnaires.[97] When the goal is to determine relationships and needs of use, the vegetation found in different places of a park, city, or landscape offers impartial information—unlike that of a human being, who is subject to social pressure when interviewed—about how often the location is fertilized, pruned, or mowed and where the space is heavily used. Depending on these factors, the numbers of this or another plant and its combinations with that or another plant will vary. This procedure is, however, not as exact as the advocates of the method would claim, because the different local characteristics of plants can overlap and influence one another.[98]

Reading clues by means of phytosociological mapping is one of the nonreactive measuring methods that have been increasingly demanded in the social sciences since the 1960s, because so-called reactive methods—including the classic survey using questionnaires—regularly contain powerful distortions.[99] For example, when asked about the maintenance of a park, municipal gardeners will tend to give an answer that reflects more their ideal notions of the necessary fiscal and time resources than the actual situation. The analysis of vegetation shows which "weed" is growing where and how much, and from that one can conclude relatively well—with the aforementioned limitation—how fertilizer is used here and how often, as well as how high the pressure is to utilize specific areas of the park.

Nonreactive measuring methods, like searching for clues, analyze physical traces that can be categorized as "erosion measurements" (abrasion) and "accumulation measurements" (residue, deposits) and interpreted accordingly.[100] This can register specific human behavior or actions. Its reasons and social context must, however, be determined by employing other methods from the social sciences. For Hard, this can be done, for example, through follow-up oral questioning, but not in the form of standardized questionnaires, as it is less susceptible to the projection of stereotypes. The interviews based on concrete clues should take the form of "spontaneous conversations" in which the interviewees can "express *their* view of things," such as their philosophy on controlling weeds, using yards, and so forth.[101]

In conclusion, Hard explained the possibilities and uses of "clue-reading" vegetation science and phytosociology, which he claims increases the ability of the landscape planner to predict, deal with, and interpret data in three steps: First, as far as the ability to predict is concerned, he knows better than before which vegetation and vegetation dynamics are to be expected, what competition and degree of competitive pressure the planted vegetation will face, and what its (usually overestimated) odds of survival are under these conditions. Second, as regards handling ability, he can then deal with spontaneous and planted vegetation better in a technical sense (and will not, for example, as has often been done to date, produce ineradicable concentrations of weeds through his maintenance efforts). Third, on the level of a general ability to interpret, he can better read open spaces with an eye to their use and utility (and even to the needs of specific users).[102]

According to Hard, this kind of vegetation science can thus be employed for analyzing situations (history of an open space, former and present use), for predicting and scheduling maintenance (especially of spontaneous vegetation or combinations of planted and spontaneous vegetation) for evaluating one's own planted open spaces (will they have a short or long life?), and finally for the analysis of role models advocated by the Kassel School, that is, "socially proven open spaces."[103]

The Order of Plants

In his enthusiasm for the methods of phytosociology, in addition to his interest in its instruments for the deep analysis of a situation, Dieter Kienast was particularly enthusiastic about implementing it for planning: in how a profound knowledge of plants concerning the dynamics of processes of succession and the expected interaction of plant communities could be used to create attractive open spaces in the city for little money. He focused his attention especially on the spontaneous vegetation of the city. Under the supervision of Karl Heinrich Hülbusch, he wrote his dissertation, "Die spontane Vegetation der Stadt Kassel," from 1976 to 1978. It was the first time the method of phytosociology was applied to urban vegetation in a large research study. Kienast's achievement consisted of, among other things, describing urban vegetation using phytosociology and especially the method of sigma-communities in order to synthesize different data on a higher level. This made the data easier to interpret and apply.

The background of this interest was historical and specifically in contemporary and not least local history":[104] after the bombing of German cities during the Second World War, which substantially destroyed Kassel as well, a variety of plant species had become established on the piles of rubble that were left behind, and they were closely studied by vegetation scientists.[105] At the same time, the constant expansion of cities led to the incorporation of villages, and the fields and meadows surrounding them, that had previously been located outside the city limits. In parallel with this development, agriculture was turning more highly industrialized, while the vegetation in rural areas was becoming increasingly homogeneous as a result of the growth of monocultures. The seeds of numerous plant species that were disappearing in the cultural landscape were nevertheless in the soil of the villages and tracts of land that had been absorbed by the cities. These plants could thus become established in vacant lots and open spaces in the city. Since the 1960s, many cities had become more biologically and ecologically interesting and had developed much more diverse habitats than the highly agricultural landscapes surrounding them. This discovery led in turn to nature conservation programs gradually being introduced in cities in the 1970s. Following the phytosociologists, vegetation ecologists now began to study urban nature and to describe the autonomy of urban vegetation. Scientists around Professor Herbert Sukopp in Berlin founded the new field of urban ecology,[106] which is now the leading method for studying and describing urban vegetation.

Kienast's interest in the spontaneous vegetation of the city was guided by various values: in addition to protecting species diversity, this included aspects of economics and user friendliness. Precise analysis and evaluation of the existing plants made it possible to base planning on the plants that were already there by considering what would go well with them. The idea was to work with nature, not against it. The next step concerned nearly natural maintenance, by which newly designed open spaces would result in cost savings and existing open spaces could be maintained more easily. That is because intense maintenance of surfaces is the second leading cause, after their complete sealing by concrete, for the destruction of plant diversity in the city.

Applied phytosociology, which Kienast wanted to help establish with his dissertation (and had already begun to do so with his thesis), has to be seen as a foundation for understanding design, not as a rejection of interventions for the sake of design. For Kienast, planning with spontaneous vegetation also included an aesthetic aspect.

His dissertation had been preceded in 1975 by the diploma thesis "Freiraumplanung Kassel Nordstadt unter besonderer Berücksichtigung der Vegetation und ihrer Verwertbarkeit für Analyse und Planung" (Planning of Open Spaces in the Nordstadt District of Kassel, Paying Particular Attention to the Vegetation and Its Utility for Analysis and Planning), which Kienast had written jointly with his fellow student and friend Thom Roelly. Phytosociological and demographic analyses of Kassel's neighborhoods were elemental to both works. Both built on the vegetation science taught by Hülbusch and connected it to social and planning questions. The goal of the thesis, in which the influence of Lucius Burckhardt's critique of planning is evident, was to provide an incontestable scientific basis for the proposed planning of open spaces. The analysis of the structures of neighborhoods and of the plant communities found there led to conclusions about the quality and use of the existing green spaces. Phytosociological mappings supported the demographic and climatological study of the neighborhoods. A specific plan on an urban-planning scale was worked out based on the data collected. His dissertation can be seen as a continuation of this approach to research.

Kienast's Diploma Thesis

Dieter Kienast and Thom Roelly submitted their diploma thesis (*Diplomarbeit*) in September 1975 at the Gesamthochschule Kassel. Their statement of topic was an analysis of the existing open spaces in various types of neighborhood in a chosen area in the Nordstadt district of Kassel from which improvements to the existing open spaces could be derived. Based on a combination of demographic, urban-planning, climatological, and phytosociological analysis of this part of the city, a map was produced on a 1:2,000 scale. Among the specific measures proposed: planting trees along streets with heavy traffic, establishing new pedestrian connections, building several pedestrian overpasses, installing community gardens, and decreasing the maintenance of some of the existing green spaces. The latter measure was based on the idea that various residents of the neighborhood would find areas that looked somewhat overgrown easier to appropriate than green spaces that were intensely maintained and designed with specific uses in mind. But the increased presence of spontaneous vegetation was not just associated with a heightened willingness to appropriate open space. Kienast and Roelly went so far as to claim that the urban climate was better where many different plant communities existed, because climate fluctuations were not as large there. They admitted, however, that the differences were not that great.[107]

Plants that prefer to be close to human habitations are known as "ruderal" or "anthropogenic" vegetation.[108] This refers to a type of vegetation found in human-dominated landscapes that is "neither planted nor maintained in a specific form by means of intentional and regulated use."[109] It is therefore called "spontaneous vegetation." The only plants which will populate a given place on their own are those for which the qualities of that location are optimal for their species to thrive. The specific combinations of plant species found together in a given location are called "plant communities."

Ruderal vegetation contrasts, on the one hand, with vegetation planned and maintained by human beings, and, on the other hand, with natural vegetation found in natural disturbances such as eroded banks or burned areas. In order to distinguish human-controlled vegetation more clearly from ruderal and natural vegetation, Kienast and Roelly introduced the term "gebaute Vegetation" (built vegetation).[110] This term includes trees along streets, flowers in beds, lawns, and so on, but excludes all kinds of spontaneous vegetation. They also made it clear that in agricultural areas (in contrast to the city) ruderal and "built" or planted vegetation nearly always overlap, because ruderal vegetation invades all areas of vegetation that have been worked by humans—namely, as a "weed." Whereas for farmers and gardeners, built vegetation is the defining feature of an area of vegetation, the vegetation scientist is interested in the weed communities and derives from their presence information about a piece of land.[111]

16
Master plan to improve the open spaces and their infrastructure in the Nordstadt district of Kassel, scale: 1:2,000, H 85 × W 70 cm. Dieter Kienast and Thom Roelly, "Freiraumplanung Kassel Nordstadt" (Planning of Open Spaces in the Nordstadt District of Kassel), diploma thesis, 1975

In connection with the instruments used for analysis in their thesis, charts of ruderal vegetation were produced and evaluated as sources of information about local and urban climatic conditions, and conclusions were drawn about the soil substrate and the intensity and frequency of use.[112] In a first step, the forms of housing and the buildings in Kassel's Nordstadt district were categorized according to criteria of building type—for example, turn-of-the-century block perimeter construction, old and new housing developments (including high-rises from 1958 onward), mixed areas of single-family homes and apartment buildings with small-scale industry, industrial and commercial areas, public buildings—and then associated with sociodemographic data, for example, population density, percentages of workers, civil servants, and self-employed. Next, they took climatological measurements and made phytosociological charts of the existing ruderal vegetation. This first section of the thesis with the analysis and assessment comprises a good 130 pages and is accompanied by many tables on the demographics of the population; climatic measurements at 8 a.m., noon, and 6 p.m.; and a number of phytosociological analyses.

They analyzed the data collected to draw conclusions about how the existing ruderal and "built" vegetation, respectively, affect the climate of the city and the use of semipublic and public spaces. Both aspects—the climate of the city and the inclination of people to use a site—were to be improved by a focused use of vegetation as laid out in a plan. Kienast and Roelly raised the question of the possibilities for using vegetation to change a particular situation.[113] The results of the analysis were integrated into a large-scale planning framework for open spaces in the area of the study on a scale of 1:2,000 (16). A follow-up project to select two or three smaller parts of the neighborhood based on the results of the measurements and analyses, and to submit plans for them that would demonstrate the possibilities of optimal use of vegetation in terms of utility and microclimate, was never realized. The plans for it never ended up being produced.

The overarching goal of this planning framework for open spaces was to increase the well-being of residents. Based on the results of previous detailed analyses, this could be achieved through infrastructural measures (new pedestrian access) and safety measures (new pedestrian overpasses). The measures involving green planning aimed, first, to improve the climate of the city by planting trees along streets and decreasing the maintenance of green spaces to permit the settlement of more weeds. Second, Kienast and Roelly proposed establishing more open spaces in which people could become personally involved. This included both setting up community gardens, which have to be jointly cultivated, and encouraging the residents to actively appropriate open spaces.

Kienast and Roelly believed that the latter could be achieved by not maintaining these areas as intensely, based on the motto that where things are overgrown, people can behave freely. They spoke of "open space that opens up new possibilities of use" and "vacant and peripheral areas with a potential for new kinds of uses."[114] Preserving them as well as creating new ones are two of the fundamental measures of planning. Open spaces are recognizable by the appearance of weeds, which should be encouraged in order to "prevent a further decline of flexible use of open space." Rules should be worked out—together with the office responsible for gardens and open spaces and with the road construction office—that clearly establish where and how maintenance is to be carried out. The "labor freed up" in the process should be employed "more meaningfully" elsewhere, for example by creating new pedestrian paths.

Setting up community gardens, in turn, was a crucial means of creating "compensatory" outdoor spaces near residential neighborhoods.[115] The planning framework for open spaces included a large number of kitchen gardens easily accessible both by the old housing developments (block perimeter developments prior to 1914 and ribbon developments from 1914 to 1925) and by those built between 1953 and 1960.[116] Setting up playgrounds also increased the number of "compensatory" free spaces. In addition to playgrounds with lawns, Kienast and Roelly proposed "construction playgrounds," where children have access to all sorts of materials that can be used creatively. The establishment of community gardens and kitchen gardens near residential areas and the realization of new pedagogical concepts for children's playgrounds had been explored by Kienast earlier between 1972 and 1975 in preparation for his thesis project in the first open-space design of his own for the grounds of a residential development in Niederhasli (pp. 172–176), a suburb of the City of Zurich. All of this was in keeping with the progressive trends for designing open spaces of the period.

Very much in the spirit of Burckhardt and the Kassel School, Kienast and Roelly's thesis was about focusing on people and their needs and developing unspectacular solutions from solid analytical work. The results of climatological studies and phytosociological charts were employed both in their written catalog of proposed measures[117] and in their planning framework for improving the residential environment. They intentionally precluded designing the neighborhood in a way that the individual style of the planner would be recognizable.

Kienast's Dissertation

In his dissertation on the spontaneous vegetation of the City of Kassel, which was published in 1978 as volume 10 in the series *Urbs et Regio: Kasseler Schriften zur Geografie und Planung* (Kassel Writings on Geography and Planning), Dieter Kienast concentrated on the phytosociological cataloging and evaluating of the plant types found in Kassel's neighborhoods, with the goal of developing a phytosociological analysis and planning tool for addressing questions related to the planning of open spaces. Knowledge and understanding of which plant communities tend to become established in which neighborhoods and with which other plant communities, as well as the grasping of processes of succession, were intended as a way of obtaining insights into the quality of open spaces (climate, diversity, use) and making planning decisions easier: Where are more open spaces needed? Which plants could be used to best support the existing ones? How can planning profit from the natural dynamics of plants? Where does the anticipated use not correspond with actual use? How can planners respond to that?

In order to provide the planning of open spaces with such a handy instrument, the presence of spontaneous vegetation in the respective neighborhoods of Kassel first had to be determined and recorded in as much detail as possible. Kienast chose the method of establishing a transect through Kassel around 11 kilometers long (ca. 7 miles) and 160 meters wide (ca. 525 feet) in which the various structures of the city's neighborhoods and open spaces would be represented at least once (17). These structures ranged from forested and agricultural areas on the outskirts of town to old village centers, industrial, commercial, and turn-of-the-century buildings, single-family homes and apartment buildings, and also to the city center, known as the "City," and the areas bordering the city. This fashionable English word was surely chosen in part because the center of Kassel had been almost entirely rebuilt after the bombings of the Second World War. Little remained of the architectural fabric of the old town center.[118] The structures of the neighborhoods in this transect were recorded on an overview map at a scale of 1:5,000.

The second methodological level was the recording and evaluating of the plant types on the higher integration level of synsociology, which is the "theory of the formation of plant communities."[119] For each location in the city, Kienast mapped not only the individual plant species and their associations, but also the plant communities and their preferences for combination with other plant communities (18). He had adopted the method of synsociology from the phytosociologist Reinhold Tüxen, who in turn had based it on the teachings of the Braun-Blanquet School and the research of the vegetation geographer Josef Schmithüsen (1909–1984).[120] Kienast was the first to apply synsociology to spontaneous vegetation in the city. Over the course of his dissertation, he developed a system of integration in which the plant communities were grouped into so-called sigma- or character-communities. That made it possible for him to take into account the fragmented plant communities (plant communities that are missing some of their typical species), which occurred very frequently in the city.[121] In sixty-two phytosociological tables, he mapped the plant communities found in the various neighborhoods according to species, number, and location and then grouped them into sigma-communities. Kienast thus not only grouped his vegetation inventories into phytosociological units but also organized them into an existing system—or rather one partially recreated by him and Hülbusch—with a structure similar to that of systematic botany.[122]

Kienast showed the results of the mapping—the synsociological classification of existing plants—in a second overview map of the selected transect. If these two maps—one of which indicates the urban structure and the other the synsociological structure—are placed side by side, then it becomes clear how the distribution of specific sigma-communities relates to the distribution of neighborhood structures. It is immediately evident that the structures of "old village centers" and "turn-of-the-century buildings"

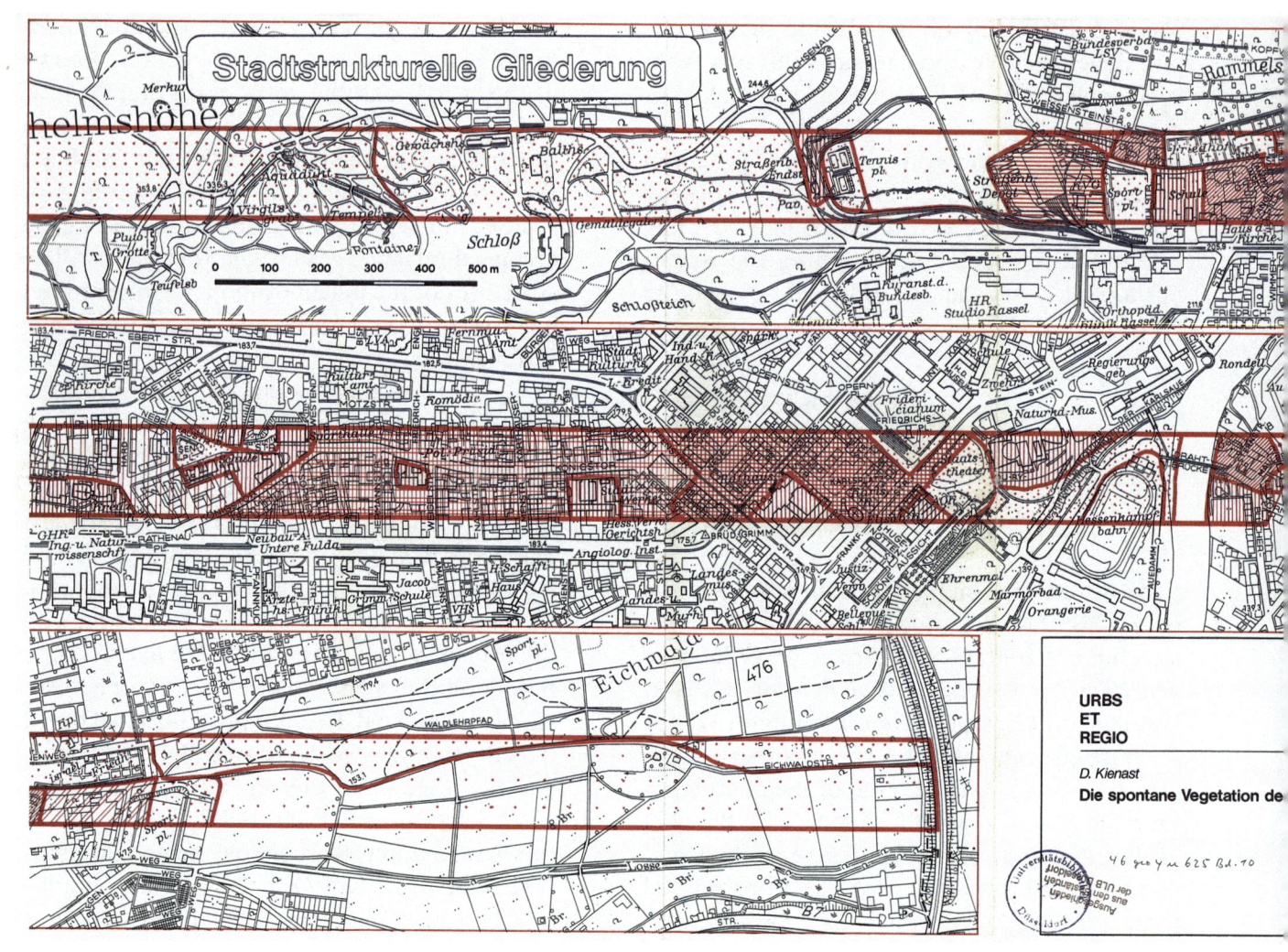

17
Categorization of the urban structure of the transect through Kassel, scale: 1:5,000, from Dieter Kienast's dissertation, "Die spontane Vegetation der Stadt Kassel" (Spontaneous Vegetation in the City of Kassel), 1978, and photographs of the plants in various neighborhoods

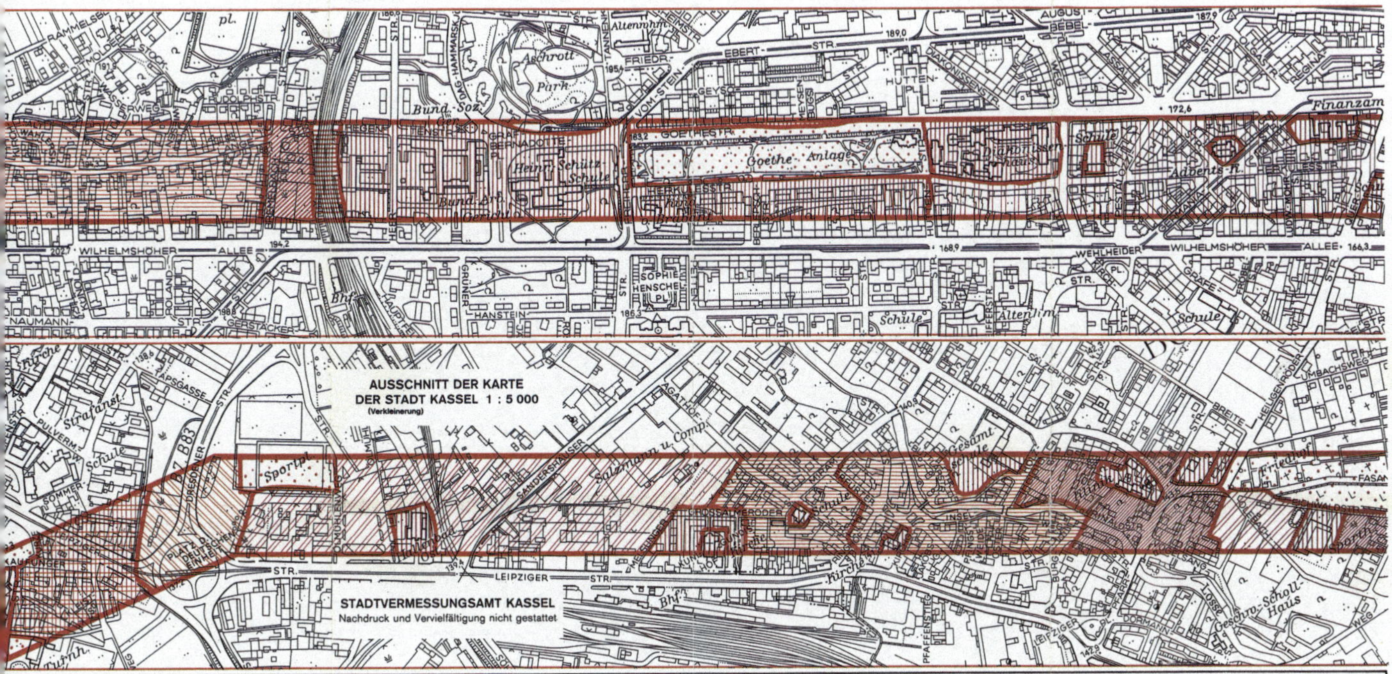
AUSSCHNITT DER KARTE
DER STADT KASSEL 1 : 5 000
(Verkleinerung)
STADTVERMESSUNGSAMT KASSEL
Nachdruck und Vervielfältigung nicht gestattet
WILHELMSHÖHER ALLEE
Goethe Anlage
LEIPZIGER STR.

10/1978

ssel

STADTSTRUKTURELLE GLIEDERUNG

- Quartierstyp 'Alte Dorfzentren'
- Quartierstyp 'Gründerzeitbebauung'
- Quartierstyp 'Geschosswohnungsbau'
- Quartierstyp 'Ein- und Zweifamilienhäuser'
- Quartierstyp 'Cityrandgebiete'
- Quartierstyp 'City'
- Quartierstyp 'Öffentliche Gebäude'
- Quartierstyp 'Industrie und Gewerbe'
- Park- und öffentliche Grünanlagen
- Landwirtschaftsflächen
- Forstwirtschaftsflächen

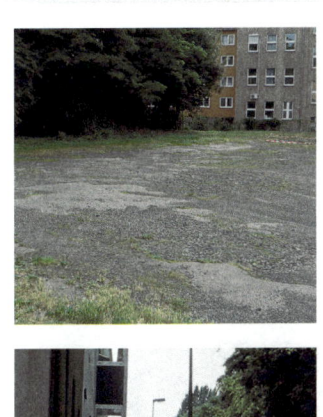

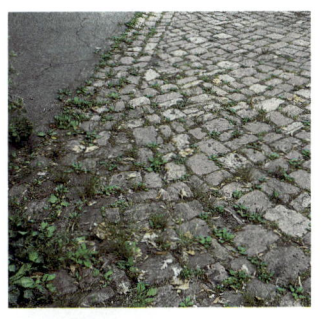

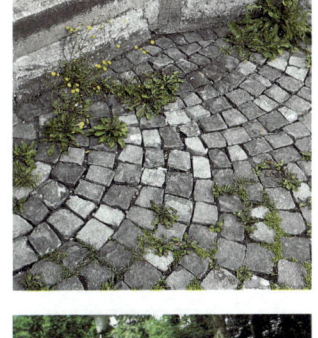

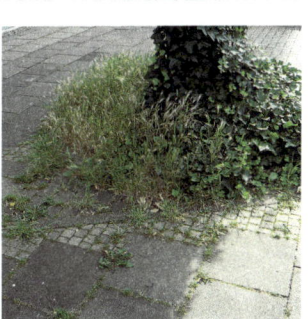

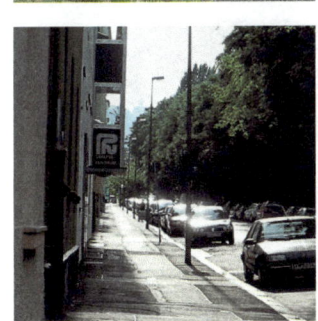

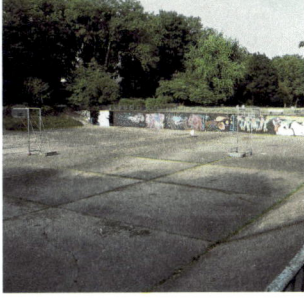

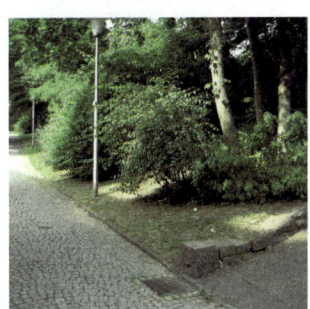

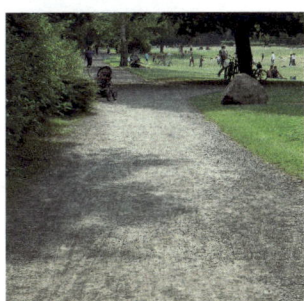

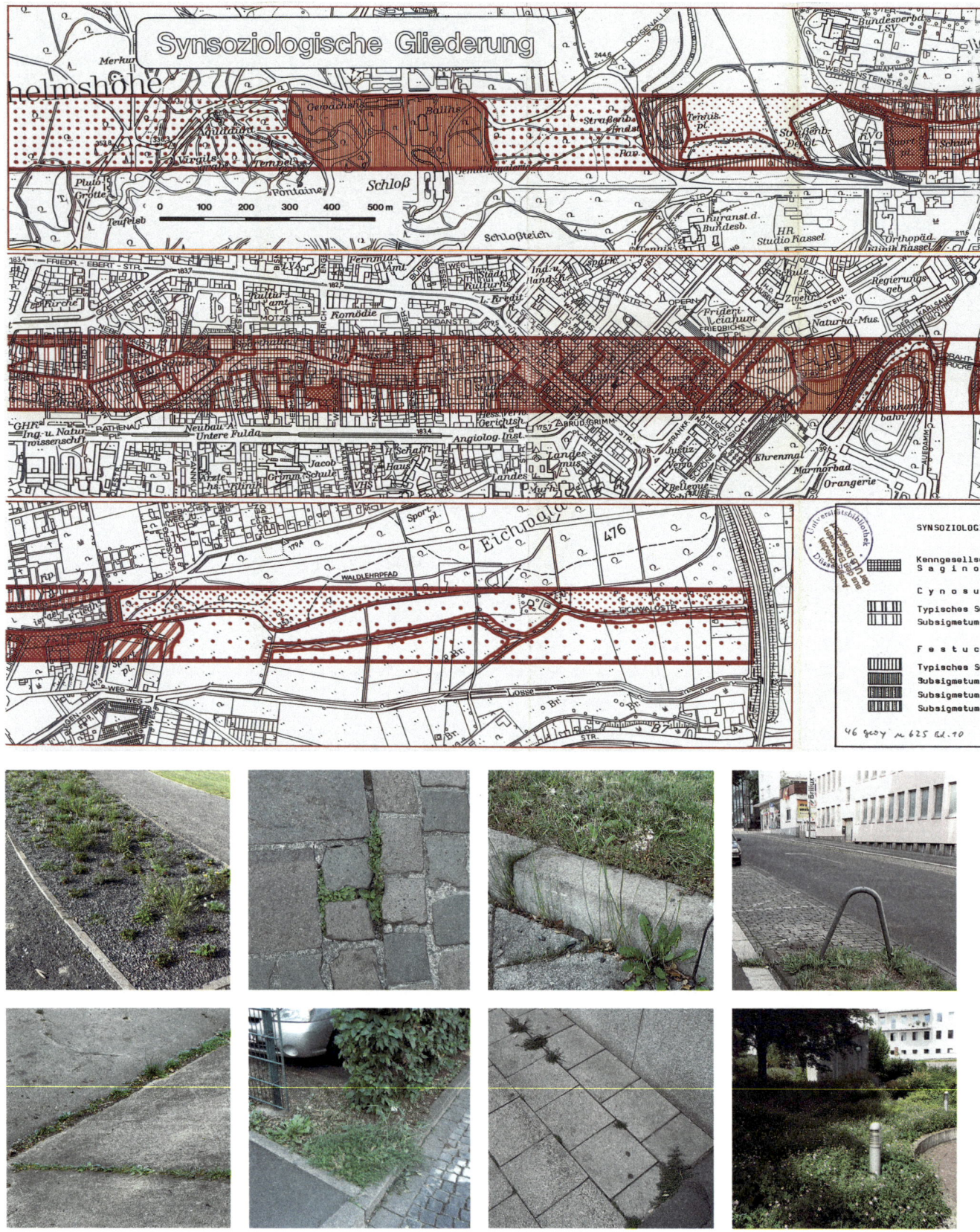

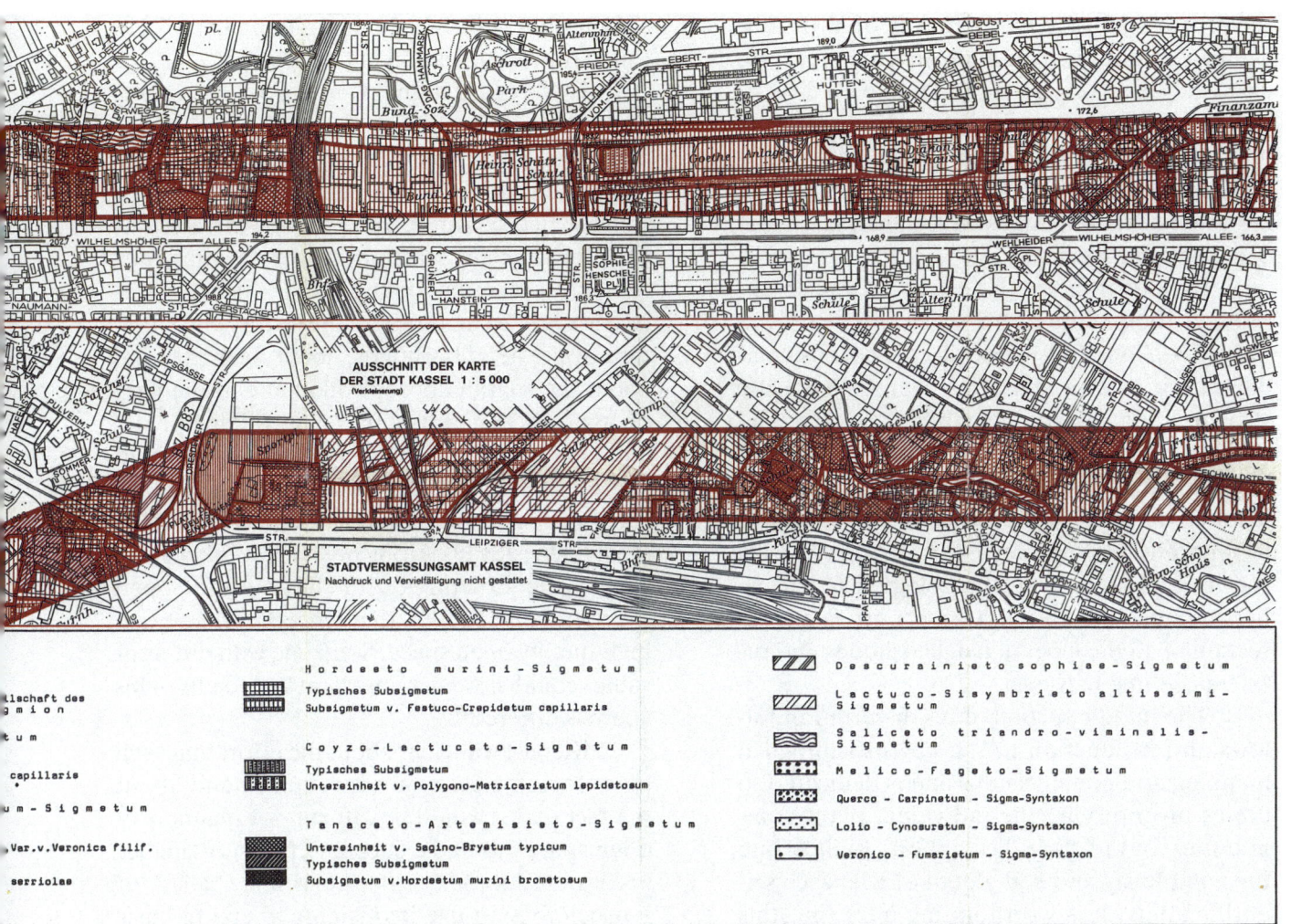

18
Synsociological categorization of the transect through Kassel, scale: 1:5,000, from Dieter Kienast's dissertation, "Die spontane Vegetation der Stadt Kassel" (Spontaneous Vegetation in the City of Kassel), 1978, and photographs of the plant communities listed (e.g., pavement-crack communities and compacted-soil communities in path areas)

have the highest concentration of different sigma-communities, which means that they have extremely diverse vegetation. The lowest diversity is found in the city areas, where Kienast found almost exclusively a sigma-alliance without a character-community, to which he and his dissertation advisor, Hülbusch, assigned the name *Sagino-Bryeto-Sigmion.*

In addition to the maps, Kienast explained the results of his inventories and the integration into the overarching system of sigma-communities in a text of rigorously scientific structure. First, he formulated the questions and the current state of research. Then he provided an introduction to the location of his study: he justified his choice of location and discussed its topography and topology and various aspects such as geology, soil, and hydrology. Then he explained how the selected transect offers a cross section of all the types of neighborhoods and rural units found in Kassel.

The middle section of his dissertation follows an introduction to the phytosociological mapping and nomenclature and is dedicated to tables inventorying the individual plant communities and to their description. To illustrate the complexity and high detail of Kienast's scientific study, it is worth noting here that this chapter of his dissertation has eighty short subchapters in which the various plant communities are briefly presented and characterized. Then Kienast integrated his observations with the aid of the synsociological method and distilled out of it the sigma-communities and sigmeta, most of which he identified and named together with Hülbusch, not only the aforementioned *Sagino-Bryeto-Sigmion* (alliance) but also the *Lolio-Plantagineto-Sigmetea* (class).

In the final section, he presented the results of his inventories, first in relation to the city in general and then separately by neighborhood type. With regard to the method of phytosociological mapping, Kienast came to the conclusion "that a sigmetum is characterized both by character and separation communities and/or by the characteristic combination of communities (of continuous communities)."[123] By doing so, he gave a boost to the method of synsociology, which was certainly controversial among phytosociologists.[124] By contrast, he saw difficulties in nomenclature and taxonomy. When identifying the spontaneous vegetation of the city, the fragmentary or derivate communities are crucial, but their names often take up entire lines of a study, as a result of the practice of characterizing a sigmetum with two community names.[125] He called for simplification in future work.

With regard to the planning of open space, Kienast concluded that the method of synsociology as a new branch of phytosociology made "methodologically cleaner analytical work" possible, which could be used to provide a scientifically relevant, causal answer to questions related to planning open spaces,[126] since statements could be made about the history and structure of the neighborhoods, about their use, and about the spatial appropriation and availability of open spaces based on the composition, spread, and distribution of the vegetation, which could be determined by phytosociological analysis.[127] From this, in turn, Kienast calculated the quality of an open space: a strong presence of plants and small-scale structures with "flexible-use" open spaces meant, to him, that a neighborhood had high-quality open space. Someone with different values could, however, interpret the results of his analysis differently.

Kienast was clear about the effort that such phytosociological analysis required and about the fact that it would be difficult for planners of open spaces to make such an effort for smaller tasks and reports. For that reason, he called for synsociology to use its insights to create basic knowledge, similar to that of sociology among the social sciences, on which planners could fall back on relatively easily.[128] Kienast established such a basis in his dissertation. First, local planners could build directly on the findings he had obtained in Kassel.[129] Second, the synsociological categorization of plant communities that he proposed could be applied to many other cities. The system of sigma-communities that he worked out was a study of basics and would make all subsequent studies of spontaneous vegetation in the city easier.

A review of Kienast's dissertation published in the journal *Phytocoenologia* by the botanist and phytosociologist Otti Wilmanns (b. 1928) in 1980 stated: "The sigma-tables are impressively clear and permit a rigorous classification that will probably prove valid in principle for numerous cities."[130] Of the results of the mapping, she emphasized especially Kienast's finding that the vegetation complexes of the city can be summed up as *Lolio-Plantagineto-Sigmetea.* According to Kienast, this class could be subdivided into "the *Sagino-Bryeto-Sigmetalia,* with communities of asphalt cracks and lawns, which are characteristic of densely built neighborhoods of intense use, and the *Tanaceto-Artemisieto-Sigmetalia,* with ruderal communities in sparsely built areas, peripheral areas, industrial areas, and old village centers."[131]

The reviewer was impressed by the mapping of categorization according to urban structures and synsociology, and by the correspondence of plant complexes and neighborhood types that it revealed. It made it vividly clear that in the city section of Kassel, which was almost completely sealed by asphalt and concrete, less than 1 percent of the open space was covered with vegetation, and that was exclusively in the form of pavement-crack communities and potted plants. By contrast, Kienast had been able to identify in the old village centers which had been incorporated into the urban structure that there was an extremely diverse and small-scale alternation of sigmeta "even including more highly developed and long-lived communities such as *Carpino-Pruneta*," despite the space "officially classified as open space" being quite minimal.[132] He was able to demonstrate by analyzing vegetation that many uncontrolled playgrounds for children had been developed in this region, specifically where there was no strict separation of function between areas for traffic and places for people to live.[133]

By looking at vegetation and urban structure as a whole, Kienast's analysis made it obvious that the city center and the areas on the periphery of the city fared worst in terms of open spaces for appropriation, presence of vegetation, and the associated climatic improvements. His work thus became a profound criticism of the urban planning measures of the postwar era applied in Germany when rebuilding many of the cities that had been destroyed by bombing during the Second World War.

Urban analyses like Kienast's opened up a new approach to thinking about urban structures, which had a lasting effect on the changes to urban planning measures in the 1980s. "Postmodern" urban planning solutions integrated existing traces and preferred structures of small units over the "wholesale redevelopment" for car-friendly cities that was popular in the urban renewal projects of the 1960s and 1970s. Kienast proved in his dissertation that open spaces can be very precisely characterized and assessed based on the presence of spontaneous vegetation and that planners can derive information from it which is important for urban planning.[134]

Kienast's dissertation continues to be a fundamental reference work for the study of urban vegetation and is required reading in the associated departments of universities. It should be added, however, that the phytosociology of the city has in general been replaced by urban ecology as a method for obtaining scientific data for sociological and geographical questions. To put it bluntly, phytosociology is "out" and had been replaced by more detailed analyses based on individual species' responses to urban stresses. Much like the method Hülbusch advocated—in which demographic data are combined with settlement structures, climate measurements, and the presence of plants—urban ecology combines measurements of findings based on different parameters. Sociogeographical studies and climate measurements play a far more important role in the latter, however, and the animals living in the city and their distribution are also incorporated into such studies.[135]

In the view of the biologist Markus Ritter, who in Basel in the early 1980s worked on the first nature atlas of a city to be published in Switzerland,[136] the phytosociological schools did not gain acceptance for the analysis of urban vegetation for the planning of open spaces because the method presumed profound knowledge of species, took too long, and was too complicated for landscape architects who lacked such knowledge.[137] Moreover, when developing their systems, phytosociologists had assumed that there were obligate interspecific controls in plant communities and that, at the historical moment when all of the synsociological connections of plant communities had been collected and described, planners would have a wonderful instrument for predicting the development of vegetation and the processes of succession. But over the course of research it became clear that there were no such determinate community types, because the forming of communities and the processes of succession were influenced by a wide variety of factors, and the character of the data collection influenced the results of the study.

The sigma leap offered by Kienast represented one possible way out of the crisis of the complexity of descriptions. The idea of entities and a universally applicable instrument was, however, untenable. In retrospect, Kienast, too, regretted that the planning component of his dissertation took second place to a scientific focus, because over the course of the work it had proven incredibly difficult to demonstrate how the phytosociological approach could be applied to planning.[138] Moreover, he doubted that it would be possible to communicate the complex specialized findings to practitioners in a way that they could apply it to landscape architecture:

The prospect of only being able to discuss a special area of plant sociology with a total of about ten international experts didn't seem very exciting. I returned to Zurich, became a partner in Peter Stöckli's office and was engaged in work of very different kinds over a two-year period; designing, mapping in the field of plant sociology, and so on. It wasn't very long before I realised that I was unable to develop in either the area of design or in plant sociology and, eventually, I decided to concentrate exclusively on working as a designer.[139]

One of Kienast's applications of phytosociological knowledge to planning was the Dry Grassland Biotope developed for Grün 80 in Basel from 1978 to 1980 and the landscape design of the Zurich–Schwamendingen streetcar line. In his own words, the ambition to achieve more drove him to start his own career as a designer in the early 1980s.[140] Kienast regarded his appointment as Professor of Garden Architecture at the Interkantonales Technikum Rapperswil (since 1998: Hochschule für Technik Rapperswil) in 1981 as a crucial step on that new path. For him, it was the beginning of an intense debate of design questions, particularly in exchange with his colleagues Jürg Altherr and Peter Erni, and in their collaborative teaching activity. Jürg Altherr (1944–2018) had completed studies in garden architecture and was working as a sculptor, whereas Peter Erni (b. 1942) was trained as an architect and possessed wide-ranging theoretical knowledge about art and architecture. Their joint teaching in Rapperswil had a crucial effect on the change in Kienast's design work. Phytosociological knowledge and the methods he had learned in Kassel were transformed in the phases of his work that followed. They contributed to the allegorizing and aestheticizing of spontaneous vegetation and influenced Kienast's way of expressing his relationship to the "nature of the city" in design.

The Third Stocktaking of the World

Lucius Burckhardt and Karl Heinrich Hülbusch were contemporaries of the structuralist theorists. In 1970, Michel Foucault was appointed to the Collège de France; in 1973, Claude Lévi-Strauss was elected to the Académie française. The influence of the debates of that period are still felt today, which demonstrates the intellectual power of semiotics but at the same time marks the ideological crisis of the 1970s, when the Marxist utopias faded in both the East and the West. "There is no socialist country, in quotation marks, to which we could appeal to say: this is how it should be done," Foucault observed at the time.[141] Subsequently, structuralists limited themselves to understanding existing semiotic systems.

Art, too, no longer functioned as a role model after the Second World War. Accordingly, the movements of the 1960s and 1970s could no longer be understood as an avant-garde—literally, "vanguard":

> Its protagonists wanted not to improve society by means of art, but rather to improve their art by exposing it to the complexity of society and its rapid changes. And in the end the divide between art and the public that was typical of the avant-gardes disappeared from the early 1960s onward. The idea of the artist as antibourgeois—"épater le bourgeois"—lost its credibility from the 1950s onward.[142]

The methods employed by the researchers and planners in Kassel could also be called "structuralist," in that they rejected the solutions of the designer as "genius" and instead based their work on sociological and scientific inventories.

They concentrated on urban structures, which they understood with the aid of phytosociology, among other things. This system of representation claimed to describe the physical reality of the urban structure and its use. Also found in the contemporaneous production of art are the evident features of the methods and themes of education in Kassel: a focus on science and rejection of the principles of the master and of intuition in design and planning; reflecting on one's own perception; precise observation as method; the scientific methods of phytosociology, including reading clues, mapping, and organizing plants according to associations, classes, and communities; studying development over time, such as succession processes and annual cycles.

The art of the 1970s was no less dystopian than the structuralist intellectuals: Whereas their precursors in the Pop and Minimalist Art in the 1960s had celebrated New York as the emblem of the American dream and faith in progress, artists around 1970 shifted their focus from the high-rises and cars, the shiny surfaces of new consumer goods, to the ground, the details of broken pavement, the poverty, the trash.[143]

In 1974, the cultural historian Günter Metken defined the technique of concealing and collecting as an artistic method in the exhibition *Spurensicherung: Archäologie und Erinnerung* (Securing Clues: Archaeology and Memory) at the Kunstverein in Hamburg. The invited artists attempted to find connections that evaded scientists because of their specialization and evidence-based rationality. In addition, their works heightened awareness of the temporality of thinking, feeling, and imagining.[144]

In the introduction to the catalog of the exhibition in Cologne called *Kunst bleibt Kunst: Projekt '74—Aspekte internationaler Kunst am Anfang der 70er Jahre* (Art Remains Art: Project '74—Aspects of International Art at the Beginning of the '70s),[145] the scientist and mathematician Wolfgang Schäfer is quoted as saying that a "third stocktaking of the world" was underway: "In doing so, we are not only revising our knowledge of physical and biological laws but also subjecting our thought patterns and the structures of our experience to a new examination."[146]

This development, which permeated all scholarly fields in the 1970s, also influenced the production of art. The time factor was, in the view of the exhibition's curators, the crystallization point for various artistic directions. It was expressed through artistic interest in phenomena of perception and natural processes, but also through scientific research and artistic methods where the knowledge acquired is elevated to a stylistic means in the production of art.

Questions about the relationship of physical reality to that which is found, to its context and to its presentation in the form of a new order, were also raised by an artwork from 1974 called *Für die Geschwister Götte* (For the Götte Siblings) by Nikolaus Lang (19), shown in the exhibition *Kunst bleibt Kunst*. Over the period of a year, the artist collected clues about the Götte family, farmers who had immigrated to the German town of Bayersoien from Switzerland around the turn of the century: pages from the land registry, statements from villagers, found objects from the since abandoned farm, including tools and clothing, but also leaves, branches, seeds, snail shells, hair, and animal bones found on the grounds of the farm.[147] Lang systematically ordered and exhibited what he had found. He inventoried the neatly sorted objects on twelve pages, supplementing them with a series of photographs of the ramshackle buildings that the grown children of the Göttes had built at a slight distance from the main farm in order to live their own lives independently of their parents.

The order brings with it its own aesthetic that results from the presentation and leads to a consolidation of the clues and found objects, which in turn inspires various conceivable narratives. On the one hand, the visual application of an epistemic order—as it had been practiced in natural history museums since the late eighteenth and early nineteenth centuries to classify and exhibit minerals, insects, and so on—suggests that a scientific, factual approach to the history of the Götte family was possible by classifying their traces. On the other hand, their presentation opens up a space of associations for stories that the viewers think up in their own imaginations when viewing the clues, which ultimately allow them to connect their own lives to those of the Götte siblings:

> Lang displays the finds as if in a museum; they appear very ancient, archaic, mysterious and slightly menacing. The objects do not conjure up the gay frivolous idyll of unspoilt nature; they are cult utensils of forgotten, pre-religious rituals. The question whether or not the Götte children

really existed in Bayersoien becomes of secondary importance; they could also be an invention of Lang's.... In this apparently objective research, he documents his real self most clearly.[148]

By ordering the clues, the artist incorporates into the presentation his own relationship to the story of the Götte siblings, and the constitution of knowledge and reconstruction that results from trying to read the clues causes viewers to be moved by the fate of the immigrant farmers as well. Finally, Günter Metken's accompanying catalog text reveals that the family was never accepted in its new location, and four of the Götte siblings built their huts far out in the countryside, where they lived the remainder of their lives alone.[149] Metken was primarily interested in the connection between the biography of the clue-seeker Lang with the lives of the people to whom his found objects point:

He preserves a situation which is both an objective detection and a subjective finding, as well as a discovery of oneself. The investigation of the terrain imparts information about himself. Should one thus describe his activity as field research or a search for personal traces? Both apply, in my opinion, because ethnology too—at least according to Levi Strauss' interpretation—involves a search for oneself in others. A description of alien communities is of significance only in relation to our own: Who could disregard himself in exploring remote circumstances?[150]

But Dieter Kienast's seeking out and mapping of plants was nevertheless somewhat different. He presented his results based on scientifically recognized methods and did not pin the plants on boards himself. Instead, he classified them and abstracted the sensorially perceived manifestations of the existing vegetation in long Latin community names up to the sigma leap of synsociology. But the method of seeking clues in space—and the always associated connection of plants to built structures produced by humans or to activities of use by people—is clearly related, gesturally, to how Lang constantly correlates objects with the stories they conceal.

The geographical location of the existing plants on the transect in Kassel can also be thought of as parallel to the process of mapping in the arts (20, 21). Geographical, geological, and social aspects are depicted and interwoven in collages of maps and photographs. In Kienast's dissertation, maps of the city with the categorization according to synsociology and urban structure indicated in red were placed side by side for comparison (17, 18).

But what is the relationship between the objective detection and subjective finding in Kienast's work? Can his "investigation of the terrain" in the Lévi-Straussian manner be linked to his own development as a planner and designer? Including his diploma thesis and other research work,[151] he spent more than five years intensely observing and mapping plants in the city. His dissertation alone has more than sixty detailed charts of existing vegetation. This kind of diligent classification, which certainly seems to correspond to the "third stocktaking of the world" that Schäfer observed, was also satirized in art, for example, by Claudio Costa's object *Mensch und Zeit* (Human Being and Time) (22).

Another feature of the art of the period that reveals parallels to the Kassel School is its focus on everyday life. The artist Allan Kaprow commented on the exhibition project *Kunst bleibt Kunst* in a letter to the curators:

> The proposal speaks at the outset of wanting... "to portray the new trends of the seventies." Excellent. But to achieve this, you plan an exhibition! Surely the committee is well aware that one of the most important developments of the last decade was a pervasive move away from object-making (icons) toward processes of all kinds; away from galleries and museums (shrines and temples) toward the daily environment and psycho-biological space of individuals.[152]

This quotation underscores that the educational reforms in Kassel have to be understood in the context of the period. Away with the genius principle in design and planning and toward knowledge acquired inductively by collecting data and then reacting to it. Away from design by the master and toward observing everyday life and the needs of the users. Away from the separation of object and viewer and toward understanding the construction of reality by one's own perception. Away from intensive maintenance of the same decorative beds and toward extensive care and design work based on processes of succession.

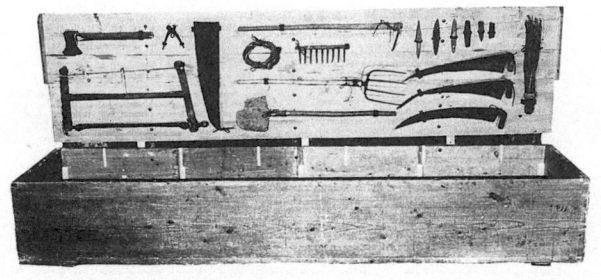

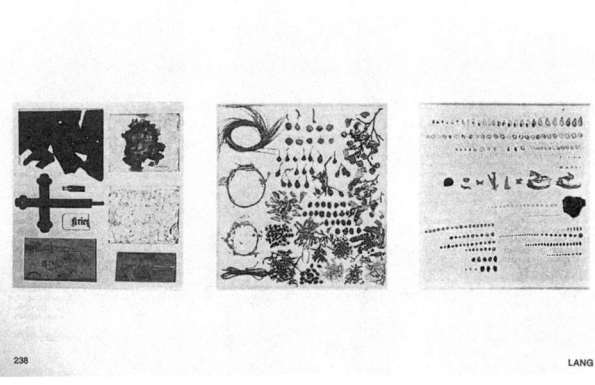

19
Nikolaus Lang, *Für die Geschwister Götte* (For the Götte Siblings), 1974. Objects and photo documentation for the exhibition *Kunst bleibt Kunst: Projekt '74* (Art Remains Art: Project '74) in Cologne

20
Newton Harrison and Helen Mayer Harrison, *On Mapping and Mixing,* 1974. Photo documentation for the exhibition *Kunst bleibt Kunst: Projekt '74* in Cologne

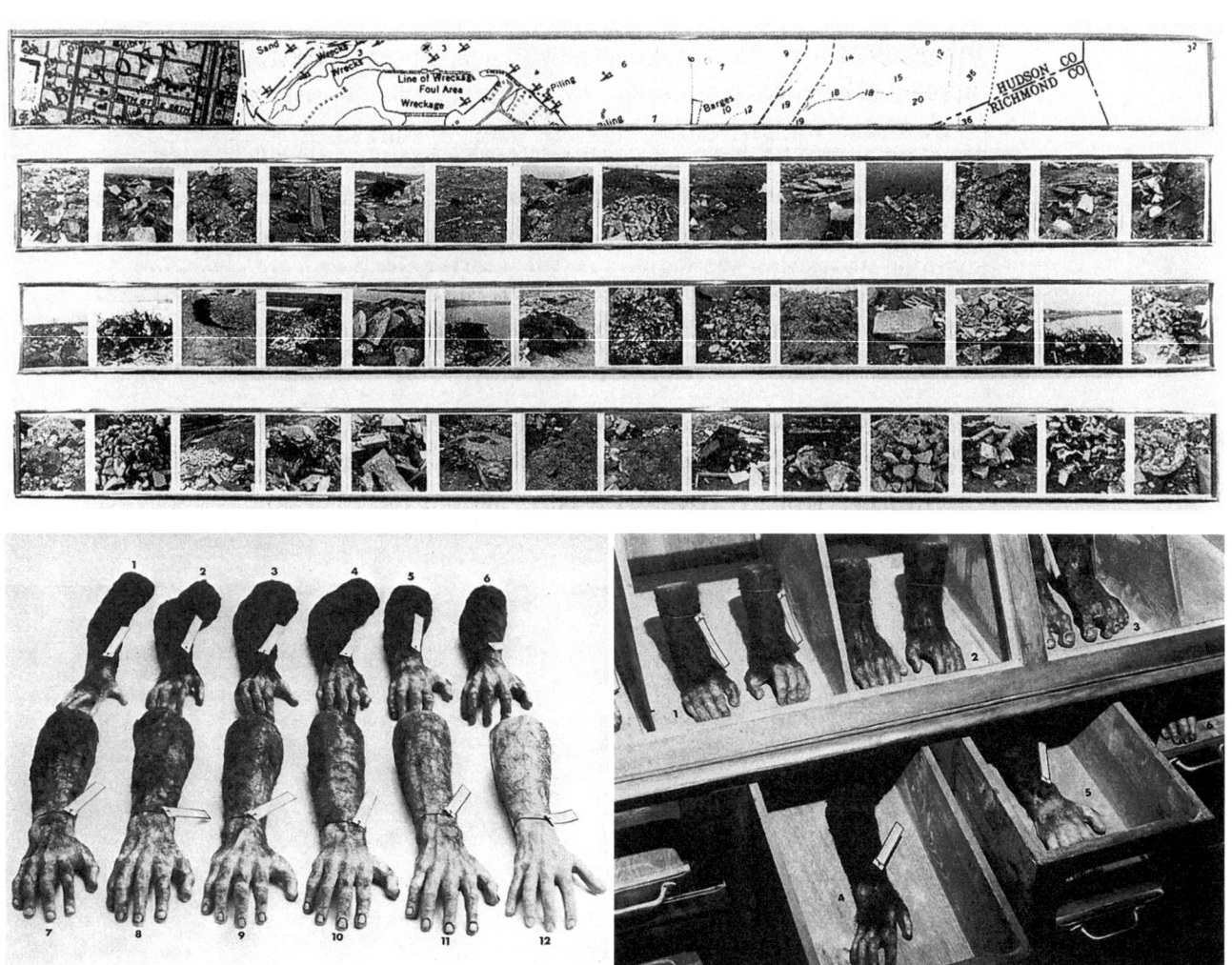

21
Robert Smithson, *Nonsite: Line of Wreckage (Bayonne New Jersey)*, 1968

22
Claudio Costa, *Mensch und Zeit* (Human Being and Time), 1974. Object for the exhibition *Kunst bleibt Kunst: Projekt '74* (Art Remains Art: Project '74) in Cologne

Placed in the context of contemporary artistic production, Kienast's phytosociological dissertation takes on new levels of meaning. First, there is the theme of searching for clues: a phytosociologist follows every trace, no matter how small, of plants sprouting between cobblestones or surviving on footpaths. The existing species and communities are identified and classified, in this case, as so-called pavement-crack communities or compacted-soil communities. They are mapped in charts that indicate how large the sample area was and how many plants and what species were found. The last-named is indicated specifically in the chart as "vegetation coverage." Kienast implemented these procedures on the synsociological level as well, noting the species that formed plant communities and indicating their quantities. If he found a large number of different plant communities per test area, he was able to conclude that this section of the city had either many different kinds of human use or provided a great variety of different spaces to the users.[153] By precisely mapping the quantities of existing plants, Kienast obtained an overview of the vegetation coverage,[154] which in turn enabled him to make detailed statements about the nature of the area and above all the forms of human use.

The plants became clues that said something about human beings and their lifestyles and behaviors. Without the presence of human beings and without the context of their use, most of the plants would not grow at all or would not remain in the state of the succession process in which they had been found.

Kienast's preoccupation with searching for plants, systematically ordering them in a vegetation chart, and then noting the results on a map of the city opened up a complex field of references to the spatial and temporal dimensions of the urban environment he was studying. The plants were read as signs of either the presence or absence of human beings, of their actions in urban space, and of the materials and construction methods they used when designing their urban space. In the way Kienast understood his phytosociological work, the plants included in the charts were interpreted as empirical, reliable data about the interplay of nature, city, and human beings.

In his text *On Historicizing Epistemology: An Essay*, the biologist and historian of science Hans-Jörg Rheinberger described how scientific systems and views about ways to obtain knowledge changed several times in the nineteenth and twentieth centuries. Epistemology therefore reflects on "the conditions under which, and the means with which, things are made into objects of knowledge. It focuses thus on the process of generating scientific knowledge and the ways in which it is initiated and maintained."[155] In that sense, phytosociology, which its aforementioned advocates repeatedly claimed had a timeless logic and empirical integrity, was also ultimately, as Rheinberger says of all sciences, a "historical development whose temporal course could be followed and whose particular conditions had to be ascertained."[156]

The obsessiveness with which Kienast pursued his vegetation inventories recalls the search for clues by artists in the 1970s. Even if phytosociology became less important in his later work, the description of the external that he pursued excessively for years left its mark inside him, for "who could disregard himself in exploring remote circumstances?"[157]

The basic stance of the Kassel School was, however, positivistic. Its methodological skepticism stopped short of the central theorems of its own disciplines. That distinguished the German planners from the French intellectuals of the second phase of structuralism—so-called neo- or poststructuralism—which made the linguistic order as such its theme.

The Natural and the Artificial: The Dry Grassland Biotope in Basel

The Dry Grassland Biotope developed in 1978 and 1979 for Grün 80 (Green 80), the second Schweizerische Landesausstellung für Garten- und Landschaftsbau (Swiss National Exhibition of Horticulture and Landscape Architecture) in Basel, was the first work in which Dieter Kienast was able to apply the knowledge he had acquired of phytosociology in a planning context (p. 68). The relationship between the natural and the artificial and the question of which images of nature can be produced in which context were the main themes of this work. As part of this exhibition of Swiss garden architecture, the appearance of the Dry Grassland Biotope stood out markedly from the other gardens (23, 24). At the same time, its themes were well suited to Grün 80, which under the motto "Vorwärts zur Natur!" (Forward to Nature!) sought to offer "a podium for the conflict of objectives between human beings and nature."[158] Apart from this commitment to environmental politics, the show was primarily about giving garden architects a chance to present their work and to encourage the transformation of the grounds into a nearby recreation area for Basel. The history of Grün 80 and the history of the Dry Grassland Biotope illustrate the conflict of objectives between garden architecture and the environmental movement.[159]

Grün 80 was held from April 12 to October 12, 1980, on the Brüglingen plain of Münchenstein in the Canton of Basel-Landschaft, and it was the second national exhibition of garden and landscape architecture, after the G 59 in Zurich in 1959. The Brüglingen plain is part of the Lower Birs Valley, which was formed by the Birs River during the Ice Age. It is the site of the St. Jakob-Park Stadium, but also of a farm, an English landscape garden, and a botanical garden, all three of which are managed by the Christoph Merian Foundation. South of this area is a major street and to the west Basel's industrial area, known as the Dreispitzareal. The new recreation area was intended especially to benefit the working classes living in that area. The organizer of Grün 80 was the Verband Schweizerischer Gärtnermeister (Association of Swiss Master Gardeners, VSG), whose president, Richard Tschan, was also the president of Grün 80. Preparations for the show began in 1975; it was decided that year that the exhibition would be held in Basel. The City of Basel organized a program of culture and entertainment under the name B 80 to accompany the show.

The first director of Grün 80 in 1976 was the historian Markus Kutter, who had coauthored with Lucius Burckhardt and Max Frisch the short book *achtung: die Schweiz* (Warning: Switzerland) in 1955, which outlined a new city as an alternative to the one planned for the Schweizerische Landesausstellung (Swiss National Exhibition) in 1964. Under the motto "Vorwärts zur Natur!" Kutter had developed a proposal that was not approved by the VSG. That same year, Kutter submitted his resignation and then, in 1977, he presented his proposal in book form.[160] The precise reasons for the rift between Kutter and the VSG are not known. Because Kutter was against the idea of a horticultural exhibition as spectacle and wanted to emphasize strongly the aspects of sustainability, the VSG may have been concerned about the festive atmosphere, attractiveness, and attendance numbers. Kutter's successor was the architect Hans-Peter Ryhiner, who would later be director of Basel Tourismus. The vice director and marketing director was Kurt Aeschbacher, who is now famous as a talk-show moderator on Swiss television. On the VSG side, Wolf Hunziker was responsible for organizing Grün 80. The Bund Schweizer Gartengestalter (Association of Swiss Garden Designers, BSG) had an advisory function, with a Grün 80 planning group and with Peter Paul Stöckli as its representative on the general planning committee.[161]

In 1976, in preparation for the competitions for the sectors of Grün 80, an open ideas competition was announced, with modest prize money.[162] Dieter Kienast and Toni Raymann, as colleagues in Peter Paul Stöckli's firm, submitted an alternative proposal for Grün 80 titled "Warum lieb ich alles was so grün ist" (Why I Love Everything That Is So Green).[163] They joined in criticism, coming from Germany in particular, of the model of the "garden show as horticultural trade show" and presented a program based on the model of a horticultural show as a driving force for urban development, an idea that was being hotly debated at the time in Germany as well.[164] They proposed, among other things, investing a good part of the funding available for the horticultural show in measures for improving the green infrastructure in the center of the city and in the neighborhoods around the grounds of the horticultural show. It received a prize and was included in the group of important proposals. But there was no move away from the basic concept for Grün 80. Rather, a large show of festive character was organized, which was criticized in the press as a "spectacle," especially by Markus Kutter and Polo Hofer, a politically active Swiss rockstar.[165]

A "Phytosociological Garden"

Only the Pflanzensoziologischer Garten (Phytosociological Garden), proposed as item 9 of Kienast and Raymann's alternative concept, was further developed. The authors foresaw this garden, based on the model of Reinhold Tüxen,[166] as part of the horticultural show that was to be presented in the Botanical Garden of the Christoph Merian Foundation. The diction of their proposal reveals the emphatic stance of Kienast, who was planting an exhibition garden with his knowledge of the formation of plant communities and of succession processes, in which he wanted to communicate to "nonspecialists" a better understanding of their environment.[167] Kienast put "natural environment" in quotation marks and explained what phytosociology could achieve to make it easier to understand the relationship between the natural and the artificial:

With its help, it is possible to convey clearly the dynamics, vegetation's competition for soil, water, and light. Here it can be shown which types of vegetation are natural and which cultural. It can be shown, for example, that without constant human intervention, the vast majority of Switzerland (up to 1,800 meters) would be forested. The lovely Alpine meadows with isolated bushes and groups of trees—widely regarded as the epitome of "nature"—are nothing other than a specific way of farming and hence not a natural area but a cultural area.[168]

This quotation clearly reflects the teachings from Hülbusch's vegetation science walks and the interaction between production methods and image of the landscape. For the biologist Markus Ritter, Kienast was presenting himself here as a "popular educator" and explainer of what nature really is, but at the same time as *homo faber*, who was able to produce these natural processes by virtue of his phytosociological knowledge.[169] They also proposed creating "the prerequisites for a natural settlement . . . by means of special preparation of the soil and the surface design."[170] It would be necessary to distinguish between natural or nearly natural and anthropogenic plant communities in the process. Their list of anthropogenic plant communities included grassland communities (*Cynosurion, Molinion*), farmland weed communities (*Chenopodieta*), compacted path soil and pavement-crack communities (*Polygonion*), shrub communities (*Artemisietea*), and edge communities (*Aegopodion*, etc.). The natural and nearly natural plant communities included marshy meadows (*Caricion*), dry grasses (e.g., *Mesobromion*), forest communities (e.g., *Querco-carpinion*), and riverbank and meadow communities (e.g., *Bidention, Alno-Padion*).

The Phytosociological Garden was intended to show the succession processes of these communities as well as the plants "fighting" for soil, water, and light (25, 26, p. 74). The intention was to make it possible to experience the different stages of succession in a grassland community up to their climax in a forest community by fencing in and ceasing to use one section of the demonstration field every three to five years—users could then borrow binoculars to observe them. The idea of fencing in was in line with contemporaneous discourses on natural preservation. In the other areas, people were supposed to interact: for example, by maintaining the trampled grasslands. In addition, they planned to integrate an open body of water in connection with the ground water improvement area planned for the Botanical Garden. Walking through the various stages of succession—which would be well explained on display panels or by specially trained guides—would, in the view of the authors, not just increase "understanding of vegetation." It was also supposed to enable

23, 24
Postcard with the symbol of Grün 80 in Basel—the dinosaur placed on a hill with decorative beds in the "Erde" (Earth) sector—and a photograph of the Dry Grassland Biotope during Grün 80

25
Dieter Kienast, Grün 80 in Basel, "Pflanzensoziologischer Garten" (Phytosociological Garden), design study. Plan no. 278-6, scale: 1:200, H 60 × W 140 cm, pencil on tracing paper, September 19, 1978, signed "Kie," firm of Peter Paul Stöckli

26
Dieter Kienast, Grün 80 in Basel, "Trockenbiotop" (Dry Grassland Biotope), sections A–F. Plan no. 287-7, scale: 1:200, H 148 × W 82 cm, pencil on tracing paper, January 24, 1979, signed "Kie," firm of Peter Paul Stöckli

A-A

B-B

C-C

D-D

E-E

F-F

visitors to observe through binoculars "the fauna belonging to each area: insects, butterflies, birds, *Charadriiformes*, etc."[171]

Kienast intended the Phytosociological Garden not only to demonstrate how well he understood and could control these natural processes, thanks to his knowledge, but also to give visitors to the horticultural show a different idea of nature and, beyond that, to demolish the equation "Nature = our Alpine landscape." The point was to present a rawer, more merciless nature (29, 30, pp. 70, 72) in the sense of denying the delightful nature that existed as an image in the heads of most of the visitors to the horticultural show. A central role was assigned to human beings as interactors with nature, controlling the processes of succession, and to the garden architect trained in the natural sciences as *homo faber* of this artificial-natural world. All of this was presented in the slightly aggressive tone of explicit didacticism in the Kassel style.

The construction site selected was a one-hectare lot in the southern part of the Botanical Garden that was cut off by the overpass of a highway from the neighboring Dreispitzareal. Kienast began with soil analysis and phytosociological mapping (27).[172] In terms of its connection to natural space, the Botanical Garden is located on the gravel terraces of the Birs River, at the transition between two alluvial terraces. In earlier times, the Birs fed into the Rhine on what is now the Dreispitzareal. Hence the gravel deposits of both rivers were intertwined. The present Botanical Garden is built on this subsoil.

On the southern edge of the grounds, Kienast exposed a cross section through the soil. This "geological exposure" revealed to Grün 80 visitors the various layers of sand and gravel from the Jura limestone vaults (28). Kienast's soil analysis uncovered all sorts of detritus and waste deposits under a layer of humus. Beneath that layer, which ranged from a half meter to a meter deep, there was a clean layer of gravel, and beneath that in turn "contaminated deposited material."[173] This composition of the soil is explained by the fact that the gravel terraces had been used as a vineyard by the owner of the estate, Christoph Merian, who during the first third of the nineteenth century had had humus placed on top of the gravel, which had in turn previously been used as a waste dump. Kienast decided to remove the first meter of soil and deposit new gravel from a gravel pit in the Dreispitzareal over the layer of clean gravel, but leaving the lower layer of contaminated soil.

The idea was to use terraces and different soils to produce as many possible kinds of plant communities and to present their interplay in the succession. Writing about the succinct relief of the grounds, with four levels and steep slopes between them, Kienast stated that these varied expositions and slope inclinations had been chosen deliberately "so that trampling, frost, and rain result in slides and erosion, which lead to greater differentiation in the development of the vegetation."[174]

Reinterpretation as a
Dry Grassland Biotope

Shortly before the planning phase began in the summer of 1978, the Phytosociological Garden was reinterpreted as a dry grassland biotope as the result of sponsorship being taken over by the World Wildlife Fund (WWF) of Switzerland and its Basel section. Two plans document the changes to the proposal.[175] Whereas the site plan dated September 19, 1978, still has the name "Pflanzensoziologischer Garten" (25), the caption of the plan for the corresponding sections, dated January 24, 1979 (26), bears the name "Trockenbiotop" (Dry Grassland Biotope). Instead of a phytosociological garden, the WWF wanted vegetation that was nearly natural and had once been typical of the location in the Birs Valley but in the meanwhile had largely disappeared or was threatened.[176] In an article published by the WWF under the title "Grün 80 Trockenbiotop: Die Natur hat Vortritt" (Green 80 Dry Grassland Biotope: Nature Takes Precedence), the destruction of nature through urbanization and increasingly industrialized agriculture was described in drastic terms, and the new one-hectare Dry Grassland Biotope was praised as a refuge where "human claims yield to threatened native nature":

> Urban agglomeration is spreading more and more. Industrial and traffic infrastructure and suburban settlements cover valley floors, creeping up the slopes and displacing the green landscape. Agriculture is also becoming progressively hostile toward nature.... All of this means a war against nature and its living diversity. This war is increasingly and often irretrievably

destroying local habitats and life forms such as hedgerows, swamps, the natural courses of streams, riverbanks and their flora and fauna. Amid this flow of "progress," this small patch of land will become a home and refuge for threatened native life and will continue to be preserved as a nature reserve even after Grün 80 closes.[177]

The WWF presented the transformation of the former gravel pit—an extremely costly transformation of a site that had been altered multiple times—as a form of compensation for the growing "cultural wastelands of urban agglomerations."[178] (29) It was hoped that a nearly natural biotope incorporated into the Botanical Garden would have an educational influence in promoting to a wide audience nature conservation and the related movement:

Reflecting on our relationship to nature and our ideology of growth and progress . . . can affect our economic and political objectives and decisions . . . in favor of nonhuman life and hence over long-term life in general.[179]

As part of this revisionist idea, a broad strip of dry grass, which had to make way for the Baselland Transport AG (BLT) line next to the Botanical Garden, was transplanted to the new Dry Grassland Biotope. Dieter Kienast integrated it into his plan for nearly natural vegetation, which was given considerable "initial assistance" in the form of seeding and planting, so that there would be something to see in time for the opening of Grün 80. During his phytosociological mapping, Kienast had found a locust (*Robinia*) forest, beech (*Fagus*) forest, and a willow (*Salix*) forest, as well as the following plant communities: an *Arrhenatheretum elatius* and *Arrhenatheretalia* community; a *Epilobio-Salicetum capreae, Tanaceto-Artemisietum,* and *Arction* fragmentary community (a fragemtary community means that it is missing some of its typical species); a *Convolvulus-arvensis* and *Calystegia-sepium* stand; a *Mesobromion* fragmentary community; and a *Lapsano-Geranion-robertiani* fragmentary community.[180] This vegetation was incorporated into the Dry Grassland Biotope. On six newly installed gravel and gravel-clay soils of different types, there were four different sowings of grasses, herbs, and leguminous plants to obtain arid areas, shaded forest borders (nitrophile), sunny borders (thermophile), and semiarid grasses with

Dry Grassland Biotope for Grün 80 in Basel, 1978–80. Photos: Georg Aerni, 2012

Page 68
The Dry Grassland Biotope was intended to illustrate the succession processes of various plant communities in their "fight" over earth, water, and light

Pages 70, 72
The same sections of the grounds shown in figures 29 and 30 three decades later

Page 74
The Dry Grassland Biotope is now a registered nature reserve with the highest ranking. That is why connecting the former grounds of the horticultural show to the adjacent Dreispitzareal via a pedestrian bridge has been prohibited. During the planning for Grün 80 that resulted in the Dry Grassland Biotope, Kienast had advocated just such infrastructure measures to improve access to the recreation areas.

Tabelle der Bodenaufbautypen

Typ	Tiefe	Deutsch	Français	English
A	0–40 cm	Neue reine Kiesschicht	Nouvelle couche de gravier pur	New pure gravel layer
A	40–90 cm	Vorhandene saubere Kiesschicht mit sehr geringen Lehm- und Tonanteilen	Ancienne couche de gravier propre à très faible proportion de terre limoneuse et argileuse	Existing clean gravel layer with very small clay and loam percentages
B	0 cm	Grasnarbe (Halbtrockenrasen) lückenhaft (Bedeckungsanteil bei Transplantation etws 40 %)	Couche d'herbe (prairie méso-xérophylique) lacunaire (proportion de couverture après transplantation environ 40 %)	Sod (semi-dry turf) incomplete (coverage about 40 % at transplantation
B	0–30 cm	Transplantationsscholle etwa 30 cm starke Kiesschicht mit organischer Bodensubstanz	Motte de transplantation, couche de gravier d'environ 30 cm mêlé à la substance organique du sol	Transplantation substrate: gravel layer some 30 cm thick with organic soil substance
B	30–70 cm	Neue eingebrachte reine Kiesschicht	Nouvelle couche de gravier pur	Newly inserted pure gravel layer
B	70+ cm	Vorhandene, relativ saubere Kiesschicht mit geringen Lehm- und Tonanteilen	Ancienne couche de gravier pur relativement propre à faible proportion de terre limoneuse et argileuse	Existing relatively clean clay gravel layer with small clay and loam percentages
C	0–60 cm	Reine Kiesschicht, locker eingebracht	Couche de gravier pur, non comprimée	Pure gravel layer, loosely laid
C	60+ cm	Verdichtetes, kiesig-lehmiges Auffüllmaterial; Oberfläche im Gefälle abgezogen, um Staunässe zu verhindern	Matériel de remplissage graveleux-limoneux comprimé; surface de la pente aplanie afin d'empêcher les retenues d'eau	Compacted gravel-clay filling, surface smoothed on gradient so as to prevent damming
D	0–30 cm	Kiesig-lehmiges Auffüllmaterial mit Humusanteilen	Matériel de remplissage graveleux-limoneux et parts d'humus	Gravel-clay filling with some humus
D	30–90 cm	Vorhandenes kiesig-lehmiges Auffüllmaterial	Matériel de remplissage graveleux-limoneux sur place	Existing gravel-clay filling material
E	0–30 cm	Kiesig-lehmiges Auffüllmaterial mit organischer Bodensubstanz	Matériel de remplissage graveleux-limoneux mêlé à la substance organique du sol	Gravel-clay filling material with organic soil substance
E	30+ cm	Vorhandenes lehmig-toniges Auffüllmaterial, verdichtet	Matériel de remplissage limoneux-argileux sur place, comprimé	Existing clay-loam filling material, compacted
F	0–30 cm	Eingebrachte Kulturerde	Terre de culture introduite	Deposited vegetable soil
F	30+ cm	Vorhandenes lehmig-kiesiges Auffüllmaterial, zum Teil gewachsener Boden	Matériel de remplissage limoneux-graveleux sur place, en partie sol originel	Existing clay-gravel filling material, partly natural soil

Table with the soil structures of the Dry Grassland Biotope

28
Visitors to Grün 80 in front of the "Geological Exposure." In the background, a section through the slope of the gravel terrace, with the various layers of sand and gravel of Jura limestone. In front of this were various rock types as numbered sample stones mounted on steel tubes; a display panel provided information about their origins.

29, 30
The Dry Grassland Biotope under construction. In the background, the highway overpass and the adjacent housing developments during Grün 80.
Photos: Dieter Kienast

the aim of integrating transplanted areas.[181] In addition, arid- and heat-loving (xerophile and thermophile) shrubs and a beech forest with appropriate plants in the undergrowth and borders were planted.[182] The terrain was given a succinct relief of terraces and levels, articulated by steps made of wooden planks, sand-strewn paths, and large limestone erratic blocks.

The idea of presenting the states of succession at Grün 80 was retained. During the show, it was primarily so-called pioneer stages that could be seen. In an article on the Dry Grassland Biotope published in the professional journal *anthos,* Kienast risked a prognosis about how the area would develop further.[183] He conjectured about the effect of the interaction between human beings and nature and briefly took position on the significance of the future "maintenance" of this area:

If the semiarid grass communities are to be preserved for the long term, then deliberate, extensive maintenance will be required. Only mowing, trampling, or pasturing can prevent these vegetation units from succeeding into bush or forest communities.[184]

The history of the genesis and design of the Dry Grassland Biotope, along with the corresponding discussions, provides insight into the complexity of how the natural and the artificial are interwoven, which remained a concern for Kienast in his entire oeuvre and manifested in different ways in his design activity. The Dry Grassland Biotope was at once a seeded and planted nature reserve, a renatured depot at the foot of a highway overpass, and a garden to display dynamic succession processes. This paradox could not have escaped Kienast.

In 1980, the recently formed firm of Stöckli + Kienast received honorable mention for the "planning and designing of the arid and desert biotope" from the Gartenbauamt (Parks Department) of the City of Zurich for the best project on the subject of "nature conservation in the community." The jury explained its decision as follows:

The object receiving the award is being assessed as a pilot project that would be conceivable in various Swiss communes. It is an example of the preservation and creation of habitats for pioneer vegetation and small fauna that has great educational value. . . . It can be achieved with relatively little funding, whereby associations with different interests could be motivated to support such projects financially. It is suited to restoring different types of soil, independently of altitude and the local situation (use as waste collection centers, railway lines, etc.). Moreover, it is independent of the size of the area available. . . . This achievement has to be praised all the more given that the authors of this good example of applied nature conservation are responsible garden architects, that is to say, representatives of a profit-oriented profession.[185]

To illustrate further the absurdity of the discourse about the Dry Grassland Biotope, I mention here an homage to it written by the biologist Marcel Amstutz under the title "Das kleine Paradies" (The Little Paradise) in 1985—revealingly, without even mentioning the garden architects as its authors.[186] Five years after it opened, Amstutz found that there were already 150 different species of plants on the 1-hectare area, and he described the "reconquest of nature" as if the people who had motivated it had only been a hindrance, whose mistakes nature, left to its own devices, may now overcome (30, pp. 72, 74):

These human interventions were necessary to better present the not-yet-green Dry Grassland Biotope at Grün 80. But over the long term, they in part have proven unsuccessful and more of a hindrance. The initially very promising transplanted semiarid grass differs only slightly today from the other pioneer areas. And the densely concentrated areas seeded with a rich mixture of seeds are now rather impoverished and look somewhat monotonous next to the uninfluenced areas, which are more sparsely developed but all the more so manifold and diverse in terms of species. Gradually, however, these parts of the Dry Grassland Biotope that were influenced by humans will be seized and redesigned by the development of vegetation that includes the entire area.[187]

The latest act in the discussion of the natural and the artificial in this early design by Dieter Kienast came to a close only a few years ago: The Christoph Merian Foundation, which still manages the Botanical Garden and the adjacent park on the Brüglingen plain as the Merian Gärten (Merian Gardens), proposed that this area be provided with better public access in the form of a pedestrian bridge from the Dreispitzareal across the train tracks and highways.[188] The overpass was supposed to lead into the border of the Dry Grassland Biotope, which, however, had since been registered as a nature reserve, so an architectural intervention such as creating access via a pedestrian bridge was no longer possible. Kienast, whose alternative proposal for Grün 80 had advocated precisely such infrastructure measures for improved access from neighboring residential areas, would surely have regarded this development as astonishing. As a dry grassland biotope, the Phytosociological Garden that Kienast had proposed back then is now a hindrance to Kienast's user-oriented approach, in which a horticultural show should primarily seek to improve the living conditions of the people in the city.

Sociopolitical Allegories

The discourse on the tension between humanity, plants, the city, and nature had specific effects on Dieter Kienast's use of plants when working with the "nature of the city." It was based on the contrast of "spontaneous" and "built" vegetation, a topic that he had first considered in detail in the diploma thesis he cowrote with Thom Roelly, and on its sociopolitical allegorizing, which Kienast later further aestheticized.

Kienast and Roelly made a direct connection between more extensively maintaining a green space and the pleasure that people, especially children, derive from their personal development as they appropriate that area as a playground:

> Ruderal vegetation—'weeds' in the jargon of gardeners—is, as it were, still the 'stumbling block' for gardeners, groundskeepers, and parks departments, since a whiff of the neglected, abandoned, unbeautiful still clings to it. Every gardener or groundskeeper who has preserved any 'professional honor' will stake everything on eradicating it by expensive means, whether mechanical or chemical. In addition to our own observations, several authors have also . . . determined that these neglected and 'unbeautiful' areas seem to be the preferred places for children and teenagers to play and spend time. For that reason, the question arises as to whether such 'wild playgrounds' cannot also be created through planning, especially as—even ignoring their outstanding suitability to appropriation—it requires only minimal investments.

In other areas, too where ruderal vegetation is spreading, it is necessary to ask whether it might not prove an essential means to minimize the enormous cost of maintenance and at the same time preserve multiuse interim spaces that permit diverse activities. This, however, assumes not insubstantial work educating the public and, not least, gardeners and planners.[189]

Kienast and Roelly were broaching a theme that had been discussed regularly by open space planners and garden designers since the 1970s. Ultimately its roots lie in the ideological debate which in the eighteenth century had played out between the advocates of gardens in the French style and those of gardens in the English style.

Style of a Garden and Form of Society

The defenders of gardens in the English style were at the same time defenders of the Enlightenment, which intertwined the concept of nature with the idea of freedom and produced a connection between nature, freedom, and morality. This was a political discussion in which the values of the Enlightenment were dovetailed with a natural right of every person. The English philosopher John Locke called for a new social contract based on "natural law," which shook the nobility's right of succession ordained by "divine right." The dominant values of absolutism were rejected as contrary to human nature, reason, and morality. Two different views in garden design illustrated this discourse: the absolutist-minded owners of gardens in the French style were discredited by making an analogy between the outward appearance of the plants presented in such gardens and the moral and human nature of the owners. The French garden, with its hedges pruned in geometric forms and topiaries, bosquets, follies, and waterworks, became the symbol of nature controlled by and even subjugated to human beings. It was hence also symbolic of a traditional social order that functioned based on hierarchies, power, and inheritable rights and therefore did not know what to do with the new ideas of freedom, since they necessarily forced open the old system.[190]

This analogy between the style of a garden and the form of society, of outward and inward nature, is strikingly captured in the painting *Maifest auf Gut Freienhagen* (May Day on the Freienhagen Estate, 1766) by Johann Heinrich Tischbein (31). The scene takes place near Kassel, at the country estate of Frederick II, Landgrave of Hesse-Kassel: whereas the landgrave and his female companion are standing in front of the conical topiary trees in their French garden—surrounded by the staff officers of the court in buttoned uniforms—their subjects are dancing around knotty deciduous trees with tall, unpruned tops so exuberantly that skirts are flying. The defenders of the English landscape garden laid claim to this connection of liveliness and freely developing nature in their polemic, even if such gardens would not be conceivable without human intervention. Human beings were treated merely as supports of a natural beauty already present in nature, though only a rational person knows how to bring it out well. The curved form, for example, preferred for the paths and lakes of the English landscape garden followed William Hogarth's "line of beauty" as a form whose irregularity was derived from nature and hence was closer to the natural and the living. All straight lines and geometric forms—such those seen in the terraces, grass embankments, and conical yews of Tischbein's painting—were rejected as unnatural.

The eighteenth-century discourse that made a direct connection between outer and inner nature, ideas of freedom and sociopolitical attitudes, flared up again under different circumstances in the early 1970s. The symbols of the new social order and its undesirable trends were the fossilized, car-friendly cities in which green space was always designed, decreed from above, and, in the case of hedges and beds, "stately" green: remnants in a thoroughly organized city. In Jörg Müller's series about the urbanization of a landscape over the course of two decades, the final image has remnants of this kind (32): between the new shopping center and the daycare center, whose building is delimited by a narrow strip of flower beds, a small lawn was reserved with a "Keep off the grass" sign, a sandbox at the corner and a bench next to it for the parents supervising their children, a horizontal concrete pipe for the children to crawl through, and a small, recently planted tree. In 1972, that was all that children had to play with in the open space. By contrast, twenty years earlier they had the entire landscape, the immediate vicinity of the home, and

31
Johann Heinrich Tischbein, *Maifest auf Gut Freienhagen* (May Day on the Freienhagen Estate), 1766. Museum Schloss Fasanerie, Eichenzell

32
The small sandbox on a lawn in front of the shopping center, which has been left for the children to play in, becomes the symbol of disciplining human beings in the newly constructed agglomeration landscapes.
Jörg Müller, *The Changing Countryside,* 1977 (detail)

the streets in the residential neighborhoods. The reorganized open space thus becomes a visual symbol of human discipline and restriction.

Several planners of open spaces and garden architects in the 1970s declared areas in the city outside the system of control to be symbols of resistance. These are areas where spontaneous or ruderal vegetation can become established: plants that have no value as status symbols. They are not used in productive ways and do not serve to beautify green spaces. Their presence is associated with the voluntary or involuntary abandoning of human control of an area.[191]

An ironic text by Kienast illustrates this polemic. It was published in 1977 in *Der Monolith,* the student newspaper for Kienast's department in Kassel, edited by Annemarie Burckhardt (33).[192] Kienast chose the form of a letter to the editor and pretended to be a vigilant citizen outraged by vegetation growing wild on the abandoned Henschelei factory grounds. The teachers and students of the architecture and urban and landscape planning department had moved there in 1976. In the style of the respectable, order-loving citizen, Kienast reported on the "weeds and wild grasses" spreading "in truly terrifying abundance" on the vacant lots of the Henschelei as a result of the "boundless negligence of authorities responsible":

Where not all that long ago the precisely maintained asphalt and paved surfaces of the factory buildings (supported by diligently manicured borders of decorative shrubs) provided a worthy—because analogous—framework, today there is vegetation growing that raises serious doubts about whether the otherwise omnipresent "caretakers" and "custodians" are still in control of the situation.[193]

After a remark that more than two hundred different plant species had been found on a visit to the site in the summer of 1976, which were there "without permission" and were "even reproducing (ick, ick)," the author concluded with three rhetorical questions. The first sought the authorities who had failed, with the result that plants were able to invade the factory grounds—Kienast used the image of a lack of "right to be here," which the plants could not demonstrate, as if talking about illegal immigrants with no residence permit. The second question makes it clear that the plants had not only populated the area spontaneously but rather that a "university professor who will remain unnamed"—this can only refer to Karl Heinrich Hülbusch—had assisted them by deliberately seeding weeds. The third question suggests a direct connection between the weeds subversively spreading unwanted and the students freely expressing their opinions in these open spaces, having monopolized the grounds much like the plants:

How long will it take for all the open spaces of the university grounds to be cleanly and hygienically paved over and the remaining areas planted with beds of roses and cotoneasters? For only such measures can offer a reliable guarantee that the unnecessary lounging about and discussion of students in open spaces can be prevented.[194]

The imaginary letter writer—the prototype of the urban philistine—adopts all of the topoi of the idea of nature as symbols for freedom or rather as an expression of the establishment. In terms of plants, this means beds of cotoneasters and roses versus ruderal plants. The spontaneously populating plants take on the role of nature as a symbol of freedom. They are the unexpected: that which comes on its own and unbidden and can only be eradicated by constant human effort. Spontaneous vegetation becomes the epitome of *natura naturans,* and the weed an image of the power of nature, at a time when the public was becoming aware of the destruction of the landscape and the lack of nature in the city.

The Recapture

In the same spirit, but in an even more drastic form, this motif appears in the story "Die Rückeroberung" (The Recapture), written in 1979 by the Swiss author Franz Hohler. Hohler imagined eagles, deer, wolves, and bears increasingly entering the City of Zurich. In the end, urban life was paralyzed by the unbridled growth of plants: "it turned out that there was another threat lying in wait over the city to make us even more helpless. It looked harmless at first, even promising, but soon it became clear that this was precisely what could finish us off."[195]

The climbing plants from front yards—ivy, clematis, white knotweed, wisteria—suddenly spread unrestrained and "started intermingling

DER MONOLITH

SEITE 2

Grenzenloser Nachlässigkeit zuständiger Stellen zufolge haben sich in den letzten fünf Jahren auf den Freiflächen der ehemaligen Henschelei Unkräuter und Ungräser in geradezu erschreckender Fülle ausgebreitet. Wo vor nicht allzu langer Zeit akurat gepflegte Asphalt- und Pflasterflächen den Werkshallen (unterstützt von emsig gehegten Zierstrauchrabatten) einen würdigen -weil entsprechenden - Rahmen gebildet haben, entwickelt sich heute eine Vegetation, die starken Zweifel darüber aufkommen lassen, ob die sonst allgegenwärtigen "Pfleger" und "Unterhalter" noch Herr der Lage sind.

Während häufiger Besichtigungen im Sommer des Jahres 1976 mussten wir über 200 (in Worten:zweihundert) verschiedene Pflanzenarten feststellen — Legion sind der Unkräuter im gesamten Werksgelände!! Unter anderem fanden wir folgende Unkräuter vor: Ackerdistel (Cirsium arvense), Acker-Gänsedistel (Sonchus arvensis), Franzosenkraut (Galinsoga parviflora), Kanadische Goldrute (Solidago canadensis), u.a.. Sämtliche der o.g. Unkräuter stehen aufgrund der §§ 2 und 3 der Verordnung zur Bekämpfung des Unkrautes vom 19.9.1960 (GVBl.S.208) (Vergl.Kasseler Wochenblatt vom 26.3.1976, 26,13,S.89) auf der Fahndungsliste und sind mit allen erdenklichen Mitteln zu bekämpfen bzw. zu vernichten. Deren Nichtbefolgen kann laut EGOWiG (Einführungsgesetz zum Gesetz über Ord-

Naturschutz: UNKRAUT IN DER HENSCHELEI!

nungswidrigkeiten vom 15.10.1970) mit Geldbuße bis zu 10'000 DM bestraft werden!

Der Verfasser sieht sich aufgrund o.g.Tatsachen verpflichtet, den Gründungsbeirat darüber zu informieren, daß sich auf dem Henschelgelände verbotenerweise Pflanzen aufhalten; ja — daß sie sich (igitt-igitt) sogar vermehren. Darüber hinaus verlangt der Verfasser vom Gründungsbeirat Antworten auf folgende Fragen:

1. Wie sind diese und andere Pflanzen, die keinerlei Daseins (Hierseins-)berechtigung nachweisen können, in das Henschelgelände

Foto:Helmut Böse

eingedrungen, bzw. wer hat dies veranlaßt oder ermöglicht?

2. Trifft es zu, daß ein hier nicht näher bezeichneter Hochschullehrer sich sogar erdreistet, der Entwicklung des Unkrautes im Hochschulgelände vermittels Ansaat weiterhin Vorschub leistet??

3. Wie lange noch wird es dauern, bis sämtliche Freiflächen des Hochschulgeländes sauber und hygienisch asphaltiert sind und die Restflächen als Rosen- oder Cotoneasterbeete angelegt werden. Denn nur derartig gestaltete Maßnahmen bieten eine verlässliche Gewähr dafür, daß unnötiges Herumlungern und Diskutieren der Studenten auf freier Fläche unterbunden werden können.

Sollte eine gänzliche Überteerung des Geländes in naher Zukunft -aufgrund fehlender Finanzmittel- nicht möglich sein, so verlangt der Verfasser einen entschlossenen, sofortigen, intensiven und nachhaltigen Einsatz von Unkrautvertilgungsmitteln, um diesem unhaltbaren - ja geradezu anarchischem-Zustand Einhalt zu gebieten.

NIEDER MIT UNKRÄUTERN UND UNGRÄSERN!!
HINFORT MIT LAUBABWERFENDEN BÄUMEN!!!
SCHLIESST DIE SCHLICKICH-SCHLAMMIGEN SCHOTTERFLÄCHEN MIT SCHÖNEN, SCHWARZEN, SCHUHFREUNDLICHEM ASPHALTSCHICHTEN!!
FÜR PASTEURISIERTE FREIFLÄCHEN IN DER HENSCHELEI!!

Dieter Kienast

33 Dieter Kienast, letter to the editor in the student newspaper *Der Monolith*, December 1977

34 Karin Widmer, illustration to Franz Hohler's story "Die Rückeroberung" (The Recapture), 1979

35 Karl Heinrich Hülbusch in front of a bed of shrubs on the grounds of the Gesamthochschule Kassel that was designed and maintained based on a knowledge of phytosociology

... to take on the battle for the streets, buildings, and subways,"[196] until the old marsh plants horsetail and pestilence wort came together in this thicket and transformed parts of the City of Zurich into an impenetrable jungle, where wild animals could cope better than human beings (34).

Kienast was impressed by Hohler's story, which was first published in *ZEIT-Magazin,* and he was still referring to it in texts written in the late 1980s. From the outset, however, what interested him about the integration of spontaneous vegetation into planning was a kind of reconciliation and synthesis. In 1989, for example, he wrote that, despite this antithesis to the urban city formulated by Hohler, a recapture of the streets had in fact occurred, albeit a modest one, in the form of pedestrian zones, public squares, and residential streets[197]—just as he and Roelly had proposed in their 1975 plan for open spaces, and just as he had presented it in his experimental model for Grün 80 (14).

This enthusiasm for the subversive character of plants spreading in the city was in keeping with the time, but its allegory remained largely rhetorical, since the practice of planning itself was primarily about the question of how these natural forces could be made useful to the planner. The associated taming of the subversive was also done for economic reasons that were likewise used to promote the use of spontaneous vegetation: in a photograph from the 1980s, Karl Heinrich Hülbusch is seen standing in front of a bed of shrubs arranged according to phytosociological characteristics; the caption points out that it cost only 50 pfennig per square meter, whereas 500 deutschmarks per square meter were being spent to create the adjoining areas (35).[198] In light of this fortunate link between the wild and the inexpensive, it is understandable that the "university professor who will remain unnamed" in the letter to the editor not only permitted weeds but deliberately seeded them to give a hand to the spontaneous.

In the theoretical discourse, however, the revolutionary energy and dynamics of spontaneous vegetation were combined. The letter to the editor accordingly evoked the image of wild plants in the backyards of the Henschelei (33) and students discussing—as a dual expression of the living. Indeed, the students of the 1960s and 1970s were without a doubt the revolutionary social force. The image of a thoroughly organized public space, by contrast, demonstrated the disciplining of design. That is why, as the letter writer indicated, only measures such as beds of cotoneasters and roses could be a "reliable guarantee that the unnecessary lounging about and discussion of students in open spaces can be prevented."[199] This linking of inner and outer nature, the appearance of open spaces and human activities stimulated by a particular look of an open space, is thus the key to understanding the planning proposals that Kienast and Roelly formulated in their diploma thesis, which also conformed to a basic attitude of the Kassel School.

For Indeterminate Spaces
in the City

Kienast and Roelly identified the places that are left out of the thoroughly designed areas and the extent of their freedom by the presence of ruderal vegetation, and they identified and classified them as open spaces that facilitate new possibilities of use.[200] At the same time, they observed that people derived more pleasure appropriating such places individually—in a liberated form of play. For that reason, planners should just leave more open spaces alone—like the weed-overgrown, paved backyards of former commercial zones—or make such places, that open up opportunities rather than define the use, available to residents by, for example, reducing maintenance.[201] Such ideas came from Karl Heinrich Hülbusch and Lucius Burckhardt, who had made the value of "dysfunctional spaces," as they are known in planning jargon, one of their themes in Kassel. In a text from 1980 of that title, Burckhardt referred to the areas left out of the planning of a city as "No Man's Land."[202] In his view, no-man's-land is interesting to people until the municipal workers burn down the bushes, grade the riverbanks, periodically mow the lawn, and install a grille. Then these places, where "Schorsch lit his homemade rocket and Anne was given her first kiss," are irretrievably lost. Green urban planning is the enemy of such no-man's-land because it cannot turn built areas into green ones but has to transform and destroy the interim areas and remnants:

> Plans to green the city by transforming No Man's Land into disciplined green spaces contribute neither to beautifying the city nor to increasing

the availability of recreational zones. Total gardening actually does not give rise to that which city gardeners expect, namely an urban landscape. On the contrary.... [203]

Kienast rejected just such total gardening, citing the same arguments that make a connection between leaving areas out of the power of planners and administrators, allowing them to grow a little wild, and then using them in playful and liberated ways. Kienast and Roelly distinguished such unmaintained areas and the ruderal vegetation that occurs there as a result of what they defined as "built vegetation," which they understood to mean trees along streets, flowers in beds, wheat fields, and lawns.[204] The term implies that the latter kind of vegetation belongs in the sphere of the city designed and hence disciplined by human beings. The municipal gardeners and green urban planners thus represent the long arm of the establishment.

The discussion of how much design urban spaces can bear is as current and volatile today as it was in the 1970s. In her programmatic lecture "The Global Street: Where the Powerless Get to Make History" in the summer of 2013, the economist and sociologist Saskia Sassen was critical of the fact that public spaces in cities are increasingly dominated by finance-driven forces and hence are losing their open character.[205] In her view, there are far too many overdetermined spaces today. Indeterminate spaces are the necessary preconditions for a nonexclusive public space that is shaped in part by the "powerless"—people without money and influence. For Sassen, this is how what we think of as "urbanism" is created:

I care about indeterminate space because I think it is one of the cases where people without formal access to politics, industries, etc. actually can "make." In a complex city powerlessness has a chance (as opposed to the plantation, where it is definite). Cities have long had much indeterminate space and messiness. The slum for example is a space for making, but our large cities are confining that space more and more. Today, in the time of mega projects, gentrification and gated communities there is not much indeterminate space left. Power grabs more and more space in today's global cities.[206]

Sassen sees architecture and landscape architecture as complex forms of knowledge that combine the competence and knowledge of different disciplines and areas to make something visible that would not otherwise be seen. Her idea of "incomplete architecture," which opens things up rather than closing them off, is satisfied by landscape architecture, because it works with dynamic forces and the living material of plants, in which change is already inherent. In this, Sassen identifies central qualities of subversion—in the sense of a possible opening up of spaces that people without money and influence can appropriate.

The discourses that preoccupied Kienast all of his life have lost nothing of their currency. He saw the necessity of both poles in the city: undesigned, extensively maintained areas and clearly formed areas—depending on the site, purpose, and use. In the final chapter of his dissertation, he praised the nearly natural gardens and public parks of Louis Le Roy (1924–2012) and also the critique of the stereotype of the Swiss garden initiated by the biology teacher Urs Schwarz (1928–2020), underscoring that the "inhospitableness of our cities,"[207] lamented by the psychoanalyst Alexander Mitscherlich (1908–1982) in 1965, was largely the result of the planning of open spaces.[208] On the other hand, already in the ideological debate over the natural and the artificial, Kienast made subtler distinctions and at the same time criticized the "pool of water in natural form" as a supposed panacea in the dreariness of badly planned neighborhoods:

> The conflict between the profession of gardeners and the guild of "natural gardeners" is built into the system: the question of the sole true form of the garden raises tempers and causes both sides to forget that, depending on the intended use, sometimes the "natural garden" and sometimes the "architectural garden" will be the better solution.... Both sides are quarreling exclusively about the look, species, and upkeep of the vegetation, while the residents' legitimate needs for the use of their open spaces are neglected once again.[209]

In Kienast's view, use is ultimately what should lend form to the vegetation—whether "spontaneous" or "built"—so that the two can have an organic relationship.

Vegetation and Use: The École cantonale de langue française in Bern

In 1984, as part of the firm Stöckli + Kienast, Dieter Kienast won the competition to design the grounds of the École cantonale de langue française in Bern, and his design reflects both the theories of the Kassel School and his turn to architecture and art since the early 1980s (36). The argument in the final chapter of his dissertation that the use of vegetation and the forms of landscape architecture should be based on use was realized here. At the same time, the project is an example of how important it was for Kienast in the 1980s to find a balance in his work between design and the use of plants, between utilization and the alignment of open spaces to urban planning and the landscape context. The competition entry was worked out in collaboration with the team of architects Somazzi Häfliger Grunder of Bern in 1983 and 1984, and the cornerstone was laid in 1988. The employees Günther Vogt, David Bosshard, and René Heer of what was by then Stöckli, Kienast & Koeppel, having expanded in the meanwhile, were involved in the construction phase, which continued until 1991.[210]

The site of the École cantonale de langue française includes a kindergarten, a schoolhouse complex, and an attached building for the administration and school cafeteria, as well as a recess yard and a large sports facility with a grass playing field, additional playing fields, and changing rooms. The school is located on the outskirts of Bern, trapped between Autobahn 6 Bern–Interlaken to the west and the looming residential high-rises from the 1960s in the Murifeld district of Bern to the east. The grounds are triangular in form and open up like a funnel between those two determinants. To the north, it adjoins a community garden colony along the full width of the property; to the south, there is a parking lot. The residential towers and the autobahn, the sounds of which are reduced by a noise barrier, give the place a decidedly urban atmosphere. Kienast and the team of architects responded to that in the choice of materials and the design of the school's exterior (37, pp. 92–108).

The "backbone" of the design and the crucial structural element that articulates the grounds of the site is a long north–south axis that runs through the entire property and connects the parking lot to the south with the community gardens to the north. It separates the athletic fields to the west from the buildings of the kindergarten and the school complex to the east, which are strung like pearls along this axis. Between the kindergarten and the schoolhouse complex, the axis widens toward the west to form a triangular recess yard. It is terminated in the north by a building that is at a right angle to the axis, which houses the school cafeteria on the ground floor and the administration and teachers' offices on the upper floor. From there is a narrow footbridge that crosses over the axis to connect to the schoolhouse.

Whereas all of the school buildings are at an angle to the axis, the playing fields of the sports complex extend diagonally from it. Kienast chose their orientation based on the course of the autobahn that terminates the property to the west. The large grass playing field is placed such that its long side runs along the recess yard that widens at an angle (p. 94). Some of the playing fields to the south of that appear to collide with the strict north–south axis or are delimited and cut off by it. Several rows of trees and shrubs cross the axis, however, without disturbing it, leading into the second zone of open spaces, which are located to the east of the access path, mostly behind the kindergarten and the schoolhouse complex (p. 96). These exterior spaces, which are followed by the "green setback" of the residential complex, look like the "backlands" of the school, where the children can play, hidden from the buildings and from view, while the open and exposed playing fields allow people to observe the matches (p. 98). A gravel path runs around the entire grounds: from the parking lot, it leads to the west of the school complex and passes between the autobahn, which is behind hedges but still quite audible, and the playing fields; to the north, it continues between the back of the administration building and the community gardens; to the east, it passes between the meadow of flowers behind the schoolhouse wings and the enormous residential towers on a

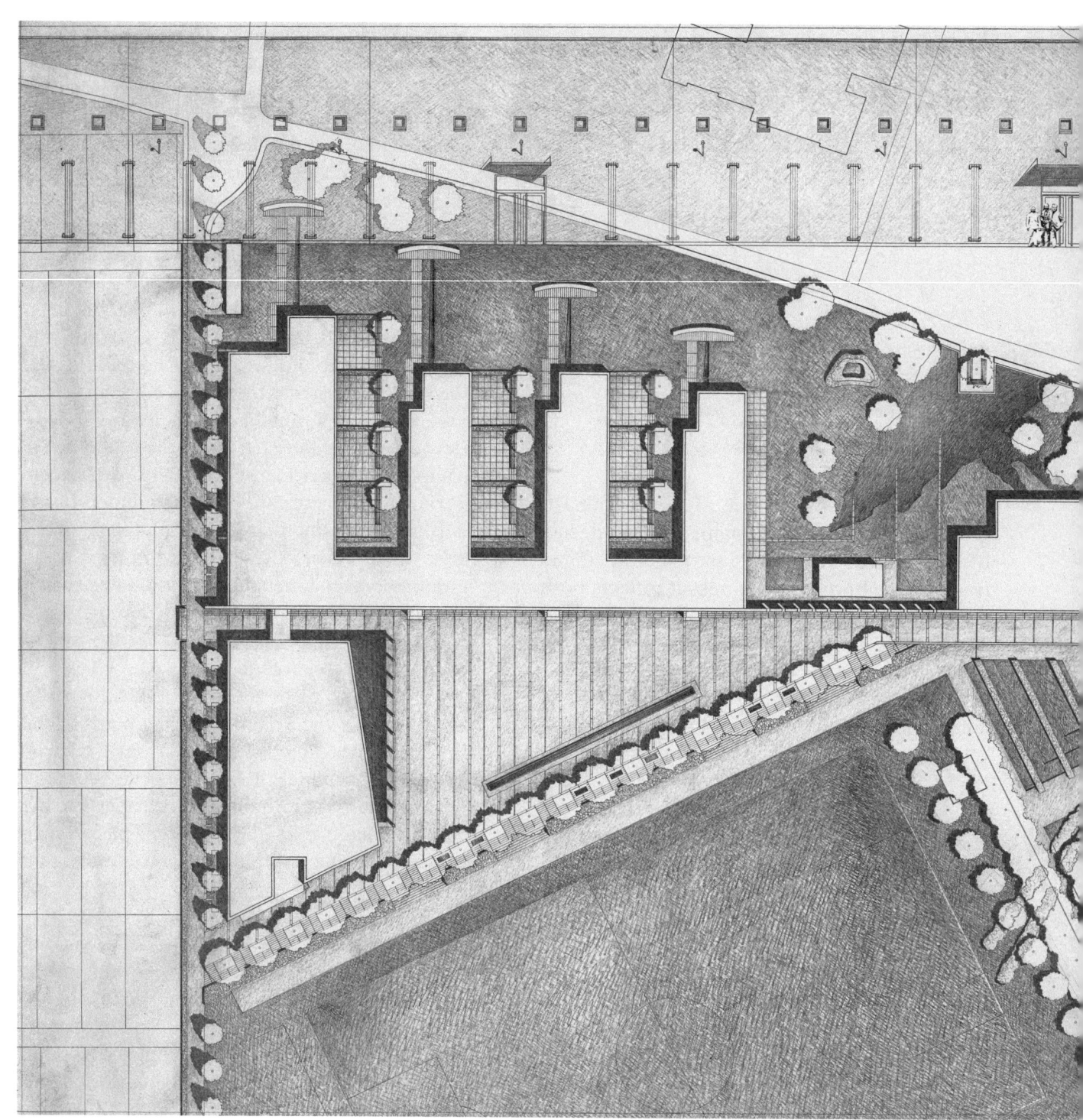

36
Dieter Kienast, design of the grounds of the École cantonale de langue française in Bern, scale: 1:200, H 174 × W 89 cm, copy of plan on thin cardboard, ink, graphite, and colored pencil, November 2, 1987, signed "Kie," firm of Stöckli, Kienast & Koeppel

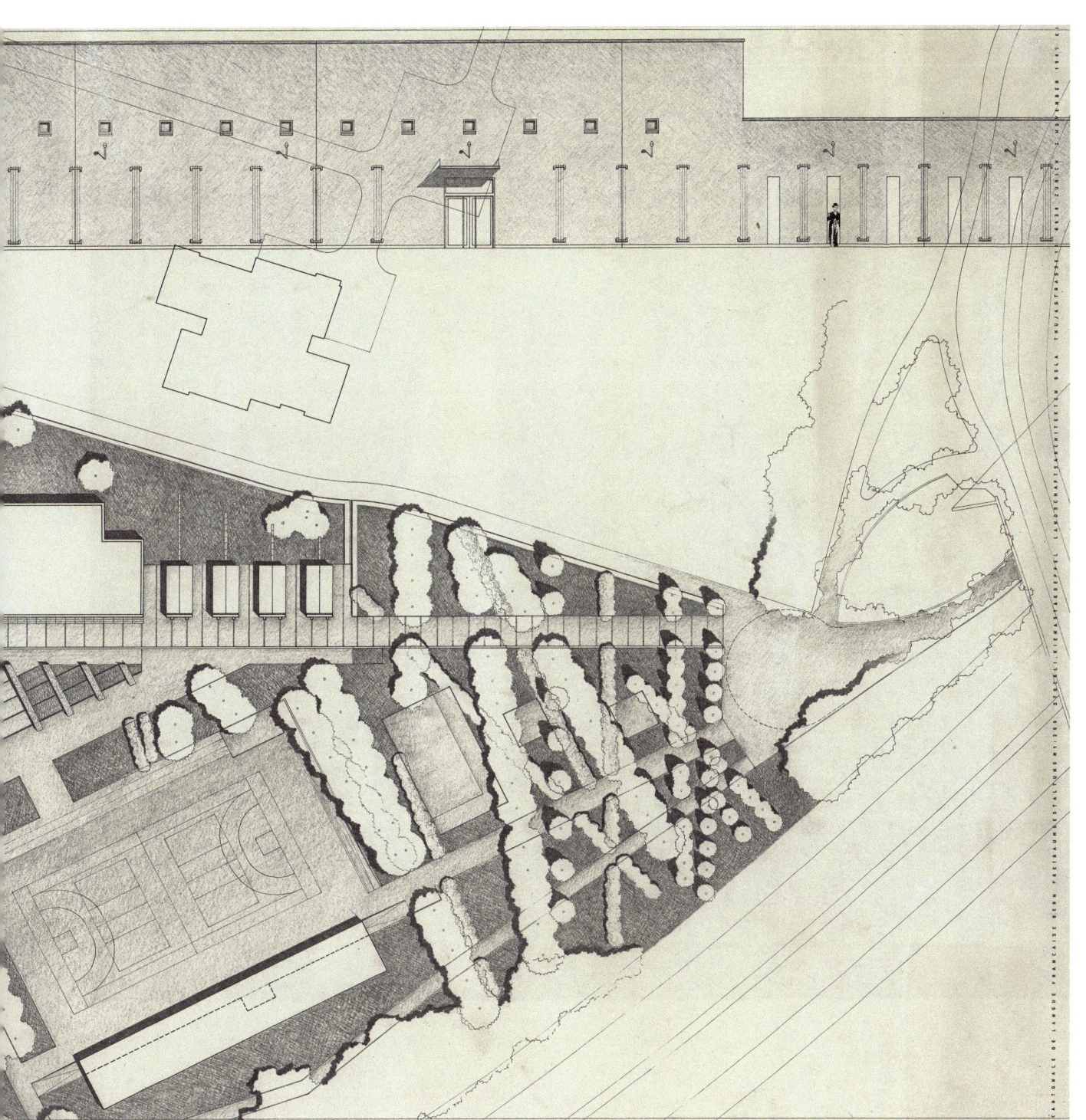

37
The basin of the École cantonale de langue française in Bern. Concrete and steel are combined with natural materials. The naturally evolved cultivated landscape on the edge of the city—with traces of a village landscape, the autobahn, and the housing development from the 1960s—was reflected in the design of the grounds. From 1991 onward, Dieter Kienast exclusively relied on Christian Vogt to document his works in black-and-white photographs.

green lawn, before finally leading back into the parking lot to the south.

The various areas of the grounds are precisely separated by design, materials, and plants, although the different manifestations of "urban nature" also play an essential role in their effect. At the parking lot, visitors are received by a small traffic island, framed with edging stones, that has birch trees and tall grass. The individual parking spaces are arranged around it, with some in chambers between the tall grass and some in chambers formed by hedges. Then visitors step onto the access path, which beyond the kindergarten widens to form the recess yard. These two central elements of the grounds are covered with asphalt, interrupted at regular intervals by strips of concrete; every other strip is wider and has a drainage channel for rainwater, since it cannot otherwise drain the sealed surface (p. 92). This creates a pattern of bright and dark that provides both order and variation. Spontaneous vegetation has appeared in several of the gaps between the asphalt and the concrete. The access path itself is bordered by a hedge to the east, as far as the kindergarten, and to the west by the rows of trees and shrubs that cut across the axis. The pattern of the ground is continued on the recess yard and related to the continuous exposed-concrete façade that connects the one-story building of the kindergarten to the two-story schoolhouse complex.

Seen from the recess yard, the exposed-concrete façade looks like a single, long, continuous wall with openings for windows and doors. The buildings immediately behind it can only be imagined. The broad open space between the kindergarten and the schoolhouse is also hidden by the façade, which does indeed become a wall here, but it can be seen through a series of tall, slender openings, through which children can slip from the recess yard to the playing areas behind the wall. In addition, the exposed-concrete façade looks like a uniform pedestal for the residential towers that loom behind it, which ties them together. This reintroduces composure into the ensemble. The prominent horizontals also offer a counterweight to the vertical towers. An elongated, narrow water basin that Kienast placed on the western edge of the recess yard, parallel to the grass playing field, is another balancing horizontal element. The residential towers reflect on the surface of the water, giving the impression that they reach vertically and upside down into the recess yard. This optical illusion brings the whole scene into an equilibrium of forces and forms.

The gray water basin fits well with the pattern of the recess yard, with its light-gray strips of concrete and dark-gray asphalt surfaces. Where the concrete strips that run across the yard meet the exposed-concrete façade, a hole has been left above each one for a wild vine (p. 100). Today, self-climbers have spread across the entire gray façade as if on a canvas, and the resulting structures can be seen from afar. Their fine ramifications and the green leaves they sprout, which turn red in the autumn, produce images that offer a pleasant contrast to the gray of the residential towers, with which the exposed-concrete façade enters into a kind of dialogue. The leaves of the vines are also reflected in the water basin, its surface constantly slightly rippled by a small drinking fountain always flowing.

In front of the plant holes and parallel to the façade, there is a transitional strip of a series of three concrete stones paved one behind the other. Kienast left the joints wide enough that all sorts of spontaneous vegetation could take root here. He did the same thing on the other side of the recess yard, where a strip consisting of a series of ten rectangular concrete stones placed one after the other forms the termination (p. 102). Their backdrop is a path lined with linden trees, each planted at regular intervals in a rectangular field with a metal tree grille. Numerous mosses, small grasses, and tiny ruderal plants are sprouting in the gaps of the grille and between the stones. Here, too, the various shades of green contrast beautifully with the bright-gray stones and rust-brown grille. There are gray benches beneath the linden trees; a low hedge of maples marks the boundary to the grass playing field. The activity in the recess yard can be observed from the benches, but not the games on the deeper-set grass playing field—to do so requires the viewer to stand. Below the boulevard, from the edge of the recess yard, one looks onto the playing field as if from spacious bleachers. One can sit on one of the steps that lead down to the playing field or simply walk down them.

The area of the parking lot, with its extensively maintained urban nature (tall grasses and young birches), and the variations on urban green in the recess yard (trimmed hedges, linden trees, spontaneous vegetation, and façades covered with self-climbers) are joined by the athletic grounds, with its classic grass playing field. It is reminiscent of those in public parks of the early twentieth century, with a grass embankment and steps for seating, as well as a basketball field with a hard surface and several smaller playing fields (p. 94). The playing fields are separated from one another by densely planted lines of trees and rows of shrubs; on the site plan, they resemble scattered pickup sticks. Their planting was not motivated by any particular use. Nor is there any evident urban-planning need for the overlapping diagonals—around the same time, Kienast achieved the same effect in the Zurzach Spa Gardens with the aid of "transparent" spatial organization (100). He seems to have been interested only in making these spaces livelier by adding flowering shrubs and rows of trees, and in disorienting the visitors. Perhaps these rows are motivated only by design and playfulness (p. 96). The planting of a belt of trees and shrubs on the side facing the autobahn and along the entire length of the sports grounds also serves to hide the nevertheless quite audible traffic and to make the walk along the playing fields more pleasant (p. 104).

The third large series of open spaces is located behind the schoolhouse complex and the kindergarten and is demarcated from the residential complex by a hedge of cornel and a sparse planting of trees (p. 98). From here visitors discover that the schoolhouse, which in the direction of the recess yard is "hidden" behind a uniform exposed-concrete façade, has four two-story wings extending out at right angles. Every classroom on the ground floor has a small courtyard paved with square concrete slabs. Each courtyard is demarcated on either side, left and right, by a hedge and by a long bench opposite the entrance to the classrooms (p. 106). A tree provides shade. Kienast left the spaces between the slabs wide enough that a bulge of spontaneous vegetation could grow up between them. Grasses and the occasional dandelion sprout on the mosses.

From the upper story of each of the four wings of the schoolhouse, a stairway leads to the open spaces located behind the complex. Kienast extended the axis of the stairs with paths of concrete stones, like those he used for the recess yard. Here, however, the vegetation growing rampantly in the broad gaps between the stones is growing over the stones themselves, which indicates that these paths are not often used; that "weeds" are not pulled; and that the area is perhaps moist. The area is designed to be a low-maintenance "meadow," interrupted only by the straight, paved paths. Kienast placed a hedge at the ends of the paths. Because each of the four paths extends a different distance into the meadow, the hedge formations appear staggered (p. 98).

The complexity of the system behind this can only be appreciated from the site plan: the diagonal axis resulting from the staggered hedge formations is parallel to the gravel path that runs along behind it marking the edge of the property (36). Hence, the hedge formation alludes directly to the tapering form of the triangular school grounds and at the same time frames the open space behind the schoolhouse complex in a gesture that is either opening or closing, depending on the direction from which it is viewed.

The area around the northernmost wing has a vegetable garden, which also serves an educational purpose. The broad open space between the schoolhouse and the kindergarten is structured by a pond, which reaches out into this interim space with several small arms. A broad belt of reeds prevents children from falling into the water. The pond is only visible to those walking along the path around the school grounds: Kienast built a rectangular viewing platform into the pond, with benches surrounding a sycamore tree (p. 108). From there one can observe the waterlilies on the water, the croaking frogs, and the yellow *Iris versicolor* on the pond's edge. The character of the pond is that of a garden, not one that has been left wild.

Those walking here can also enjoy views into the open spaces behind the school, since the adjacent hedge of cornels has been pruned at a height that does not block views entirely. At the same time, the children playing are not exposed to views: they can hide in the tall grass or behind the formations of hornbeams; the dense belt of reeds around the pond conceals the playing fields behind it from peering eyes; and small metal igloos are strewn across the grounds, with copper beeches growing on them.

The exterior spaces of the school complex were influenced by the ideas of the Kassel School in many ways, but at the same time they demonstrate that, since returning to Switzerland and coming into contact with the architecture and art appreciated there, Kienast was beginning to apply what he had learned in his own way: the low-maintenance "meadow" in the "hinterlands," where the children are offered a number of different ways to play, without it being obviously marked as a "playground," is in keeping with the idea of the "vacant and peripheral areas with flexible use"[211] which Kienast had called for in his student days. In front of the classrooms are small courtyards for spending shorter breaks, and children can walk there in their slippers thanks to the slabs. This offers children immediate contact

École cantonale de langue française in Bern, 1983–91.
Photos: Georg Aerni, 2012

Page 92
A continuous concrete wall ties the kindergarten and the schoolhouse complex together. It looks like a pedestal for the adjoining residential towers of the Bern-Murifeld housing development and thus synthesizes them into a "landscape." The long horizontal water basin in the recess yard provides a counterweight to the vertically looming residential towers.

Page 94
The grass playing field is separated from the recess yard by a maple hedge and a row of lindens. Steps that double as bleachers link the two areas.

Page 96
Rows of trees and shrubs run across the central access path of the school complex and connect the athletic fields to the meadows behind the kindergarten. The step-like modeling of the terrain is not motivated by use but is purely aesthetic—a playful detail.

Page 98
Behind the kindergarten and the schoolhouse complex, meadows with tall grass stretch out, with isolated fruit trees as relics of the once agrarian landscape. This "hinterland" is a place for the children to play. There are hedges at the ends of the paths leading from the wings of the schoolhouse into the "hinterland."

Page 100
The recess yard with the elongated water basin designed by Dieter Kienast has an asphalt covering with a striped pattern of paving stones and drainage channels. Where they meet the concrete wall, a wild vine was planted. It spreads over the wall and creates its own vegetational pictures.

Page 102
The zone beneath the row of lindens that demarcates the recess yard was designed with upright rectangular, bright-gray concrete stones and rectangular, rust-red tree grilles. Mosses, grasses, and other ruderal plants in highly varied shades of green sprout in the large gaps between the stones and in the cracks of the grilles. Kienast placed great store by such subtle contrasts in the "nature of the city."

Page 104
View from the recess yard across the water basin toward the grass playing field. The autobahn to the west is hidden by the gently rising modeling of the terrain at the other end of the playing field and by dense plantings of trees and shrubs.

Page 106
Each classroom on the ground floor has its own separate front recess yard with direct access. Hedges and benches terminate the yards; one tree provides shade in each of them. Stairs lead from the classrooms on the upper floor and along a path paved with concrete stones to the meadows of the "hinterland" where the children can play in peace. Spontaneous vegetation is growing over the concrete stones; the rainwater from the roofs is fed into a drainage channel lined with gravel. Considerations related to ecology, design, and use complement one another.

Page 108
In the basin of the school grounds, there is a viewing platform that is publicly accessible from the outside. Residents of the neighborhood can observe aquatic birds, frogs, fish, and plants from there. The design of the entire grounds was based on a clear consideration of the degree of openness and closure of the spaces vis-à-vis the surroundings, of their specific use, and of the dialogue with or retreat from the outside world.

with open space and a view of the meadow. At the same time, they engage with the spontaneous vegetation: they can observe which plants, when, and how will grow in the spaces between the slabs and get to know, as Kienast had in Kassel, the "beauty of weeds" in a playful way. The design elements of the courtyards are simple and developed based on use: a surface covering, two hedges to demarcate the courtyards of the other grades, a tree, a bench—and that's all.

Children can wander out from the courtyards and gradually conquer the entire grounds. The tall, isolated "hedge walls" can be introduced as elements that add to the excitement of hide-and-seek; the overgrown igloos may be used as caves or as starting points for other games and discoveries. The design of the outdoor spaces is harmonious in terms of material, form, and function, for example when rainwater from the roofs is collected in broad gutters lined with concrete stones and filled with gravel and diverted to seep into the ground of the "hinterlands" behind the school. Or when the central recess yard and the exposed-concrete façade, behind which the residential towers of the Murifeld housing development loom, are left entirely gray and employ "poor" materials. The recess yard provides both access to the school and a place where other games can be played, traces of which are found in the yard, such as the painted patterns for chess and Nine Men's Morris or the ping-pong tables. The interplay of material, form, and function produces meaning, which helps us to read the given site and understand its use and its context.

The gray recess yard and its diverse urban green—vines growing over the gray façade, the row of lindens, the maple hedge with the grass playing field behind it, and the green veil of the spontaneous vegetation growing between the concrete stones and in the tree grilles—provide an alternative design world to the dreary functionalism of the residential towers and their standardized setoff greenery. The design reacts to its urban environment and is urban in character. Nevertheless, it liberates an open space that in turn opens up space to think. Kienast commented on this search for nature in the first of his "Zehn Thesen zur Landschaftsarchitektur" (Ten Theses on Landscape Architecture):

> Our work is a search for the Nature of the City, whose color is not solely green but also gray. The Nature of the City rests in its elements: the tree, hedge, lawn; but equally the water-permeable

hard surface, broad square, rigid street-gutter line, high wall; and the unobstructed fresh-air or visual axis, the center and the periphery. [212]

This statement seems to have evolved out of his experience with the École cantonale de langue française. As does the thesis in which Kienast remarked about the significance and capability of experiencing his open spaces:
The city, with its outdoor spaces, cannot be planned as a whole. We trust in mosaic-like interventions in the hope that they will result in meaning and the ability to be experienced, not just of the particular place but also of the whole.[213]

Kienast's work for the École cantonale de langue française in Bern reveals the transformation from use value to experiential value, which results from that use value and which emerges when open spaces obtain a legible meaning through design, and especially through the use of materials and plants. Another idea formed in Kassel and clearly evident on the school grounds is the improvement of "green infrastructure" by creating open spaces in a neighborhood. It is clearly visible in the spatial disposition of the school complex: the semipublic spaces of the school that are not accessible to everyone are protected from the surroundings by keeping the path by which the school is accessed from the parking lots narrow; its very design triggers a hesitation to walk on it. The administration building on the north side shields the recess yard from the community gardens. The access path here, too, is open but also narrow, and the bridge element between the administration building and the schoolhouse feels like another physically palpable constriction of access.

Thanks to the path around the entire complex and the design of the edges of the sports grounds and of the open spaces east of the school, people from the surrounding areas can nevertheless enjoy the design of the school's open spaces as they stroll past. In the form of the viewing platform at the pond, Kienast even created on the school grounds a place to spend time that can only be accessed from outside. There are, of course, also regulations that permit the residents of the neighborhood to use the school's grass playing field and the other sports fields as well. This results in a kind of zoning, through forms of use and through those people with the right to utilize these open spaces, that is evident in the physical design as well. Those who do not have the right to access are not cut off completely but can at least enjoy the open spaces on the school grounds by observing them. These open spaces have very "different shades of green," as Kienast used to say, and thanks to the growth of spontaneous vegetation they convey even in the very artificial-looking areas, such as the recess yard, an openness to development that parallels the symbolism of the openness desired in the artificially and culturally shaped world of educating growing children.

New Images of Urban Nature

We have already discussed in detail the allegorizing of various species of vegetation. Over the course of his development, however, Dieter Kienast also aestheticized vegetation with an eye to make it possible to read a place by the vegetation used there and to begin to make it perceptible by the senses. The beginnings of that development can be found in the design of the outdoor areas for the École cantonale de langue française in Bern.

Ever since Plato, there has been a topos in philosophical discourse about the enjoyment of nature, expressing that this state can only occur if there is a distance between the viewer and the nature being viewed, that is to say if the viewer is not also working with nature, such as a farmer. Nature that is enjoyed by viewing it as a landscape has to be exempted from the viewer's own categories of production.[214] In this perspective, a city dweller can find the surrounding agricultural landscape beautiful but a farmer cannot,[215] since he or she associates it with the efforts and sweat expended to cultivate it, with milk quotas, or with the weather for a good harvest, and so on.

When the discussion about spontaneous vegetation as a revolutionary force for the planning of open space began, what Kienast called the "built vegetation" of the city was linked to a paradoxical discourse on aesthetics: designed, stately green was associated with the aesthetic criteria of the beautiful and functionless and was at the same time criticized for that. Separating it from functional needs or from the reality of the city—since such design was intended to create a kind of alternative image to the city—turned greenery into mere "prettifying" or, in the words of the philosopher Wolfgang Welsch, "surface aestheticization."[216] Planning using ruderal vegetation offered a way out of this dilemma: the repopulating plants do not fit in any categories of production and use but are, moreover, directly connected to the given location and to the people who spend time there. As a result, they do not fall under the categories of a nonfunctional "prettifying" either. As a consequence, spontaneous vegetation has the inherent possibility of an aesthetic that provides adequate images of the urban character of different locations. In keeping with the use of the word in philosophy, "adequate" means following reality as closely as possible and in unity with the object depicted.[217] The corresponding passage in Dieter Kienast and Thom Roelly's diploma thesis reads:

> Whereas ruderal vegetation is still stubbornly negated in the planning of open space today, the latter has worked all the most intensely to come to terms with built vegetation. Containers of flowers, beds of decorative shrubs, and monstrous lawns ensure that attention is paid primarily to the aesthetic function of vegetation, rather than to the functional (physical). Another characteristic of the traditional planning of open spaces is the tireless zeal with which everything that looks technical—e.g., parking lots, streets, industrial buildings, and more—is "greened away," veiled, hidden, in a green belt.... Thus, there is a constant effort to preserve the image of the city oriented around the symbols of agriculture.... It has to be regarded as a typical feature of an unresolved nature conservation and heritage movement. So it is hardly surprising that these "ideologues of greenery and concealing"—that is to say, landscape planners—are still struggling in vain to be equal partners of urban planners.[218]

Kienast and Roelly do not directly state that planning with spontaneous vegetation is also about the possibility of producing new images in the planning of open spaces. Their goal is not to make the city more bearable by creating rural images but rather, on the contrary, to create images that represent a "nature of the city." For them, the coexistence of nature and city has to be made visible, whereby a crucial role is played by the untamable, unordered, raw components of nature versus the city completely organized

38
Front yard of the M.-M. House in Erlenbach, renovated in 1988 to 1989. Kienast sowed plant seeds in the joints between the concrete slabs that cover the front yard to accelerate the growth of vegetation.

39
The access path to the L. House and Garden in Riehen, with strips planted with spurge, 1995. Photos: Christian Vogt, ca. 1991 and ca. 1996

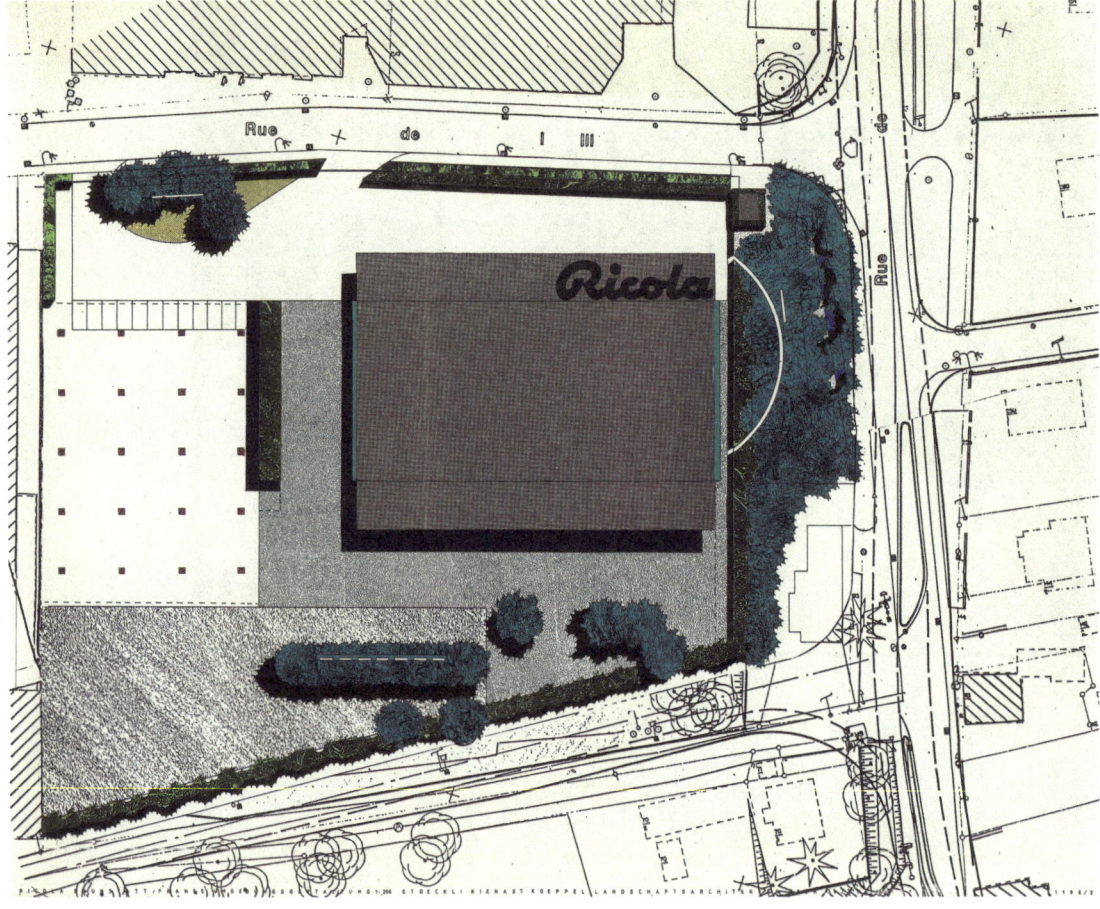

40
Eichbühl Cemetery in the Altstetten district of Zurich (1963–66) by Fred Eicher. Forecourt of the Verabschiedungskapelle (Farewell Chapel, architect: Ernst Studer) with planted strips in the asphalt

41
Design of the grounds of the packing and distribution building for Ricola AG in Brunstatt, France. The squares on the plaza next to the building, which were meant to be cut out of the asphalt, were originally intended to be planted with herbs. From here, the herbs were to grow wild and spread across the grounds and conquer them, so to speak. The concept was not realized. Plan no. 1198-2, scale: 1:200, H 71 × W 85 cm, colored collage on copy of plan, undated (1994), firm of Stöckli, Kienast & Koeppel

by human beings. While the aesthetic quality of spontaneous vegetation was only indirectly formulated by Kienast and Roelly, it was later explicitly employed by Kienast in his projects. Kienast sometimes even sowed spontaneous vegetation himself, preferably in the gaps between paving slabs outdoors:[219] for example, in the front court and in the small "secret garden" for potted plants in front of the bathroom of the M.-M. House in Erlenbach, a renovation by the ARCOOP architectural firm (38), or a garden seating area in front of the living room of the L. House in Riehen, a redesign by Herzog & de Meuron.

The image of the pavement-crack communities observed and identified during the mappings of vegetation in Kassel is also used, however, when "cultivated plants" are placed in the cracks, such as the strip of thyme in the B. Garden in Gockhausen or the strip of spurge along the access road through the garden of the aforementioned house in Riehen (39). The references for these interventions can be sought in the reservoir of images from vegetation studies, both spontaneous and "built." In addition to the vegetation maps of Kassel, Fred Eicher's strips of plants for Eichbühl Cemetery (1963–66) were an important model for the crack images (40).

The dialectic of the formal and the wild, of geometric form created by human beings and dynamic plants that break open these forms, which in Kienast's early years was allegorized in a sociopolitical way, became an aesthetic game of its own over the course of his work. In his design of the grounds for Herzog & de Meuron's packing and distribution building for Ricola AG in Brunstatt, beds of herbs were planned on a square (41). Small squares were to be cut out of the asphalt at regular intervals to plant herbs, which were supposed to spread from there across the entire area. In the L. & L. Garden on the Zürichberg, Kienast placed squares of narcissus (pp. 116, 118) and in the garden courtyards of the PTT/Swisscom Administration Building in Worblaufen squares of primulas, which after several years are barely recognizable as such, because in the meanwhile they have spread and the entire lawn is covered with flowers. They are offshoots of the subversive potential of vegetation and of the discourse between form and freedom. This slightly subversive way of highlighting the free and vital nature of plants which oppose human form corresponds with the postmodern attitude granting all forms of "play" their own energy and cognitive power. In the case of postmodern philosophy, however, this play concerns signifiers in language.[220]

L. & L. Garden in Zurich, 1993–94.
Photos: Georg Aerni, 2013

Pages 116, 118
On the small patch of lawn in the L. & L. Garden on the Zürichberg in Zurich, Kienast planted squares of narcissus that are visible once a year and will eventually cover the entire lawn.

The whole development is not without a certain irony, since weeds can be eradicated in two ways: by tearing them out or declaring them to be an "heirloom." Kienast's reinterpretations of spontaneous vegetation increasingly neutralized the subversive character of weeds. This was already inherent in allegorizing, aestheticizing, and utilizing weeds, as the Kassel School insisted. The way Kienast used the tension between "spontaneous" and "built" vegetation when designing urban nature in the last decade of his work can be illustrated using two projects in Zurich as examples.

Processes of Reinterpretation: The Front Yard and Rear Courtyard for Ernst Basler + Partner in Zurich

In 1995 and 1996, the headquarters of Ernst Basler + Partner in the Stadelhofen district of Zurich were renovated. The façade was redesigned, and Kienast Vogt Partner planted a garden in front of it. Romero & Schaefle designed a small building with a roof terrace behind the office building where employees could spend their breaks. Between the office building and the western wall of this break building, a garden courtyard was added based on the idea of the classic rear courtyard (42). The way that Dieter Kienast used the plants in both areas continues the division into "built" and "spontaneous" vegetation defined in his early years. He employed both for deliberate aestheticizing: the tufa wall in the garden courtyard is constantly dripped with collected rainwater or sprayed with it through built-in nozzles. Moistened in this way, the slightly fissured surface of the wall was populated first by algae and then by lichens and mosses, until the wall transformed into a kind of hanging garden with cranesbill and ferns (p. 122). In front of that is the second essential element: an oversized concrete pot that collects the rainwater flowing from the roofs channeled through subterranean plumbing. The tufa wall is both a termination point and an eye-catcher in the intimate courtyard covered with grayish-green gravel. The company's conference rooms were placed opposite it in the renovated office building. From there the tufa wall looks like an oversized canvas on which nature is painting its images itself, so to speak. The other plants in the courtyard include wild vines climbing the long wall to the left of the tufa wall and the *Iris versicolor* that Kienast often employed.[221] Within this triad, the tufa wall displays the wild and subversive "non-garden" vegetation of the city that settles anywhere it is not constantly combated.

Already during his time in Kassel, Kienast had shown enthusiasm for the freedom of spontaneous vegetation and for its ability to serve as an information-aesthetics record of local conditions, climatological circumstances, and human interventions. In the courtyard garden of Ernst Basler + Partner, he exaggerated the aestheticizing of spontaneous vegetation. Thanks to the austere frame of the wall, the image of the vegetation exhibits itself (p. 124). The result is an aesthetic reinterpretation that juxtaposes this form of vegetation with other, more domesticated forms. The small front yard of the office building—in front of the main entrance on the side facing the street, making it the pendant to the rear courtyard—is an example of the latter.

Today, the ribbon of the front yard accompanies the entire street façade. When Kienast and Vogt first worked on the site, they planned a garden accompanying one third of the street façade. After Kienast's death in 1998, the firm of Günther Vogt transformed the remaining parking spaces into front gardens as well (p. 126). The artificiality of the arrangement is expressly emphasized: it is designed to look as if a steel drawer had been pulled out of the building, and a garden unexpectedly appeared there. The first, as it were, "original" front garden, which is framed by steel plates on the base of the façade, is subdivided into several strips of different ground cover. One strip of green rubble along the façade adjoins a strip covered with bark mulch where a row of fifteen lindens was planted (43, 44). Their tops are pruned flat, forming a continuous roof, as has been traditional for lindens in the city since the eighteenth century. Wild garlic is growing beneath the trees, a plant that flowers in the spring in sparse deciduous woods and exudes a garlic-like scent. This is followed by a water basin as the third strip, which is delimited by a shin-high steel wall. Between the basin and the strip with bark mulch, a narrow drainage strip with coarse gravel was added for the water that overflows.

The basin of water, the outermost element of the "garden in a drawer" (*Gartenlade*), is, like the concrete pot and the damp tufa wall in the rear courtyard, fed with rainwater collected from the roofs of the building and directed by a gutter into the basin. The steel-clad gutter is inserted into the concrete wall that delimits the garden on the western side.[222] The low wall with the steel gutter spans the site like a bracket. On the opposite short side, an entrance pavilion with the name of the company terminates the front yard.

Front yard and rear courtyard of Ernst Basler + Partner (since 2016: EBP Schweiz) on Mühlebachstrasse in Zurich, 1995–96. Photos: Georg Aerni, 2012

Pages 122, 124
Dampened by rainwater, the tufa wall in the courtyard was populated first by algae, then lichens and mosses, until the wall transformed it into a hanging garden of sorts, with ferns and geraniums.

Pages 126, 128
The front yard on Mühlebachstrasse was planted in a plinth demarcated by a steel plate and placed in the street space. It makes the slight slope of the street evident. Wild garlic (*Allium ursinum*) was planted beneath the linden trees; it blooms in the spring and smells strongly of garlic. In 2000 and 2001, the firm Vogt Landschaftsarchitekten extended the front yard by adding ten lindens in front of Mühlebachstrasse 9. Greenery was added to the original layout in the form of a strip of boxwood, all of the strips of the new front were likewise planted with boxwood. The concrete-and-steel inflow gutter for rainwater has meanwhile vanished between the two parts of the garden.

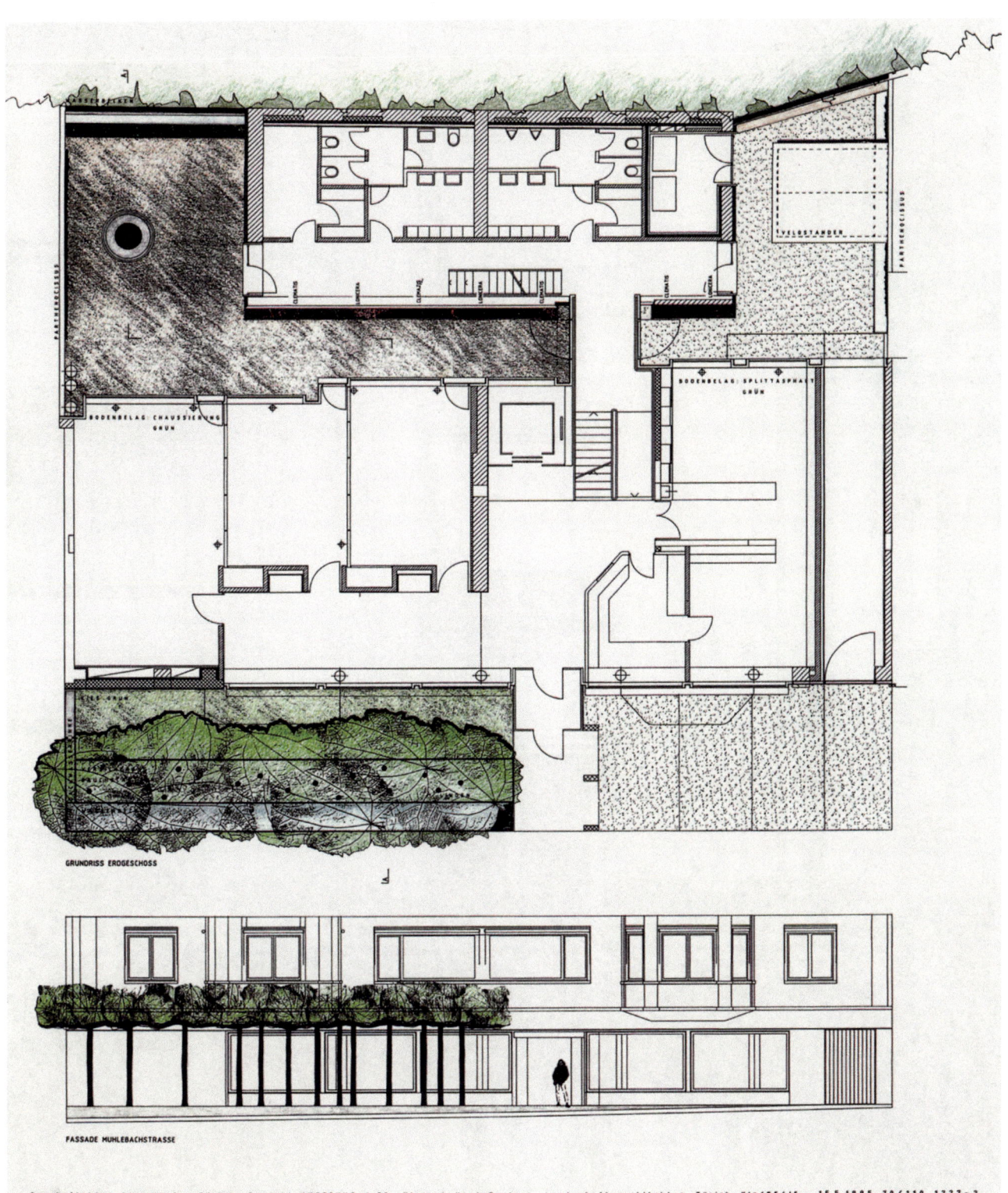

42
Design of the grounds for the headquarters of Ernst Basler + Partner in the Stadelhofen district of Zurich, floor plan and sections. Plan no. 1227-2, scale: 1:50, H 113.5 × W 70.5 cm, colored pencil on copy of plan, May 15, 1995, signed "Kie/BE/JS," firm of Kienast Vogt Partner

43, 44
The front yard for Ernst Basler + Partner in Zurich shortly after its completion in 1996. It looks as if it had been planted in a steel drawer pulled out of the building. On the side facing the street there is a water basin that receives rainwater from the roof. When it overflows, the water drains into a strip lined with gravel. It leads to another strip, lined with a steel plate and covered with bark mulch, where fifteen lindens with tops pruned into boxy shapes are planted Nearer the building, are two additional strips, one with bark mulch, the other with gravel. The front yard looks very artificial as a result. Photos: Christian Vogt, ca. 1996

Kienast himself described the design as a symbolic reminder of the tradition of front yards on this street, which has been lost.[223] Over the years, all of the front yards on this section of the street had been replaced with parking lots. This disappearance of the old front yards in the city of Zurich is a widespread, undesirable trend that can no longer be stopped. When the company moved into this office building, Ernst Basler + Partner initially restricted themselves to three parking spaces (which were later also transformed); the other planned spaces were abandoned in favor of the redesigned front yard.[224] The pedestrian sidewalk is the border zone between the sphere of the former "useless" gardens and the city of the street. As the epitome of built vegetation, the small garden in the steel drawer becomes the exaggerated symbol of the conditions for a front yard in an urban space where economizing is everything.

In an interview, Dieter Kienast said of this project:
We are trying . . . to achieve a coherence of meaning, form, and material. . . . We have to accept that there are ever more buildings, and then design on the scarce space in an urban way and break free of the rural model.[225]

He also spoke of the ecological effect of the garden on the urban climate and of the connection between the ecologically important collection of rainwater and contemporary garden design. Arguments based on improving the climate seem a little exaggerated for this tiny site, though they may be indisputable on the micro level. The quality of this garden in a drawer rests more in its powerful symbolism for the conditions of designing with nature in urban space—especially in an urban space like that of Zurich where economic considerations weigh so heavily, where every square meter costs a small fortune and hence has to be capitalized on as much as possible. In such a context, sacrificing four parking spaces for a front yard represents a small revolution, though this says less about the courage of the client than about the conditions in Zurich. This front yard is, however, the entrance to an engineering and planning company whose portfolio includes traffic infrastructure, flood protection, energy technology, and resource and climate protection, as well as site development.[226] The garden is a kind of green calling card for a company that has an important influence on the design of the environment, but focusing more on the built environment and construction of all kinds.

This artificial garden in a drawer can be interpreted in two ways: as a green gem in a jewelry box, a work that displays the exclusivity of its design through both its artificiality and its simple beauty; or as the duality of front yard and rear courtyard that belongs to a tradition of cultural history. In the latter interpretation, the row of topiary lindens in the metal drawer represents nature ordered in the Apollonian spirit—the face of the house—while the damp, shady rear courtyard, with its tufa wall covered with spontaneous vegetation in a hidden spot, expresses an otherwise repressed, wild Dionysian nature.

The aestheticizing and symbolizing of nature that is so strikingly evident in Kienast's final creative phase, which is like a conscious display of the different manifestations of nature in urban space, was present in embryo in the allegorizing of vegetation during his Kassel period. Kienast adopted allegories, images, and meanings and employed them in altered form in his designs. In his early work, the function of open space—as truly being *open* space that people can use freely in different ways—was placed above everything. In his later work, this concern receded behind the effort to produce strong images through design. Neither the front yard nor the rear courtyard of the headquarters of Ernst Basler + Partner is a place to spend time, which results not only from the design but also from the available space. Kienast's design corresponds to the scarcity of space and to the desire of the company to have a calling card, a contemporary "design of nature."

Nevertheless, these visual garden spaces are not without function. Kienast assigned to them not only the function of the pleasure of viewing them but also an educational function, since they show the forms of nature in the urban context and under the conditions of designing for a small space (43, 44). This is achieved primarily through the use of plants and their "preparation," such as the box-like pruning of the tops of the trees or the framing wall for spontaneous vegetation to settle. At the same time, the nature of urban space is explained, in that the flat patch of soil in the front yard in which the lindens are planted is made physically palpable by means of the fifty-centimeter-tall steel drawer. The drawer sits above the asphalt of the sidewalk and street. It was Kienast's emphatic intention to exhibit precisely that, not to deceive passersby into thinking that this small urban plot still consisted of actual topsoil, of "Mother Earth." Instead, there are probably various cables to supply energy beneath it. The drawer also clarifies the microtopography of the terrain: following the slight rise of the sidewalk and street, it is lower to the east; the surface of the water basin in front indicates the horizontal plane (p. 130). The process of making this visible is also a reference to the fact that garden or landscape does not always mean that the terrain and topography follow the spirit level of the planner.

In its reduction, the front yard presumably did not seem natural—that is, "green"—enough to the company's management. When, after Kienast's death, they commissioned his partner, Günther Vogt, and the firm he had founded in 2000, Vogt Landschaftsarchitekten, to expand the front yard to include the parking places of the adjoining building, Mühlebachstrasse 9, the existing design was also altered slightly: the strip with green rubble was replaced with a boxwood, connecting to a strip covered with grayish gravel that leads to the façade of the building (43, p. 126). Günther Vogt had been involved in the first design of the exteriors. The occasion for the extension and redesign was the company's expansion into the neighboring building. The renovation by Romero & Schaefle included adapting the façade to the company's original building. Vogt Landschaftsarchitekten extended the steel drawer to the west to add ten lindens. Rather than lengthening the water basin, they added a boxwood hedge the same width. The adjoining strip with drainage gravel was repeated in the new section of the front yard, but all of the other strips were planted with boxwood. In subsequent years, Vogt Landschaftsarchitekten added a third section in this style.

Today, this front yard looks more like a green oasis in a dense urban space than a critique of this urban density and of the loss of all the other front yards to parking spaces. The "legibility" of his interventions that was so essential for Dieter Kienast has been diminished by such extensions: the gutter that guided the rainwater from the roof to the basin, which rested on the wall that terminated the old garden, originally turning the entire ensemble into a graspable whole, has now disappeared behind the leaves of the wild garlic in the old section and behind the branches of the boxwood in the new section of the front yard.

Conditions for Designing Urban Open Space:
The Garden Courtyard for the Schweizerische Rückversicherungs-Gesellschaft in Zurich

The principle of the declension of natural forms applied to the front yard and rear courtyard of Ernst Basler + Partner can also be found in another, older work by Kienast Vogt Partner: the garden courtyard of an office building for the Schweizerische Rückversicherungs-Gesellschaft (called Swiss Re since 1999), built by Stücheli Architekten on Gotthardstrasse in Zurich in 1994 to 1995 (45, 46).[227] The conditions for designing with nature in an urban space are on display in this garden courtyard as well. Its themes are urbanity and possible forms of nature, whereby the relationship between the natural and the artificial focuses largely on the context that in this case the courtyard is situated above an underground parking garage.

The garden courtyard has an irregular ground plan and is framed by a wall 3 meters tall. It extends behind the building directly in front of the cafeteria on the ground floor, from which it can be entered through three doors. Its elements include a circular water basin, seven katsura trees, and a series of drainage channels where irises bloom in the spring. As a reference to the hollowed-out subterranean space, the surface of the courtyard is slightly inclined toward the building. The entrance to the underground garage is on the south side.

The wall of the courtyard is set back slightly from the edge of the property and is perforated in several places. Beyond the wall, a number of cornels were planted and then later replaced by Lombardy poplars (still later replaced by cypress oaks)—to the south, the trees now frame the access to the parking garage, with ostrich ferns planted beneath them. The treetops loom over the wall. This results in a dialogue between inside and outside, especially since the trees can

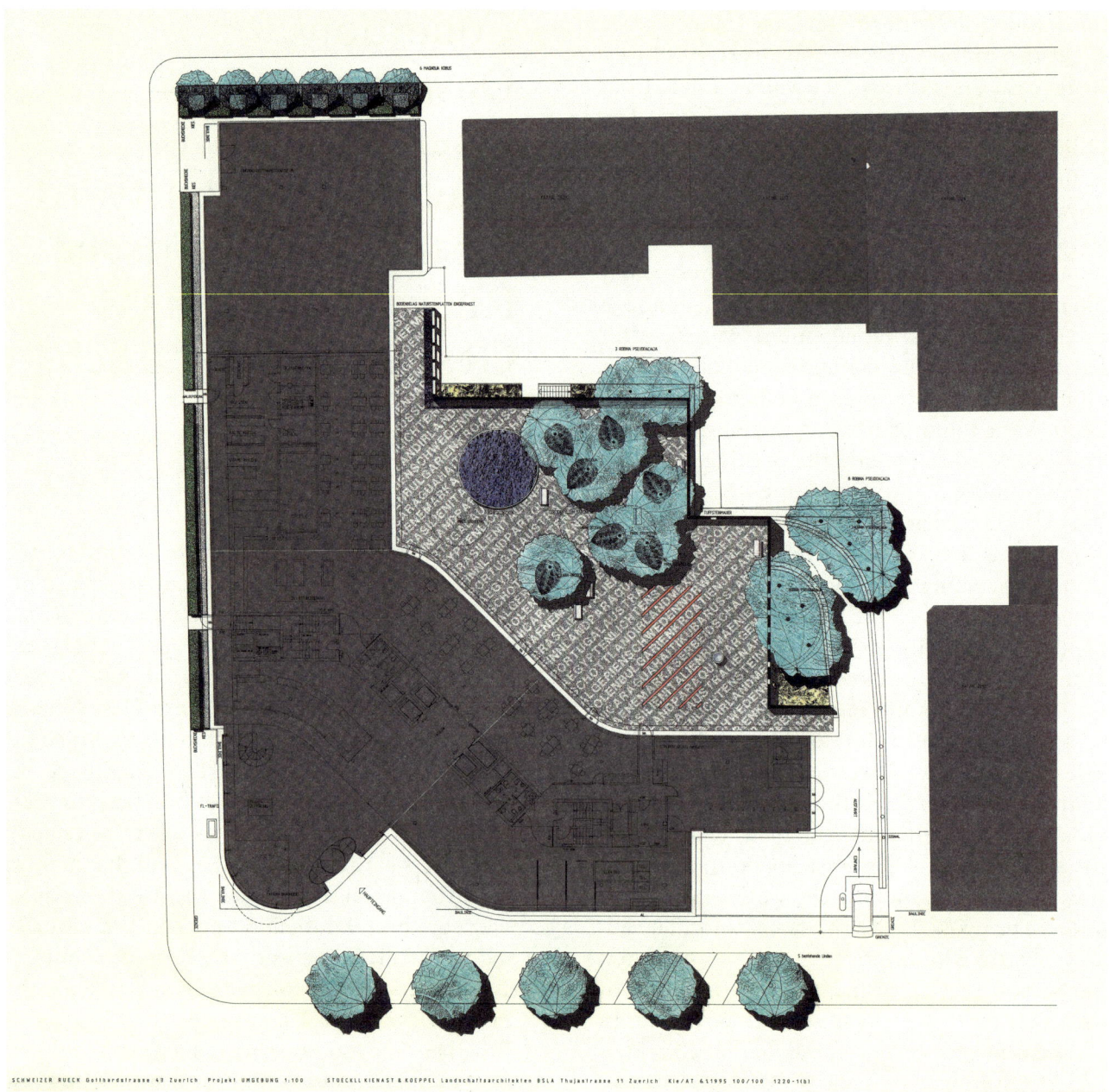

45
Design of the grounds of the Schweizerische Rückversicherungs-Gesellschaft (Swiss Re) office building on Gotthardstrasse in Zurich, floor plan. Plan no. 1220-1b, scale: 1:100, H 100 × W 100 cm, collage and colored pencil on copy of plan, January 6, 1995, signed "Kie/AT," firm of Stöckli, Kienast & Koeppel

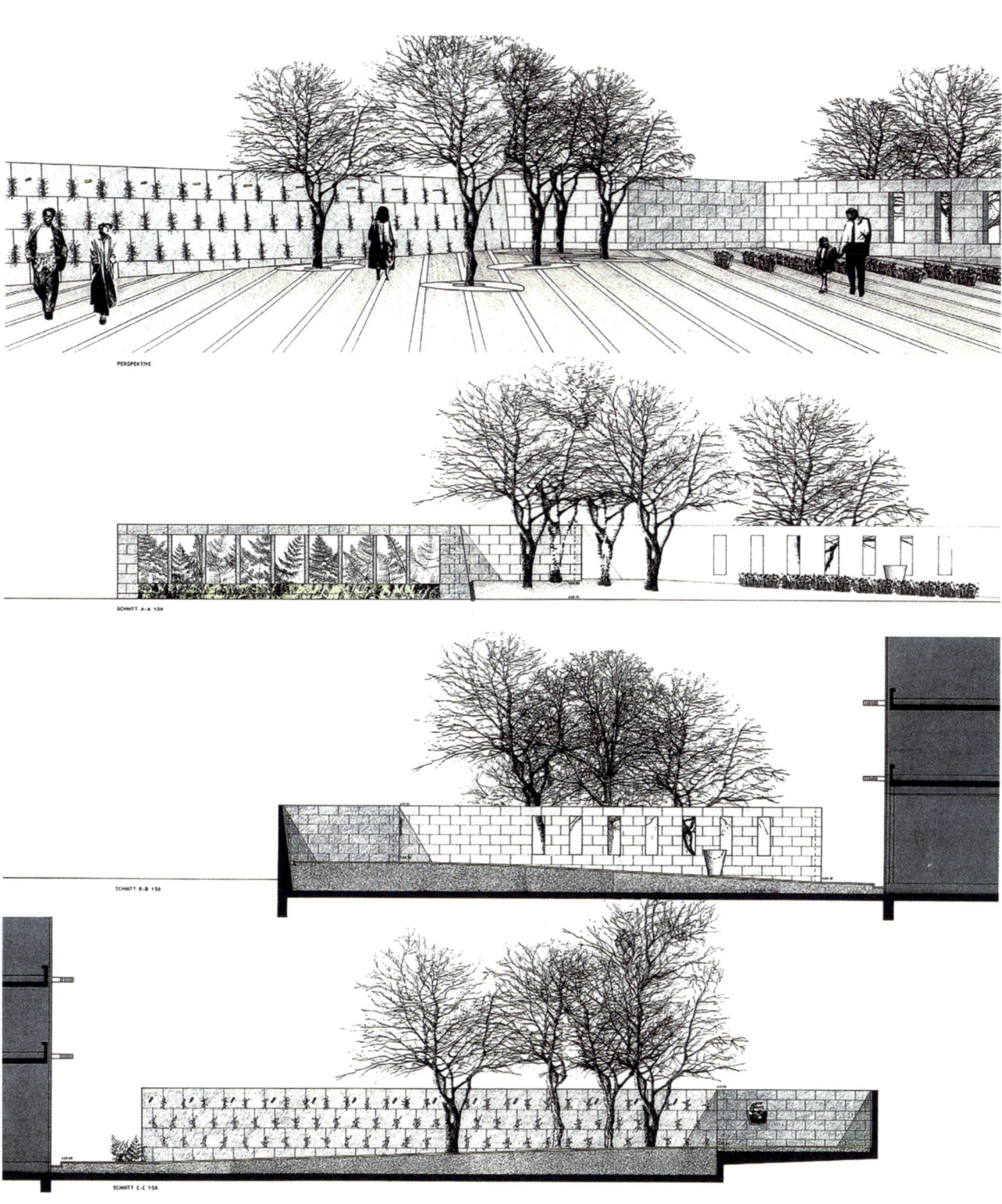

46
View and sections, plan no. 1220-2, H 100 × W 80 cm, collage and colored pencil on copy of plan, November 9, 1994, signed "Kie/AT," firm of Stöckli, Kienast & Koeppel

47
Cornel was planted over the entrance to the underground parking garage and beyond the boundary wall of the garden courtyard. Later it was replaced by Lombardy poplars, then by cypress oaks. The garden courtyard seems to reach over its walls. There is a dialogue between inside and outside.
Photo: Christian Vogt, ca. 1996

also be seen through the vertical, rectangular openings in the wall (46, 47). The courtyard seems to reach beyond itself and thereby establish a connection to its surroundings. To the east, the wall follows an entrance to the courtyard that is kept open as an escape route. Five large windows with silkscreens of irises were inserted into the wall; they look like real plants flowering in the courtyard.

The wall bears depictions of nature in other ways as well: tufa slabs have been applied on one side. According to the presentation plans, vegetation was supposed to sprout from the cracks here, as if resisting the other austere forms of the courtyard and the materials used (concrete, asphalt, steel). The wall of tufa slabs is located in the shadiest part of the courtyard, where the katsura trees are standing. No spontaneous vegetation has developed in the hanging garden, like that on the wall in the courtyard of Ernst Basler + Partner—either because, as some would have it, the caretaker often scrubbed off the weathering or because too little water was sprayed on the wall. Instead, a wild vine was planted that has now grown over nearly the entire wall.

In a niche in the wall, in front of the irises, a grotesque has been mounted as a symbol that this part of the courtyard is "damp" (p. 140). Grotesques are found in histoical gardens, especially in grottos, in the context of bossage work and in cool, humid parts of the garden. In the garden courtyard on Gotthardstrasse, the grotesque becomes an ornament that opens up a space of associations. Here, too, Kienast was playing with associations from the history of culture and reinterpreting images of vegetation: first, the strips of irises echo the image of the numerous pavement-crack communities that Kienast identified as part of his phytosociological studies in Kassel; second, the plants themselves and the grotesque above them identify a damp location and thus even on dry days point to the drainage of water that occurs there. The courtyard was maintained less intensively from 2007 to 2008, before the new occupant of the building, the Bank Vontobel, began caring properly for it again, resulting in the establishment of a series of other plants in the strips that then overgrew the austere, slender openings. The connection of the drainage channels to the vegetation spontaneously growing in pavement cracks—the wild nature of the city—thus became clearly visible for a time.

The fact that the garden courtyard is located above an underground garage, which is not hidden but rather explicitly expressed in the design, contributes to the overall complexity of the images of the natural and the artificial and underscores the challenge of designing with nature in the city.[228] It is not earth that fills the foundation of the courtyard but concrete. Between the concrete slabs, long strips of coarse asphalt were poured. The concrete slabs themselves have a granular pattern from the crossbeams. The surface treatment of the courtyard in general is closely related to the glass façade of the office building. It looks like a symbolic mirror or an imaginary cast shadow of the façade, and yet it is its broken counterimage. The façade is characterized by emphasis on the horizontals as a result of the brise-soleil above every floor. The guide rails of the shades in front of the windows, which look like slender supports, articulate the bands of the floors vertically. The pattern of the ground in the courtyard responds to the façade but inverts its emphasis: the asphalt strips run directly into the vertical lines of the façade and are the dominant element in the courtyard itself. By contrast, the horizontality that dominates the façade is absorbed and split up, as it were, by the imprinted crossbeams on the surface of the courtyard. The slope of the ground toward the building establishes a spatial relationship between the façade and the courtyard in addition to this symbolic mirroring effect (2).

It is, however, possible to read something else from the specific pattern and the slope of the courtyard. The slight diagonal illustrates the artificiality of the subterranean. Moreover, small curvatures of the ground indicate the concrete structures of the underground garage. At the same time, the sloping surface and the materials of the individual strips thematize the runoff and drainage of rainwater, as well as the extreme sealing of the ground in the city. The asphalt strips inserted between the concrete slabs can take up some rainwater, and as a result of both their sloping and the drainage function, they direct the excess water more quickly to the drain openings in front of the building, from which the rainwater flows into the sewer system. There are only two places in the courtyard where the rainwater is actually collected and absorbed by plants rather than being led on into the drainage: in the southern section, the asphalt strips are replaced by drainage channels with plants, and the katsuras in the center have broad tree grilles in the form of leaves. In that sense, the use of water-permeable asphalt for the strips that mirror the verticals of the façade is an ambiguous act: the strips should ultimately be seen as ornamental or symbolic, since the underground prevents the water from draining into the groundwater—yet this drainage process is exactly why asphalt is used in urban areas. Kienast was presumably interested precisely in addressing these ambiguities and playfully juxtaposing the natural and the artificial until that dichotomy itself is reduced to the absurd.

Seen in this way, the imprinted crossbeams on the concrete slabs also have a level of symbolic meaning that goes beyond the ornamental: the imprinting of the concrete with hammer and chisel produces the surface structure of natural stone. Visually, it transforms the material in the sense of a metamorphosis. The artificial surface of the concrete is visually transfigured partly into natural stone by being worked in this way—a complex, vexing play, like that of the asphalt strips that lead the water into the drainage because the courtyard is located above an underground garage.[229]

By his own account, Kienast placed the large, round water basin in the courtyard "like a spirit level,"[230] so that the surface of the water symbolically attempts to straighten out the slope of the ground (p. 142). The basin reflects the sky and thus establishes contact to a natural element above, whereas the underground area has a parking garage instead of earth.

After this multiple demonstration of the artificiality of the site—and Kienast's design points almost obsessively to this—he is relieved of the worry of using green as a concealing element in the city[231] and can instead concentrate on how it appeals to the senses: Kienast chose the seven katsura trees in the center of the courtyard because their leaves change color and also because of their scent.[232] In the spring, the leaves are reddish-bronze; in the summer, green; and in the autumn, first yellow and then scarlet. In the autumn, the katsura leaves smell sweet, like baked goods, which gave them their nickname "gingerbread trees."[233] The trees are also brought out well in the otherwise geometrically strict and tonally restrained design, while still providing lively intensity as a counterpoint. The tree grilles emulate the heart-shaped leaves of the katsura trees and thus evoke associations of their leaves even in winter when the trees are bare (p. 144).

These two approaches to designing with nature—the intellectual one, always reflecting on the conditions and the context of activity, and the direct, romantic, sensual one—are what distinguish Kienast's work: as a result of his personal development, Kienast united the gardener,

Swiss Re (now Bank Vontobel) garden courtyard on Gotthardstrasse in Zurich, 1994–95. Photos: Georg Aerni, 2012

Page 140
The grotesque in the wall niche and the drainage channels planted with irises
Page 142
The water basin and the katsura trees, and between them the tables and chairs where employees spend their breaks
Page 144
The grove of katsura trees. Even in the limited space of the garden courtyard it is possible to work with contrasts of bright and dark, damp and dry, sunny and shady zones. A reduction of means emphasizes the opportunity to experience the individual plants with the senses: the leaves of the katsura trees change color several times a year, and they smell like gingerbread upon falling to the ground.

the critical thinker shaped by Kassel, the scientist, and the phytosociologist. The courtyard of Swiss Re is clearly an expression of this synthesis: his design results from an absolutely intellectual and extremely controlled approach with many references, ambiguities, and rifts. In distinction from the "surface aestheticization" by means of beautifying and hiding, this method can be characterized by what the philosopher Wolfgang Welsch has defined as a "deep-seated aestheticization." According to Welsch, "genuinely aestheticized culture" permits "ambiguity"; it breaks up hierarchies and brings out the suppressed.[234] It is "sensitive to differences and exclusions—and not only in relation to the forms of art and design, but equally in daily life and towards social forms of life." Seen in this way, the garden courtyard certainly has an "ascetically hidden . . . richness," as Theodor W. Adorno observed in his *Aesthetic Theory* when he wrote that aesthetic unity obtains its dignity from diversity itself: "It does justice to the heterogeneous."[235]

The garden courtyard displays the conditions of designing with nature in urban spaces, emphasizes the artificiality of the site, and at the same time counters it with natural elements. Ecological and technical necessities such as the drainage of rainwater or the location of the garden above an underground garage are turned into powerful images by the design and become a cultural statement about the sealing of the soil in the city. By exposing urban complexity and the ambiguity of the natural therein, Kienast can indulge in his sensualist approach to nature without being suspected of hiding with green. On this site, he runs through the various possible manifestations of nature in urban space: from the spontaneous algae and mosses to the silkscreens of irises and the leaf-shaped tree grilles. The visual character of the individual plants and their symbolism, as well as their sensory manifestation—their scent—and the dynamics of their transformation by spontaneous growth or the changing of the seasons, all play a crucial role. This brings together Kienast's express effort to design in sparse urban space with a coherence of meaning, form, and materials[236] and his call for the rediscovery of plants as an urban element, in order to appreciate their sensuousness and discover their symbolism.[237] It results in fragments of the natural in a controlled and symbolic world.

The "social relevance" that Welsch hopes to gain from the "potential for sensibility" of aesthetic work is only partially achieved by the garden courtyard. The employees who eat lunch and drink coffee here are hardly likely to grasp the full complexity of the levels of meaning of their surroundings. If there is a social relevance, then it consists in reflecting on the themes of the courtyard and the consequences one draws from them. In its deliberately inhospitableness as a whole and in the seductive beauty of its details, it is like a thorn in the side of the viewer as a rational being and also as someone affected by the design. Whether it is because the site has aged well in the meanwhile and the trees have grown large, or because the relationship to nature and the understanding of the artificial versus the natural has transformed again, the garden courtyard today is, according to the employees of Bank Vontobel, a frequently used, very popular outdoor space that is accepted without criticism.

1. Dieter Kienast, "Vom naturnahen Garten; oder, Von der Nutzbarkeit der Vegetation," *Der Gartenbau* 100, no. 25 (1979): 1117–22, here 1122.
2. "Our work is a search for the Nature of the City, whose color is not solely green but also gray." The first of the "Zehn Thesen zur Landschaftsarchitektur" (Ten Theses on Landscape Architecture) was first formulated by Kienast in 1992 and published posthumously in 1999: Dieter Kienast, "10 Theses on Landscape Architecture," in *Dieter Kienast: In Praise of Sensuousness*, exh. cat. (Zurich: gta Verlag, 1999), 84–94, here 85. The ten theses have been newly translated for this volume from the last amended version from November 1998, as published in *Dieter Kienast: Die Poetik des Gartens; Über Ordnung und Chaos in der Landschaftsarchitektur*, ed. Professur für Landschaftsarchitektur ETH Zürich (Basel: Birkhäuser, 2002), 207–1. See appendix, 403–5. Kienast probably took the phrase "die Natur der Stadt" (the nature of the city) from Heide Berndt, who published a book of that title in 1978. See Heide Berndt, *Die Natur der Stadt* (Frankfurt am Main: Neue Kritik, 1978).
3. Kienast, "Zehn Thesen zur Landschaftsarchitektur," 209; see appendix, 404. Most of the theses were formulated in the lecture "Von der Notwendigkeit künstlerischer Innovation und ihrem Verhältnis zum Massengeschmack der Landschaftsarchitektur" (On the Necessity of Artistic Innovation and Its Relationship to the Mass Taste of Landscape Architecture), held in Berlin in November 1991. See Dieter Kienast, "Von der Notwendigkeit künstlerischer Innovation und ihrem Verhältnis zum Massengeschmack der Landschaftsarchitektur," in *Choreographie des öffentlichen Raumes,* ed. Jürgen Wenzel (Berlin: self-published, 1994), 52–64. Republished in Professur für Landschaftsarchitektur ETH Zürich, *Dieter Kienast*, 103–15.
4. Dieter Kienast, "Longing for Paradise," trans. Bruce Almberg and Katje Steiner, in *Dieter Kienast* (Boston: Birkhäuser, 2004), 84–90, here 89.
5. Viktor Shklovsky, "Art as Device (1917–1919)," in *Viktor Shklovsky: A Reader,* ed. and trans. Alexandra Berlina (New York: Bloomsbury Academic, 2017), 73–96, here 80.
6. Johannes Stoffler has pointed out that when Kienast was appointed Professor of Landscape Architecture at ETH Zurich, the first book he asked the library to acquire was the German translation of Simon Schama's cultural history *Landscape and Memory* (New York: Knopf, 1995). See Johannes Stoffler, "Gegen die Vereinfachung der Gartenbotschaft," *Stadt und Grün* 53, no. 1 (2004): 16–20, here 16.
7. Thom Roelly in conversation with the author, May 28, 2010.
8. Günther Grzimek, "Leitbild für das Studium am Lehrstuhl für Landschaftskultur an der Hochschule für bildende Künste, Kassel," in *Hochschule für Bildende Künste Kassel, Lehrstuhl für Landschaftskultur* (Kassel: Lehrstuhl für Landschaftskultur, 1968), 101–2.
9. Ibid., 101.
10. This and the following quotations, ibid.
11. Ibid., 102.
12. This clarification was made by Thom Roelly in conversation with the author, January 11, 2011.
13. See Karl Heinrich Hülbusch, "Anmerkungen zu diesem Notizbuch: Krautern mit Unkraut," *Krautern mit Unkraut; oder, Gärtnerische Erfahrung mit der spontanen Vegetation,* Notizbuch der Kasseler Schule 2 (Kassel: Arbeitsgemeinschaft Freiraum und Vegetation, 1986), 1–15, here 9.
14. See Jesko Fezer and Martin Schmitz, "The Work of Lucius Burckhardt," in Lucius Burckhardt, *Writings: Rethinking Man-Made Environments; Politics, Landscape & Design,* ed. Jesko Fezer and Martin Schmitz (Vienna: Springer, 2012), 7–26, here 21.
15. The reason that a "teaching sofa" was needed was because the school administration would not allow Burckhardt to teach a design studio alone, as he was trained as an economist and sociologist. Throughout his tenure, a second person, a trained architect, had been appointed to join him in his teaching activity. See Silvan Blumenthal, "Das Lehrcanapé und sein Architektenbild: Lucius Burckhardt an der ETH Zurich, 1970–1973" (diploma thesis on an elective subject, ETH Zurich, Professor Laurent Stalder, 2008); Silvan Blumenthal, "Das Lehrcanapé," *archithese* 39, no. 2 (2009): 96–99; Silvan Blumenthal, *Das Lehrcanapé: Lucius Burckhardt und das Architektenbild an der ETH Zürich, 1970–1973,* Standpunkte Dokumente 2 (Basel: Standpunkte, 2010).
16. See H. Brockelmann, "Das Kasseler Modell der integrierten Gesamthochschule im Studiengang Architektur, Stadt- und Landschaftsplanung" (seminar paper, Architecture and Urban and Landscape Planning, Gesamthochschule Kassel, 1992), Archiv Grauer Raum, docu:lab, University of Kassel, Department of Architecture, Urban and Regional Planning, Landscape Architecture and Landscape Planning.
17. See Jens Brömer, ed., *Portrait einer Hochschule: Universität Gesamthochschule Kassel,* 2nd ed. (Kassel: Gesamthochschule, 1994), 38. In 1978, the organizational unit Architektur, Stadt-, und Landschaftsplanung was split into the departments of architecture and of urban and landscape planning.
18. See Gesamthochschule Kassel, ed., *Kunst und Architektur* (Kassel: Gesamthochschule Kassel, 1990), 7.
19. This is certainly also conditioned by the discipline of landscape architecture as such, which in general draws on these two fields. Kienast made these interests very explicit themes, however, in his texts and in his projects.
20. Students in Kassel who no longer fell under the regulation of the IAP received two diplomas under the new regulations for studies: after nine semesters, the Diploma I, which was recognized as equivalent to a degree from a *Fachhochschule,* and after three additional optional semesters of in-depth study, the Diploma II, which corresponded to a degree of in engineering from a technical university. In between those two stages, studies could also be interrupted for an extended period of professional practice. On this and

what follows, see Heide Hoffmann, "Das Studium der Stadtplanung: Studiengänge im Vergleich: Kassel, Dortmund, Berlin, Kaiserslautern, Hamburg-Harburg" (seminar paper, Architecture and Urban and Landscape Planning, Gesamthochschule Kassel, winter semester of 1996–97), Archiv Grauer Raum, docu:lab, University of Kassel, Department of Architecture, Urban and Regional Planning, Landscape Architecture and Landscape Planning.

21 See Martin Schmitz, "From Critical Urban Studies to the Science of Walking," in Lucius Burckhardt, *Who Plans the Planning? Architecture, Politics, and Mankind,* ed. Jesko Fezer and Martin Schmitz (Basel: Birkhäuser, 2019), 7–11, here 7.

22 Hoffmann, "Das Studium der Stadtplanung," 56. Ironically, twenty years after the introduction of the interdisciplinary course of study in Kassel, the author came to the conclusion that as a result of self-responsibility and self-regulation, the students were suffering from "lack of orientation." The interdisciplinary education in the basic studies continued to be welcomed, but in some cases it occurred at the cost of a solid foundation for a specialty. Moreover, the students were said to have the feeling that they could not adequately evaluate their own achievements because of the lack of written exams.

23 In the sixth of the "Zehn Thesen zur Landschaftsarchitektur," Kienast wrote: "For us, working with our co-disciplines of architecture, engineering, and the visual arts is not an unwelcome obligation but a welcome axiom. Working together generates mutual innovations." Kienast, "Zehn Thesen zur Landschaftsarchitektur," 208; see appendix, 403.

24 This feature of Kienast's pedagogy was addressed on November 6, 2008, in a podium discussion at the Hochschule für Technik Rapperswil (HSR, until 1998 Interkantonales Technikum Rapperswil, ITR) with his former colleagues Bernd Schubert and Jürg Altherr, his former assistant Hans-Peter Burckhardt, and a graduate from 1986, Sybille Aubart. In addition, thesis projects supervised by Kienast at the ITR were exhibited.

25 On Kienast's teaching, see "*Gartenarchitektur— Freiraumgestaltung: Nachdiplomstudium 1989/90,*" Schriftenreihe Abteilung Grünplanung, Landschafts- und Gartenarchitektur, Interkantonales Technikum Rapperswil 1 (Rapperswil: ITR-Ingenieursschule, 1992); Ulrike Rothe, editorial staff, Universität Karlsruhe (TH), Institut für Landschaft und Garten, Prof. Dr. Ing. Dieter Kienast, "Dokumentation von Lehre und Forschung Sommersemester 1992 bis Wintersemester 1996/97," Karlsruhe, 1997, Archiv Fachgebiet Landschaftsarchitektur, Karlsruhe Institute of Technology (KIT) (I am grateful to Ulrike Rothe and Wolfram Müller for calling my attention to and lending me this book); Lehre des Instituts für Landschaft und Garten, Lehrstuhl für Landschaftsarchitektur und Entwerfen, "Dieter Kienast," in *Ein-Blick, Aus-Blick: Universität Karlsruhe, Fakultät für Architektur; Forschung und Lehre,* ed. Fakultät für Architektur, Universität Karlsruhe (Tübingen: Wasmuth, 1999), 120–25.

26 Exceptions include the (unrealized) green urban planning of the Scheibenschachen neighborhood of Aarau (1978, firm of Stöckli + Kienast), the lakeshore design of the Triechter in Sursee (1991–95, firm of Stöckli, Kienast & Koeppel), the residential grounds and riverside park at Moabiter Werder in Berlin (competition 1990–91, firm of Stöckli, Kienast & Koeppel), the urban and landscape planning for Kronsberg in Hanover (1994–99, firm of Kienast Vogt Partner), and the urban planning analysis of the Neustädter Feld prefabricated building development in Magdeburg (study, 1994).

27 Harald Szeemann, ed., *Besser sehen durch documenta 5: Befragung der Realität; Bildwelten heute* (Kassel: Bertelsmann, 1972), n.p.

28 Ibid. The formulation "individual mythologies" came to Szeemann in 1963 while installing a work by Étienne Martin in the Kunsthalle Bern.

29 Ibid.

30 Ibid.

31 See Joseph Beuys interviewed by George Jappe, "Not Just a Few Are Called, But Everyone," trans. John Wheelwright, in *Art in Theory 1900–2000: An Anthology of Changing Ideas,* new ed., ed. Charles Harrison and Paul Wood (Malden, MA: Blackwell, 2003), 903–6, here 905.

32 Szeemann, *Besser sehen durch documenta 5,* n.p. Szeemann concluded his "Individuelle Mythologien" with the summary that everyone who did not want to see that the "great disorder" was being countered by an order of the artist's own is brought into an accepted "great order" and would "again succumb to the seductive, edifying, and persuasive image." This was true, according to Szeemann, the same as ever for both the left and the right.

33 The 1972 contribution to the catalog is one of Szeemann's few written statements on documenta 5. In an interview in 2002 on the occasion of an exhibition of the archive of documenta 5, Szeemann remarked that he was constantly preoccupied with organizing it and wanted to leave the texts reflecting on it to others. See Gabriele Mackert, *Skandal und Mythos: Eine Befragung Harald Szeemanns zur Documenta 5 (1972)*, exh. cat. (Vienna: Kunsthalle, 2002), 20.

34 Szeemann explained the motto "Seeing better through documenta 5" by noting that art is an offering to the sense of vision, in the hope that the other mechanisms of the machine would then engage: the brain, the heart, and the stomach. Ibid., 11.

35 Heinz W. Hallmann, "Die Entwicklung der Landschaftsarchitekten und ihre Ausbildung in Deutschland," part 2, *Das Gartenamt: Stadt und Grün* 41, no. 3 (1992): 165–70, here 166.

36 See Karl Heinrich Hülbusch, "Notizbuch der Kasseler Schule: Programmatische Anmerkungen," in *Krautern mit Unkraut,* 158–63, here 158.

37 See Eike Schmidt, "Der Naturgarten: Ein neuer Weg?" (The Natural Garden: A New Direction?), *Garten + Landschaft* 91, no. 11 (1981): 877–84, here 881.

38 Helmut Böse, *Die Aneignung von städtischen Freiräumen: Beiträge zur Theorie und sozialen Praxis*

des Freiraums, Arbeitsbericht des Fachbereichs Stadtplanung und Landschaftsplanung 22 (Kassel: Gesamthochschule Kassel, 1981); Jürgen Milchert, Review of Helmut Böse, *Die Aneignung von städtischen Freiräumen,* in *Das Gartenamt* 32, no. 2 (1983): 116–17.

39 Milchert, Review, 116. Karl Heinrich Hülbusch wrote something along the same lines in the afterword to Böse's publication. See Böse, *Die Aneignung von städtischen Freiräumen,* 221–22.

40 Böse, *Die Aneignung von städtischen Freiräumen,* 100, quoted in Milchert, Review, 117.

41 Milchert, Review, 117.

42 On this, Böse wrote: "Originality is not a problem of planning. Originality does not solve any problems either. Originality results from appropriation (by the users)." Quoted in ibid.

43 In contrasting the *bricoleur* and the engineer, Lévi-Strauss emphasized in particular their different ways of working and techniques. The engineer works toward a goal; he always has the necessary tools at hand and a clear objective in mind. The *bricoleur* improvises; he has to find other solutions by detours when he lacks the right tools. In doing so, he falls back on an arsenal of the already existing and repurposes it. Lévi-Strauss uses these models for, among other things, his theory of culture and analysis of intertextuality. See Claude Lévi-Strauss, *The Savage Mind* (Chicago: University of Chicago Press, 1966), 16–36.

44 For example, Bernard Tschumi won the competition for Parc de la Villette in Paris in 1982, with a similar program in terms of content and with architecture intended to provoke "events" that renounced the English landscape garden style of the nineteenth century. This project, which was excessively instrumentalized architecturally, had nothing to do with the concerns of the Kassel School despite similarities in their respective discourses.

45 On this, see, among others, Hülbusch, "Notizbuch der Kasseler Schule," 159 and 161; Helmut Böse, "Vorbilder statt Leitbilder," *Garten + Landschaft* 96, no. 11 (1986): 28–33; Gerhard Hard, "Schwierigkeiten beim Spurenlesen," in *Freiraum und Vegetation: Festschrift zum 60. Geburtstag von Karl Heinrich Hülbusch am 21. Mai 1996,* ed. Helmut Böse-Vetter, Notizbuch der Kasseler Schule 40, no. 2 (Kassel: Arbeitsgemeinschaft Freiraum und Vegetation, 1996), 39–51, here 50.

46 Hülbusch, "Notizbuch der Kasseler Schule"; Böse-Vetter, *Freiraum und Vegetation,* 2.

47 Quoted in Böse-Vetter, *Freiraum und Vegetation,* 2. It is above all a debate over who represented the pure teaching, whereby the external specialists looking at the Kassel School are clearly unable to differentiate so precisely the value of the purity advocated by the insiders.

48 In 2006, I received an e-mail from Henning Schwarze, a member of the Arbeitsgemeinschaft Freiraum und Vegetation, a working group supporting open spaces and vegetation. He was interested in Kienast's photographs of turf grass (*Festuco-Crepidetum capillaris*) and wrote that neither the Verein nor the Kassel School had officially existed in Kienast's day, but that his dissertation clearly revealed the approaches of this school and represented "even today one of the central works on vegetation in open spaces in the city." He found it astonishing that Kienast struck "notes of the profession" in the dissertation and in his later works that were completely different.

49 See Reto Mehli, "Das Lei(d)tbild 'Landschaft': Zur Kritik ästhetischer Leitbilder in der Gartenarchitektur," in *Reise oder Tour?,* Notizbuch der Kasseler Schule 26 (Kassel: Arbeitsgemeinschaft Freiraum und Vegetation, 1992), 128–56.

50 See Jürgen von Reuss, "Lassus in Kassel: Ein Blick auf die Gartenkunst," in *Die Kunst, Landschaft neu zu erfinden: Werk und Wirken von Bernard Lassus,* ed. Andrea Koenecke, Udo Weilacher, and Joachim Wolschke-Bulmahn, CGL Studies 8 (Munich: Meidenbauer, 2010), 85–115. While still a student, Kienast had attended a guest lecture by Bernard Lassus. Thom Roelly in conversation with the author, April 6, 2011.

51 Thom Roelly in conversation with the author, April 6, 2011, and Mehli, "Das Lei(d)tbild 'Landschaft.'"

52 See Lucius Burckhardt, "Kritik der sechziger Jahre," *Werk* 60, no. 12 (1973): 1588–90.

53 Jörg Müller, *Alle Jahre wieder saust der Pressluft-hammer nieder; oder, Die Veränderung der Landschaft* (Aarau: Sauerländer, 1973); translated as *The Changing Countryside* (New York: Atheneum, 1977).

54 Burckhardt, "Kritik der sechziger Jahre," 1588.

55 Ibid.

56 Lucius Burckhardt, "Der Architekt in der Gesellschaft von morgen," *Werk* 52, no. 11 (1965): 234*–44*, here 243*.

57 Ibid.

58 In his text "Das Ende der polytechnischen Lösbarkeit" (The End of Polytechnic Solvability) of 1989, Burckhardt subsumed the methodology of "enlightened academicism" under the formula ZASPAK: "Benenne das Ziel, analysiere das Problem, mache daraus eine Synthese, zeichne den Plan, schreite zur Ausführung, kontrolliere den Erfolg" (name the objective, analyze the problem, synthesize the analysis, do the planning, implement it, and monitor the outcomes). For Burckhardt, the trick with such an approach is that a clear goal be stated from the outset, and this pushes all further thinking in a certain direction. He mentioned a task presented to students in his class during their first month at ETH Zurich: "analyze the leisure requirements of the small city of Biel and thereupon plan a theater for Biel on the site of a former freight yard." By formulating the goal in this way, the theater is established from the outset as the answer to satisfying the needs of the citizens of Biel. This makes a study unnecessary, and a test of its success would surely not call into question the whole enterprise given the initial situation. Burckhardt, "The End of Polytechnic Solvability," in *Who Plans the Planning?,* 110–18, here 110–11. For an earlier text that discusses "enlightened academicism" and the formula ZASPAK and has been translated into English, see Lucius Burckhardt, "From Design Academicism to the Treatment of Wicked Problems (1973)," in Burckhardt, *Writings,* 77–84, here 79.

59 Grzimek, "Leitbild für das Studium," 101.
60 The discussion that follows is a synthesis of several of the texts in Burckhardt, *Who Plans the Planning?*.
61 See Burckhardt, "From Design Academicism." Burckhardt referred to Horst Rittel, "Systematik des Planens," *Werk* 54, no. 8 (1967): 505–8. See also Horst Rittel, *Planen, Entwerfen, Design: Ausgewählte Schriften zu Theorie und Methodik* (Stuttgart: Kohlhammer, 1992), 20–32. During his time in Kassel (1973), Peter Latz, too, studied Rittel intensely. See Udo Weilacher, *Syntax of Landscape: The Landscape Architecture of Peter Latz and Partners* (Basel: Birkhäuser, 2008), 80.
62 Burckhardt, "The End of Polytechnic Solvability," 117–18.
63 On this, see Silvan Blumenthal, "Das Lehrcanapé" and Blumenthal, *Das Lehrcanapé*, with additional bibliographical references. Rolf Gutmann was succeeded in the second year of the *Lehrcanapé* by the architect Rainer Senn. The Germans Jörn Janssen, Hans-Otto Schulte, and Heinrich Zinn, all of whom were appointed as guest lecturers at the same time as Burckhardt, were tolerated by the school administration for only two semesters and then dismissed, because, in the words of the Dean of the Division of Architecture, Bernhard Hoesli, one could no longer speak "of training in the traditional sense" and there were few projects but lots of discussion and writing. Quoted in Bernhard Hoesli, "Die Abteilung für Architektur," in *Eidgenössische Technische Hochschule Zürich, 1855–1980: Festschrift zum 125jährigen Bestehen*, ed. the Rector of ETH Zurich (Zurich: Neue Zürcher Zeitung, 1980), 253–90, quoted in Blumenthal, *Das Lehrcanapé*, 14. As the former ETH student Jan Verwijnen recalled it, the students only agreed to the hiring of Aldo Rossi as guest lecturer in 1972 because the Division of Architecture assured them that Burckhardt's guest lectureship would be extended one year until the summer semester of 1973. Jan Vewijnen, "Politische Radikalität und poetische Präzision," in "Viele Mythen, ein Maestro: Kommentare zur Zürcher Lehrtätigkeit von Aldo Rossi," *Werk, Bauen + Wohnen* 84, no. 12 (1997): 39–41, here 39.
64 *Canapé News* 29 (1973): 70 (emphasis in the original).
65 *Canapé News* 20 (1972): 14, quoted in Blumenthal, *Das Lehrcanapé*, 32.
66 Burckhardt's critique of planning, which thus far has only been explicit here in connection with his teaching, had its origins in his critical study of the correction plan for Greater Basel presented in 1949, which was to redesign the city to make it friendlier for cars and would have completely destroyed Basel's old town. Two publications triggered a nationwide debate over planning: *Wir selber bauen unsere Stadt*, edited by Lucius Burckhardt and Markus Kutter, with a foreword by Max Frisch (Basel: Handschin, 1953); and *achtung: die Schweiz*, edited by Lucius Burckhardt, Max Frisch, and Markus Kutter (Basel: Handschin, 1955). At the conference "Der Stadtplan geht uns alle an" (The Urban Plan Affects All of Us) held in Dortmund in 1955, Burckhardt demonstrated the possibilities of participation processes in planning. From 1962 to 1972, he was editor in chief of the journal *Werk*, which became a platform for these discussions of the critique of planning. He himself described planning and design as being in a triangle of politics–environment–mankind. He worked out the political dimensions on which every space and every design are based and developed from his findings a "critique of the static logic of planning based on clean solutions to problems." Burckhardt interpreted the shape of the environment as a result of the interaction of space and a changing society, with new practices, production methods, and ideologies. Within these tensions, he argued, architectural concepts met urban and rural realities, whereby the great dynamics of economic expansion and social changes after the Second World War would lead to a rapid transformation of the environment that could scarcely be controlled. As the third factor in this interaction, Burckhardt analyzed the daily lives of people and their everyday practices and related them to their built and landscape environment. He believed this perspective would make the social character of spatial contexts understandable. Jesko Fezer has pointed to the connections in terms of subject matter and method between Burckhardt's analytical triangle and Henri Lefèbvre's three conceptions of space as presented in *La production de l'espace* in 1974. On this, see Jesko Fezer, "Politics—Environment—Mankind," in Burckhardt, *Who Plans the Planning?*, 12–16.
67 Burckhardt and Lassus organized the conference "L'intervento minimo" in Gibellina, Sicily, in September 1981. Burckhardt published the results, among other places, in Lucius Burckhardt, "Der kleinstmögliche Eingriff," in *Die Kinder fressen ihre Revolution: Wohnen, Planen, Grünen, Bauen*, ed. Bazon Brock (Cologne: DuMont, 1985), 241–47.
68 On the individual actions, see Lucius Burckhardt, *Warum ist Landschaft schön? Die Spaziergangswissenschaft*, ed. Markus Ritter and Martin Schmitz, Textsammlung Lucius Burckhardt 2 (Berlin: Schmitz, 2006); translated by Jill Denton as *Why Is Landscape Beautiful? The Science of Strollology*, ed. Markus Ritter and Martin Schmitz (Basel: Birkhäuser, 2015).
69 See Annette Ulbricht, ed., *Von der Henschelei zur Hochschule: Der Campus der Universität Kassel am Holländischen Platz und seine Geschichte,* Kasseler Semesterbücher Studia Cassellana 15 (Kassel: Kassel University Press, 2004), 100.
70 Lucius Burckhardt, "A Matter of Looking and Recognizing: In Conversation with Thomas Fuchs," in *Why Is Landscape Beautiful?*, 282–87, here 284. For further information on the "Voyage to Tahiti" action, see the following essays in the same book: "The Science of Strollology," 231–66, here 249–50, and "What Do Explorers Discover?", 267–70.
71 Lucius Burckhardt, in *das Zebra streifen*, ed. Helmut Aebischer, with texts by Gerhard Lang et al., Schriftenreihe des Fachbereichs Stadtplanung und Landschaftsplanung 20 (Kassel: Infosystem Planung, 1994), 1. See also Gerhard Lang,

72 "Der mobile Zebrastreifen" (1993), http://www.gerhardlang.com/d_work_zebra.html.
On this, see also Hannah Stippl's dissertation on Burckhardt's watercolor illustrations for his theory of landscape: Hannah Stippl, "Nur wo der Mensch die Natur gestört hat, wird die Landschaft wirklich schön: Die landschaftstheoretischen Aquarelle von Lucius Burckhardt," University of Applied Arts, Vienna, 2011.
73 Burckhardt, "A Matter of Looking and Recognizing," 287.
74 Lucius Burckhardt, "Strollogy—A New Science (1998)," in *Why Is Landscape Beautiful?*, 288–94, here 290.
75 Ibid., 290–91.
76 Ibid., 291.
77 Ibid.
78 Published in 1980 by the Deutscher Werkbund, Arbeitsgruppe Kassel.
79 Discussed in detail in Fernando Groener and Rose-Marie Kandler, eds., *7000 Eichen* (Cologne: König, 1987).
80 Quoted in Patrick Werkner, *Kunst seit 1940: Von Jackson Pollock bis Joseph Beuys* (Vienna: Böhlau, 2007), 228.
81 Armin Zweite, ed., *Joseph Beuys: Natur, Materie, Form,* exh. cat. (Munich: Schirmer/Mosel, 1991), 42.
82 In addition to their robustness, oaks have been "a form of sculpture, a symbol for this planet ever since the Druids, who are called after the oak. Druid means oak. They used their oaks to define holy places." "Richard Demarco Interviews Joseph Beuys, London (March 1982)," *Studio International* 195 (September 1982): 46–47, here 46.
83 On the interests of the Swiss German architects with whom Kienast worked from the late 1980s onward, on the work of Beuys, and on these questions in general, see Martin Steinmann, "The Presence of Things: Comments on Recent Architecture in Northern Switzerland," trans. Ingrid Taylor, in *Construction, Intention, Detail: Five Projects from Five Swiss Architects / Fünf Projekte von fünf Schweizer Architekten,* ed. Mark Gilbert and Kevin Alter, 2nd ed. (Zurich: Artemis, 1994), 8–25.
84 In a lecture in 1991, Kienast stated his position on the issue of participation. He continued to support open planning, the participation of those affected in the process of planning and building, and to that end it was necessary to "break up the stable crusts of the narrowmindedness of specialist disciplines" and to "prevent everyday life from being determined by others." He then formulated in six points his experiences with citizen participation, and a certain disillusion comes through: citizens showed relatively little interest in dealing with issues of open spaces. Those who were able to prevail had a say, while marginal groups were ignored; nor, as far as he could tell, was the common good ever the goal. Moreover, the wishes expressed were usually on a very banal level: a responsible planner was going to worry about safe passageways, trees, barbecue pits, slides, and benches anyway. Citizen participation only succeeded, in his view, when a true dialogue developed, when the citizens actively participated in the process of developing ideas, and when the planner did not simply take orders and help carry out the wishes of lobbying groups. "Under current practices," Kienast concluded, "citizen participation in issues of public space seems to be full of obstacles that will require great efforts to remove. We have to continue to search for suitable forms." Kienast, "Von der Notwendigkeit künstlerischer Innovation," 62–63.
85 See the first sentence of the first of Kienast's "Zehn Thesen zur Landschaftsarchitektur," 207; see appendix, 403.
86 Ibid., 207; see appendix, 403.
87 Karl Heinrich Hülbusch, "Vegetationskundige Spaziergänge," in Böse-Vetter, *Freiraum und Vegetation,* 417–20, here 417.
88 Hülbusch, "Notizbuch der Kasseler Schule," 158.
89 Georg Simmel, "The Philosophy of Landscape" (1913), trans. Josef Bleichner, *Theory, Culture & Society* 24, nos. 7–8 (December 2007): 20–29.
90 Joachim Ritter, "Landschaft: Zur Funktion des Ästhetischen in der modernen Gesellschaft" (1963), in *Subjektivität: Sechs Aufsätze* (Frankfurt am Main: Suhrkamp, 1989), 141–90, here 150–51.
91 See Hülbusch, "Vegetationskundige Spaziergänge," 419.
92 Ibid., 418.
93 Ibid., 419.
94 For more detail, see Robert H. Whittaker, ed. *Classification of Plant Communities* (The Hague: Junk, 1978); the various schools and traditions are presented on pages 6–10.
95 This presentation is very simplified. On the tasks, goals, methods, and use of phytosociology, see Rüdiger Knapp, *Einführung in die Pflanzensoziologie: Pflanzengesellschaften, Vegetationskunde, Vegetationskartierung und deren Annwendung in Land- und Forstwirtschaft, Landschaftspflege, Natur- und Umweltschutz, Unterricht und anderen Gebieten,* 3rd ed. (Stuttgart: Ulmer, 1971). I am grateful to Norbert Kühn, Professor of Vegetation Technology and Plant Use at the Technische Universität Berlin, for checking and commenting on the abbreviated definition presented here.
96 Dieter Kienast, *Die spontane Vegetation der Stadt Kassel in Abhängigkeit von bau- und stadtstrukturellen Quartierstypen.* Urbs et Regio: Kasseler Schriften zur Geografie und Planung 10. PhD diss. Gesamthochschule Kassel (Kassel: Gesamthochschul-Bibliothek, 1978), here 3–4.
97 Hard, "Schwierigkeiten mit dem Spurenlesen."
98 I am grateful to Norbert Kühn for pointing this out.
99 On this, see Hard, "Schwierigkeiten beim Spurenlesen." Hard referred to the study Eugene J. Webb et al., *Unobtrusive Measures: Nonreactive Research in the Social Sciences* (Chicago: Rand McNally, 1966) and criticized several of the studies published in Munich in the series *Arbeiten zur sozialwissenschaftlichen Freiraumplanung* by Gert Gröning, Maria Spitthöver, and Joachim Wolschke-Bulmahn, whose questionnaire-based results Hard called into question.
100 Hard, "Schwierigkeiten beim Spurenlesen," 47.
101 Ibid., 48.
102 Ibid., 50–51.
103 Ibid., 51.

104 I am grateful to the biologist and vegetation scientist Markus Ritter, Basel, for an introduction to this context in a conversation on March 19, 2009. In the 1980s, he coauthored with Martin Blattner the *Basler Natur-Atlas,* one of the first atlases of this type for a city, for the Schweizerischer Bund für Naturschutz (Swiss Association for Nature Conservation). See Martin Blattner, Markus Ritter, and Klaus C. Ewald, *Basler Natur-Atlas,* ed. Basler Naturschutz, Sektion des Schweizerischen Bundes für Naturschutz (Basel: Basler Naturschutz, 1985).

105 On this, see the works cited by Kienast in his dissertation: Rolf Weber, *Die Besiedlung des Trümmerschutts und der Müllplätze durch die Pflanzenwelt,* Museumsreihe 21 (Plauen: Vogtländisches Kreismuseum, 1960), and H. Pfeiffer, "Pflanzliche Gesellschaftsbildung auf dem Trümmerschutt ausgebombter Städte," *Vegetatio* 7 (1957): 301–20. The latter compares the development of plant species on rubble dumps to an "enormous experiment with nature" whose "scale can best be compared to the settlement of the biotope created by volcanic eruptions." Pfeiffer, "Pflanzliche Gesellschaftsbildung," 301, quoted in Kienast, *Die spontane Vegetation,* 16a.

106 See Herbert Sukopp and Rüdiger Wittig, eds., *Stadtökologie: Ein Fachbuch für Studium und Praxis,* 2nd ed. (Stuttgart: Fischer, 1998).

107 Dieter Kienast and Thom Roelly, *Standortökologische Untersuchungen in Stadtquartieren – insbesondere zur Vegetation – unter dem Aspekt der freiraumplanerischen Verwertbarkeit.* Gesamthochschule Kassel. Schriftenreihe OE Architektur, Stadt- und Landschaftsplanung 3, no. 2 (Kassel: Gesamthochschule Kassel, 1978), here 106. This view meets with great skepticism from urban ecologists such as Norbert Kühn. In his view, this thesis is untenable. Norbert Kühn, e-mail to the author, March 1, 2011.

108 Kienast and Roelly, *Standortökologische Untersuchungen in Stadtquartieren,* 11.

109 Reinhard Bornkamm, "Die Unkrautvegetation im Bereich der Stadt Köln," *Decheniana* 126 (1973): 267–85, here 268, quoted in Kienast and Roelly, *Standortökologische Untersuchungen in Stadtquartieren,* 11.

110 Kienast and Roelly, *Standortökologische Untersuchungen in Stadtquartieren,* 11.

111 Ibid., 12.

112 Ibid., 14.

113 Ibid., 15.

114 Ibid., 127.

115 Ibid., 127–28.

116 On this, see map 4.3 in ibid., and the outline plan for open spaces submitted with their thesis.

117 Ibid., 125–37.

118 See Kienast, *Die spontane Vegetation,* chap. 4.

119 On the definition and terminology of this method, see ibid., 253–54.

120 Ibid., 251–52.

121 A review of Kienast's dissertation published in the specialist journal *Phytocoenologia* emphasized that the compiling and mapping of typical community complexes as Kienast proposed were probably the only meaningful and practical method to describe the vegetation of settlements. Moreover, Kienast's work is said to have created a crucial foundation for urban sigma-sociology by mapping to identify minimal areas, taking into account initial and degradation states, and considering habitat fragments when identifying sigmeta. Otti Wilmanns, Review of Kienast, *Die spontane Vegetation,* and Hülbusch et al., *Freiraum- und landschaftsplanerische Analyse, Phytocoenologia* 8, no. 1 (1980): 151–52.

122 Norbert Kühn, e-mail to the author, March 1, 2011.

123 Kienast, *Die spontane Vegetation,* 352.

124 Ibid. By his own account, Kienast thereby refuted the view of Gisela Jahn that communities of defined species composition are very difficult to find and that the method of synsociology must therefore be called into question. Gisela Jahn, untitled, *(Sigmeten) und ihre praktische Anwendung,* Berichte der Internationalen Symposien der Internationalen Vereinigung für Vegetationskunde 21 (Vaduz: Cramer, 1978), 529–32, here 530.

125 Kienast, *Die spontane Vegetation,* 352.

126 Ibid., 354.

127 On this, see also Karl Heinrich Hülbusch et al., *Freiraum- und landschaftsplanerische Analyse des Stadtgebietes von Schleswig.* Urbs et Regio: Kasseler Schriften zur Geografie und Planung 11 (Kassel: Gesamthochschulbibliothek, 1979), 2, and Wilmanns, Review, 151.

128 Kienast, *Die spontane Vegetation,* 355.

129 Ibid., 355–56: "After the extensive and detailed inventory material and the results of this work, a mapping of the entire settlement area of Kassel could be carried out without difficulty. A mapping on different 'rankings' would be conceivable. As with the transect, it could only be carried out according to sigmeta and their subunits, according to sigmeta without subunits, or only according to previously known alliances, orders, and classes. Depending on the planning questions, sometimes a detailed mapping and sometimes a generalized mapping will make sense. . . . Based on the material now available, which was produced within the framework of methodological and sigma-systematic basic research, the objection to practical results can be reduced to an extent that would make a sigma-sociological study possible for planning or for a report."

130 Wilmanns, Review, 151.

131 Ibid., 152.

132 Ibid.

133 See Kienast, *Die spontane Vegetation,* 367–69.

134 The botanist who reviewed Kienast's dissertation, Otti Wilmanns, saw it as a promising area of work for the future: "Combining detailed analysis based on vegetation science, urban ecology, human sociology, and planning will open up an entirely new, promising area of work that is desperately needed in practice." See Wilmanns, Review, 152.

135 See Sukopp and Wittig, *Stadtökologie.*

136 Blattner, Ritter, and Ewald, *Basler Natur-Atlas.* The first work to employ the basic concepts of ecology, such as "ecosystem" and "biotope," to urban space was that of the Belgian scientist Paul Duvigneaud, who studied the city of Brussels

from this perspective in 1974. Paul Duvigneaud, "L'écosystème 'urbs': Études écologiques de l'écosystème urbain Bruxellois 1," *Mémoires de la Société Royale de Botanique de Belgique* 6 (1974): 5–35. See also Herbert Sukopp, *Rückeroberung: Natur im Großstadtbereich,* Wiener Vorlesungen 102 (Vienna: Picus, 2003).

137 Markus Ritter in conversation with the author, March 19, 2009.

138 Dieter Kienast, "Cultivating Discontinuity," trans. Felicity Gloth, in *Between Landscape Architecture and Land Art,* ed. Udo Weilacher (Basel: Birkhäuser, 1996), 137–56, here 141.

139 Ibid.

140 Ibid.

141 Michel Foucault, in Knut Boesers, "Die Folter, das ist die Vernunft: Ein Gespräch Knut Boesers mit Michel Foucault," in *Die Sprache des Großen Bruders: Gibt es ein ost-westliches Kartell der Unterdrückung?,* ed. Nicolas Born and Jürgen Manthey, Literaturmagazin 8 (Reinbek bei Hamburg: Rowohlt, 1977), 60–68, here 60.

142 Philip Ursprung, *Die Kunst der Gegenwart: 1960 bis heute* (Munich: Beck, 2010), 18.

143 Ibid., 39.

144 See Günter Metken, *Spurensicherung: Archäologie und Erinnerung,* exh. cat. (Hamburg: Kunstverein in Hamburg, 1974).

145 The exhibition at the Kunsthalle Köln, in Cologne, was curated by Evelyn Weiss, Wulf Herzogenrath, Horst Keller, Manfred Schneckenburger, Albert Schug, and Dieter Ronte.

146 Evelyn Weiss, "Die wiedergefundene Zeit," in *Kunst bleibt Kunst: Projekt '74; Aspekte internationaler Kunst am Anfang der 70er Jahre,* exh. cat. (Cologne: Josef-Haubrich-Kunsthalle, 1974), 14–25, here 15.

147 Weiss, "Die wiedergefundene Zeit," 21, and Günter Metken, "Nikolaus Lang: Für die Geschwister Götte," in *Kunst bleibt Kunst,* 238–41.

148 Ibid., 21.

149 Metken, "Nikolaus Lang," 240–41.

150 Ibid., 240.

151 The last big phytosociological analysis in which Kienast participated was Hülbusch et al., *Freiraum- und landschaftsplanerische Analyse.*

152 Quoted in Weiss, "Die wiedergefundene Zeit," 11.

153 This is called the "quantitative feature" because it refers to the diversity of the existing vegetation units. It indicates the average number of species in a selected area determined for the test. Kienast, *Die spontane Vegetation,* 360 and 391.

154 Ibid., 391. The synsystematic category of the plant communities can also be derived from the vegetation coverage. This is called the "qualitative feature" of the existing vegetation.

155 Hans-Jörg Rheinberger, *On Historicizing Epistemology: An Essay,* trans. David Fernbach (Stanford, CA: Stanford University Press, 2010), 2–3.

156 Ibid., 3.

157 Metken, "Nikolaus Lang," 240.

158 Quoted in Richard Tschan, "Grusswort des Schweizer Gartenbau-Präsidenten," *Deutscher Gartenbau* 34, no. 15 (1980): 666.

159 Annemarie Bucher discussed in detail the aesthetic of nature presented at Grün 80 and its social context as part of her dissertation. Annemarie Bucher, "Naturen ausstellen: Schweizerische Gartenbauausstellungen zwischen Kunst und Ökologie" (DSc diss., ETH Zurich, 2008). The analysis that follows therefore concentrates on several facts of the story leading up to Grün 80 and the Dry Grasland Biotope. My source was the documents on Grün 80 in the Archiv SKK Landschaftsarchitekten in Wettingen.

160 Markus Kutter, *Vorwärts zur Natur: Was war damit gemeint? Ein Ausstellungsprojekt und seine Hintergründe* (Niederteufen: Niggli, 1977). See also Markus Kutter, "Die Folgen, die eine Gartenbau-Ausstellung hätte haben können," *Tages-Anzeiger Magazin* 3 (1978): 14–18.

161 See "Gesammelte Protokolle," Archiv SKK Landschaftsarchitekten, Wettingen, boxes 254 and 312.

162 The names of the sectors were: "Schöne Gärten" (Beautiful Gardens), "Land und Wasser" (Land and Water), "Säen und Ernten" (Sowing and Harvesting), "Grüne Universität" (Green University), "Markt" (Market), and "Thema Erde" (Theme of Earth). The last of these sectors was intended to address environmental themes. Kienast's Dry Grassland Biotope became part of the "Schöne Gärten" sector and was later incorporated into the Botanical Garden of the Christoph Merian Foundation. As an employee of Peter Paul Stöckli, Kienast worked with him on the competitions for the "Land und Wasser," "Thema Erde," and "Grüne Universität" sectors. For the "Thema Erde" sector, Stöckli's firm produced an outdoor project, and Dieter Kienast was responsible for planting it with summer flowers (see "Aufwandskontrollblätter," Archiv SKK Landschaftsarchitekten, Wettingen). Kienast was also tasked by the BSG (Association of Swiss Garden Designers) with the content of several educational display rooms in the same sector. In collaboration with the group oekomedia, he produced, among other things, a planning model in which visitors could playfully alter a neighborhood by raising or lowering curtains: a balcony with flowers or an empty one, a parking lot or front yard, and so on. Kienast was also involved in a display room with a "before and after" series of photographs that showed how human beings were destroying the environment and how they could harm themselves in the process (see Archiv SKK Landschaftsarchitekten, Wettingen, box 312). After Grün 80, Kienast worked for the Christoph Merian Foundation as technical director of the Botanical Garden in Basel from 1981 to 1985 (see Archiv SKK Landschaftsarchitekten, Wettingen, box 360).

163 Dieter Kienast and Toni Raymann, "Grün 80; oder, Warum lieb' ich alles was so grün ist? Alternatives Konzept zur Durchführung der Gartenschau 1980 in Basel," idea competition for Grün 80, typescript, 1976. Archiv SKK Landschaftsarchitekten, Wettingen.

164 See, among other sources, the debate between Lucius Burckhardt and Karl Heinrich Hülbusch and their respective supporters over the Bundesgartenschau in Kassel. Deutscher Werkbund, Arbeitsgruppe Kassel, ed., *Durch Pflege zerstört:*

Die Kasseler Karlsaue vor der Bundesgartenschau (Kassel: n.p., 1980). On Basel and Grün 80, see Lucius Burckhardt, "Gärtnern: Kunst und Notwendigkeit," *Basler Magazin* 21 (1977): 1–5.

165 Markus Kutter, "Queerbeetein durch die Grün 80," *Tages-Anzeiger Magazin* 32 (1980): 13–21, and Polo Hofer, "Kein grüner Salat an der Grün 80," *Basler-Zeitung,* August 23, 1980, 10.
166 The botanist and phytosociologist Tüxen planted a phytosociological garden in Hanover.
167 Kienast and Raymann, "Grün 80," 22.
168 Ibid.
169 Markus Ritter in conversation with the author, March 19, 2009.
170 Kienast and Raymann, "Grün 80," 22.
171 Ibid., 23.
172 See Dieter Kienast, "Botanischer Garten Südteil: Naturnahe Biotope" (Botanic Garden Southern Section: Biotopes Close to Nature), *anthos* 19, no. 1 (1980): 56–65.
173 Ibid., 62.
174 Ibid.
175 Archiv SKK Landschaftsarchitekten, Wettingen, plan nos. 287-6 and 287-7.
176 Kienast, "Botanischer Garten Südteil," 57.
177 Rudolf Schenker, "Grün 80-Trockenbiotop: Die Natur hat Vortritt," *Panda Nachrichten: Mitteilungsorgan der Stiftung WWF Schweiz* 13, no. 2 (1980): 15.
178 Ibid.
179 Ibid.
180 Kienast, "Botanischer Garten Südteil," 58–62, with additional explanations.
181 For the details of the intended plants, see Kienast, "Botanischer Garten Südteil," and Dieter Kienast, "Botanischer Garten Südteil: Naturnahe Biotope," manuscript, January 4, 1980, Archiv SKK Landschaftsarchitekten, Wettingen.
182 On the details, see Kienast, "Botanischer Garten Südteil" (published version).
183 Ibid., 65.
184 Ibid.
185 Quoted in the award certificate from the Zurich city council, October 8, 1980, Archiv SKK Landschaftsarchitekten Wettingen.
186 Marcel Amstutz, "Das kleine Paradies [Dry Grassland Biotope, Brüglingen]," *Basler Magazin,* April 27, 1985, 1–5.
187 Ibid., 3.
188 I am grateful to Christophe Girot for this information, who in 2008 sat on the jury, together with representatives of the Christoph Merian Foundation, for the redesign of the Dreispitzareal.
189 Kienast and Roelly, *Standortökologische Untersuchungen in Stadtquartieren,* 12.
190 The moral philosopher William Shaftesbury wrote about this in *Characteristicks of Men, Manners, Opinions, Times* (London: John Darby, 1711): "A Princely Fancy has begot all this; and a Princely Slavery, and Court-Dependence must maintain it." Like Locke, Shaftesbury posited nature as a primarily moral force. In his view, it was expressed in the harmony of the cosmos and in the morality that is inherent to human nature. Because for Locke and Shaftesbury the new social order was based on natural law, and morality was manifested by human nature, their ideas were taken up by a series of intellectuals who opposed the existing social order in their critique of formal gardens and proposed the English landscape garden as an alternative. On this, see Adrian von Buttlar, *Der Landschaftsgarten: Gartenkunst des Klassizismus und der Romantik* (Cologne: DuMont, 1989), 9–13.

191 This tradition includes, for example, Gilles Clément's "jardin en mouvement" (1994) in the Parc André Citroën in Paris. It is a part of the park that has been allowed to grow and is low-maintenance in a kind of "three-field system" with minimal interventions. The "mouvement" of the name refers to the dynamic processes of nature that are visible in the succession processes. In a text about it, Clément speaks of a biological abundance ("un plein biologique") that can occur where there is no architecture or design ("le vide architectural"). Clément also refers to the subversive character of plants that grow wild, whose appearance can be read as a sign of human power receding ("un recul du pouvoir lisible à l'homme") and hence as a grave defeat ("une grave défaite"). See Gilles Clément, *Le Jardin en mouvement: De la vallée au parc André-Citroën* (Paris: Sens & Tonka, 1994), 4 and 18.
192 Dieter Kienast, "Naturschutz: Unkraut in der Henschelei!," *Der Monolith: Studentische Zeitschrift der Organisationseinheit Architektur, Stadt- und Landschaftsplanung, Gesamthochschule Kassel* 12 (December 1977): 2. In the typescript of this text, Kienast signed his middle name, Alfred, in order to underscore the alienation and irony. See gta Archives (NSL Archive) / ETH Zurich (Dieter Kienast Bequest).
193 For this and the following quotations, see ibid.
194 Ibid.
195 Franz Hohler, "The Recapture," trans. Diane Dicks, in *At Home: A Selection of Stories* (Basel: Bergli, 2009), 62–77, here 74.
196 Ibid., 75.
197 Dieter Kienast, "Gestalt des öffentlichen Raumes," in *Stadt, Stadtraum, Raumqualität: Referate zur SWB-Veranstaltungsreihe im Kunst- und Kongresshaus Luzern,* ed. Schweizerischer Werkbund, Ortsgruppe Innerschweiz (Lucerne: Schweizerischer Werkbund, 1989), 40–46, here 42.
198 Böse-Vetter, *Freiraum und Vegetation,* 411.
199 Kienast, "Naturschutz."
200 Kienast and Roelly, *Standortökologische Untersuchungen in Stadtquartieren,* 125–27.
201 Ibid., 127–29.
202 Lucius Burckhardt, "No Man's Land (1980)," in *Why Is Landscape Beautiful?,* 126–27, here 126.
203 Ibid., 127.
204 Kienast and Roelly, *Standortökologische Untersuchungen in Stadtquartieren,* 11.
205 Saskia Sassen, "The Global Street: Where the Powerless Get to Make History," lecture at the conference "Thinking the Contemporary Landscape: Positions and Oppositions," Hanover-Herrenhausen, June 21, 2013, https://video.ethz.ch/conferences/2013/ila/06_friday/c10211a6-3d35-4393-a809-d4223782ae07.html.
206 Ibid.

207 The title of one of his books: Alexander Mitscherlich, *Die Unwirtlichkeit unserer Städte: Anstiftung zum Unfrieden* (1965; enl. ed., Frankfurt am Main: Suhrkamp, 2008).
208 Kienast, *Die spontane Vegetation,* 392.
209 Ibid. This debate, like the linking of garden styles and forms of society, is very old. It can already be seen in the controversies between the advocates of the "wild garden" in the circle of William Robinson and those of the architectural garden at the turn of the twentieth century. It is always about different ideas of nature.
210 See gta Archives (NSL Archive) / ETH Zurich (Dieter Kienast Bequest, box ECLF).
211 Kienast and Roelly, *Standortökologische Untersuchungen in Stadtquartieren,* 127.
212 Kienast, "Zehn Thesen zur Landschaftsarchitektur," 207; see appendix, 403.
213 Ibid.
214 Compare one of the central passages in Joachim Ritter's 1963 essay: "Landscape is nature that becomes aesthetically present when seen by a feeling and perceiving viewer: the fields in front of the city, the stream as 'border,' 'trade route,' and 'problem for bridge builders,' and the mountains and steppes of shepherds and caravans (or those searching for oil) are not already 'landscape.' They only become so when human beings turn to them without a practical purpose, to view them in 'free' enjoyment, in order to be in nature as themselves. When they go out in it, nature changes its face." Ritter, "Landschaft," 150–51.
215 This hypothesis, made by philosophers from Plato through to Joachim Ritter, is perhaps doubtful but is presented here uncontested to further the book's argument, which ultimately leads in another direction.
216 See Wolfgang Welsch, "Aestheticization Processes: Phenomena, Distinctions and Prospects," in *Undoing Aesthetics,* trans. Andrew Inkpin (Thousand Oaks, CA: Sage, 1997), 1–32, here 1. Welsch subsumes all forms of superficial embellishment under the term "surface aestheticization."
217 On the concept of "adäquat" (adequate), see Rudolf Eisler, *Wörterbuch der philosophischen Begriffe* (Berlin: Mittler & Sohn, 1904), http://www.textlog.de/1426.html.
218 Kienast and Roelly, *Standortökologische Untersuchungen in Stadtquartieren,* 12–13. In this passage, the following literature is referred to in parentheses (omitted in the quotation): "Buchwald, 1962," "Hülbusch, 1973," and "Bierhals, 1972." The bibliography at the end of the work clearly identifies only one of these titles: Erich Bierhals, "Gedanken zur Weiterentwicklung der Landespflege," *Natur und Landschaft* 47, no. 10 (1972): 281–25.
219 Sowing spontaneous vegetation is paradoxical by definition, which underscores its reinterpretation through deliberate design.
220 In one of the central passages of *De la grammatologie* (1967), Jacques Derrida wrote: "The advent of writing is the advent of this play; today such a play is coming into its own, effacing the limit starting from which one had thought to regulate the circulation of signs, drawing along with it all the reassuring signifieds, reducing all the strongholds, all the out-of-bounds shelters that watched over the field of language." Jacques Derrida, *Of Grammatology,* trans. Gayatri Chakravorty Spivak (Baltimore: Johns Hopkins University Press, 1976), 7.
221 The irises were later removed. They are still visible in the photographs by Christian Vogt published in 2000: *Kienast Vogt: Parks und Friedhöfe / Parks and Cemeteries,* trans. Felicity Gloth (Basel: Birkhäuser, 2000), image p. 147.
222 The front yard was expanded in 2000 and 2001; since then, this supply gutter has disappeared between the original part of the front yard and the new part. A third strip has been added to the front yard.
223 Dieter Kienast, "Funktion, Form und Aussage" (Function, Form and Statement), interview by Robert Schäfer, *Topos* 18 (1997): 6–12, here 8.
224 Over the past fifteen years, all of the other parking spaces in front of the company have been removed to extend the front yard.
225 Kienast, "Funktion, Form und Aussage," 6 and 12.
226 See EPB's firm profile under https://www.ebp.ch/en.
227 Today the building is leased by Vontobel Holding AG. During the renovation, the garden courtyard was maintained only minimally for several years. In the meanwhile, the weeds were removed and the irises replanted in the drainage channels. The garden courtyard is thus unchanged and has aged well.
228 On the discussions that follow, see also Anette Freytag, "Bildkritik: Ästhetisierungsstrategien in der zeitgenössischen Landschaftsarchitektur," *Die Gartenkunst* 15, no. 2 (2003): 204–24.
229 I am grateful to my colleague Albert Kirchengast of Christophe Girot's chair for landscape architecture, ETH Zurich, for discussing these themes and coming to shared conclusions.
230 Dieter Kienast, "Courtyard Design: Offices of the Swiss Re, Zurich," in *Kienast Vogt: Aussenräume / Open Spaces,* trans. Felicity Gloth (Basel: Birkhäuser, 2000), 184.
231 On this, see the critique formulated in Kienast and Roelly's thesis of the "traditional" planning of open space, which tries with "tireless zeal" to "green away" everything that looks technical and to maintain an image of the city that is oriented around the symbols of an agrarian landscape. See Kienast and Roelly, *Standortökologische Untersuchungen in Stadtquartieren,* 11.
232 See Kienast, "Courtyard Design," 184.
233 I am grateful to Norbert Kühn for this reference.
234 Welsch, "Aestheticization Processes," 1, 9, and 25.
235 Theodor W. Adorno, *Aesthetic Theory,* ed. and trans. Robert Hullot-Kentor (Minneapolis: University of Minnesota Press, 1997), 191.
236 Kienast, "Funktion, Form und Aussage," 6.
237 Kienast, "Zehn Thesen zur Landschaftsarchitektur," 209; see appendix, 404.

II
Forms of Use

Aesthetic Experience and Coping with Everyday Life

In 1979, Dieter Kienast's programmatic text "Bemerkungen zum wohnungsnahen Freiraum" (Remarks on Free Spaces Close to Habitations) was published in the professional journal *anthos,* which at the time was the official organ of the International Federation of Landscape Architects (IFLA).[1] The principles for designing the surroundings of housing colonies and apartment buildings that he formulated there had been tested by him in the prior years for a housing development in Niederhasli (a suburb of Zurich) and in his private garden in Zurich.

Kienast's central thesis was that quality of life is crucially dependent on the design of the open space surrounding a home,[2] and that landscape architects are therefore called on to improve it.[3] Kienast substantiated this thesis with a statistic that people spend as much as 80 percent of their nonwork time in the immediate surroundings of their home.[4] Nonwork time does not mean leisure time, however: in open space close to residential areas, according to Kienast, the primary task is coping with everyday life.[5] The categories of use should be correspondingly diverse:

First, use has to be defined more broadly than walking in the park, observing the beautiful flowers, or slightly tortured play on the playing field. For me, the use of open space means first coping with everyday life: for example, going shopping with children, the route to the streetcar stop, waiting for the streetcar, cooking while keeping an eye on the kids, the route to school, working in the allotment garden, contact with the neighbors, activity after work. For children, however, coping with everyday life also means playing: experiencing self-realization in play, practicing social behavior, having fun.[6]

The "use value" of open space—a concept from the Kassel School—is the focus for Kienast. In the same paragraph, he laments that as a rule open space is dominated by a "beautified environment," which is also the responsibility of architects and landscape architects, and which for that reason today's open spaces near residential areas are "in some cases rather badly equipped" for the everyday life he describes and its demands for use.[7] Kienast attacked in particular the design of the grounds of Swiss residential complexes of the 1960s and 1970s, including Volketswil, Spreitenbach, and Bern-Bethlehem.[8] He identified in these housing developments located "auf die grüne Wiese" (on the green meadow), as the topos has come to be known in German, great shortcomings in terms of differentiated opportunities for play, sports, and encounters for residents of various ages.[9]

His critique focused on playgrounds where everyone, apart from the small children playing in the sand, is bored—especially the teenagers and adults. On the one hand, in his view, the right spaces are necessary for various age levels so as to create relationships between children and adults. Here the primary question is from what age children are permitted to play how far from home so that they can gradually break free of parental supervision. On the other hand, Kienast argued for giving children new substances with which to play, for example, leftover construction materials such as compound concrete bricks, gravel, soil, sand, boards, and posts.[10]

The second line of argumentation for improving open spaces near residential areas concerns community gardens, which should be located in the immediate vicinity of housing. Kienast was not referring to allotment gardens but rather "tenants' gardens" on communally cultivated or individual lots. For each six units in a complex, the tenants would be provided with a lot of at least a 100 square meters. Kienast believed that making that possible represented an important contribution to the "emancipatory planning of open spaces."[11] Tenants' gardens could do more than just ensure self-sufficiency: working in the garden, chatting, and sharing supervision of children would improve relationships between neighbors. Apart from this social function, the necessary work in the garden done in the evening, on the weekend, or "in between" has lots of value as recreation.[12]

In addition to his argument for tenants' gardens and to pointing out the benefits of designing not just playgrounds but also the entrances of each housing complex and the surrounding grounds to be suitable for children to play, two other passages in this text are revealing. Kienast called for "using, rather than the term 'green urban planning' common in Switzerland, . . . the more apt term 'open-space planning'"[13]—an early reference to the themes of landscape architecture that can and should be "not just green but also gray," as would become so essential in his later work.[14] "Green urban planning," Kienast wrote in 1979, "strongly suggests that it is primarily about green areas—meadows, lawns, plantings—but that is by no means accurate."[15]

The second striking statement refers directly to the theme of use and its influence on design. Kienast ended his manifesto on the design of open spaces near residential areas with a section on the possibility of users altering open spaces—a fact that until then had generally been more tolerated than welcomed by landscape architects but which Kienast elevated to a program:

Open spaces that meet the above requirements should, for obvious reasons, dispense with the usual "orderliness." It must also be accepted that they change—along with their users. Something that the landscape architect in particular has (thus far) not been pleased to see since it changes the design. I believe, however, that it is infinitely more important that open space be used intensively, changed, and even made disorderly . . . . Open spaces in which everything is artificial, planned, intended, paved, indestructible, and neatly maintained prevent self-determined use and evolution. Thus, they become unusable and gleam in an aseptic form as a sign of elite architecture.[16]

The final sentence of this manifesto for usable open space, with its landscape architecture changeable by people, is particularly remarkable in view of Kienast's strictly formal works, which dominate his oeuvre from the mid-1980s onward. Indeed, such "unalterable" grounds, its aesthetic having delighted many architects and landscape architects, were, on the other hand, called into question by many critics. They were unalterable because the works are precisely composed, so that any deviation from the original concept represents a disturbance. The cultural historian Ilse Helbich commented skeptically in a radio essay in 1998:

> The same Dieter Kienast who says "what matters to us is the people" has produced grounds of which one asks how they are intended to be used, because they cannot be "occupied"—in the literal sense—nor played nor walked on with pleasure. Further indications of this are the bodies of water that serve no function. Often they are shadow mirrors stamped out in rectangles. They are far removed from the idea of a swimming pool or a biotope. What is their point? The Swiss Dieter Kienast explains that he wanted to depict the rifts in our relationship to nature today. And one does see them: by consciously building transverse paths and disharmonies into his gardens, he wants to make the painful contradictions with which modern human beings have to live visible and palpable. Gone are the pleasant aspects of an aesthetic view of nature emphasizing its painterly qualities. The discrepancy "here nature"—"here human-constructed environment" is presented as a problem.[17]

Just under twenty years lie between Kienast's argument for uses that alter open spaces and Helbich's critique of Kienast's "inaccessible gardens," the title of her radio essay. In terms of time and subject matter, the two positions stake out the area of tension in which Kienast's open spaces stand in terms of their design and use. The postulated inaccessibility points to a characteristic that over the years became more crucial in Kienast's work: the phenomenological dynamic between open space and viewer.

Kienast's concept of use changed in the 1980s and 1990s. Whereas initially the practical was foregrounded, the contemplative gradually became more important. Use is always connected to form, space, and material—for the landscape architect, the last of the three includes plants as his very material.

"Using Open Space Means Coping with Everyday Life": Kienast's First Design for the Grounds of a Housing Development

The grounds of a housing development in Niederhasli, planned and executed in several stages from 1972 to 1975, was the first larger project that Dieter Kienast developed while working in the firm of Peter Paul Stöckli and in constant exchange with him (48). Kienast had met Stöckli, who was four years older, during his internship for the landscape architect Albert Zulauf in Baden, Switzerland, in March 1967. Since February 1972, he had been working regularly for the firm that Stöckli had founded in 1970. On January 1, 1980, Kienast was made his official partner.[18]

When designing the grounds of housing developments, one is primarily dealing with semi-public spaces that are freely accessible but used mainly by the residents of the complex and in particular by their children. Kienast spent more than four years working on designing the grounds for Niederhasli. First, his project enabled him to apply the subject matter of his studies in Kassel; second, he intensely studied the traditional Swiss design of the grounds of residential complexes. In 1978, he was accepted into the Bund Schweizer Gartengestalter (Association of Swiss Garden Designers, BSG; later Association of Swiss Landscape Architects, BSLA) for this work, among other things.[19]

In designing the grounds, much of which was Kienast's own work, several principles of open-space planning were formulated that still hold true today:
— Designing open space means, first and foremost, organizing the space. When designing the grounds of a residential complex, it is also about zoning a social space.
— It should take into account the private areas such as the interior living space and the gardens, the public and communally used areas, and the transitional zones that connect the interior and exterior spaces, such as entrances and passageways.
— Ideally, the design of the open space should initiate social processes, such as games, conversations, and communal child care.
— The goal is—as formulated by one of Kienast's younger colleagues, the landscape architect Stefan Rotzler (b. 1953)—that the specific design of the open space should offer its users an atmosphere in which they enjoy spending time and where it is possible to experience something like "producing meaning" or "moments of individual happiness."[20]

In 1972 and 1973, the grounds of the section of the complex on Lindenstrasse were realized first, followed by the Huebwiesen section. In addition to a distinct topography adapted to the site of the residential blocks, Kienast's achievement at Niederhasli was above all a detailed analysis of the user groups and their needs. The results of this analysis were reflected in the design of the topography and the network of paths, as well as in the features and programs of the related sites.

"On the Green Meadow"

Niederhasli is located in the Glatt Valley, 18 kilometers north of the center of Zurich, just before the entrances into the Wehn Valley between Dielsdorf und Rümlang. All three municipalities belong to the urban agglomeration of Zurich, which had been growing rapidly since the 1960s. The history of the town and of this housing complex is representative of the architectural development of numerous municipalities of Zurich during the 1970s.

Niederhasli is a typical commuter town. Most of its residents work in Zurich. While the number of residents increased by just 333 to 1,375 from 1950 to 1960, over the following decade it more than doubled to 2,838 as a result of the large influx; today the population is over 9,200.[21] More and more land of this agricultural community was rezoned. Whereas in neighboring towns it was primarily single-family homes being built alongside the traditional farmhouses, Niederhasli was dominated by apartment complexes "on the green meadow"

(49, 50).[22] The Lindenstrasse-Huebwiesen housing development, part of the expansion of the town, was located on a plot delimited by three streets that bordered on an agricultural zone with meadows and fields. The lot tapers to a point in the southwest. Later, additional single-family homes and apartment buildings were built around this complex.

The approach was always similar: first, the architects planned multistory apartment buildings, and then the landscape architects were asked to design the grounds and plant greenery. In an unpublished and undated recollection found as a manuscript in his papers, Kienast described his confrontation with the faits accomplis in Niederhasli:

The problems of separating planning tasks by profession arose from the outset. All of the buildings, underground garages, entrances, and so on, were already designed by the architects. The landscape architect was then let loose on the remaining areas. The landscape planning work began with that premise. In a euphoric mood, its author began to analyze the floor plans of the units, which hardly offered any feasible approaches to planning the open spaces—how could they?[23]

Kienast complained that the floor plans were too small and incorrectly standardized and were not well placed within the overall grounds. Because no changes could be made to the architecture, "the possibilities to compensate for it in the open spaces were very limited." Kienast observed, moreover, that in retrospect he would have planned the grounds of the housing development differently, but several approaches to the planning had indeed been successful and should still be pursued. He did not identify specific examples, so that we can only guess what still worked and what did not from his later perspective.

Peter Paul Stöckli's firm was hired to design the grounds of the Lindenstrasse section of the complex early in 1972, and several months later the grounds of the Huebwiesen section as well.[24] Initially, the design was based on a plan dated January 20, 1971, by the Zurich architect Ernst Nüesch for four residential blocks with L-shaped, slightly curved ground plans. The plan indicates the number of floors of each building and also the ground-level and underground parking spaces. An orthogonal system of paths between the blocks provides access to the grounds, and loose rows of trees had been placed between them. Taken together, these elements provided the minimal version of the grounds of a residential complex, which the landscape architects subsequently altered considerably. A series of photographs sent to Stöckli's firm in February 1972 illustrate the original situation of the Lindenstrasse and Huebwiesen complexes:[25] the former complex had been built by the general contractor Wanner, for the Milchbuck Building Cooperative of Zurich,[26] and the latter complex by the MOBAG construction company of Zurich.[27] One of the photographs shows the flat terrain of meadows and fields for the future development; another shows a path through the field, on the edge of which is the soil excavated to construct the housing blocks. The rural surroundings are evident, as is the fact that this area will soon be built up with housing blocks and single-family homes, like those seen on the horizon (49, 50). The area to the right of the path through the field is the site of the future Huebwiesen section of the development, adjoined by the Lindenstrasse complex. The excavated soil is used to design a slightly hilly topography. The photographs do not show anything of the future development. It is reasonable to assume that it was already under construction when Dieter Kienast began planning the grounds, coordinating with Stöckli and the garden architect Georges Boesch of Zurich, who is also listed on the plans as the garden architect responsible for the project.[28]

Comparing Kienast's sketches of ideas and initial designs to the preliminary projects makes the course of the construction clear. Three residential blocks on the western section of the lot, the Lindenstrasse buildings, were realized first. They were followed by the Huebwiesen section of the development, which was composed of three parts: three residential blocks on the pointed eastern section of the lot, two blocks to the south on the opposite side of Langackerstrasse, and two blocks and a kindergarten to the east opposite Lindenstrasse. The residential buildings with the L-shaped, slightly curved ground plans vary in height. Depending on the size of the ground plan, each block has two to four sections of different eave heights; starting with a minimum of four floors, they have as many as seven. The development has around 350 apartments in all.[29] The estimate of the cost for the grounds, dated April 3, 1972, indicates a terrain of 15,500 square meters with total costs of 43.50 Swiss francs per square meter.

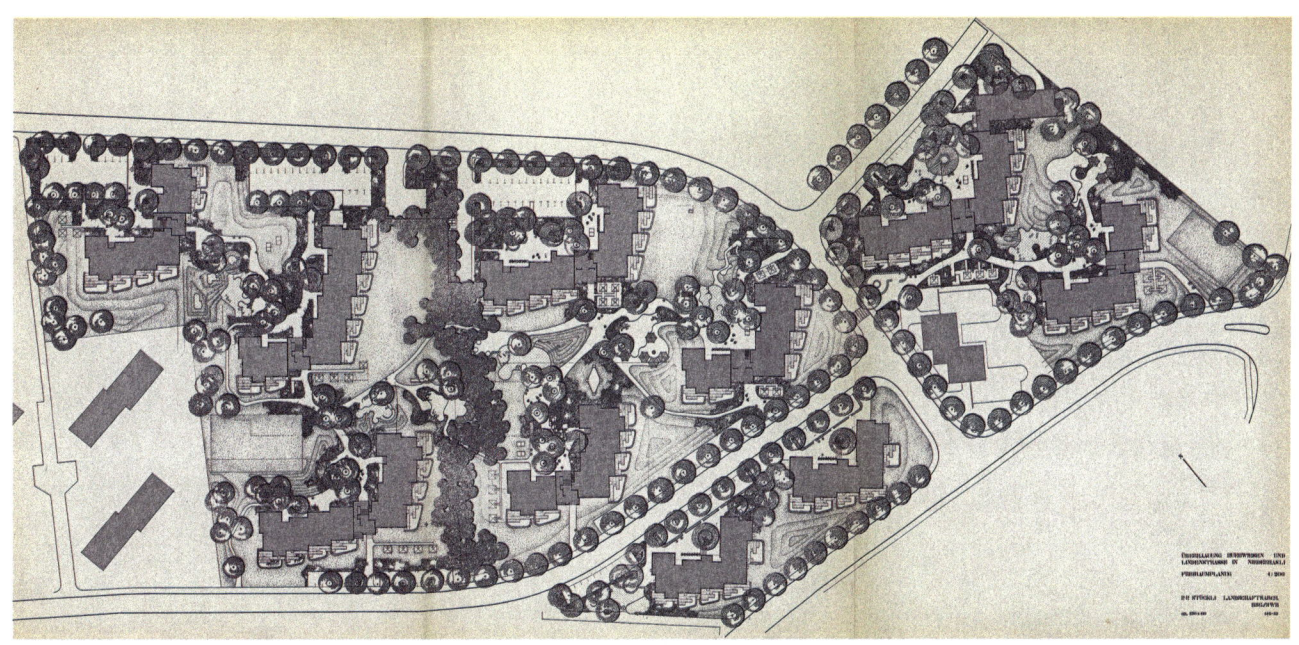

48
Grounds of the Huebwiesen and Lindenstrasse housing development in Niederhasli, presentation plan no. 142-33, scale: 1:200, H 110 × W 230 cm, copy, undated, firm of Peter Paul Stöckli

49, 50
The development was built on an open field next to an existing housing complex. The excavated soil, which would later be used to model the terrain for the landscape design, was stored on site. Photos: 1972

51–54

Dieter Kienast, sketches and initial designs for Niederhasli, undated, ca. February–March 1972 (detail views)

Sheet 1: Sketch of the grounds of the Lindenstrasse development; assigning of areas and functions, with legend

Sheet 2: Topography and system of paths and areas of the Lindenstrasse development

Sheet 3: Design drawing of the topography, paths, pools, and sand playgrounds of the Lindenstrasse development

Sheet 4: Design drawing of the surroundings of the Lindenstrasse and Huebwiesen developments, with detailed legend of the program and the organization of the areas

Kienast's undated sketches of ideas and preliminary drawings must have been produced between early February and mid-March 1972 (51–54).[30] The first three drawings are of the Lindenstrasse section alone; the other two include the Huebwiesen section. The plan of the preliminary project for Lindenstrasse is dated March 16, 1972, and the one for Huebwiesen July 19, 1972. An undated presentation plan published in *anthos* in 1983, along with a series of black-and-white photographs, shows the complete grounds including the complexes of buildings opposite Langackerstrasse and Lindenstrasse.[31] There are several photographs of the completed complex, very probably taken in 1975.[32]

The Design Process:
Moving from the Use
to the Design

Five sketches of ideas and initial drawings for the design of the grounds of Lindenstrasse and Huebwiesen have been preserved. Although they are undated, it is possible to order them chronologically, shedding light on the design process.

After the zones of use had been roughly subdivided in the initial sketch (51), the next sketch (52) provides more detail about the topography and paths. The next two sketches codify the design (53, 54): the style is calmer; the linear is employed; programmatic zones are expressed in the design; and on the fourth drawing there is a detailed legend. The latter also points to a special feature: the Lindenstrasse section of the development, which has a detailed program for the design of the grounds, appears with the first part of the Huebwiesen development on the right side, for which only the basic features of the topography and the most important paths are marked. In the fifth sketch, the topography of the first part of the Huebwiesen grounds is drawn with more details, mimicking the drawing style of the Lindenstrassen grounds, making it difficult to determine the exact chronology. The crucial thing, in any case, is that the design process happened nearly simultaneously: a summary definition of the areas of use, from which a special order derives, is followed by embedding and codifying the design program into the landscape that derived from the topography and the system of paths designed for it. The parameter for designing spaces according to their use is their closeness to or distance from the houses and the question of whether they should be used privately or communally and by which residents or groups of residents.

According to Kienast's maxim, it should be possible both to use all of the areas of the grounds and to play there. As Kienast later admitted, however, this turned out to be a "pious wish":

> I too succumbed to the idea of filling up as many areas as possible with sensible and nice activities, at least according to plan. In addition to a forum, which fortunately was later eliminated, everything a garden architect can imagine was planned: from a bocce court, shallow pools, playgrounds, to fire pits, pergolas, and roof gardens. In the various meetings of different committees, however, the program was reduced again.[33]

The sketches of ideas, the drawings of initial designs, and the preliminary project plans enable us to reconstruct the ideal planning, which incorporated all of the needs of the users, as well as the most important concerns in terms of their (non) implementation in the context of other designs for the grounds of residential developments in the 1960s and 1970s.

Sheet 1 (51) shows the rough subdivision of the separate zones based on the parameters "close to the building" versus "far from the building," "user groups," and "privately versus publicly used." The point of departure for the design was the three L-shaped, slightly curved residential blocks, from which the separate areas were developed. Three blue zones, also L-shaped, were assigned to the building entries. Another blue zone is located on the edge of the development. A legend on the right-hand side of the page briefly explains what the colors used represent. On the left side of the sheet, Kienast noted more specifically of the color blue that it marked entrance zones with rest areas where residents of all ages could meet as "pedestrians."[34] For that reason, this area had to be attractive for all ages and be provided with benches, playground equipment, and scooter parking. The blue area on the outer edge is a parking lot that can also be used as a place for children to play, with trees to provide shade and protect the development from noise and dust. On the other side of the building, marked in brick red, are private areas along the balconies where Kienast planned to place "garden seating protected from the public area." A line in brown felt-tip pen indicates a soil bank that terminates the private areas and also provides "noise reduction and privacy."

Around the two main zones near the residences—the public zone at the "front" and the private zone at the "rear" of the building—Kienast marked in yellow "central areas" for "as many age groups as possible." The adjoining areas in purple indicate "opportunities for infants and children to play." The bands with green hatching are the only ones that are not explained. They very probably represent trees and hedges, which are natural and customary in landscape design as structuring and framing components.

In the second sketch of ideas (52), the principle of subdividing areas is retained and further differentiated. The design of the grounds now includes part of the Huebwiesen development as well. The most significantly elaborated areas graphically are the entrances in front of the residential blocks and the "central areas for as many age groups as possible" already laid out in the first sketch. The zones in front of the entrances were filled with brick-red felt-tip pen on Sheet 2, which suggests some kind of surfacing, since the paths between the buildings are also in this color. Individual trees and seating surrounded by hedges or walls are also evident. Otherwise, it is difficult to get further detail from this sketch. Some of the entrance zones transition directly into the "central areas." The highly expressive lines can be read as topographic elevations and depths. Between them are surfaces of various shades of yellow and blue, which indicate areas of sand and water. Snaking lines and individual squiggles mark playing equipment, with notes such as "sand and water," "puppet theater," and "fortress" added.

The areas for infants and children have the fewest details; Kienast's indications are "construction playground" and "badminton." Another area—the future grass playing field—is indicated only by the dimensions 25 by 30 meters; another is defined by the label "sunbathing lawn." In contrast to that of the southern edge of the development along Langackerstrasse, the design of the northern edge on Lindenstrasse is worked out in detail. Kienast placed two rows of trees across the two large existing parking lots. Where the clotheslines for drying laundry were planned, he interrupted the inner row of trees so that the trees do not cast shadows on the laundry.

On Sheet 3 (53), the individual areas are even further differentiated. In contrast to the first two sketches, which are drawn in an expressive style, Kienast worked also with a ruler here. The dominant elements are the entrance zones in front of the residential blocks and the paths that lead from there through the grounds of the development. The uniform surface covering is emphasized with yellow. The connection paths and forecourts, which are also intended for roller-skating and tricycle riding, are adjoined by a hilly area equipped with sandboxes, playgrounds, and bodies of water. The topographic formation is arranged such that the playgrounds are always located between the entrance of one residential block and the back of the one to the north of it, so that they still fulfill the "protective wall function" indicated on Sheet 1 for the private gardens for the ground floor units that are located by the rear façade. The areas in front of the building entrances, which are filled in with green, indicate shrubs and trees, which are planted in concrete cylinders because of the underground parking garage beneath them (55). They provide shade for the planned seating and articulate the forecourts to create spatially more intimate areas in which to spend time.[35]

Kienast's notes on the forecourt of the northernmost residential block indicate "entrances," a "roller-skating area," "hopscotch courts," and "tricycle riding." In front of the center block there is a small forum intended as a theater, but it was later eliminated. Northeast of the form is a grove-like area enclosed by walls and hedges, where ping-pong tables and seating are evident, labeled by Kienast as "reading corner," "storytelling," and "jass" (a card game).

The other playing areas are distributed around the hilly landscape: sweeping, meandering sandy playgrounds can be seen, as well as two curved bodies of water, one of which has a small island, and snaking partitions in the form of yellow plastic tubes. There is a place for a "puppet theater" and a "playing hill with wooden fortress." The areas left blank in the design sketch, which are surrounded by hills and walls, are grass playing fields. The area to the east labeled "sunbathing lawn" on Sheet 2 now has three ping-pong tables and, connected by a path, a bocce court. The field on the southwest edge of the development marked "25/30 m" on Sheet 2 has a note "ball fence combined with climbing tower, etc." The last area lacks an explanation and is located east of the southernmost residential block: it is framed by hills and is marked by yellow dots. Its function is evident only from the legend on Sheet 4: Kienast was planting a seating area with a fire pit and steps.

166

On Sheet 3, it is still somewhat unclear how the development would be terminated facing the street: Kienast left off the tree-lined boulevard depicted to the north on Sheet 2 and planned a broad green area in the south labeled "badminton" and "sunbathing." On Sheet 4, its use changed: the grassy area for games and relaxation was replaced by a plaza with rotary washing lines to hang laundry.

The fourth drawing is the only one that combines both parts of the development, Lindenstrasse and Huebwiesen, on one sheet (54).[36] The spatial arrangement, the program, and the details of the design are laid out most completely here. A color legend explains the material composition of the surfaces; a number legend indicates the function assigned to the separate spaces. The three preliminary stages of the Lindenstrasse grounds are integrated into the design and further differentiated. Kienast retained the spaces, the topography, and the elements he had developed earlier. Specific components such as the network of paths, the mounds of soil, the trees, the entry areas, and the lawns have been transferred to the Huebwiesen grounds in the same colors and the same style. As in the design process thus far, the colors of the individual areas are assigned to specific uses: analogously to Sheet 1, yellow is for "areas for all age levels" and purple for areas reserved for children and teenagers to play. Here, too, Kienast planned an extended grass playing field and private gardens behind the residential blocks, though this time without the protective soil banks. Like the Lindenstrasse section of the development, the Huebwiesen complex is terminated at the street with tree-lined boulevards or green zones.

It is striking that all of the ideas for play and use recorded in the drawings thus far reappear in compressed form on Sheet 4. It does not seem to have been an exaggeration when Kienast said in retrospect that in Niederhasli "everything a garden architect can imagine" had been planned.[37]

The legend for the forum speaks of a "place for diverse events (tenant meetings, celebrations, dances)" and of a "puppet theater." The two large "playing hills" were to be equipped with a wooden fortress with slides, climbing ropes, and "concrete pipes to crawl through"; the sandy playgrounds with walls of wood or concrete for painting that are also suited as "niches for role-playing." One unusual feature is the construction playground located behind the northernmost residential block, which first appeared on Sheet 2. Here the children could build huts, dig holes, and play hide and seek, and there was even supposed to be "an old car to disassemble." Smaller playing equipment, such as horizontal bars and merry-go-rounds, was distributed along the paths throughout the grounds, to encourage children not just to stay within the playgrounds but to "conquer" the entire open space.

The courtyards in front of the buildings and the paths provided spaces for roller-skating and bike riding. In the final design, the center entrances to the residential blocks were designed to be open on both sides of the building, so that even today it is still possible to ride a bike through the buildings without dismounting. All of the exterior stairs have a bike ramp. On the asphalt and paved forecourts, trees and shrubs in concrete cylinders form intimate areas with seating in the shade. Kienast took advantage of the structural requirement to place plants in concrete cylinders in order to create a special design that was fashionable at the time. For example, the Swiss garden architect Fred Eicher, where Kienast completed an internship in 1969 and 1970, had used such cylinders at the entrances to a housing development designed by Ernst Hiesmayr in the Nussdorf district of Vienna.[38] Ernst Cramer (1898–1980), whose work Kienast greatly admired, also placed tall, slender concrete cylinders in the entrance to his Heuried housing development in the Wiedikon district of Zurich.[39]

The pools for children are not specifically included in Kienast's legend for the program on Sheet 4, but they are in the legend for materials and surfaces. In the project as realized, they were wading pools in which small children could splash about while the people looking after them can sit under broad umbrellas and watch (56, 57). Moreover, the development has large sand playgrounds near the entrance (58). In addition to communal areas, the tenants in the ground-floor units have small gardens they can design themselves; the other apartments have balconies. Such private areas, of 10 to 15 square meters each, were uncommon in developments of multistory residential blocks at the time. Individual garden seating areas with very small gardens were common, as illustrated by an atrium development in Reinach, near Basel, built in 1959 to 1960 and published in the journal *anthos* in 1968: each of its houses, comprising four and a half rooms, had a spacious courtyard garden (atrium) and a private garden.[40] Private gardens in housing developments were still the exception in the 1960s.[41]

In 1983, the Swiss landscape architect Toni Raymann, who had been employed for many years in Stöckli's firm, called for more open space for private use in apartment buildings. As a positive example of this from the early 1970s, he mentioned the development in Niederhasli and also the Sonnhalde development in Adlikon and the Wygarten development in Fällanden.[42]

Niederhasli Today

Kienast's design of the grounds of the development in Niederhasli underwent numerous changes in the design, planning, and implementation phases. Compared to the sketches of ideas and the preliminary designs, the first project drawings considerably reduced the types of playgrounds and displayed a much less differentiated topography. In the presentation plan produced later, the playgrounds were simplified even further; the system and paths and the topography were retained, albeit slightly reduced.

The most striking change was the elimination of the hill landscape that was supposed to connect the Lindenstrasse and Huebwiesen sections of the development. Instead, there was a much smaller hill on the Lindenstrasse side and a plaza with a large pool on the Huebwiesen side. Between the two parts, a hornbeam and oak forest with low border trees was planted.[43] Kienast saw the broad belt of shrubs and trees as another area where children could play. But the reason for the planting was probably a desire to have a spatial separation between the residential blocks, which were under different management. Other playgrounds were also eliminated, such as the construction playground with its unusually "free" program and the sunbathing lawn, the bocce court, and the two playing fields. Kienast commented on this with resignation: "was cut—the children were not allowed to play soccer."[44] The playgrounds were built in a slightly different form, for example a terrace with steps of hewn quarry stone instead of the hill topography. The system of paths, the entrance plazas, and the pools (now without islands) were largely implemented as planned.

The number of trees increased. First, the entire site was surrounded by a boulevard lined with silver lindens (*Tilia tormentosa*) and plane trees (*Platanus acerifolia*). Second, the number of solitary trees and groups of trees on the grounds was increased; they were planted in concrete cylinders on the entrance plazas and at intersections of the paths or directly on the hills (55). They included oaks, maples, hornbeams, and Lombardy poplars. The shrubs visible on the first photographs of the finished Weiden development (primarily in the sand playgrounds) included hazelnuts, cornels, and privets. The bottoms of the low, broad, concrete cylinders were covered with cotoneasters, snowberries, and ivy. The tall, narrow cylinders were filled with hanging plants that "poured out": ivy and jasmine, as far as can be read from the photographs. The planting of the borders of the pools consisted of reeds, grasses, and catnip with blue flowers (56). The edges of some of the paths had shrubs, such as *Astilbe*. The private gardens were lined by hornbeam hedges, their leaves turning brown and dropping off in winter, which gives the hedges a diaphanous look during the colder months.[45]

Even if, in retrospect, Dieter Kienast somewhat ironically distanced the planning from his own wealth of ideas, and his original program was cut by the building cooperative, his concept for the open space was remarkable in several respects. In collaboration with Peter Paul Stöckli, Kienast had conceived a spatial arrangement based on a program aiming to initiate social processes. A lot of attention was paid to creating a good and, in this case, elaborate system of access to the development based on short paths. The paths are relatively broad and are thus suited for children playing, on roller-skates or bicycles, without blocking adults walking on the paths. Thanks to entrances to the buildings being open on both sides and the bike ramps on all of the stairs, children can ride and run through the individual residential blocks.

The generous access, the distribution of playing equipment at regular intervals along the paths, and the hilly landscape that conceals more play spaces, turning the environment into an area to be explored, all encourage children to appropriate the whole terrain as a playground (pp. 172–177). The hilly terrain adds excitement to a development in which all of the blocks look alike, and the changing topography, which is specific to each part of the development, helps to provide orientation within the monotonous complex.

The placement of sand playgrounds and pools near the entrances to the buildings was based on the idea of making it easier to keep

55
Ping-pong tables, sand playgrounds, and concrete cylinders for shrubs and trees were placed near the entrances to the buildings of the Lindenstrasse development. Photo: probably Dieter Kienast, 1975

56–58
Wading pool for small children, with large umbrellas

Seating area between pool and sand playground

Large sand playground near the entrance to the Lindenstrasse development. Photos: ca. 1975

an eye on children. In his "Bemerkungen zum wohnungsnahen Freiraum" (Remarks on Free Spaces Close to Habitations), Kienast advocated designing open space so that mothers preparing food in the apartment can look out the kitchen window to keep an eye on their children playing outside.[46] This was intended to ease their burden, very much in the sense of giving mothers more autonomy, so that they can gradually emancipate themselves from their children. Common areas in front of entrances and playgrounds near the buildings also offer an opportunity for children of different age groups to adapt to one another or for one grown adult to look after several children, and thus relieve other parents. Thanks to the easily navigated system of paths and the hilly topography, older children and teenagers can move through the entire development without obstacles. Kienast wanted to provide them with areas where they could avoid being observed by adults. Given the density of construction and the multistory apartment buildings from which it is possible to observe the entire grounds well, this idea of retreat was probably more wishful thinking than reality. The only thing that could protect teenagers from the eyes of adults was probably the bushes of the belt trees between the sections of the development.

Kienast regarded the spaces for leisure-time recreation created especially for adults as less successful, because they were barely adopted by the users.[47] Neither the fire pits nor the ping-pong tables, and not even the benches, were well received by adults. The small gardens for residences on the ground floor, a very special offer compared to other housing developments of the time, were assessed very differently by the residents, according to Kienast, and sometimes not appreciated at all. Kienast attributed this to the situation of the housing market at the time.[48] The apartments of Niederhasli were, he said, by no means inexpensive; at the same time, the complex had all of the disadvantages of suburban developments (Kienast did not specify what he meant by that). For that reason, he argued, many renters moved to Niederhasli only temporarily, in anticipation of later finding a better location near the city. Kienast regarded this "waiting on call" as one of the reasons why the open space had not been adopted and used as he had planned. Moreover, his original concept had been implemented only in a highly simplified form.

From today's perspective, the development seems above all very child-friendly and decidedly advanced in terms of how the planning addressed the needs of the users and the desire to set social processes in motion. From an aesthetic perspective, the large sand playgrounds with rings and concrete walls, the yellow plastic tubes placed snakingly in the sand, the blue lamps, and the yellow horizontal bars look modern, cheerful, and light, as does the kidney-shaped pool painted bright blue, with red umbrellas and mobile benches with bent steel frames, seats painted red or green, and wooden armrests. The metal play equipment—such as horizontal bars, merry-go-rounds, and slides—distributed throughout the grounds was sometimes designed especially for the site. Swings and wooden climbing fortresses, which often look heavy and unwieldy on playgrounds, are nowhere to be found. The areas with water are well embedded, thanks in part to the shrubs, grasses, and reeds planted along the edges. The use-oriented design in Niederhasli placed value on the harmony of forms and on producing atmospheres by varying the greenery. Such approaches were stepped up even more in Kienast's later works as part of the search for a perfected aesthetic.

The decision to separate the small private gardens in front of the ground-floor units from one another with hornbeam hedges, which from autumn to spring are translucent and hence do not block views entirely, is interesting in that this would become Dieter Kienast's "favorite hedge" in the future: wherever he could, Kienast would replace conifer hedges with hornbeam or beech or plant a strip of these deciduous hedges in front of an existing conifer hedge.[49] The reaction of some tenants in Niederhasli was precisely the reverse: some of them clearly did not like the translucence much, since they replaced the hornbeams with evergreen conifers to provide more privacy.

Cotoneasters, by contrast, were taboo in Kienast's later works, and groundcover such as ivy and snowberries also largely disappeared; for shrubs, he continued to employ cornels often and hazelnut now and again. Otherwise, the use of native, robust plants, as dominates the design of the grounds in Niederhasli, gave way to a selection in which Kienast was explicitly concerned about the specific appearance during the changing seasons. Evergreen conifers soon came to be generally frowned on in his plans. When choosing a tree, for example, Kienast decided based on how its leaves rustle, the color of its flowers in the spring, and the color of its leaves in the autumn, but also based on its symbolic or historical significance in the context of a certain project.

It speaks for the quality and modernity of the design of the grounds in Niederhasli, completed in 1975, that large parts of it survived unchanged until the end of 2006 without becoming a target of vandalism (pp. 172, 174). A visit to the site at the time left the impression, however, that the playground equipment was not regularly maintained, nor were the slowly weathering concrete and plastic elements kept up or replaced when worn out. Grass had overtaken the sand play areas, or they had been covered with bark mulch, and it had probably been a long time since a child had splashed about in the wading pools. The red umbrellas had yellowed; the border plants were overgrown or had disappeared entirely. The belt of trees and shrubs between the two sections of the development had been given a firebreak, and an asphalt path had been added that intersected with the paved plaza for sitting by the pool of the Lindenstrasse section. The original division of the two sections of the development, which looked natural thanks to the greenery, was destroyed by this forceful intervention. The belt of trees and shrubs had been intended as a hideaway for children, and Kienast had wanted it to grow wild to make it even more attractive as a place to play. Instead, it was thinned out, a path cut through it, and a chain-link fence built to separate it.

The slight neglect of the open spaces reflected a social change: these days, the Lindenstrasse-Huebwiesen development is a kind of bedroom community. Leisure-time behaviors have also changed: it seems that no one wants to play jass outdoors anymore. In the 1980s, there was a retreat indoors. The increasing interest in designing one's own apartment minimized social life in exterior spaces. Dieter Kienast, who repeatedly referred to his work as "determined by society,"[50] reacted to this development and, especially in the 1990s, began creating exteriors that took into account society's need for design.[51]

Between 2007 and 2010, as part of a renovation in which the residential blocks of the Lindenstrasse sections of the development were insulated with light and dark pink thermal insulation panels—without regard to their modernist forms—the design of the grounds was also updated in an unfortunate way. The concrete elements of the cylinders for the trees and shrubs at the entrances and the concrete "reading and playing corners for adults" and some of the ping-pong tables were not refurbished. The deciduous trees, the flowering shrubs, and the hanging climbers in the concrete cylinders were replaced by ill-suited

Niederhasli housing development, 1972–75. Photos: Georg Aerni, 2008

Page 172
The gentle hills of the topography help to differentiate the landscapes between the repetitive housing blocks. Playground equipment and seating areas are nestled between them here and there. Thermal insulation panels were installed in the housing blocks in 2007 and 2008.

Page 174
Playground prior to the 2009 renovation

Page 176
Example of the plants near the building entrances: privet hedges, cornel, and hazelnut bushes, as well as the hilly topography with large trees, differentiate the exterior.

plants such as palms and conifers. Several trees were not replanted, so that the staggering of the spaces of the separate parts of the development was lost. On the meadow, next to the completely weathered round concrete seats, new play equipment was installed: small sandboxes, swings, and seesaws on a red artificial surface that was supposed to guarantee a soft fall for children. All of the large sand playgrounds were removed and covered with gentle hills of bark mulch on which cornel bushes grew. The yellow, tubular plastic elements disappeared from the site entirely; the blue lamps were dismounted or surrounded by columns of artificial stones, which look rather out of place in the context of the other materials used: concrete, cement, and plastic.

The hill playgrounds were also given new equipment: clumsy wooden towers and wide swings with wooden supports, which had been deliberately avoided in the original design because they had become standard props. Of the playground equipment specially designed for the site, the small merry-go-rounds were all that was retained, with new seesaw animals added. All of the objects have plastic mats under them. These round, blue islands, which really stand out visually against the green lawn, are also framed with border stones. One cannot describe it as the aesthetically well-considered embedding into the hilly topography as Kienast had intended. The playground equipment on its blue plastic carpets is distributed willy-nilly across the site, just like the square sandboxes.

The idea of giving the users a little freedom to alter the design, in the form of mobile benches that they could take wherever they wanted to sit and spend time, was abandoned. Little by little, the mobile benches gave way to fixed ones. All of these interventions are foreign bodies in a design from the 1970s of which only the rudiments have been preserved. This is particularly glaring around the pools: now there are new grill stations with bulky benches and tables; a single, oversized umbrella provides the only shade.

It is regrettable that the quality of the design of the grounds in Niederhasli was not recognized and preserved or judiciously adapted to today's needs. Updating it with renovated concrete elements, an appropriate replacement of missing or unsuitable plants, and thoughtful additions of new playground equipment could have prevented the current unsightly mishmash of styles. It is understandable that the weathered concrete and plastic parts seemed unattractive and old-fashioned to the building cooperatives and the residents. But there was no coherent concept for their renovation. Clearly, those responsible had not given any thought to the spatial or design qualities of the existing site. That is evident not least from the efforts of the residents to demarcate and set up separate spaces by adding benches, tables, and grill stations. Kienast's elaborate zoning of areas close to and areas far from the building in keeping with the needs and demands of different age groups has been undone by the later redesign. The quality of the original design has been made truly evident by comparing it to the development's outdoor spaces now. For playing children in particular, the present site is much less attractive. Leaving the larger concepts aside, even considerations like planting trees in the parking lots so children can use them as a place to play while remaining comfortable in the shade are passé these days. From the perspective of those who grew up in a housing development like the one in Niederhasli, and who spend their days there, it would pay to study the concepts of the 1970s, to update them, and to refine progressive approaches to designing the grounds.

Swiss Housing Developments
during the Boom Years

The programmatically motivated, comprehensive topographic design of the terrain between residential buildings repeated in series, as demonstrated by Dieter Kienast for the housing development in Niederhasli, was widespread in the Swiss-German landscape architecture of the late 1960s and early 1970s. One pragmatic reason for this was that removing the soil excavated for construction sites on such a scale was very expensive, and so the soil was instead used on site.[52] Apart from that, however, modeling the terrain in combination with the use of plants, the system of paths, and the use of water as a design element are basic components of landscape architecture. Elevations and depressions in the terrain make it possible to define separate spaces, which as a rule are further distinguished by planting trees and shrubs. This modifies the perception of the space: for example, subtle topographical modeling can make an open space between two residential blocks look more spacious than if it simply extended out flat.

59
Willi Neukom, design of the grounds for a retirement and residential development in Dübendorf, 1966. Photos: Fritz Maurer
60
Ernst Cramer, design of the grounds of the Heuried housing development, 1969–74. Photo: Erwin Küenzi
61
Eduard Neuenschwander, design of the grounds of the Friedau housing development in Aadorf, 1971–74. Photo: Heinrich Helfenstein, 2008

The modeling of the terrain occupies an important place in the discourse on the significance of landscape design in improving the mass construction of housing. In fact, housing construction using prefabricated parts became the central task of the booming building industry in the 1960s and 1970s. The larger the serially erected developments, the cheaper they could be produced and built in general. The landscape architects hired to design the grounds of such housing developments saw their task as an opportunity to establish themselves as experts in the design of the other areas. They wanted to counter the "endless monotony" of repeating residential blocks with "improved surroundings made livable" and to offer residents "the wealth of experience that such construction projects otherwise lack in terms of the buildings," as the landscape architect Christian Stern (b. 1935) expressed it in 1968, as a representative for his professional colleagues, in a programmatic article titled "Die Bedeutung und Entwicklung der Freiraumgestaltung im industriellen Wohnungsbau" (The Significance and Evolution of the Design of Open Spaces in Industrial Dwelling Construction).[53]

The spaces modeled with excavated soil were intended to create a varied landscape between always identical residential blocks. "Diversity," "orientation," and "individuality" were the slogans of the landscape architects of the period. The individual solutions, which depended on the originality of the designers and on the amount of the budget, did not differ very much from one another because the basic conditions of construction were similar, for example, the staggering of the residential buildings or the fact that tall mounds of soil could not be placed over underground garages for structural reasons. Whereas the areas near the buildings were flat and usually paved, the hilly landscape was, as a rule, placed further from the buildings and moved out toward the edges of the development—in part to block street noise and provide privacy.

Typical examples of the period include the grounds, completed in 1966, for the retirement and residential development in Dübendorf by Willi Neukom (1917–1983) (59), and Ernst Cramer's design of the grounds for the In Surinam development in Basel, realized from 1968 to 1970.[54] The developments of the 1960s that featured somewhat more restrained modeling of the terrain included Volketswil (Zurich), Müllerwis (Greifensee), and Adlikon-Regensdorf, all of whose grounds were designed by Christian Stern. Examples contemporary with Kienast's grounds in Niederhasli include Ernst Cramer's design of the grounds for the Heuried housing development (1969–74) (60) and the grounds for the Sonnenberg (1970–74) and Friedau (1971–74) developments, designed by Eduard Neuenschwander (1924–2013) (61).[55]

In his monograph on Ernst Cramer, Udo Weilacher presented Cramer's landscape designs under the heading "Sculptural garden zones"[56] and situated them within a history of sculptural works, the most prominent of which was the Garten des Poeten (Poet's Garden) for the G 59 horticultural exhibition in Zurich. In their book on Neuenschwander, Claudia Moll and Axel Simon praise Cramer for a style of design that had "little in common with the setback greenery of many comparable developments of the time"[57] and emphasized his skill. If one compares the central works of these years,[58] it becomes clear that Cramer, Neukom, Neuenschwander, Stern, and other Swiss-German colleagues were indebted to related design principles.

With his first more significant, autonomous landscape planning project, Kienast joined their ranks but also developed his own conceptual profile. For example, the development in Niederhasli scarcely differed at all from other landscape designs of the 1960s and 1970s in terms of the selection of materials. The paving was made of stones of concrete, cement, or cement composites. Playgrounds and smaller walls were made of prefabricated concrete elements; and there was a generel reliance on industrially produced materials.[59] In that sense, Niederhasli certainly conforms to the design trends of Swiss landscape architecture at the time. Kienast's personal priorities are evident above all in the elaborate zoning of spaces for different user groups, the detailed program for use, and the enormous importance that he attributed to children's play.

In terms of the structural subdivision of spaces, all of the developments mentioned followed the same principles:[60] hilly modeling and curving paths to produce a particular perception of the space as one moved through the development; no steps in the entrances or only if combined with ramps, so that residents could ride bicycles through the entire development; areas for small children to play near the entrances; the integration of playgrounds or flat areas of grass for ball games into a hilly landscape somewhat removed from the residential buildings;[61] the use of trees and shrubs to demarcate spaces for different purposes; the use of plants to create intimate spaces that cannot be seen from the residential buildings;[62] and distinguishing the developments from their surroundings by placing a hilly topography on their borders and/or planting a belt of trees around the development.

Children's Play as a Driving Force behind the Emancipatory Planning of Open Spaces

No other Swiss-German landscape architect of the 1960s and 1970s worked with self-designed utilization programs, to be realized in spatial formulations, in as intense and concentrated a way as Dieter Kienast. He paid particular attention to children's play. His open spaces for children, such as the unrealized playing field for ball games in Niederhasli, were clearly inspired by the *skrammellegepladsen* (adventure playgrounds) of the Danish landscape architect Carl Theodor Sørensen (1893–1979), who developed the idea in 1931 and was the first to implement it.[63] From the mid-1950s onward, this playground concept became popular in Switzerland as well under the name *Robinsonspielplatz* (Swiss Family Robinson playground), albeit deviating from the original idea of making as much material available to the children as possible so that they could work and develop individually: wood, soil, and plants, but also iron, dishes, old machines, and junk cars.[64]

Kienast also called for playgrounds offering a spectrum of materials to challenge the creativity of children. The grounds of a residential complex should have enough areas where children can transform everything: "beaten paths, holes, dirt piles, wild vegetation in open spaces . . . have to become as natural as the weekly mowing of the grass is today," he wrote in an essay from 1980 with the utopian title, borrowed from Maxim Gorky, "Ein Spiel ist der Weg der Kinder zur Erkenntnis der Welt, in der sie leben, und die zu verändern sie berufen sind" (A Game Is the Way that Children Get to Know the World in which They Live and which They Are Called Upon to Change)."[65] Kienast explicitly mentioned that screened-off areas, like the belt of bushes in Niederhasli, make the ideal sites for children to play, and they should be allowed to grow wild:

> In keeping with the location and its use, a vegetation develops here that probably contradicts the ideals of beauty shared by the residents up to now, but it sends a clear signal to children: this area belongs to no one, is maintained by no one, and here we will not be disturbed.[66]

Well-maintained parts of the development, with expensive plantings such as beds of shrubs, signal conversely for Kienast that "Here we have to behave carefully, here nothing can be changed." Expansive sand playgrounds, with ball fences, screened-off grass fields, and areas growing wild, by contrast, demonstrate to children that they are somewhat "released" from the control of adults. One special element that Kienast takes up in his text, following Lucius Burckhardt and Karl Heinrich Hülbusch, is the proposal to integrate "no-man's-lands" on the edges of developments—dysfunctional places that are not designed at all, with which children can do what they want.[67]

By working out his concepts of use in detail, Kienast arrived at an insight that the landscape planner's personal will to design has to be subordinated to the needs of children to ensure their self-realization in play. He regarded traditional children's playgrounds as visible signs of a failure to meet the needs of children, because areas had been reserved for children to play with the idea of banning them from other areas as much as possible.[68] In 1980, this attitude led him to call for a design to establish a rough spatial-topographical framework, that is to say, the connecting paths and places to encounter others. In his mind, it should be possible for all of the places and areas in between to be reinterpreted according to the needs of users:

> It is not therefore our brief to build everlasting monuments as a sign of our virtuosity in design. Rather, we have to create open spaces in which the essential components, such as trees, connecting paths, plazas, small gardens, are established but, more than that, also permit every change: the composition of residents changes over the years and demands that the open spaces be capable of evolving. Only in this way can they meet the changing needs of the residents over the long term and permit new interpretations. Perhaps one day a

development will have no infants, bicycles are being repaired or soapbox cars built in front of the buildings; perhaps, because there are no cars, part of the parking lot can be used for roller-skating. This sort of growth and maturing of open space—along with its residents—makes it unnecessary to renovate open space, as is done today, especially in backyards.[69]

In the texts that Kienast wrote over the course of the 1980s, he continually speaks of making open spaces available for different forms of use. Yet Kienast's understanding of use had changed fundamentally. Whereas until then aspects of coping with daily life and active self-realization shaped his designs, increasingly the emphasis was placed on a more contemplative human development: atmosphere and aesthetics became more important, and finding consolation in beauty and the experience of nature implicitly meant for Kienast a new way of better coping with daily life through the use of open spaces. Kienast's idea of use, of dealing with daily life, and of what is necessary to do so shifted between the 1970s and 1990s, but human beings and their needs remained the center of his reflections.

Form as the Antithesis of the Natural Garden: The Brühlwiese Municipal Park in Wettingen

On the occasion of a symposium in 2008 on the relevance of Dieter Kienast's work, the architect and filmmaker Marc Schwarz and the landscape historian Annemarie Bucher made the documentary film *D. K.: Eine Spurensuche* (A Search for Clues)[70] In it, Kienast's colleagues spoke, including the Swiss landscape architect Rainer Zulauf (b. 1953), among others, who explained why the Brühlwiese Municipal Park in Wettingen was a public nuisance for many residents—and a powerful impulse for a new generation of landscape architects, at once vexing and inspiring (62). According to Zulauf, it was simply necessary to recall what was going on in Switzerland in the early 1980s: the natural garden movement, influenced by the biology teacher Urs Schwarz and the Solothurn School, had become the dominant movement in garden architecture.[71] Kienast's redesign of the park was diametrically opposed to their ideals of "naturalness." Accordingly, the uncertainty of his professional colleagues was intense:

> And then suddenly this park took shape there! At the time I was commuting between Rapperswil and Baden and had a look at it and thought only: Impossible! And, despite that, precisely this fractured relationship was fascinating: someone makes a path with three rows of trees; someone comes and propagates a grove, a forest, in a park! Someone makes two pyramids as part of the architecture of a garden! That was something unimaginable at the time, and we didn't recognize that all the components of this reservoir were our history and our garden culture. That was not perceived,

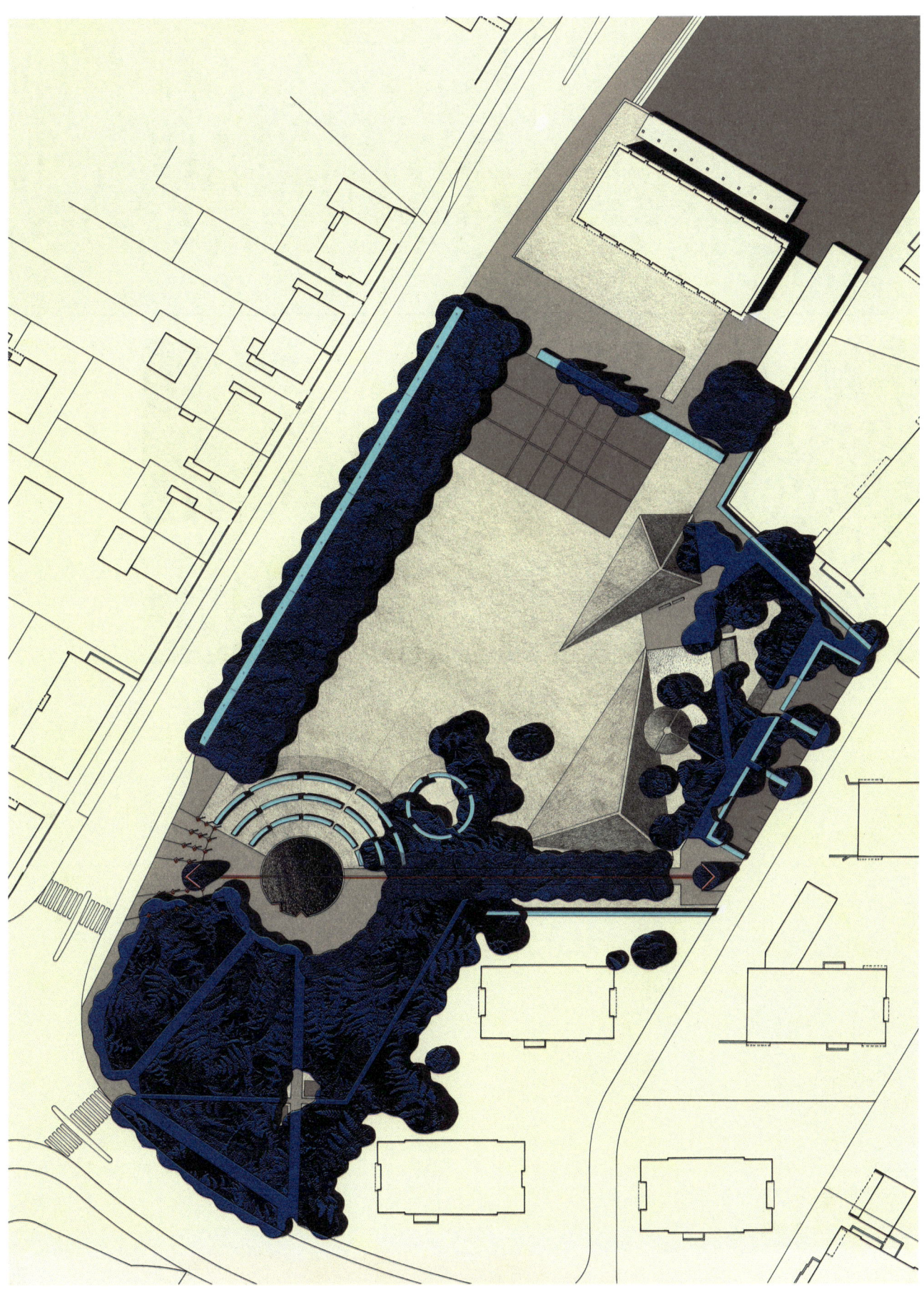

62
Brühlwiese Municipal Park in Wettingen, revised presentation plan of 1991, scale: 1:200, H 83.5 × W 121 cm, pencil and colored film on copy of plan, undated, firm of Stöckli, Kienast & Koeppel

63
Pool in the L. Garden in Riehen, 1995. Photo: Christian Vogt, ca. 1996

64
The "ground wave" at the Allenmoos Outdoor Swimming Pool in Zurich, 1995–99. Photo: Johannes Stoffler, 2004

because we were all involved in our own thing: we designed and made and regarded designing more as an easy thing. And then someone comes and makes such a fragmentary thing, like a kind of collection of props from the history of the garden and puts it together and says: That is a park!⁷²

Of the elements listed by Zulauf, it was above all the earth pyramids, with their unusual formal rigor (p. 200), that caused feelings to run high among the residents of Wettingen. Peter Paul Stöckli saw the pyramids as a logical consequence of Kienast's intense preoccupation with form as such and with the topographical relief of designed grounds.⁷³ Kienast's effort to come to terms with the relief can be followed in numerous works: in very delicate topographic forms, so-called microtopographies, and in decidedly powerful ones. In order to make slight gradients visible, Kienast repeatedly used pools of water. In the L. Garden in Riehen, for example, he inserted a shallow, extended pool into the ground, its edge marked with a steel plate and projecting upward (63). As a result of this austere, straight form, the gentle slope of the terrain along the pool up to the border of the garden can be experienced. At the same time, Kienast loved to design with striking forms of earth, including terraces, as in the Dry Grassland Biotope in Basel (p. 70) or the L. & L. Garden in Zurich (p. 118); pyramids, as in Wettingen, the Bad Münder Spa Gardens, and the Berggarten (Mountain Garden) at the Internationale Gartenschau (International Horticulture Show) in Styria, Austria, in 2000; or "ground waves," as at the Freibad Allenmoos (Allenmoos Outdoor Swimming Pool) in Zurich (64) and at the Expo 2000 in Hanover.

This interest in designing with reliefs was heavily influenced by the conception of the Dry Grassland Biotope in Basel, which in terms of its visual impression was much closer to the aesthetic of the Naturgarten (natural garden) than its apparent antipode, the Brühlwiese Municipal Park in Wettingen, also known as the Brühlpark. According to Jürg Altherr, who was enthusiastic about the Dry Grassland Biotope in Basel, Kienast distinguished himself from other landscape designers who were part of the natural garden movement precisely thanks to this interest in relief and concise forms.⁷⁴ In contrast to those of his colleagues who left nature "alone," Altherr argued, Kienast from the outset subordinated this act of leaving alone to a form that made observers conscious of it in the first place. Indeed, the form underscored dynamic, natural growth and thus made it possible to experience the connection between the human will to form and the power of nature.

In orientation and layout, neither the Dry Grassland Biotope nor the Brühlwiese Municipal Park has anything to do with the natural garden movement. The two works were developed at nearly the same time. Kienast worked on the Dry Grassland Biotope from 1976 onward; the drawings and the lists of plants date from 1978, and the site opened in 1980. The conversion of the Brühlwiese into a municipal park had been under discussion since 1975; in 1979, the Municipality of Wettingen commissioned Peter Paul Stöckli's firm for a preliminary project, and Dieter Kienast handled its planning from 1979 to 1982, in collaboration with Stöckli and Walter Vetsch; the earth pyramids were added to the project in the spring of 1982, and the park opened on June 30, 1984.

Reading Kienast's description of the earth pyramids in the Brühlpark, it is clear that he was repeating the principle of the Dry Grassland Biotope to present an intertwining of the natural and the artificial, but reversing it: his decision to place the earth hills in the municipal park was a reaction to the desire of the residents of Wettingen to have slopes for sledding.⁷⁵ With their pyramidal form, he was alluding to the ridge of the Lägern, the foothills of the Jura Mountains, which rose up behind the town (72). The users of the park, he said, could now perceive the Lägern as a parallel to their "sledding hills." The "geometrically shaped mounds of soil," as Kienast preferred to call them, were planted with a calcareous meadow:

> The hills thus become 'readable' on different levels. The geometrical shape makes the artifact clear; the nearly natural vegetation overlays the artificially designed hill and demonstrates that nearly natural vegetation does not depend on naturalizing topography.⁷⁶

This gesture suggested by the earth pyramids of relating use, form, and vegetation to one another in such a way that a tension results was characteristic of Kienast's oeuvre for an extended period—and at its core it never really disappeared again entirely. Kienast preferred to develop his projects from contrasts. The design process of the municipal park and the changes made between 1979 and 1982 document his growing interest in the principles and methods of postmodern design as well as in collage, working with the fragmentary, and quoting from the history of the garden. Moreover, over the course of designing and building the municipal park in Wettingen, Kienast arrived at a synthesis in his approach to design that would play a crucial role in his subsequent designs: garden and landscape architecture has to cover ecological, social, and aesthetic aspects, and the landscape architect has to see to it that all three of these aspects are introduced.[77]

A New Park for
a New Center

The project for the Brühlwiese Municipal Park began when the Municipality of Wettingen altered its approach to its green spaces. The population of the town on the Limmat River near Baden had been growing since the 1950s and reached the scale of a city. In 1959, the new town hall opened; by 1970, a new center had grown up around it. In 1975, the district school southwest of the town hall, dating from 1955, was expanded; three residential towers were built to the west of the school and the town hall, which are still the tallest buildings in Wettingen. A market has been held weekly since 1960 on the paved plaza in front of the main entrance to the town hall. Stretching out behind it is the only large green space in the center of Wettingen: the so-called Brühlwiese. By the end of the 1970s, it was being used as a soccer field; already since the 1960s, there had been efforts to turn this green space into a park for the community.

The first project proposed to the town council in 1975 was by the garden architect Albert Zulauf of Baden, in whose firm Dieter Kienast and Peter Paul Stöckli had met while interning in 1967. But the loan required to implement Zulauf's proposal was not granted.[78] In January 1976, the town council applied to the residents' council for permission and financing for a municipal planning authority project to convert the soccer field into the "Gemeindepark Brühl" (Brühl Municipal Park) (66).[79] The plan was to convert the red-clay-soil plaza in front of the town hall into a hard court, to slightly alter the small playground north of the soccer field that had evolved, to divide the soccer field into a "network of cement-compound paths for strolling and walking" measuring 2 meters across, and to plan three groups of three trees each: a copper beech, a plane tree, and a *Tsuga* (hemlock) conifer.[80]

Although the residents' council approved the application, loud protests soon followed. Critics regarded both the plan submitted and the project description as sound evidence of the municipality's amateurish and uncoordinated policy on green spaces, in which the planning department had been granted authority to which it was not entitled. "A park does not result from planting nine trees and building a path through a meadow!" proclaimed an anonymous text that fiercely attacked the town council and the residents' council already in January.[81] This six-page protest letter documented with texts and photographs a series of sins committed by the planning authority regarding the design of open spaces and criticized the lack of an overall concept for green sites and sports facilities in Wettingen. As a consequence of this deplorable state of affairs, it concluded by calling for the Brühlpark project to be sent back to the town council for it to work out a guideline for the planning of green urban areas and sports facilities as well as a green pedestrian axis through the entire municipality—with the recommendation that "a proven BSG garden architect/green planner" be consulted.[82] These demands got a hearing from members of the residents' council. They submitted to the planning authority of the Municipality of Wettingen an application for the reversal of the planning commission and called for a discussion of the principles of urban planning with regard to the design of the city center.[83] In response, the residents' council decided to have urban planning studies done for the Center and Brühlwiese areas.[84]

The actual commission was not awarded until 1979, to the Baden architects Burkard, Meyer, Steiger whom the landscape architecture firm Peter Paul Stöckli had hired as consultants on the Brühlwiese.[85] As part of its urban planning study, Burkard, Meyer, Steiger worked out three variations for the redesign of the Center and Brühlwiese areas, including a "Green Variation" (67). Parallel to this, Stöckli, working in concert with Kienast, developed two variations of their own, labeled "A" and "B." Variation A (65) was integrated into the "Green Variation" of Burkard, Meyer, Steiger, which shows the town hall with its forecourt, a new park, the municipal school, and the three residential towers. At first glance, the concept can be described

186

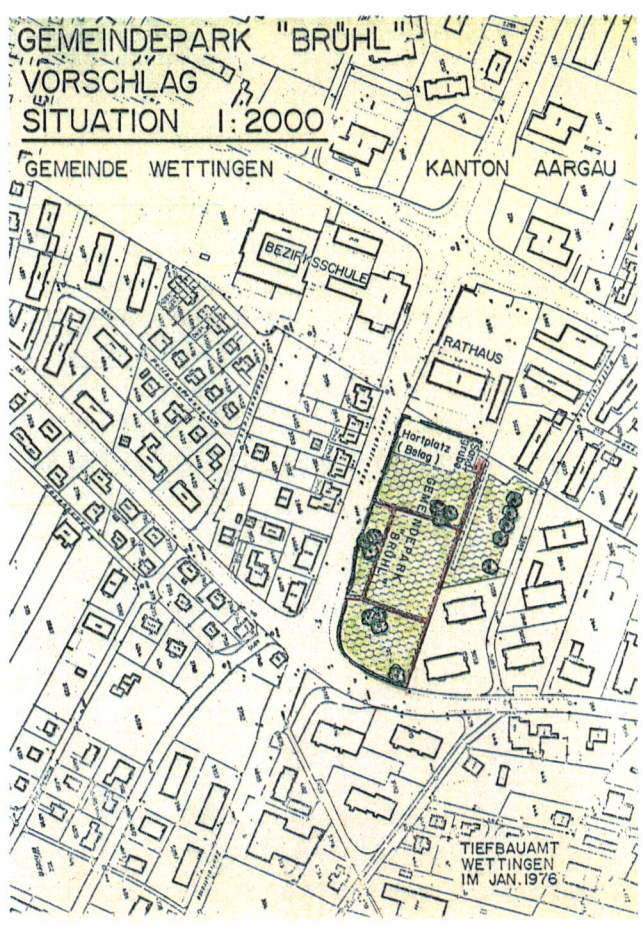

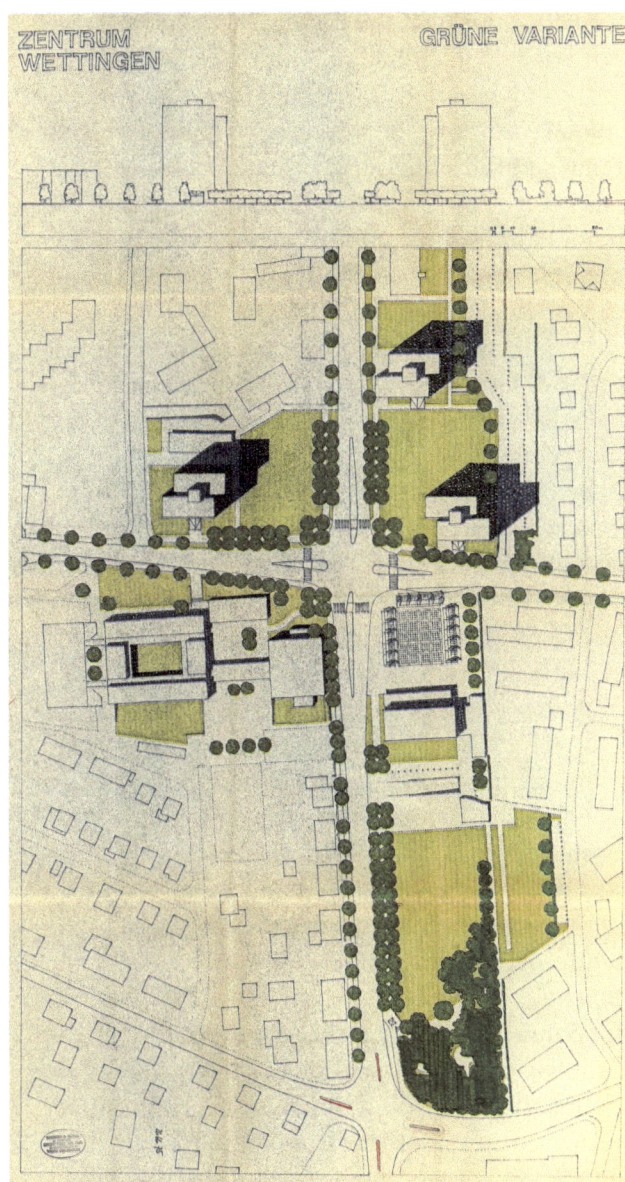

65
Brühlpark Wettingen, site plan Variation A, plan no. 316-1, scale: 1:500, H 52 × W 96 cm, copy of plan, undated, firm of Peter Paul Stöckli

66
Brühl Municipal Park, site plan, scale: 1:2,000, 1976, Municipal Planning Authority of Wettingen

67
"Center of Wettingen: Green Variation," urban development plan, undated, Burkard, Meyer, Steiger + Partner, Baden

as what critics of such layouts at the time called "buffer greenery" (67): Lawns extend around the residential towers and the school building; boulevards with one and two rows of trees shield the pedestrians from the streets. The trees are arranged in number, location, and pruning so that they contribute to a differentiated spatial division of the center, which points to the influence of landscape architects and their specific skills.

The section of the plan showing the new part illustrates Stöckli and Kienast's concept for the redesign: The spacious lawn of the disused soccer field is retained and lined with boulevards on both sides. Between the lawn and Zentralstrasse to the south, Stöckli and Kienast planned a boulevard with two rows of trees, and between the lawn and the small playground to the north one with a single row. At the eastern border of the site, they planned to plant a small forest to mark off the playing field from Bahnhofstrasse, which had previously only been separated by a hedge. The small forest provides a spatial counterweight to the town hall, to which it relates at the same time: a semicircular clearing has been cut into the forest along the axis of the town hall, in the center of which the landscape architects placed a solitary tree—precisely in the central axis of the town hall.

The spatial separation of lawn and playground determined by their use was reinforced by Stöckli and Kienast with a boulevard. The playground itself is not further detailed on the plan; it is closed off to the north on Ryffelstrasse by a row of trees. One essential urban planning component is the proposal formulated in the plan to introduce an east–west diagonal to connect Zentralstrasse and Ryffelstrasse, oriented around an existing housing development, with buildings located diagonally to the lawn.

The "Green Variation" of the various urban planning proposals had a backstory: in contrast to Niederhasli, where the garden architects were brought in only afterward, the responsible parties in Wettingen took the ambitious approach of involving the landscape architects in the urban planning measures from the outset. After the first advisory meeting between the architects and the landscape architects in January, the Stöckli firm was commissioned by the town council on February 15, 1979, to work out a preliminary project for the new park, including an estimate of costs.[87] The first studies were presented in the Wettingen town hall in March.[88] By September, Kienast, who was working with Stöckli, had developed two variations, which were discussed with the architect Adrian Meyer, who had been entrusted with the design study for the Center area, on September 12.[89] Their reflections on the downtown area and the new park occurred in parallel and in constant exchange. Nevertheless, there are differences between the "Green Variation" of the architects and the solutions of the landscape architects.

Variation A (65) corresponds in rough outlines to the parts sketched on the general urban development plan. A "small urban forest with footpaths and beaten paths," as it is labeled on the plan, forms the eastern termination of the city. Kienast redesigned the clearing with the solitary tree placed in the axis of the town hall as a plaza with a circular pool, which is framed to the west by steps, like an arena or the orchestra of an ancient theater.[90] He reinforced the diagonally connecting path as a design element by lining it with two rows of "tall trees with small tops." He broadened the already planned boulevard to the south with two rows of trees, as a termination that provides protection from the heavy traffic on Zentralstrasse, into a boulevard with three rows of "tall trees with large tops."

Kienast retained the open lawn as a "playing field" and the central element of the new park. But he found different solutions for the other areas. In front of the town hall, he arranged three plazas of varying use: a rectangular plaza with a strict grid of planted trees, which transitions into a rectangular plaza labeled as a "dry plaza," lined with trees and open to the sky; a third area serves as a parking lot, with a hedge-lined entrance for a future underground garage for the municipality. The plaza with the grid of trees is directly linked by a path to a children's play area, which Kienast placed to the north in a second small forest to be planted.

Variation B is organized spatially in a similar way, but only a few formal components are retained. The striking difference from Variation A raises the question of whether Kienast and Stöckli intended this "rural" variation, which corresponded to the planning practices of the time, as a way of ensuring they would get the commission even if their bolder proposal was rejected. Following a presentation of both variations at the meeting of the town council on June 7, 1979, and a consultation with the architect Adrian Meyer, Stöckli and Kienast submitted a further refined version of Variation A as their preliminary project on September 26, 1979.[91] The town council submitted the proposed urban planning solutions of Burkard, Meyer, Steiger and Stöckli/Kienast to the residents' council for a decision at a meeting of the municipal parliament on October 25.[92] On December 13, 1979, the loan for the first stage of Brühlpark, based on the Stöckli/Kienast plans of the preliminary project, was approved by the residents' council.[93]

On the site plan and the comparable section of the preliminary project (68, 69), the name "Stadtpark Brühlwiese" appears for the first time; the "municipal park" of the title was more than just a phrase—it was a program. In his handwritten notes on Variation A, Kienast argued that a municipal park was not a miniature countryside but rather had to have an urban look.[94] It should not be one-sidedly functional and not be an "overly gardened park." Its design had to be spacious, getting by with just a few elements and based largely on articulation by means of trees and boulevards. This approach recalls the spacious and clearly structured plans of Kienast's teacher Fred Eicher[95] and those of his role model Leberecht Migge (1881–1935). In connection with use, in a report to the municipality in 1979, Kienast called for an "open design grid"—that is to say, for arranging the areas without assigning a specific use to them.[96] This was intended to encourage overlapping uses by different groups of residents and, with an eye to the long term, make it possible to reinterpret the areas and their use.

Another striking feature concerns the presentation of the site plan: in the preliminary project, it is no longer Zentralstrasse that is used as a line of orientation for the plan but rather the buildings of the housing development in the northeast, which are diagonal to Zentralstrasse. The diagonal of the connecting path through the park thus becomes a straight line and hence a kind of backbone for the grounds as a whole. The lawn, the streets, and the town hall look slightly tipped on the plan. The spatial organization looks clear: the axis with the boulevard lined by trees with small tops and the pool become a separating element between the small urban forest in the east and the open lawn in the center of the park.

Kienast commented on the individual components of the park and their possible use: "large field ... for ball games, but not marked as a soccer field"; "dry plaza ..., partially covered with a grid of trees, partially sunny; rollerskating plaza, exhibitions, festivals"; "parking lot ... left clear at the entrance, conceived as a possible expansion of the dry plazas during festivities"; "children's play area ... not a miniplaza for small children but a generously laid-out area; large sandy areas, hills for sledding ...; equipment for moving, role play, and creative play"; "boulevard along Zentralstrasse"; "urban promenade with wide sidewalk (6 m); hedges on the street side and triple-tree-row boulevard"; and "small urban forest ... with trees, clearings, paths, and seating."[97] The little forest opens up onto a semicircular plaza with seating, and in its center a pool with a diameter of 20 meters.

The sledding hills were incorporated into the preliminary project at the suggestion of the residents' council. They were to be formed from soil excavated to build the planned underground garage and on the site plan are indicated by contour lines.[98] The aesthetically radical form of earth pyramids was first incorporated into the project between March and June 1982, after receiving approval from the Municipality of Wettingen in December 1979. The construction of the earth hills was part of the second stage of the project, for which the firm, now operating under the name Stöckli + Kienast, submitted a second estimate for its implementation on March 8, 1982,[99] enclosing the same plan that accompanied the submission in 1979.[100] This application, too, was approved. Just nine days later, on March 17, 1982, the firm drew up a site plan with sections, presumably based on sketches and preliminary drawings by Kienast that have not survived,[101] showing the new municipal park in forms that perplexed both many of his colleagues and many of the residents of Wettingen (70, 71). The plan and the three sections of the project were submitted three months later with the application to build the "pyramidal hills," which was approved on August 5.[102] Already by the autumn, however, there were questions about and objections to the new design of the park.

From Design to Use:
Strategies for Composition
and Effect

In contrast to the preliminary project, Kienast developed a formal language pointedly focused on contrasts during the construction of the Brühlpark, especially in the areas of the earth pyramids, the diagonal, the pool, and the urban forest. The earth pyramids form the boundary between the spacious, rectangular grass playing field and the

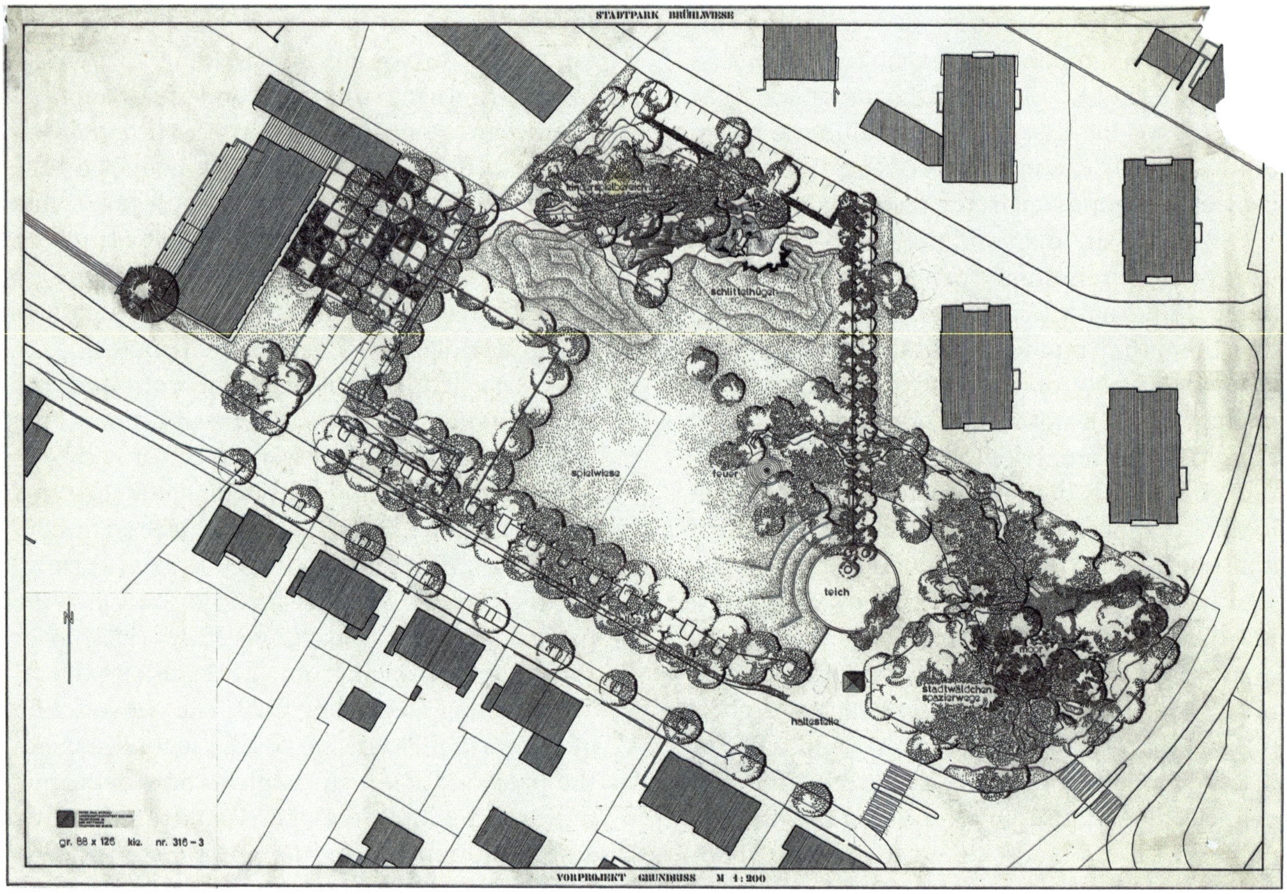

68
Preliminary project for the Brühlwiese Municipal Park, site plan no. 316-3, scale: 1:200, H 86 × W 126 cm, pencil on tracing paper, undated (1979), signed "Dieter Kienast," firm of Peter Paul Stöckli

69
Section A–A, plan no. 316-4, scale: 1:100, H 30 × W 115 cm, pencil on tracing paper, October 23, 1979, signed "Kie, CM," firm of Peter Paul Stöckli

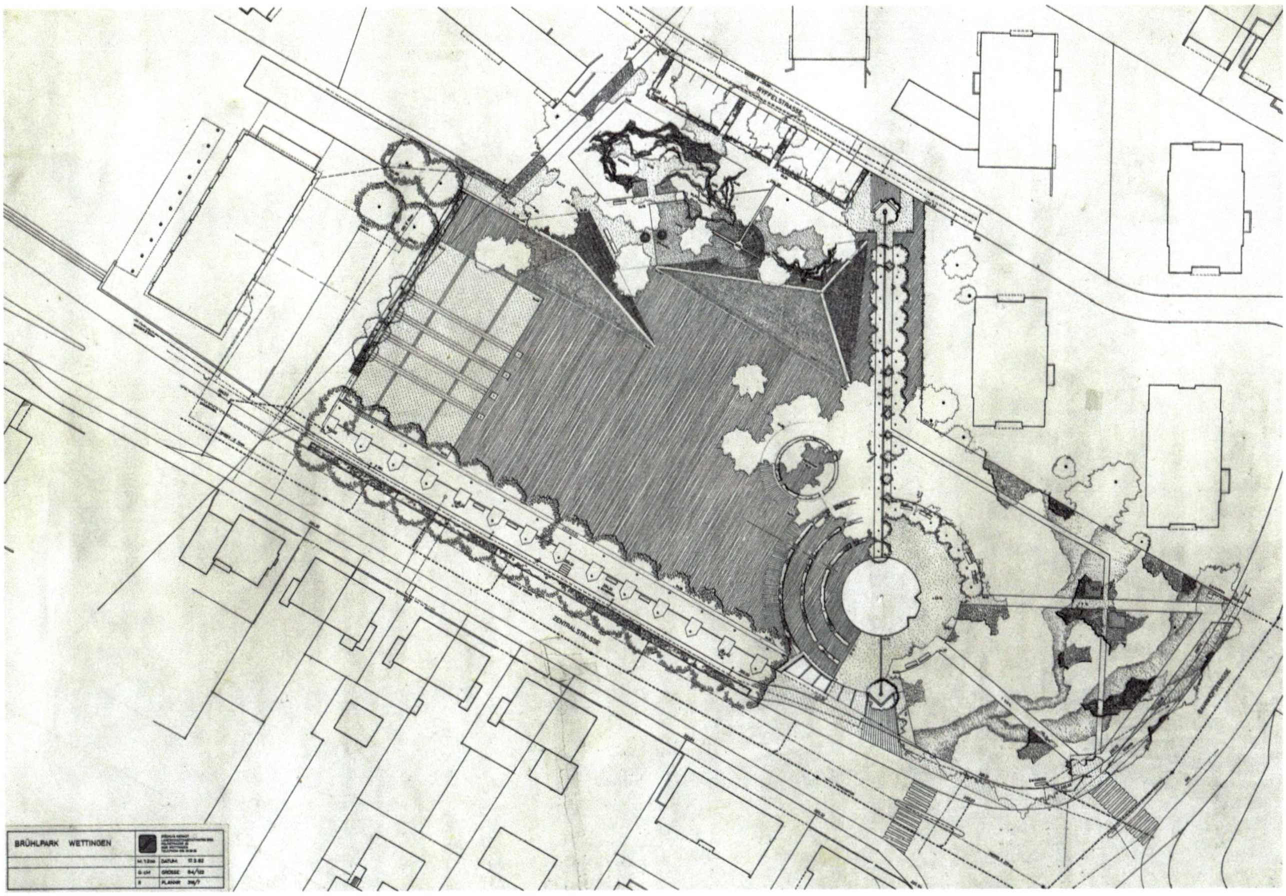

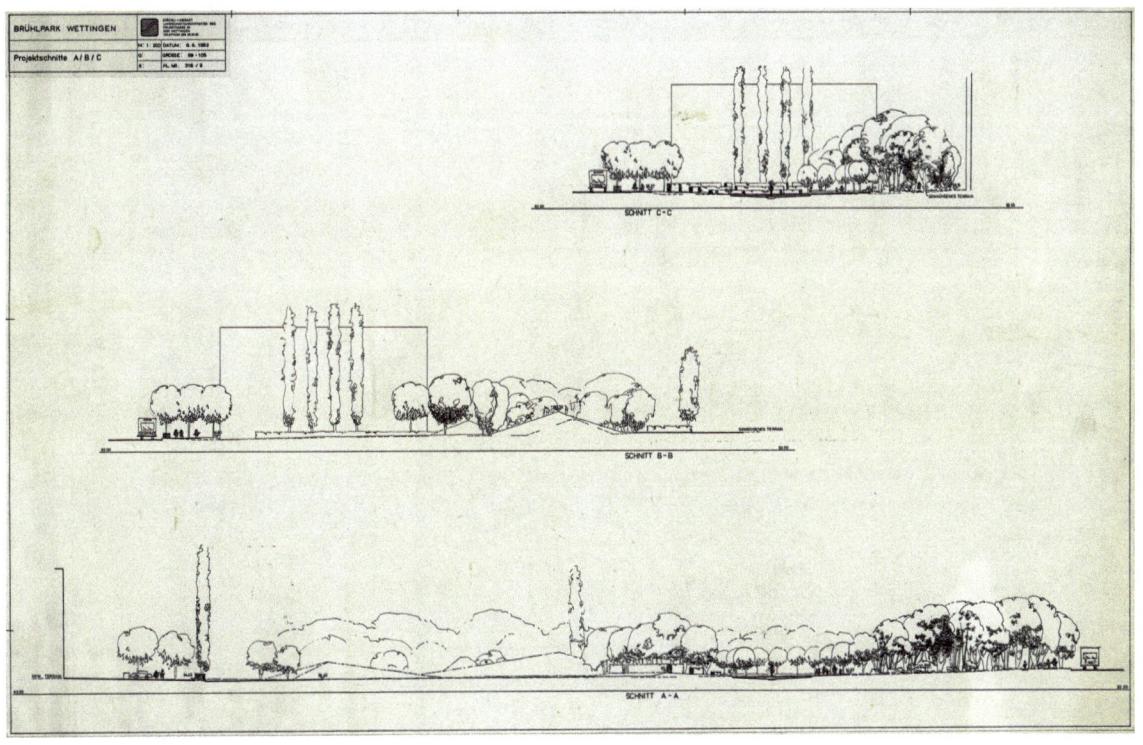

70
Brühlwiese Municipal Park,
site plan no. 316-8, scale: 1:200,
March 17, 1982, signed "CM,"
firm of Stöckli + Kienast

71
Sections A–A, B–B, and C–C,
plan no. 316-9, scale: 1:200,
June 8, 1982, signed "CM,"
firm of Stöckli + Kienast

72
The rectangular grass playing field—the central element of the park—is bounded to the northeast from the playground by the earth pyramids of the sledding hills. Kienast attributed their appeal in part to the way their form seems to echo the Lägern mountain chain, stretching out as a backdrop to the park and following the Limmat Valley to the northeast. Photo: Christian Vogt, 1991

73
Entrance in the southeast. The newly planted urban forest frames a semicircular plaza with a circular pool. A hedge formation completes the circle on the opposite side. The pool lies in the axis of the diagonal path connection of Zentralstrasse and Ryffelstrasse. The Lombardy poplars and one of the little walls formed into a triangle mark the southeastern end point of the diagonal path. They are surrounded on the left and right by steel trellises overgrown with Virginia creeper and clematis. A bus stop is located in front of it. Photo: Christian Vogt, ca. 1991

playground (72). Together, they create a kind of miniature massif, with a backdrop formed by the Lägern mountain chain stretching out and following the Limmat Valley in the northeast. The flat grass playing field also seems like a projection screen for the shadows cast by the trees along the boulevard to the southwest (p. 200). This area already reveals how the form and function of the grass playing field and of the sledding pyramids serve, first, their use, and, second, the perception of natural circumstances, such as a cast shadow or a surrounding landscape, by heightening a staged aesthetic experience.

In the eastern section of the park, Kienast expanded the diagonal between Zentralstrasse and Ryffelstrasse, which had originally been introduced to improve access, into an axis route; its course can be read from, among other things, the materials inserted into the ground (73, p. 208). At each of the two entrances, a small wall in the form of a triangle opening into the park was erected and a Lombardy poplar placed in the space surrounded by them. The two looming trees now mark the ends of the axis. Several rows of paving stones were placed from each entrance wall to the street, thus visually emphasizing the ends of the axis. The spontaneous vegetation preferred by Kienast has populated the cracks.

Three strips of paving stones run the full length of the axis toward the park. Where the axis and the pool intersect, we encounter an unusual solution to a detail: the paving stones do not stop in front of the concrete edge of the pool but run over the edge straight through the pool and continue over the opposite edge. This marking of the axis with very visible material is clear testimony to Kienast's will to make his works "legible." In retrospect, however, Kienast found this excessively clear reference to the axis superfluous:

> For a time, I found this strip to be incredibly important to demarcate the entire width of the park. Today, this need seems mysterious to me, since it merely conveys the illusion of logic. The insistence that people absolutely have to notice certain things is no longer contemporary today.[103]

The pool located in the axis is round and at one point shaped like the stem of an abstract water lily leaf. This "stem" was originally emphasized by a small wall that has since vanished (p. 206). In the preliminary project, Kienast had intended an area planted with reeds, which would grow rampant into the pool and seem to dissolve its edge (68). The entire area around the pool is characterized by the collision of geometric forms with freely growing plants, as the plans show: the urban forest extends out to the east; it was designed on a grid with modules of 2.5 by 2.5 meters and was intended to thin out over time like a forest.[104] The plan indicates native woods such as oak, ash, beech, maple, and various acacias, as well as a marshy bed and an area with a mixture of sand and marl, to encourage more diverse vegetation in the underbrush. The mixed forest is composed of layers of trees, shrubs, and herbs; the ground vegetation was intended to develop in harmony with the location and its use.[105]

The idea of planting in a grid and an overlapping of the grid with different soils is already an indication of the tension of formal and informal logic that Kienast sought. This was reinforced by his desire to introduce in the forest dead straight roads, on the one hand, and seemingly random beaten paths, implemented along with the gridded planting. All of the forest paths lead to a semicircular plaza east of the pool. According to Kienast, visitors step out of the cool of the forest into the bright, central plaza.[106] Once again, with his concept for the urban forest, he was trying to achieve a synthesis of ecological, user-friendly, and aesthetic-sensuous park design.

The semicircular plaza surrounding the pool with a fluid transition into the forest was already conceived in the preliminary project. In the final plan, one of the two rows of maple trees with small tops that follow the diagonal north of the plaza opens up toward the plaza and surrounds it up to the central axis of the pool. The second row of trees leads straight to the edge of the pool (68). This refinement of the original idea produces a strange spatial effect. In the preliminary project, both rows of trees lining the boulevard still lead up into the pool. The trees are standing in the pool, presumably as an unmistakable symbol of an overlapping of two contrary forms and logics—a big theme in "postmodern" designs of this period.[107] In the preliminary project, the west side of the pool is framed by three steps covered with grass that recall the cavea of an ancient theater. The cavea would have offered additional resting places with a view of the pool and of the urban forest, while at the same time reinforcing the spatial demarcation of this area of the park and diversifying the relief of the park. Northwest of the theater, a circular fire pit was planned for communal grilling.

Brühlwiese Municipal Park
in Wettingen, 1979–84. Photos:
Georg Aerni, 2012 and 2013

Page 198
View from the diagonal boulevard, the backbone of the entire site, toward the earth pyramids and the town hall

Page 200
The broad grass playing field of the municipal park is also a projection screen for the shadows cast by the three rows of trees of the boulevard to the southwest.

Page 202
View along the boulevard lined with three rows of trees and across the grass playing field toward the urban park that borders the park to the east. The detailed rendering of the boulevard shows what Kienast meant by the "legibility" of nature in the city.

Page 204
Seen here is the terraced hedge formation, which echoes the cavea of an ancient theater originally planned in this location, and behind it the hedge rondel, which recalls the form of the fire pit originally planned, the earth pyramids of the sledding hills, and the diagonal boulevard of maple trees with small tops.

Page 206
The pool of the municipal park has the stylized form of a waterlily leaf. In the summer, visitors step out of the dark and cool urban forest into a light-flooded, hot plaza, where the pool offers refreshment. Behind it extend the cavea-shaped rows of hedges. The boulevard lined with three rows of trees that borders the park to the southwest connects the urban forest to the Wettingen town hall.

Page 208
Hedge formation and hedge rondel seen from the diagonal boulevard

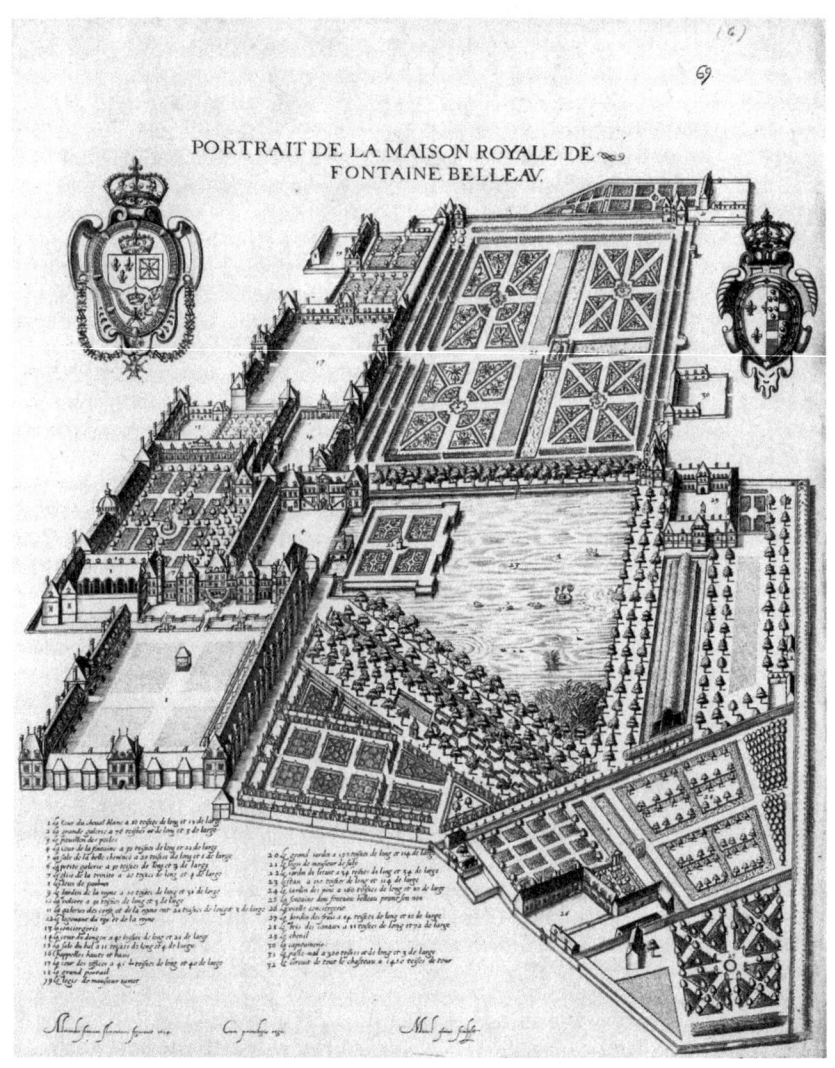

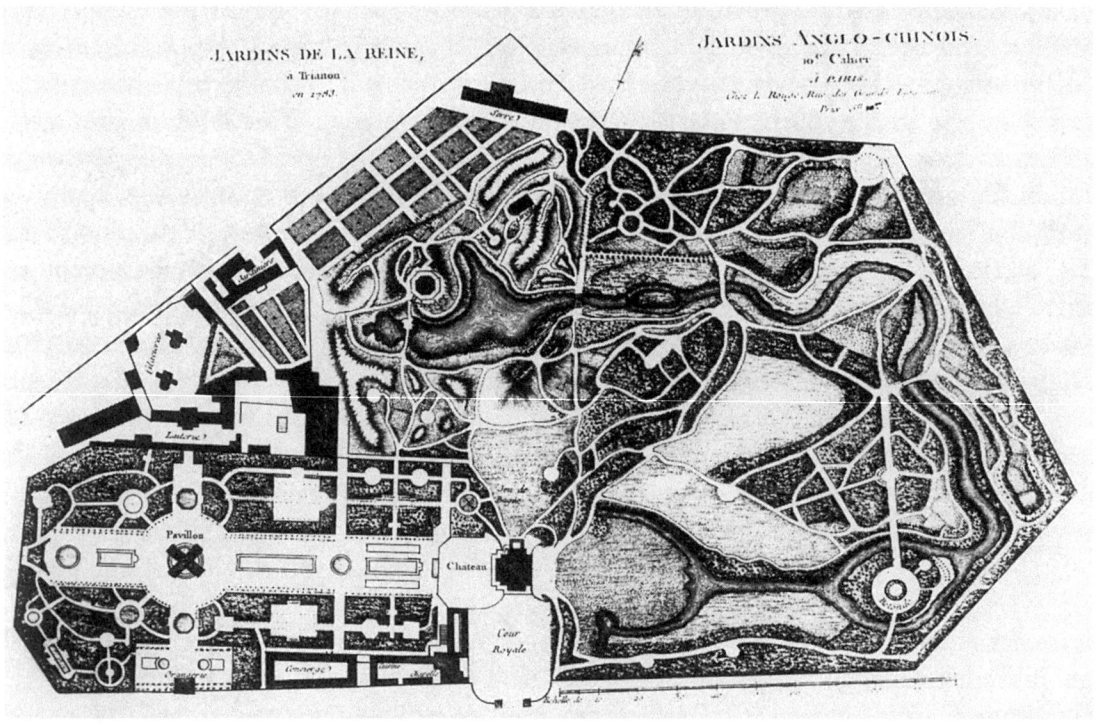

74
Château de Fontainebleau and gardens documenting its state in the early seventeenth century. Engraving by Michel Lasne after a drawing by Alessandro Francini, 1614

75
Château de Versailles, Petit Trianon, plan of the garden by Georges-Louis Le Rouge, 1783

In the final plan, Kienast dispensed with these possible uses. But he still retained their forms: the *cavea* became a semicircular formation of three rows of hedges of geometrically pruned field maple (pp. 206, 208). The effect of steps is produced by trimming the hedges at different heights, since the terrain was only slightly banked. The fire pit has transformed into a hedge rondel, also consisting of geometrically pruned hedges of field maple. Between these two geometric areas, several deciduous trees were planted. Clearly, the form was more essential here than the use. Rather than a fire pit for grilling, children can play hide-and-seek or tag in the hedge rondel, and lovers can hide and kiss. The forms of the planted objects are divorced from a clearly defined function, though this opens up new uses.

When decorating the municipal part with flowers, as the residents had desired, Kienast also took paths that would not have been expected of him: despite repeated requests, he refused to include flower beds.[108] Instead, in the southeastern entrances, east and west of the Lombardy poplars that mark the beginning of the diagonal axis, he placed eighteen steel trellises. The two outer rows were to have clematis, then a row each of Virginia creeper, and finally honeysuckle.[109] Kienast was clearly trying to use different types of flowers and the autumn colors of the vines to make this area interesting in the changing seasons. He took into account the people's desire to have flowering plants, but at the same time the steel trellises were intended to underscore the urban note of the park.

With his fragmentary combination of the individual components of the park, Kienast created an assemblage of the traditional repertoire of the art of the garden, citing and reinterpreting it in an urban context. He had been inspired to do this by the logic of Baroque and late Baroque gardens, for example Alessandro Francini's plan of Fontainebleau documenting its state in the early seventeenth century: a composition of garden spaces both connected and separated by axis and boulevards (74). The boulevards and axis create the structural framework for the coherence of the separate parts. One detail that recurs in Kienast's work is the overflowing zone of reeds in one corner of the water parterre at the center of Fontainebleau: in the preliminary project for Brühlpark, the pool is superimposed with a reed zone, although here the form itself is taken up by the formlessness of the reed and at one place dissolved by it: "[We] were excited . . . by the relevance and unexploited potential of late Baroque garden plans, their geometric order dissolved, as it were, by the acid of an organically determined formal system."[110] There are such aspects in the park at Versailles as well. A copy drawing of André Le Nôtre's bosquet "Les sources" in the Trianon de Marbre reveals an informally designed small forest with winding paths and streams and geometrically shaped pools, next to a stately, diagonal boulevard. The plan for the Petit Trianon that Le Rouge produced in 1783 illustrates how this garden space changed over the decades, including the landscapes added at the request of Queen Marie Antoinette (75). The castle and the central axis of the garden extending out before it represent thresholds between the landscape and the formal areas, though they also interlock in places. This kind of meeting of organic and geometrically shaped spaces also occurs in the Brühlwiese Municipal Park, with the diagonal axis as the threshold that separates the informal part of the urban forest from the geometrical areas such as the hedge theater, the hedge rondel, and the earth pyramids.

Materiality and Poetics

Another telling example of the formal language that Kienast developed in Wettingen is the boulevard of three rows of linden trees, which follows the length of the grass playing field and delimits the park to the south facing Zentralstrasse. In front of the town hall, Kienast placed four slender Lombardy poplars, which loom into the sky like enormous flagpoles and, as vegetal elements, contrast with the cubic-architectonic ones. In Marc Schwarz's film *Dieter Kienast: In Praise of Sensuousness*, the camera captures the tops of the poplars moving in the wind in front of the modern building, thus illustrating the dynamic that Kienast initiated by placing the trees as components that shape the space and the structure.[111] The poplars are standing in a hedge that separates the two parking lots on the western part of the grounds. The dry plaza immediately to the north is much smaller than in the preliminary project and covered by just four trees.

Lindenallee is the connecting link between the town hall and the urban forest. Kienast saw the boulevard with three rows of trees as an urban promenade.¹¹² The significance he attributed to it is reflected in the particular attention he devoted to its details (p. 202). The boulevard is an early example of his idea to design the "nature of the city," and of his call for "legibility" by means of design:¹¹³ every element of the transition from the street to the grass playing field was thought out and harmonized with the elements around it. The southernmost of the three rows of trees is framed by geometrical hedges pruned to two different heights; the one facing the road is lower than the one facing the footpath. This hedge-and-tree strip is lined with gray curbstones along the street and with brownish-red paving stones along the asphalt road. When the sun shines in winter, the bare branches of the hedge cast picturesque shadows on the asphalt at midday.

The middle row of trees is in the sidewalk zone. The planting pits in the asphalt are covered with steel tree grilles. They are not the usual rounds discs, but rather an oblong, rectangular grille with an especially unusual triangular end.¹¹⁴ This tip extends far into the footpath, while the other end of the grille extends into the adjoining grass field. The grille becomes the hinge between two zones of the boulevard, while the trees loom upward at the point of intersection between the triangle of the tip and the nearly square body of the grille. Facing the lawn, the sidewalk and the tree grilles that extend into the embankment are framed by a meandering band of paving stones of the same brownish-red color, forming the border along the hedge that demarcates the first row of trees that is parallel to the one on the other side of the sidewalk—although here it is completely straight. The boundaries between the natural element of the lawn and the hedge and the urban elements, such as the asphalt footpath and the steel tree grilles, are thus particularly clearly staked out. At the same time, the hewn paving stones transition from the urban zone to the vegetal one—by means not only of their materiality but also of their form, being industrially produced and yet still irregular.

Between the trees in the middle row, benches were placed, sometimes facing the park, sometimes the footpath. The function of the linden-lined boulevard as a mediator between the city and the park is reinforced in this way, since it is up to the visitors to decide whether they prefer to observe the bustle on the street or the activity on the lawn.

The trees in the third row were planted directly into a strip of grass, which is connected to the grass field by a gentle downward slope. Here, too, there is zoning, this time topographical in nature: the level of the footpath has been extended a bit by means of fill dirt. First there are the trees, and then the terrain transitions with a slight curve into the large grassy area below. In summer, when the trees have leaves, the grassy field is well shaded; in late autumn and winter, the sun passes through the bare branches of the treetops at a lower angle, turning the field into a projection screen for cast shadows extending well into it (p. 200). Kienast counted on and anticipated such images of "urban nature." They are the picturesque pendant to the architectonic rigor that characterizes the design of the park. Its material and form mark it as an urban park, as an urban-planning element, without the poetry fading into the background.

Urban Planning,
Not Ideal Landscape

In an article published in the local paper in Wettingen on June 30, 1984, shortly before the municipal park opened, Dieter Kienast emphasized that the Brühlpark "in its present form" also expressed the intention of adding a component missing from the design of the town center in 1956, in that the town hall had "finally" been given an adequate park:

> These urban planning goals were achieved above all through the large boulevard lined with three rows of lindens. It separates Zentralstrasse from the park and at the same time joins them, forming the connecting link between the town hall and the urban forest, which in turn forms the eastern termination of the series of spaces: the town hall plaza, the town hall, and Brühlpark.¹¹⁵

In referring to the municipal park "in its present form," Kienast meant the change that the project had undergone between 1979 and 1982 up to the realization of his intention to create a park with a decidedly urban character in terms of materials, form, and use. Kienast regarded the comparatively low budget available to the firm Stöckli + Kienast for the project—the construction costs were 41.80 Swiss francs per square meter (as opposed to approximate costs of 100 francs for comparable sites in Zurich)[116]—as a "guideline for the economical approach to the existing and modest use of elements of garden architecture."[117]

Against the backdrop of the heated discussions of the park, Kienast wrote in his text about the opening that the active participation of residents in its creation had been a pleasure, and after all everyone had his or her own idea of a park: Some may be a little disappointed by it. We want to tell them again that the goal of the design is not an ideal landscape in the sense of the English landscape garden but rather a site with an urban look and diverse uses for all seasons, ages, and occasions.[118]

The most striking components of the municipal park are the two earth pyramids, which are placed like an aesthetic caesura between the area for children to play located in a grove and the broad, open area of the grassy field. The northern, three-sided pyramid is more than 4 meters tall; the southern, four-sided one is 3 meters tall. The asymmetrical pyramids face each other; the longest and flattest arm of each is pointed at the other pyramid, while at the same time they are slightly pushed together. In this way, owing to the space between their tips and the gentle ascent of the two sides that face each other, they form a kind of "gate" to the children's playground—a topographic arrangement that is at once boundary and opening. This gatelike situation was further reinforced right at the entrance to the grove: according to the final plan, two fountains were to be placed symmetrically in front of the entrance; instead, two stone pillars were erected. An existing hill with two slides was integrated into the new landscape; on the plan it hugs the northern pyramid. The contrast in design between the naturalistic topography of the children's play area—the forest and its beaten paths with bushes and erratic boulders—and the smooth surfaces of the geometrically strict pyramids was deliberately chosen and to the same degree even continued retrospectively, in that Kienast did not plant lawns on the pyramids, but rather a calcareous meadow as nearly natural vegetation on an artificial form (p. 198).

If it had been up to Kienast, the grove that provides shade for the playground would have been more like an orchard, with berry bushes, including currants and gooseberries, nut trees, and stone fruit such as apples, cherries, damson plums, and apricots.[119] The children were supposed to enjoy themselves beneath the trees, climbing them and eating their fruit. This again reflects Kienast's effort to design sensuous experience and pleasures such as eating, touching, tasting, and smelling, yet simultaneously dovetailing the natural and the cultural in the municipal park: cultivated fruit was supposed to be planted on a naturalistic topography, while a calcareous meadow was sown on the geometric hill.

The town council rejected the proposal. It called instead for "modest trees,"[120] and so several types of ash, field maple, various willows, elder, hazelnut bushes, and blackthorn were planted.[121] In the center of the grove, there was a well pump and two wooden ramps for the water, from which it fell into meandering grooves in concrete and seeped into the soil at the northern edge of the forest. Other than that, the town council approved only common playground equipment such as swings and seesaws.[122] Kienast's wish not to have a fixed playground, but rather a participatory project in which the northern part of the park could have been designed by children and young people together in an evolutionary process of several years, likewise went unfulfilled.[123]

The community also protested the pyramids and their "unsuitable edges." Their usefulness as sledding hills was also doubted.[124] This time, however, Kienast and Stöckli were able to prevail by arguing that the pyramidal hills were no steeper than originally planned and had already been very popular with children when unfinished.[125] Their geometric form, they repeatedly argued, explicitly underscored that the soil had been deposited there by human beings. Like the grounds of the Niederhasli development, the idea of the free child was combined with aesthetic ambitions.

Kienast's preliminary project of 1979, with its earth hills in a naturalistic topography, and the final plan of 1982, with its geometric earth pyramids, were separated by three years that were crucial for his stance as a garden architect. In 1981, he had been appointed Professor of Garden Architecture at the Interkantonales Technikum Rapperswil, which led him to work "right through the history of garden art." He "realized that many historical conceptions were not outdated, but still have relevance for us today."[126] Kienast's recourse to history is reflected in many of the elements of the municipal park in Wettingen, which can be read as an assemblage of quotations from the art of the garden, probably the most prominent of which was the pyramid.

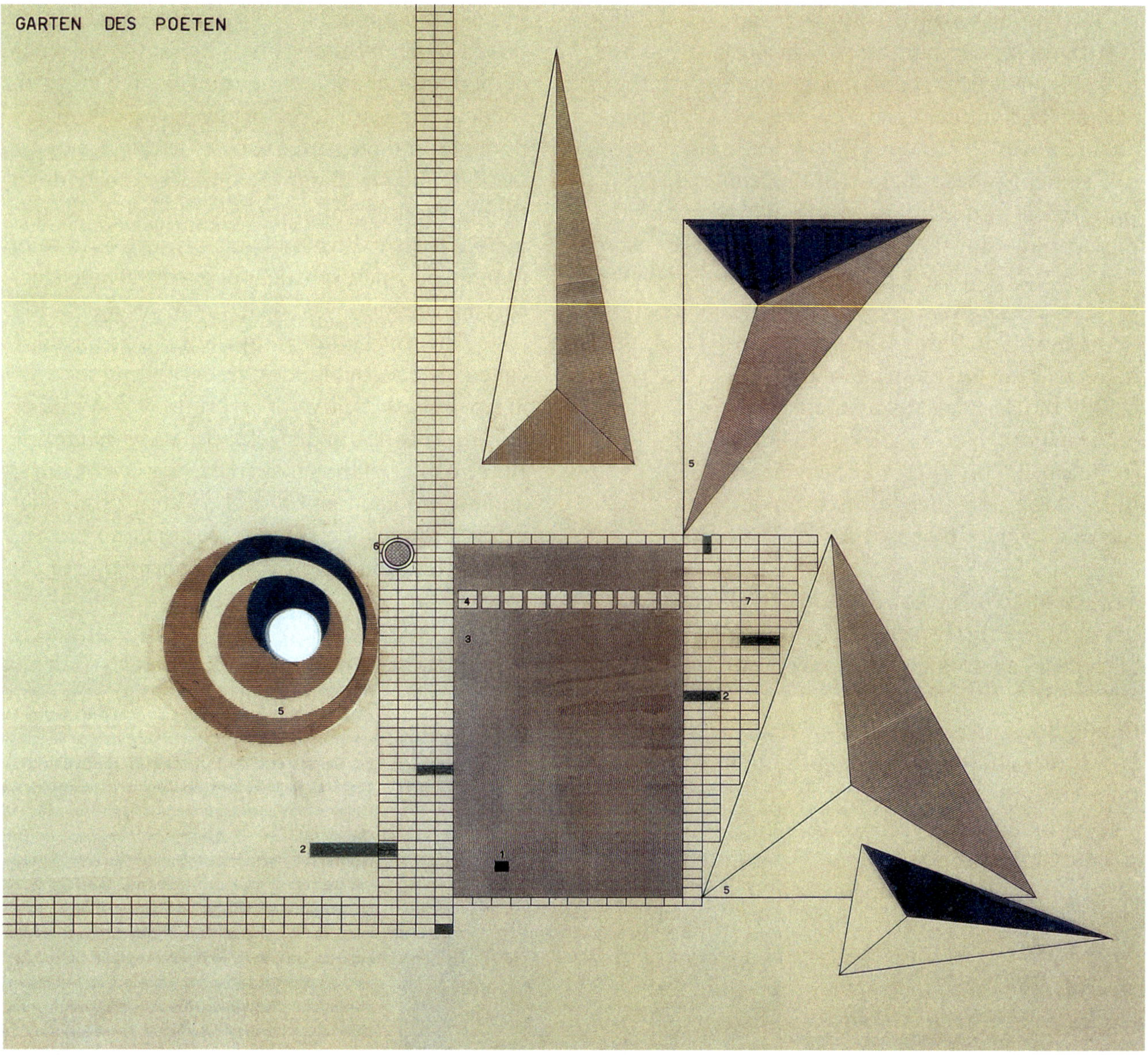

76
Ernst Cramer, plan of the design of the Garten des Poeten (Poet's Garden), at the G 59 horticultural show in Zurich. Copy of plan on tracing paper, film

77
Fred Eicher, grounds of a housing development in the Nussdorf district of Vienna, 1964. Photo: Fred Eicher

In the mid-nineteenth century, the German garden designer Prince Hermann von Pückler-Muskau (1785–1871) had two grass-covered pyramids built in his landscape garden in Branitz, near Cottbus. One of them housed his own grave and the other that of his wife. The peak of the latter had an iron crown stamped with a quotation from the Koran: "Graves are the mountaintops of a distant, new world."[127] The pyramids belong to a symbolic context here, whereas Kienast was pursuing a different, formal interest: his pyramids were not intended to express a "distant world" but rather make it possible to experience this one in a new way. The omnipresent form thus represented a deliberate contrast with the principles of the so-called natural garden in which form was to be avoided.

When conceiving the Brühlpark, Kienast did not just have the "agreeable simplicity" of the work of Fred Eicher in mind;[128] another crucial inspiration came from the radical clarity of Ernst Cramer, as represented, for example, by the Garten des Poeten (Poet's Garden) that he realized for G 59 in Zurich, the first Swiss horticultural exhibition (76). The Garten des Poeten, which was composed of four earth pyramids from 2 to 4 meters tall and a terraced cone, was "an independent work of art, an abstract interpretation of landscape and an architectural ensemble" in one.[129] The pyramids and cone were grouped around a central, rectangular pool, while the plaza surrounding the pool, the access paths, and the seating furniture, with cable and manhole covers, were all designed in concrete. This material, and even more so the gesture of using serially produced standard products of concrete, such as manhole covers, in a garden, was still completely frowned upon in 1959.[130] He dispensed with flowers in the ensemble, apart from a round concrete container planted with red geraniums as a sly wink in view of the other grounds featured at the horticultural show, which was popularly known as the Blumen-Landi (Federal Flower Show).[131]

Without a doubt, Kienast's reinterpretation of the earth hills as pyramids was inspired by the Garten des Poeten. Cramer's influence is also evident from other details of the municipal park, such as the use of urban materials like concrete and concrete slabs, the lack of flowerbeds, and the preference for straight lines for the connecting paths and boulevards. The garden architect Cramer was aware, according to Udo Weilacher, "that a straight line in plan would never be as hard when it crossed the landscape; vegetation in its inherent vitality, unlike architectural building materials, admits neither merciless severity nor paralysis in abstraction."[132]

Long before Kienast adopted its pyramidal forms, the Garten des Poeten had left its traces in other works by Swiss landscape architects, including Fred Eicher's design of the grounds of a housing development in Nussdorf, near Vienna, in 1964 (77). Here, the pyramids are used for the first time not as purely artistic landscape elements with no active use, but rather as a place where children could play. As in Cramer's garden, they are combined with flat, rectangular pools, and Eicher paved the connecting paths and plazas with concrete slabs. The few trees planted in between them are at once linchpins and spatial components.

Kienast's introduction of the earth pyramids was related to both of these projects. In terms of the topography and relief, Cramer's pyramids were striking elevations on the broad plain of the lakeshore at the transition to the Zürichhorn. This formation of plane and elevation is mirrored in mitigated form in Eicher's design of the grounds. In Kienast's case, this aspect becomes relevant again, because in front of the pyramids is the elongated grassy area of the former soccer field, which he retained as a spacious void in the center of the park. "The hills were intended as a contrast to the flat surroundings and the mountain ridge of the Lägern,"[133] Kienast explained in an interview in which he described the relationship to Cramer's pyramids "not intended": it "nearly led to a new formal design with a lot of hills. But, in the end, I left it as it was as I felt that the park could only take a very small number of simple elements."[134]

In his Garten des Poeten in 1959, Cramer was trying to counter the Blumen-Landi with a design of form and material that expressed modernity. This rebellious spirit returned in a different context in Kienast's decision.[135] The deliberate placement of dominant geometric elements, such as pyramids, can be seen as a demonstration of a new aesthetic to confront the principle of hiding designed forms practiced by the natural garden movement.

No Garden without People

In the article "Vom Gestaltungsdiktat zum Naturdiktat; oder, Gärten gegen Menschen?" (From the Dictates of Design to the Dictates of Nature; or, Gardens against People?) of 1981, Kienast laid out his position on the natural garden movement.[136] Starting out from the use of open space by people, he developed an anthropocentric stance on dealing with nature. A key passage in the text is a quotation from Theodor W. Adorno's lecture "Funktionalismus heute" (Functionalism Today, 1965) on the relationship of purpose, space, form, and material: How can a certain purpose become space; through which forms, which materials? All factors relate reciprocally to one another. Architectonic imagination is, according to this conception of it, the ability to articulate space purposefully. Conversely, space and the sense of space can become more than impoverished purpose only when imagination impregnates them with purposefulness.[137]

Starting out from the question of use and of the human being as a designing creature, Kienast developed a kind of "third way" to distinguish himself from the natural garden movement, which in his view had introduced the dictates of nature to replace the design dictates of modernism. He analyzed and differentiated the various types of natural gardens accordingly. Whereas he admired the work of the Dutch architect Louis Le Roy,[138] he rejected the postulates of Urs Schwarz.[139] The Swiss biology teacher's book, *Der Naturgarten* (The Natural Garden, 1980), was required reading for landscape architects in the context of the environmental movement. For Schwarz, knowledge of nature was more important when designing natural gardens than knowledge of design, since he was primarily concerned about natural cycles. In the words of the landscape historian Annemarie Bucher: "Gardens should be understood as symbioses, as ecosystems with native plants and animals, which will regulate themselves when provided with nearly natural plantings."[140]

The reason that Kienast felt obliged to contradict Schwarz's understanding of how to approach nature is best understood from the latter's definition of "weed," which was diametrically opposed to Kienast's view, as a gardener, phytosociologist, and defender of spontaneous vegetation. Schwarz advocated the motto: "Above all, however, we are changing our attitude toward nature. We define nonnative, alien plants as weeds and the native, domestic ones as plants. And then we carefully begin to make room for the plants by removing the weeds."[141]

Kienast also resisted Schwarz's term "natural garden,"[142] which he replaced with "nearly natural garden," because there is no "garden in which direct human influence is eliminated."[143] He distinguished between several types of nearly natural gardens: The nature-conversation type, in which plants and animals are granted primacy over human beings and propagated as an ecological compensation area, which Kienast regarded as an "unsuitable effort by unsuitable means ... to solve the problems of our civilization." He understood the nature-imitation type of garden to be a "pseudo-natural" garden design in which an artifact is deliberately focused on an artificially produced naturalness. The third type, and the one he preferred, was the user type. As examples, he named not only the works of Louis Le Roy but also Gleisdreieck in Berlin and the art actions organized there. He wrote that the user type animates, "thanks to its seeming lack of professionalism, to specific action of one's own, playful construction, creative development of the residents." "In the open spaces of this time, there is no imitation: the artifact can stand out clearly. Construction, buildings, and nature are not antipodes in their appearance."[144] From Gleisdreieck in Berlin, he went on, one could learn "that design, cultural expression, can be outstandingly realized with nearly natural forms of vegetation":

> The condition for deliberate design with nearly natural forms of vegetation is precise knowledge of them, otherwise it becomes be necessary, as in the traditional garden, to work too hard with pruning shear, spade, and saw. This type of open space is outstandingly well suited to appropriation by users. Enduring, stable vegetation evolves in harmony with the site and its use and hence does not place any limits on use but rather represents the true form of an "open space."[145]

By adopting this attitude, Kienast was trying to produce a synthesis of the Kassel School principles of the emancipatory planning of open space

78
Kienast Vogt Partner, Berggarten (Mountain Garden) at the Internationale Gartenschau (International Horticultural Show) in 2000 in Styria, Austria.
Photo: Udo Weilacher

and designing with spontaneous vegetation and the need to give form to purposes. From here it is a short step to the earth pyramids covered with a calcareous meadow in the municipal park in Wettingen. The unusual form for sledding hills and the planting of them, which slightly goes against their form, was clearly anchored in Kienast's logic, even though their visual appearance could not have been more different from what one imagines a "nearly natural garden" to be.[146]

The pivotal point in Kienast's turn from the natural garden movement was the importance he attributed to the use of open space. Just how strongly use was tied to the space designed for it was already demonstrated by his planning the grounds of the housing development in Niederhasli, where the form was derived from a detailed concept of use. The focus was on creating attractive, nonstandard spaces for children to play. This process was repeated for the Brühlwiese Municipal Park, but under the opposite conditions: the sledding hills requested provided an opportunity to develop the pyramidal form, but the use had to be adapted to it. Kienast's priorities had shifted in the meanwhile.

The critique of the earth pyramids referred to their supposed impracticality: children had difficulty handling their sleds on the peaks of the pyramids; there were no flat areas for turning or for sitting before starting down the slope.[147] Kienast and Stöckli countered this by arguing that the point was to challenge the dexterity of the children and repeatedly emphasized that the children were being provided with an extraordinary landscape for playing, and that they had already conquered it for themselves in its unfinished form.

Kienast's recurring idea of an "alternative" playground could be realized at least in part in the topography of the children's play area in the Brühlwiese Municipal Park. This kind of artificial landscape gave children more freedom and creativity than average playgrounds do.[148] In the meanwhile, a so-called adventure playground has been installed between the pyramids and the grove—a compensation for the failure to realize Kienast's original concept for the playground.

Form without practical use was inconceivable to Kienast in the early 1980s, but with the need to venture more in formal terms, the relationship between form and use changed. Finally, with the Berggarten conceived in 1996 for the Internationale Gartenschau 2000 (International Horticultural Show 2000) in Styria, Austria, he designed a folded landscape that gave rise to a previously unknown sense of space (78). The concerns of urban planning or broadly sociopolitical effectiveness that Kienast had formulated in 1976 in his alternative concept for Grün 80 in Basel now receded behind the experiential quality of the new landscapes for horticultural shows: the use of open space as a way of coping with daily life meant something entirely different in the 1990s than it had just two decades earlier. Playful interaction with open spaces and the potential for aesthetic experience had become increasingly important over time.

The question of the later use of this extraordinary, artificial landscape became a problem, until the Styrians decided to present the contemporary sculpture collection of the Steirisches Landesmuseum Joanneum in the Berggarten and to celebrate a large garden party there twice a year.[149] The fact that Kienast's folded landscape can stand up to the new use despite being outright overloaded with sculptures, that its "inherent value" does not vanish behind the art objects exhibited there, says something about the strength of its design. The Berggarten is an example of one of several public landscapes designed in the 1990s by Kienast and the firm of Kienast Vogt Partner, founded in 1995, that needed the context of a festival in order to continue being used.

Transparency and Collage: Making City and Countryside Legible

The Brühlwiese Municipal Park in Wettingen was the first project in which Dieter Kienast tried out postmodern design principles: complexity, ambiguity, and integration of an existing situation into a new context of form and meaning. Recourse to the history of the garden played just as central of a role in this as did the composition principles of collage and transparency. The point of departure for Kienast's design was the urban planning situation of the park, the question of required uses, and the recourse to forms from the history of the garden, which he wanted to employ and interpret in a contemporary way.

Over the course of the 1980s, Kienast underwent a change in approach to his designs, trying out different modes in parallel. In 1983 to 1984, he was working simultaneously on the competition for the École cantonale de langue française in Bern and on the spa gardens in Zurzach. In the school project, his grappling with the theories of Kassel is present, including the analysis of requirements, the integration of spontaneous vegetation, and design using forms derived from urban nature. Its integration into the city seems to have been especially relevant to the design of the spa gardens, in keeping with the theory of the "transparent organization of space," which was very popular at the Division of Architecture at ETH Zurich—Eidgenössische Technische Hochschule (Swiss Federal Institute of Technology) at the time.

With the municipal park in Wettingen, Kienast had been interested in a language of form that drew on the historical vocabulary of garden art and used contrasts as design tools to experience both the dichotomy of the natural and the artificial, as well as the topography of the site. The Zurzach Spa Gardens represented another turning point: Kienast was no longer developing his designs by starting out from the practical demands of the users, but now shaping the space based on methods of analysis and design, such as the transparent organization of space, which he had taught himself. Even more so than with the municipal park, in the case of the spa gardens he was applying formal principles from the history of the garden and of architecture to a specific terrain. Nevertheless—and this is the paradoxical aspect—the qualities of the place in question are clearly worked out and made tangible precisely by means of the defamiliarizing interventions.

In the Zurzach Spa Gardens, visitors have a sensual experience of diverse spaces designed with nature: the cool forest, the orchard, a large clearing, a bank covered with reeds, a quay-like promenade, or an uncovered stream that expands to form a pond in the park, and much more. The stimulation of aesthetic experience is the theme of the park, and its thoroughly composed contrasts make it Kienast's first site where every removal or addition of an element disturbs the overall effect: the decade of the "unalterable spaces," and of concepts that emphasize form, begins with this park.

Transparent Spatial Organization and the Formal Principle of Collage

When Dieter Kienast returned to Switzerland from Germany in 1979, Swiss architects were grappling with new ways of representing landscape architecture projects that would distinguish them from the then common organic forms. The Swiss architect Bernhard Hoesli (1923–1984)—who, in the 1950s, had reformed the teaching approach at the School of Architecture at the University of Texas in Austin and then later brought the subject matter he had developed there to ETH Zurich in the 1960s—has been quoted as saying in this context: "We have to draw straight lines; God will see to it that they become crooked."[150]

Clear examples of this reorientation are the competition plans produced by Hoesli with the Swiss architects Arnold Amsler and Arthur Rüegg and the garden architect Ruedi Siebrecht for the design of the lakeshore in Biel (1977) and the "Land und Wasser" (Land and Water) sector of the Grün 80 horticultural show in Basel (1977) (79, 81).

The revision of the preliminary project for the municipal park in Wettingen in 1982, during which Kienast's language of form underwent an obvious transformation, may have been inspired by these plans. In the Biel lakeshore and Brühlwiese Municipal Park projects, boulevards and bosques created the spatial framing. Geometric and organic forms oscillated. In the center of each, there was an extended grassy area with a few groups of two to three solitary trees. Void and mass—mass understood here as a porous, spatial caesura created by trees—alternate. On the plan for the Biel lakeshore, existing traces had been integrated into the new design; on the plan for the municipal park in Wettingen (70), which was based on a smaller and less complex terrain, that is only true of the existing children's playground with the hill for slides.

The competition entry for the "Land und Wasser" sector of Grün 80 received an award, and it differed fundamentally from all of the other plans submitted and awarded a prize (81).[151] Following the model of late Baroque gardens, in this plan a site is developed in which areas of strictly formal design—with large axes, boulevards, and formal gardens so typical of the Baroque—alternate with informal areas of small forests with curving paths, rural zones with meadows, and reeds along the banks. Next to the overview plan, the office placed a detailed axonometric drawing, instead of—like most of the other competitors—leaving open this area that corresponds geographically to a site with sports facilities next to the grounds for Grün 80 that was not a part of the competition. Above the plan, in small square fields, the most important principles of the plan are indicated by small diagrams. The conscious contrast of the natural and the artificial is emphasized here again.

Arthur Rüegg (b. 1942) has described this period as a struggle over the reinterpretation of landscape design: over forms and over the visual representation of design ideas.[152] How could an axis be defined without depicting it entirely? Or how could existing traces of the landscape be integrated into the design and thus emphasized? In terms of presentation, architects in particular turned against the widespread naturalistic drawing style, as represented, for example, by Willi Neukom's plan of 1963 for the lakeshore design in Zurich (80).

In Switzerland, plans like those for Biel and Grün 80 functioned as role models for the new methods of presentation in landscape architecture. Influenced by such plans, from 1980 to 1985 Dieter Kienast gradually parted ways with a drawing style à la Neukom and Grzimek, the latter having been his first teacher in Kassel. Another important influence for this change was evidently the theoretical-methodological texts *Transparency* by Colin Rowe and Robert Slutzky (1963/1968)[153] and *Collage City* by Colin Rowe and Fred Koetter (1978),[154] which were translated into German by Bernhard Hoesli. The changes in presentation also suggest a shift in the methodology of the design. Kienast introduced the "transparency" principles of finding and organizing form into landscape architecture, and he was one of the first to work with the technique of collage. The presentation plans he submitted for competitions in the 1980s and 1990s won first prizes. They were initially hand-drawn, as was the case with the Zurzach Spa Gardens (82), then a mix of hand drawing and collage technique, as with the plans for the Moabiter Werder in Berlin (83, 129) and Günthersburgpark in Frankfurt am Main (84), both developed in collaboration with Erika Kienast-Lüder and Günther Vogt. Thanks to such plans and the design methodology they represent, European landscape architects derived crucial inspiration for a reorientation of the profession: from the mid-1980s onward, issues of form and its organization, the historical and urban planning context of a design, and the way symbolism can be produced by a design intervention became just as important themes as ecology and participation.

In his texts, Kienast repeatedly referred to the programmatic writings *Transparency* and *Collage City*.[155] From the time of his project for the municipal park in Wettingen, they shaped his design practice as well—his lifelong search for forms to provide a design language for social and ecological concerns. In the Zurzach Spa Gardens, these theories and the postmodern search for and integration of urban planning traces and contexts can be clearly verified.

The design method of transparency had been explored between 1953 and 1958 as part of the teaching of the Texas Rangers—as Bernhard Hoesli, Colin Rowe, John Hejduk, Robert Slutzky, and Lee Hirsche were later called—at the School of Architecture at the University of Texas in Austin. The resulting text by Rowe and Slutzky is one of the only tangible theoretical results of that teaching experiment. Hoesli later

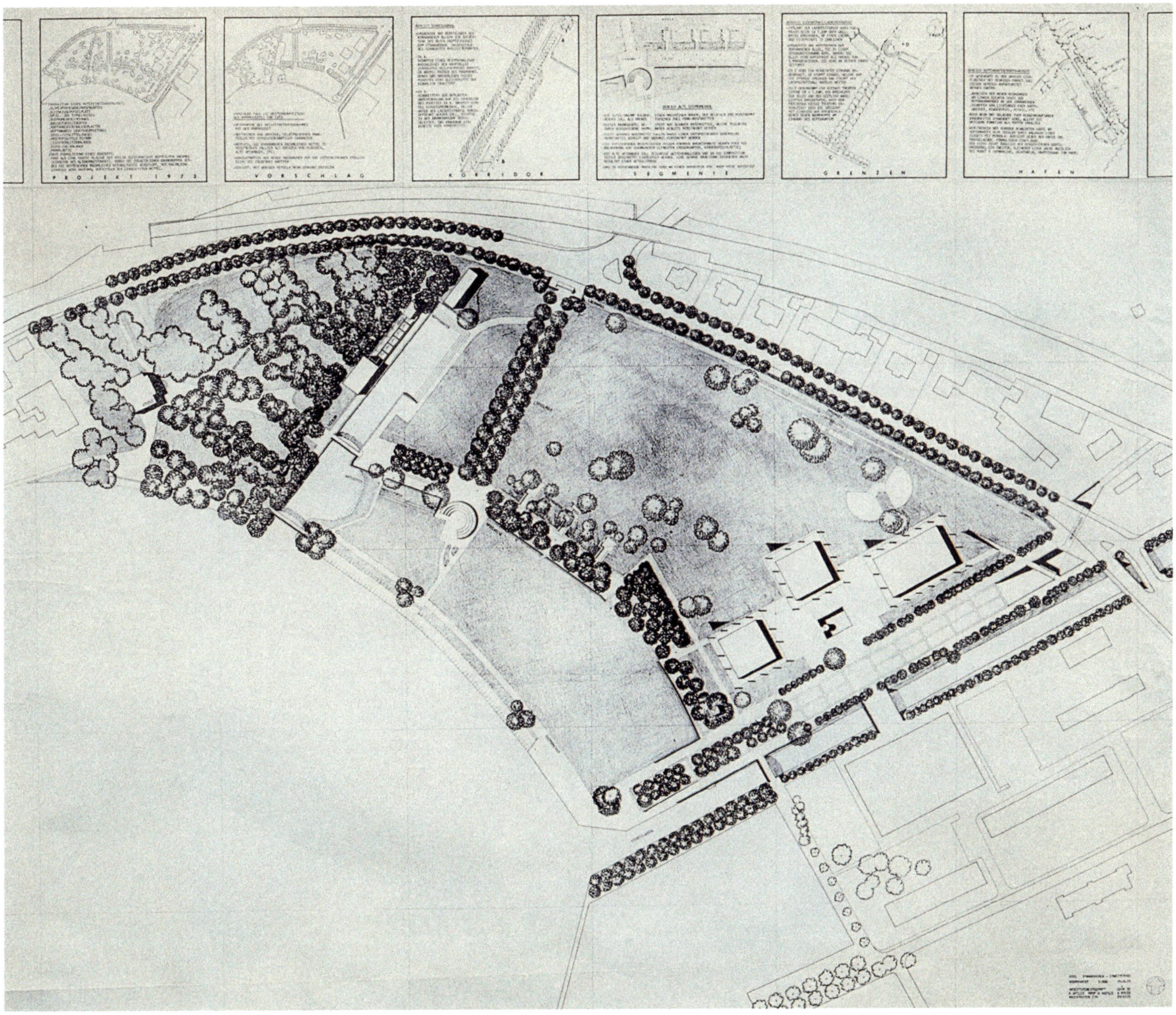

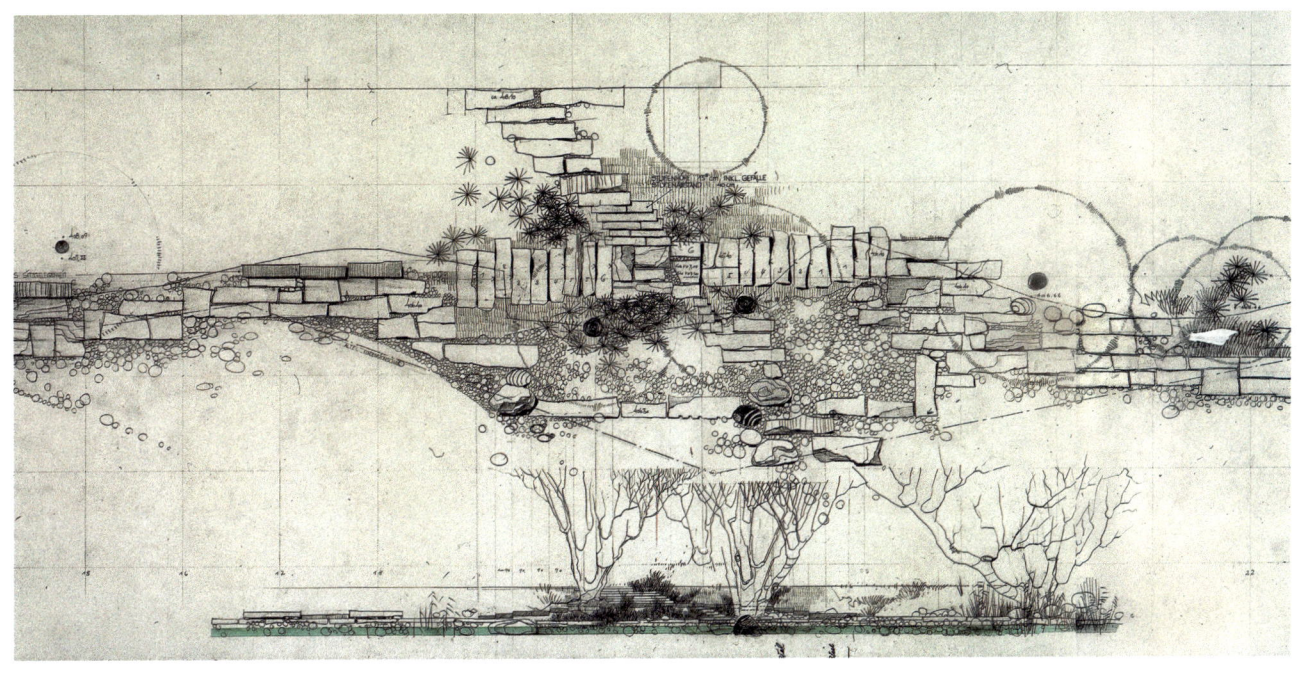

79
Bernhard Hoesli, Arnold Amsler, Arthur Rüegg (architecture), and Ruedi Siebrecht (landscape design), design of lakeshore in Biel, competition entry, 1977

80
Willi Neukom, design of lakeshore in Zurich, Zürichhorn, 1963 (detail)

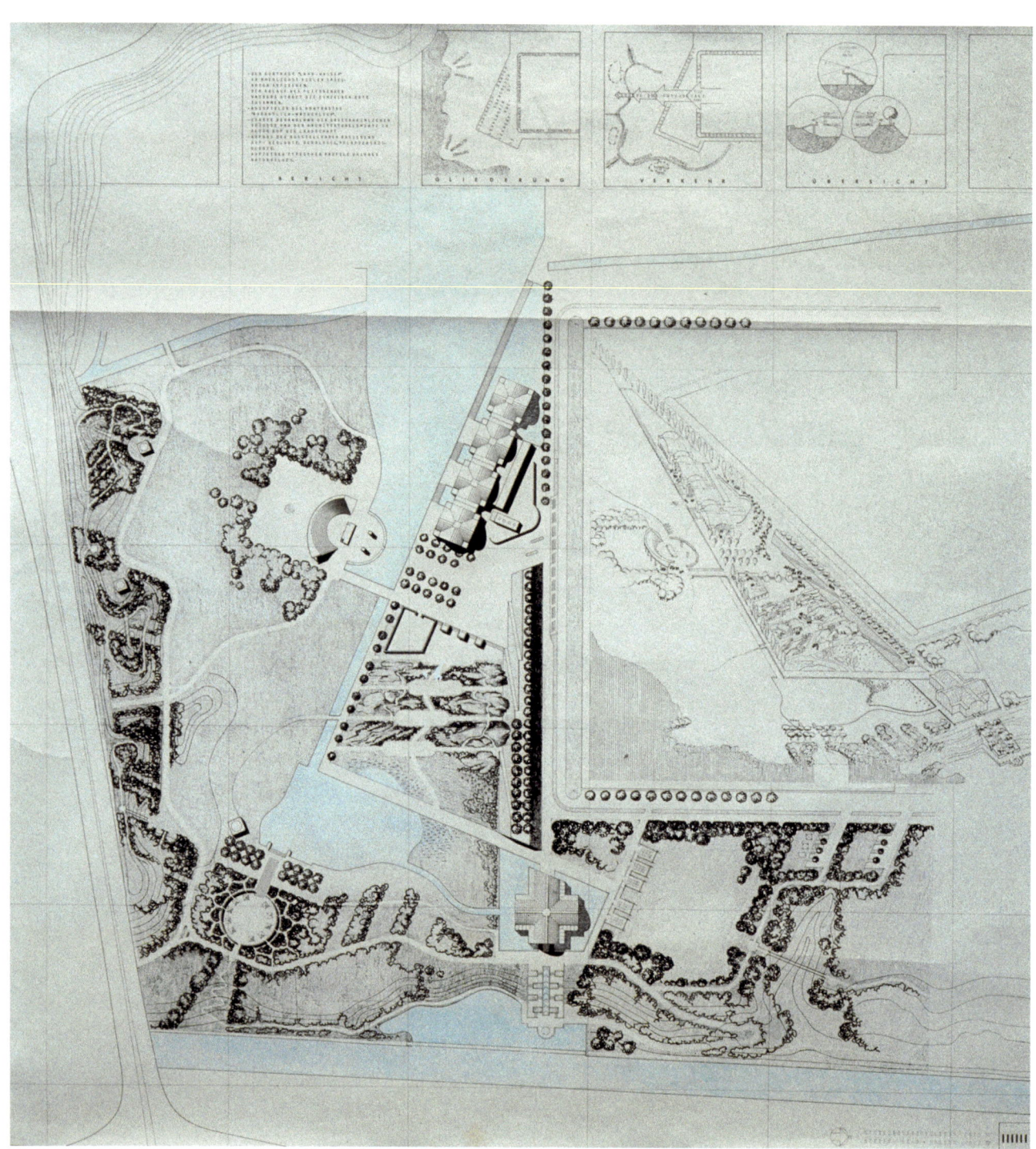

81
Bernhard Hoesli, Arnold Amsler,
Arthur Rüegg (architecture),
and Ruedi Siebrecht (landscape
design), Grün 80 in Basel, "Land
und Wasser" (Land and Water)
sector, competition entry, 1977

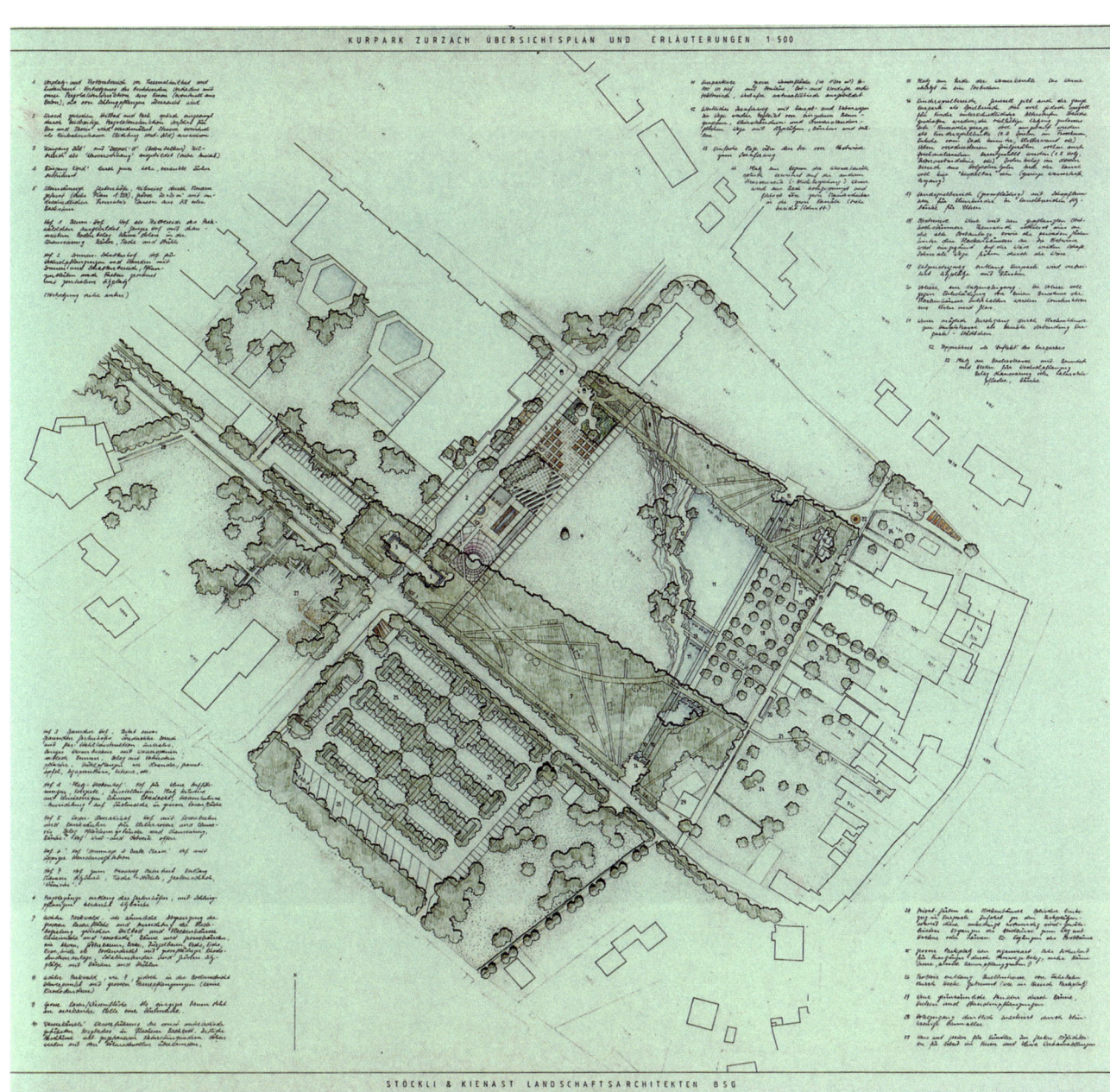

82
Zurzach Spa Gardens, overview plan and explanations, scale: 1:500, H 100 × W 100 cm, undated (1983), firm of Stöckli + Kienast

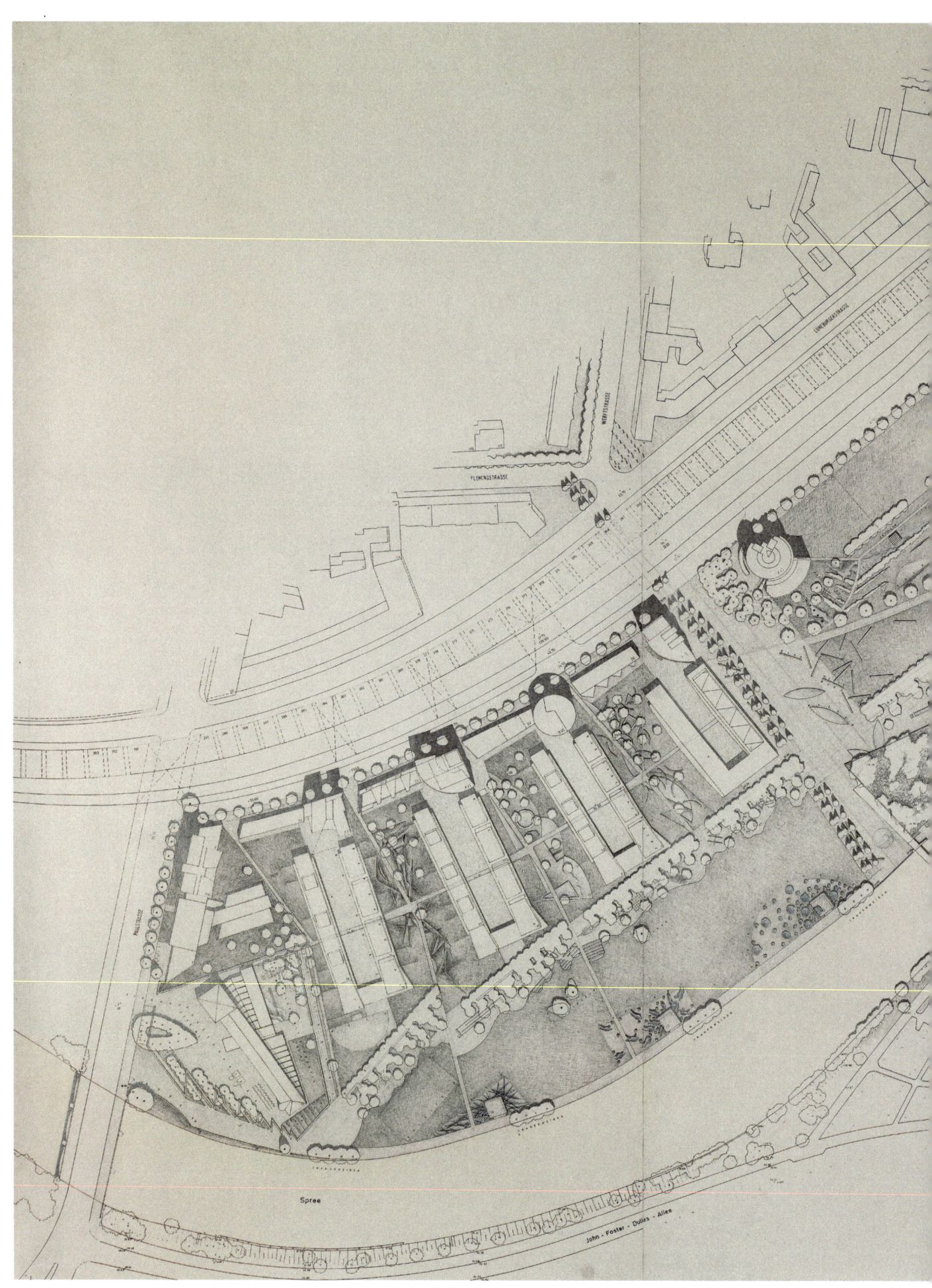

83
Moabiter Werder competition in Berlin, second prize, master plan, scale: 1:500, H 147 × W 211 cm, copy of plan on lightweight cardboard, graphite and colored pencils, 1991, firm of Stöckli, Kienast & Koeppel

84
Competition for the expansion of Günthersburgpark in Frankfurt am Main, project description, perspectives, and sections, H 89 × W 85 cm, copy of plan on lightweight cardboard, graphite and colored pencils, 1991, firm of Stöckli, Kienast & Koeppel

tried to establish the concept of transparency as a reliable design method in the introductory course on architecture at ETH Zurich. In the battle over the reorientation of architectural education fought at ETH in the 1970s, Hoesli was one of those who wanted to put an end to the discussion and question of all architectural tasks and have the design classes again concentrate more on specific designs and finding of form. In 1982, he wrote retrospectively about this period in an addendum intended for a new English edition of the Transparency essay:
Soon after the publication of my commentary [on the German translation of the Transparency essay] schools of architecture entered the rapids of "la contestation." Architecture is a form of sociology, we were told, and, if concerned with buildings at all, a kind of social engineering at best. There could not possibly be an interest in architectural form, which was declared of no importance at all or "unmasked" as a device of oppression to the advantage of the interest of a ruling class and to the detriment of the common good. Interest in problems of architectural form was held in contempt. Space was denounced as architect's fiction.[156]

At the transition from the 1970s to 1980s, interest in form increased again, but, he continued, now functionalism was being attacked by claiming that it saw form merely as a result. By contrast, now form was being seen as a component of a typology or as a historical model to be used at will.[157] Hoesli emphatically defended himself against these side effects of postmodernism: "The idea of form as neither an end in itself nor as a result of design but as an *instrument of design* seems still quite difficult to grasp."[158]

The principle of transparency, which Rowe and Slutzky defined using analyses of Cubist paintings and buildings by Le Corbusier as examples, was for Hoesli an important *"means of organizing form."*[159] Rowe and Slutzky distinguished between two types of transparency: an inherent quality of materials, such as the transparency of a glass curtain wall on a building, and an inherent quality of organization. They defined the former as literal transparency and the latter as phenomenal transparency.[160] In their treatise, they were particularly interested in the latter. It referred to an organization of space that divides the space into real and implicit layers, and to the alternation between deep and shallow spatial positions. "Layer" and "depth" are keywords crucial to understanding this second concept of transparency, as are "suggestion" and "ambiguity." In his commentary, Hoesli summarized these features in a definition:
> In general: Transparency arises wherever there are locations in space which can be assigned to two or more systems of reference—where the classification is undefined and the choice between one classification possibility or another remains open.[161]

Hoesli specifies the kind of design principle he means using the example of Mark Jarzombek's thesis project at ETH Zurich. The brief was to integrate a cultural center into an urban structure. Jarzombek interpreted the building itself as a small city in which all of the areas of life and work are interlocked (85). In Hoesli's view, this design succeeded completely in organizing the space transparently:
> Old and new, public and private areas, collective and individual use, are inseparably interwoven in a many facetted, rich, texture—and all meanings mentioned above are stated in terms of the geometric property of belonging to the one or the other orthogonal system of directions that generate the plan. There is identity of meaning and geometry. The sequence of the plans indicates progressively how transparent form-organization can be used to unify and differentiate within a complex yet clear organization, how meaning is present in terms of space.[162]

Following the principle of transparency, enables the practitioner, first, successfully to react to complex contexts and, second, to introduce oversight and order in complex contexts. For that reason, transparency as a means of organizing form is particularly well suited to urban planning projects with an established context. Because landscape architecture projects are also faced with addressing the existing fabric, Kienast clearly quickly realized the potential of this method to renew the forms in landscape architecture.[163] He first employed the method in 1982 in the design for the municipal park in Wettingen: for example, the pool is located in the axis of the town hall and of the semicircular clearing but it is also a crucial

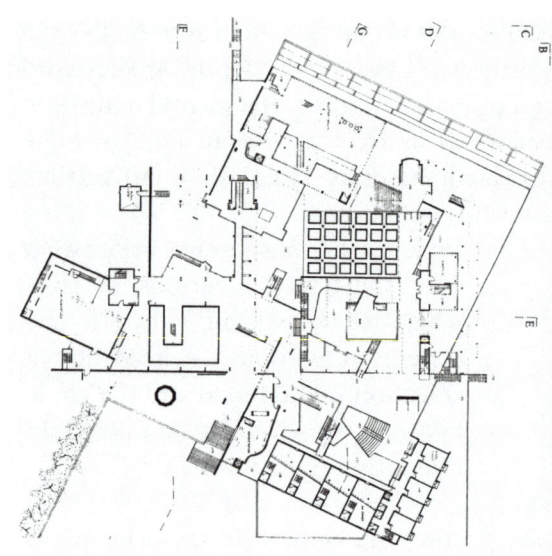
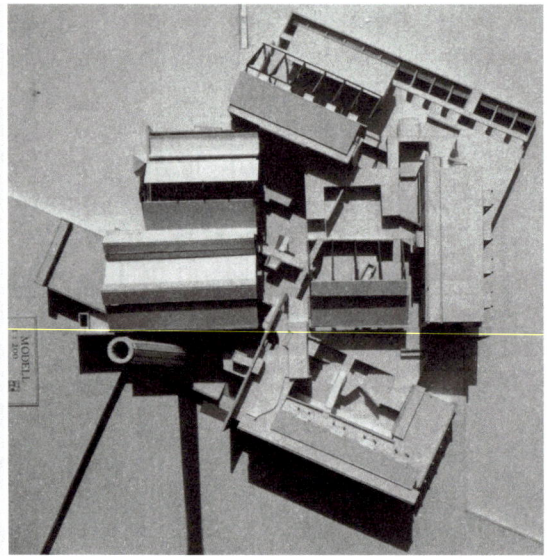
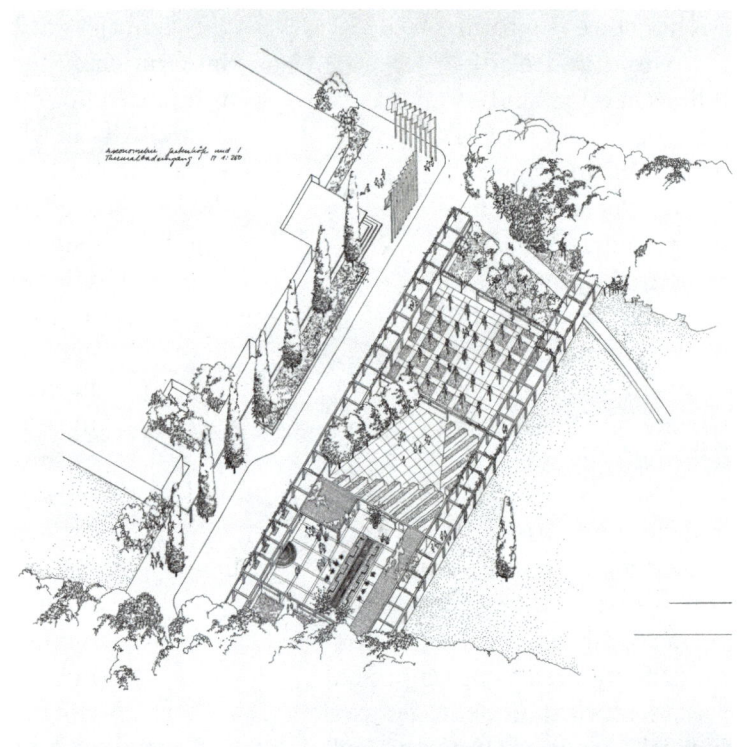

85
Mark Jarzombek, diploma thesis project for Bernhard Hoesli's chair at ETH Zurich, winter semester 1979–80, floor plan and model

86
Zurzach Spa Gardens, axonometric drawing of the thematic gardens, plan no. 471-2, scale: 1:250, H 100 × W 100 cm, ink and pencil on tracing paper, July 26, 1985, signed "CM, KIE," firm of Stöckli + Kienast (detail)

87
Zurzach in the Aargau, 1975

component of the diagonal. In the case of the Zurzach Spa Gardens (known as Bad Zurzach since 2006), realized only a little later, after the firm of Stöckli + Kienast won the invitation-only idea competition in 1983, Kienast applied the principle of transparency consistently throughout the design in order to reorganize a disparate space in terms of urban planning and landscape (82).[164] In both projects, he tried to make the park an urban planning link that follows both the context and its own logic.

The Park as Link
between Natural and Urban

The fully documented design process for the spa gardens in Zurzach enables us to understand in detail how Dieter Kienast applied the form-organization of transparency and the principles of collage. In a way similar to Bernhard Hoesli's description of Jarzombek's thesis, Kienast superimposed several grids of different orientation on his design, playfully attributing different ordering systems to the individual parts so they belong to two or more systems of reference. In this way, he managed to create a differentiated urban planning framework and to introduce zones that mediate between the park and the adjacent parts of the city.

An aerial photograph from 1975 illustrates the initial situation for the design (87): the terrain to be transformed into a park extends between the thermal bath in the northwest (groundbreaking ceremony in 1968), the core of the village in the southeast, part of which dates from the Middle Ages, a large parking lot in the southwest, and a meadow with trees in the northwest, where already at the time of the competition there were plans to construct a shopping center. The site is bounded on three sides by streets; on the fourth, southeastern side, it transitions into the backyard gardens of the houses of the historical village. The terrain itself is subdivided into three strips: in the northwest, adjoining the thermal bath, there is a small parklike area with a few trees and angular paths paved with stone slabs; to the north of it stand various houses. The middle zone of the terrain includes a meadow with several dozen fruit trees. To the southeast of that, there is another, fenced-in meadow, at the southern end of which lies the youth center, a former manor with a garden and later the schoolhouse of Zurzach. Apart from this one, all of the buildings on the property were to be demolished for the new park.[165] The spa gardens were to become a centrally located recreation area that mediated between the modern thermal bath and the medieval silhouette of the town. The mountain stream that had previously been fed underground through the village and flowed into the Rhine north of Zurzach was to be exposed and made one of the attractions of the grounds. The theme of water, an elementary one for a spa town, would thus be present in the design of the open space.

Kienast extensively analyzed the existing situation and the demands of the park. He had taken a lot of time to reflect and find responses to problems that at first seemed to him very difficult to solve:

> What will the transformation of an agricultural area measuring two hectares into a park mean for the residents of Zurzach and the guests of the spa? How can the required connection between the old space and the new small town be achieved? What does a park in village/rural surroundings look like? What are the requirements on the part of the spa guests and the residents and can they be reconciled? What are the unmistakable features of the place?[166]

All of the sketches from the design process for the Zurzach Spa Gardens have been preserved. The twelve drawings that Kienast made in three phases between July 17 and August 17, 1983, document the different attempts to organize space and forms (88–99).[167]

The first sketch is a formal inventory of the buildings surrounding the park. The existing forms were mirrored in the park like positive and negative. Kienast arranged the two dominant axes of the park in parallel to the direction of its northwestern and southeastern boundaries: one at a right angle to the southwestern street and running parallel to the thermal bath complex, and one extending diagonally to that street and echoing the course of the southeastern boundary. He retained these two systems of reference in nearly all twelve sketches and in each case developed a grid from it that is superimposed on the various systems of reference. Sometimes it is a simple grid and sometimes a tartan grid, with a form that can be traced back to the eponymous

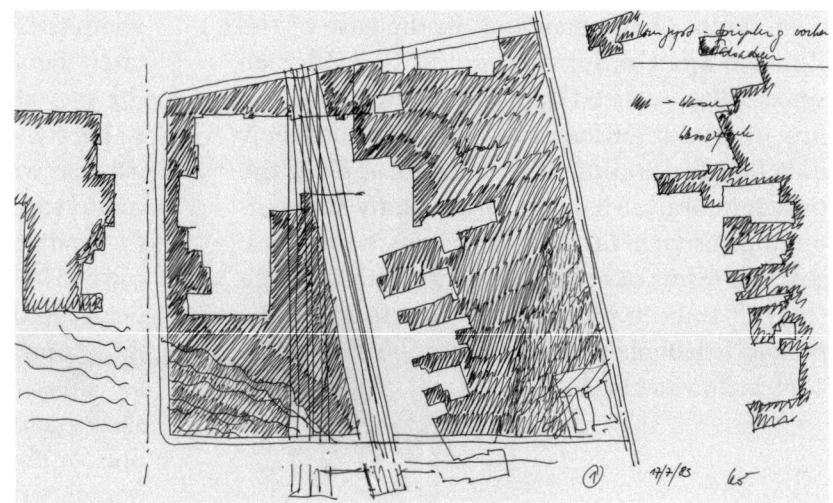

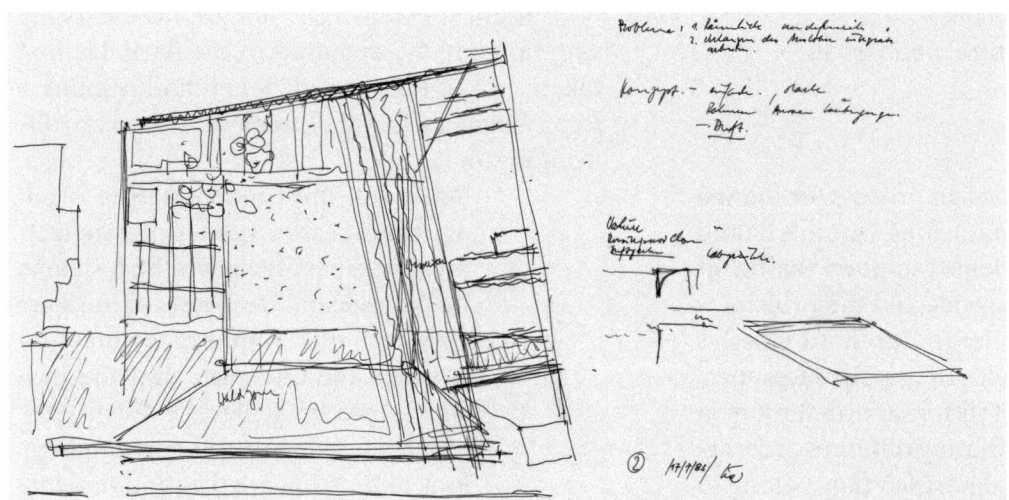

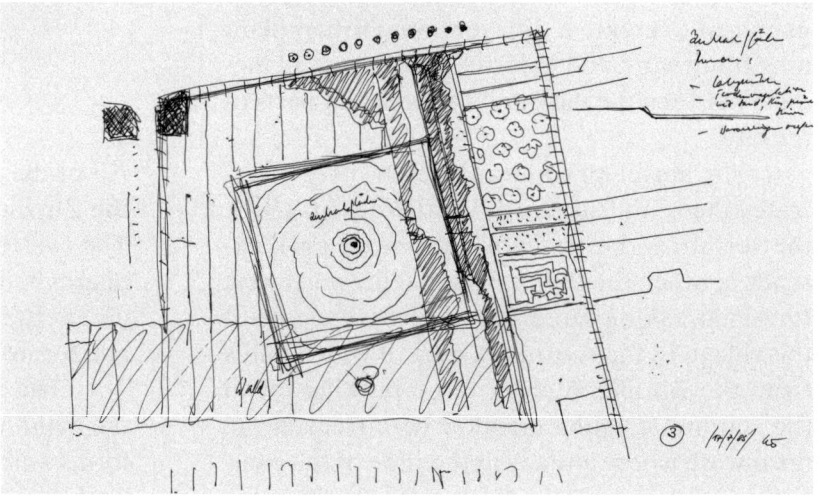

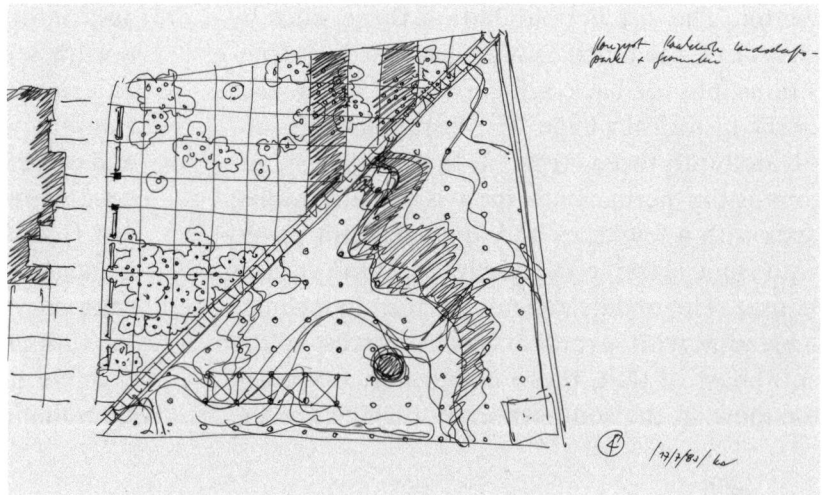

88–95
Dieter Kienast, twelve sketches for the Zurzach Spa Gardens, ink on tracing paper
  Sketch nos. 1–7, July 17, 1983
  Unnumbered sketch (presumably no. 8), August 2, 1983

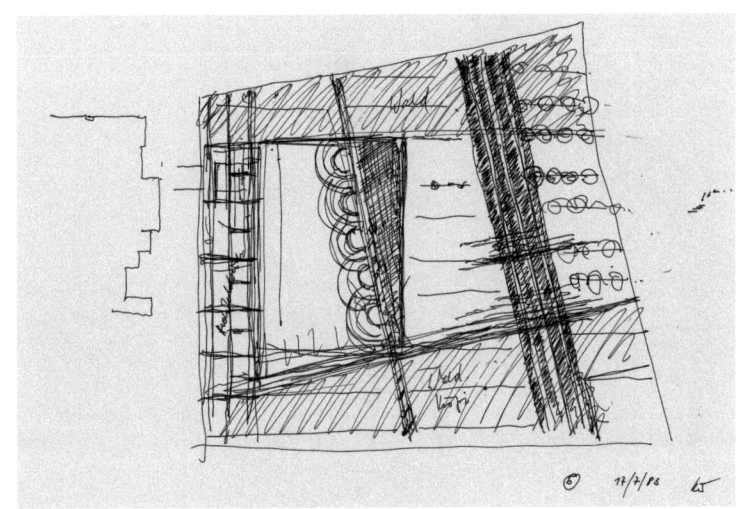

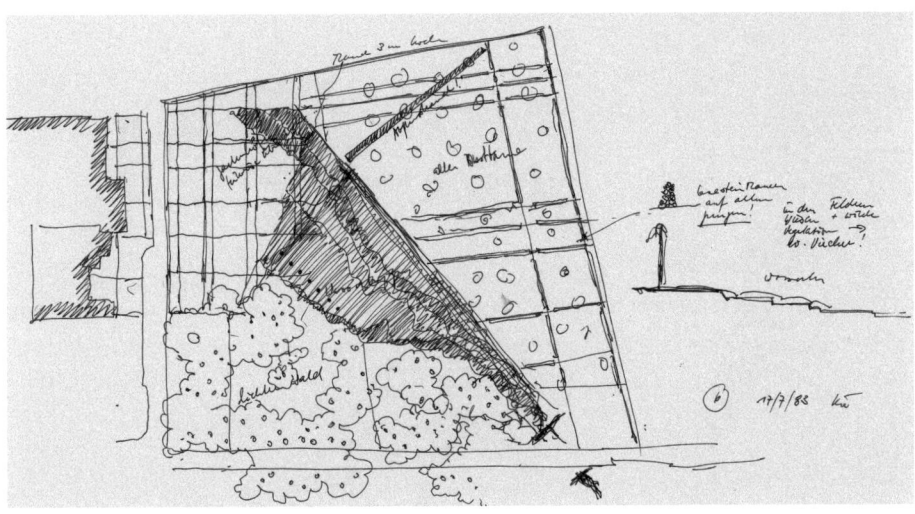

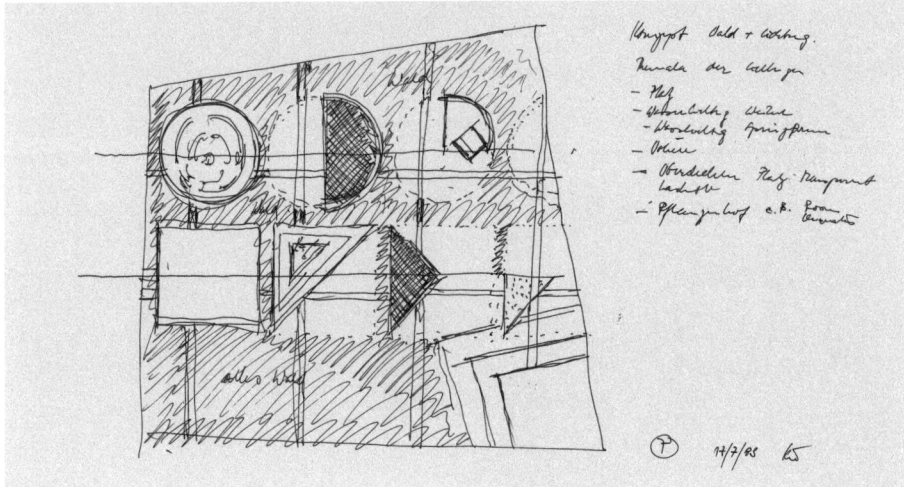

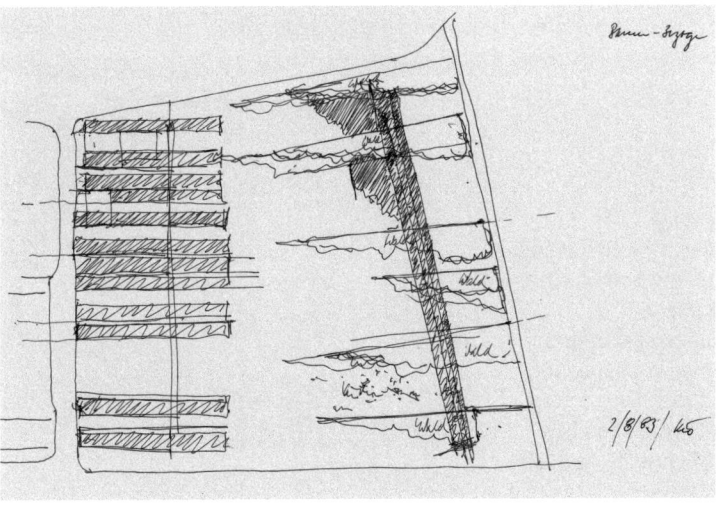

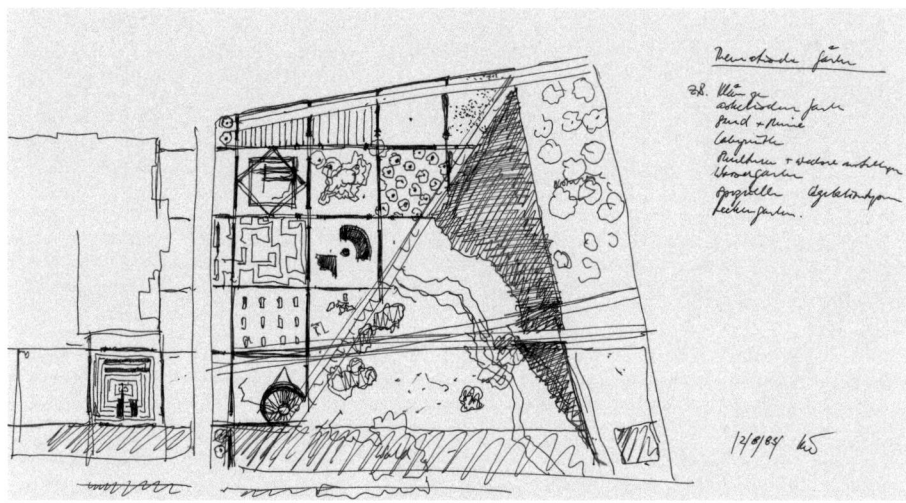

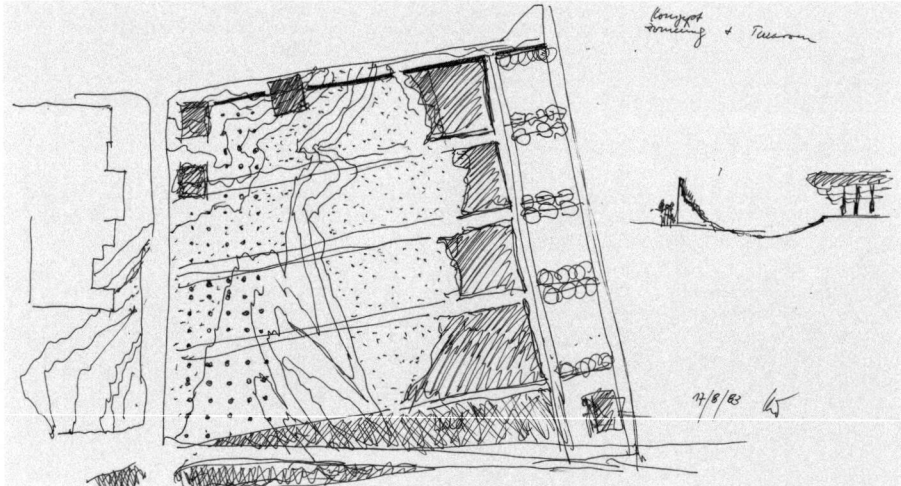

96–99
Dieter Kienast, twelve sketches for the Zurzach Spa Gardens, ink on tracing paper
    Unnumbered sketches (presumably nos. 9 and 10), August 2, 1983
    Unnumbered sketches (presumably nos. 11 and 12), August 17, 1983

pattern of Scottish fabrics: two vertical and horizontal bands close together and in a color distinct from the ground are superimposed to create a grid.[168]

The zoning of the terrain takes into account the existing three-part division—green space with paving slabs near the thermal bath, the orchard, and the smaller grassy area with existing buildings—while a series of sketches shows a strip parallel to the southeastern and/or the northwestern boundary. The widths of these zones are varied. As the third system of reference for a tartan grid, Kienast chose the course of the northeastern street. Some of the axes parallel to that reach out like fingers toward the park, but then they either do not continue or they cut through the park, either entirely or through just one zone.

Toward the center of the park these various ordering systems overlap—or Kienast only extends them far enough into the park that the center remains open for areas of grass or water (Sketch nos. 2, 3, 5, 9, 12). In one sketch, a diagonal cuts through the park from south to north (Sketch no. 6); in two others, from west to east (Sketch nos. 4 and 10). These diagonals divide the site into two triangles. In both designs with the diagonals from west to east, square thematic gardens face an organically designed "natural" area, separated by the diagonal of the path. In the drawing with the still little differentiated diagonals from south to north, Kienast introduced two grids whose geometry is disturbed by the landscape and by organically flowing parts. One early idea that he retained and varied during the design process was to frame the park with a forest zone. In the end, he settled on placing this zone in the southwest, then in the last of the twelve sketches labeled it "Projekt Waldlichtung" (Project Forest Clearing), before coming to the solution ultimately implemented of bordering the park with a small forest in the northeast and southwest.

In Sketch no. 12, the transparent organization of form based on interlocking different systems of reference is pushed to the extreme, while at the same time finding an elegant solution to integrating the spa gardens into the city (99): Kienast planned four zones on the boundaries of the site, some separating the park from its surroundings and others connecting it to them. As a result, entering the park becomes a different experience depending on the direction from which the visitors approach the park and the side from which they enter.

To the northeast and southwest, the site is framed by a small forest in each case, and thus shielded from the new shopping center and the large parking lot, respectively. Visitors cut through a forest zone before arriving at the opposite of the urban world: the natural sphere of the park. These forest areas have clear boundaries drawn with a ruler. Their course is determined, on the one hand, by the immediately adjacent street and, on the other, by the street on the opposite side of the park. Hence the two streets are mirrored in each case in the side of the corresponding forest that faces the park. As a result, not only the two forest areas but also the clearing in the center of the park feature a trapezoidal form. This presentation of the forest zones as framing and blocking the city is supported by extended hedges (103). Here Kienast was imposing the principle of transparency by which space is built up in layers.

To the northwest, toward the thermal bath, the park is bounded by a strip of rectangular thematic gardens. Kienast noted on the plan of the sketch "Themengärten" (Thematic Gardens), and on the upper right edge "Sand, Steine / + Geröll" (sand, stones / + pebbles) and "Labyrinth / Skulpturen / Ausstellungen" (labyrinth / sculptures / exhibitions). The thematic gardens were inteded to be connected via a pergola structure to the sidewalk and the plaza in front of the thermal bath, and this is still retained in the design plan of 1984 (100). In the project plan of 1985, this motif has disappeared, and the forecourt is lined by Lombardy poplars, which, as previously in front of the town hall in Wettingen, were supposed to rise up into the air like flagpoles, as the axonometric drawings, also from 1985, show very well (86). The thematic gardens become a transition zone between the complex of buildings and the nature of the park. They are provided with nature designed like a garden and provide amusement for the spa guests entering the park from the bath.

Kienast designed the southeastern border of the spa gardens as an orchard. In the sketch, he still placed the trees freely in the meadow, whereas in the design plan he employed a grid, though one that also has blank spots. In the process, he created a flowing transition to the gardens of the historical houses that surround the park. The contextualization is very clear from the design plan: the orchard seems to flow into the meadows of the gardens, with a few trees scattered in front of the buildings. The zone labeled "Obstwiese mit Schafen" (orchard with sheep) on Sketch no. 12 (99) alludes the rural character of Zurzach and the medieval structure of the village. It is an evocation of the nature here that was once used for agriculture. The existing paths were also incorporated and extended into the center of the park.

The climax of the spa gardens is the exposed mountain stream. Kienast guided it through the park parallel to the southeastern boundary. The stream comes to the surface in the forest at the southern end of the park where Kienast designed a straight canal to guide the stream toward the northeast. At the height of the clearing, the stream gradually widens into a pond: whereas its southeastern bank continues as a straight line, the water splays out to the west, toward the center of the park, where it is lined by a flat bank (101). Kienast had gravel poured there so that ruderal vegetation would develop;[169] a belt of reeds was added later. On the northeastern boundary of the park, the water flows back into a channel and "falls," as Kienast put it in his description of the project, "clearly visible and audible, back into the canalization."[170] The appearance and disappearance of the stream, which transforms into a pond in the park, is staged like an event. As he had earlier with the diagonal line of the municipal park in Wettingen, Kienast emphasized the beginning and endpoint of the channel's axis by using distinctive materials. The design plan reveals his intention to create small plazas that extend across the adjoining streets. On the project plan (100), he abandoned this element again: the stream comes to the surface in a small forest clearing. Kienast framed its site of entry into the park with a colorful mosaic basin—a response to a trip to Barcelona that year and the enthusiasm for Antoni Gaudí it had awakened.[171]

Other aspects of the design are made more precise on the project plan—Kienast's first "gray plan":[172] the bank of the pond that runs to the zone of reeds and ruderal vegetation to the west is accompanied by a path in the shape of a segmental arch that frames the central lawn of the park. A few trees arranged in small groups—known as "clumps" in the terminology of English landscape gardening—line the path. The grassy field creates a generous expanse. As a sweeping clearing, it contrasts with the two forested areas in the northeast and southwest (102, 103). With a single poplar placed near the thematic gardens, Kienast wanted to mark this plane and at the same time make it look optically even more spacious.[173] The thematic gardens that terminate the grassy area to the northeast are framed by wide pergolas.

Postmodern
Design Principles

The planning of the Zurzach Spa Gardens unites different postmodern design methods, above all spatial organization based on the principle of transparency and also as a reaction to existing traces and structures. They enabled Kienast to turn the park into an urban-planning element that lends form and structure. The individual components of the site take on meaning in that Kienast uses them to run through different forms of nature, some of which respond to traces and structures: in the southeastern part, for example, he worked with the agrarian nature of the orchard, which recalls the village character of Zurzach and transitions into the historical core of the village; in the northwestern part, the thematic gardens respond to the urbanization of Zurzach by the modern complex of the thermal bath and its looming tower. The nature of the thematic gardens is visibly artificial; its urban use and its amenity are brought to the fore. In his approach to the exposed stream as well, Kienast played with naturalness and artificiality by having the stream enter the park through a mosaic basin. On its bank facing the agrarian nature of the orchard, Kienast gave the pond a straight "backbone" in the form of its line. Facing the thematic gardens, it was supposed to develop a belt of reeds and ruderal vegetation. But even this "natural nature" is immediately contrasted by a segmental arch. In the northeast, an extended staircase with three stone steps covered with gravel marks off the pond. The broad grassy area and the clumps cite elements of the English landscape garden, whereas the straight axes, the thematic gardens, and the bosque zone derive from the Italian and French garden styles (105). But other grounds likely also inspired Kienast's design for the Zurzach Spa Gardens, such as Leberecht Migge's public garden of 1910 in the Fuhlsbüttel district of Hamburg (104), with a spacious central lawn, its expanse accentuated by the placement of a single, large chestnut tree, and framed by pergolas, boulevards, and two gridlike groves (maple and birch) with an identical box layout. Migge likewise drew on the traditional repertoire of forms in garden history for his design.

The collage, the quotation, and the fragment are characteristic of postmodernism. Kienast alluded many times to *Collage City* by

Colin Rowe and Fred Koetter and pointed to the sentences in which its authors spoke of the "suggestiveness" of the garden "for the 'planner' or the 'designer' of cities" and of the "garden as a criticism of the city and hence as a model city."[174] Because Kienast left open in his texts how much these ideas were brought to bear in his work, he may have employed these quotations strategically, using Rowe's authority to underscore the relevance of garden and landscape architects for urban development, which was a subject barely addressed otherwise in the discourse on architecture of the time. Kienast's designs clearly reflect Rowe and Koetter's call for a "city of composite presence" and a "metropolis of loosely organized sympathies and enthusiasms."[175] Kienast's interest in detecting role models in the history of garden art and the influence that contemporary landscape architecture had on his projects both find expression in his "composite presence." Particularly striking examples of the latter are his borrowings from two prizewinning projects from the competition for the Parc de la Villette in Paris, which was the most important international competition for landscape architects in the early 1980s and had an enormous influence on the evolution of landscape architecture in Europe:[176] whereas Sheet nos. 4 and 7 of Kienast's sketches for Zurzach make clear reference to Bernard Tschumi's competition entry (107), motifs from Bernard Lassus's design have obviously found their way into Sheet nos. 9 and 12 (106).

Kienast's "composite presence" also drew on personal memories from his travels, such as the mosaics inspired by Gaudí that frame the mountain stream's emergence into the park. The various motifs and quotations in Kienast's work are always integrated into a concept based on precise analysis of the site. Other crucial components of his plans are designing with contrasts and presenting various forms of nature.

Just how much Kienast's design for the Zurzach Spa Gardens of the mid-1980s differed from the methods of other Swiss landscape architects is clear from looking at the plans of the competition entries that received the second and third prizes (108, 109). Kienast's ideas were not, however, implemented completely, and the qualities of the spa gardens were only partially appreciated and accepted by the residents. For example, the thematic gardens were never realized, so that attractions were lacking, which were then integrated later into the finished park at great cost: a music pavilion was placed on the lawn and showy ornamental beds with canna were installed. Both elements disturb and undermine the sweeping, spacious effect of this part of the park. Moreover, the paving slabs of the old spa gardens were never completely removed, so that the grassy area is even more fragmented. A set of spray nozzles for waterworks was placed in the reed belt next to the pond and surrounded by decorative plants, which destroys the belt and looks trivial next to the parts of the bank where the reeds and ruderal vegetation have been preserved.

Inaccessible, Unalterable Gardens?

The indisputable refinement of its concept also reveals the supposed deficit of the Zurzach Spa Gardens: as soon as even one element is altered, the overall impression is lost, and the composition is destroyed. Ilse Helbich's criticism quoted at the beginning of this chapter alludes to Kienast's gardens being perceived as "inaccessible" and sees their quality lying solely in the possible reflection on the rifts and tension of our day—seeing sharp COR-TEN steel edges and concrete steps contrasting with the natural elements.

Whereas in the 1970s Kienast still argued for the appropriation and possible altering of a site by its uses, in many of his later works this is only possible at the cost of aesthetic sacrifices, since the compositions have been thought through in such detail. It must, however, be mentioned here that Kienast's projects realized in Switzerland are all for relatively small sites. The successful competition entries for the large urban parks developed in collaboration with Erika Kienast-Lüder and Günther Vogt in Berlin (Moabiter Werder, 1990–91) and Frankfurt am Main (Günthersburgpark, 1991–92), which were based on the same design methods as the Zurzach Spa Gardens, were never or only partially carried out (83, 84).[177] They would have been a better test of the user-friendliness of Kienast's postmodern designs in terms of receptiveness to change and processes of appropriation and participation. Kienast himself emphasized user-friendliness in a lecture at the symposium "Die Choreographie des öffentlichen Raums" (The Choreography of Public Space) in Berlin in 1991, by illustrating his reflections

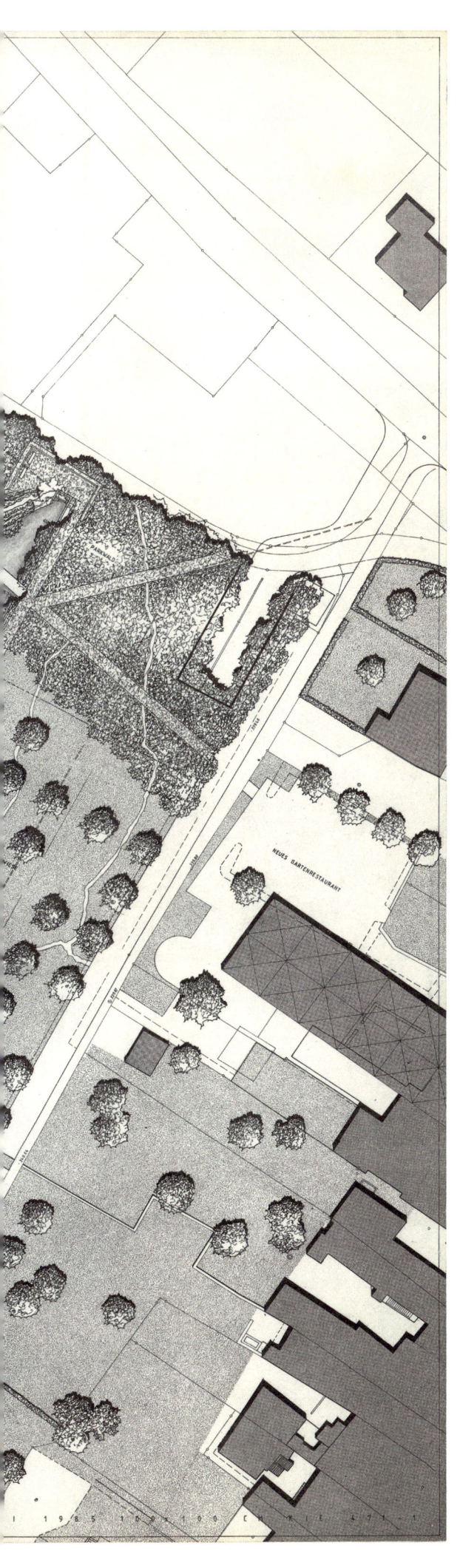

100
Zurzach Spa Gardens, project plan no. 471-1, scale: 1:250, H 100 × W 100 cm, copy of plan on lightweight card-board, ink and pencil, July 26, 1985, signed "CM, KIE," firm of Stöckli + Kienast

101
The mountain stream widens into a pond, which is bounded to the southeast by a hedge, to the northeast by elongated steps, and to the west by a flat shore zone

102
View along the northeastern shore. The pond is accompanied by atmospheric willows. To the left is urban forest in the northeast of the park. Photos: Christian Vogt, ca. 1991

103
Another small forest borders the park to the southwest. A hedge separates the forest from a street that runs along here. Beyond the street, a second hedge was planted to protect pedestrians from the traffic. Photo: Christian Vogt, ca. 1991

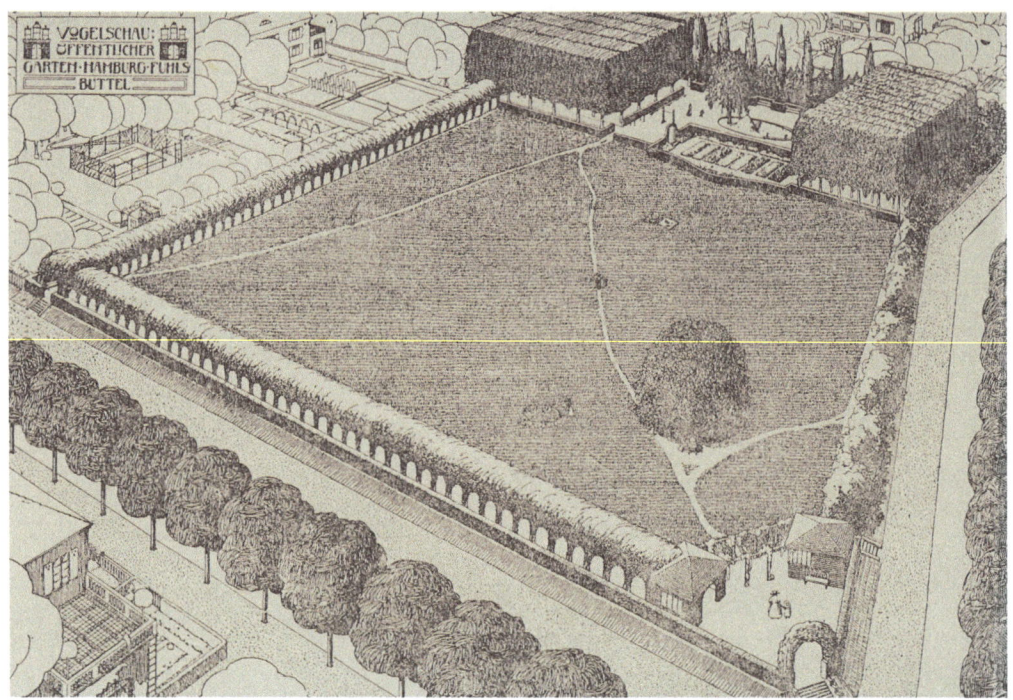
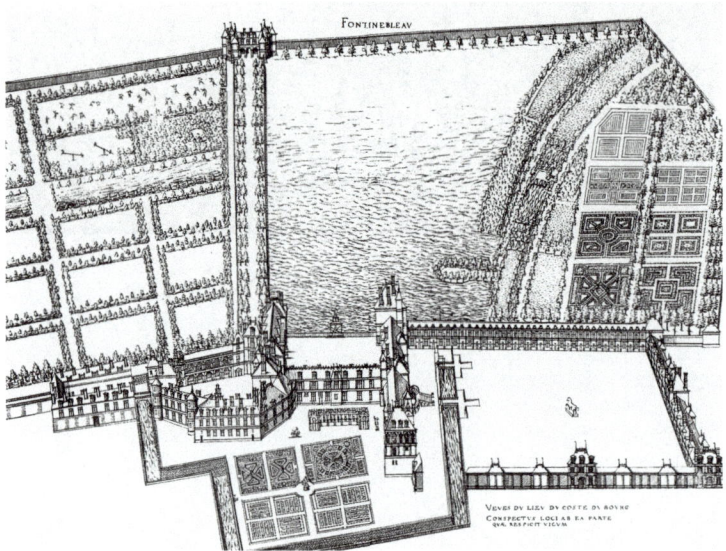
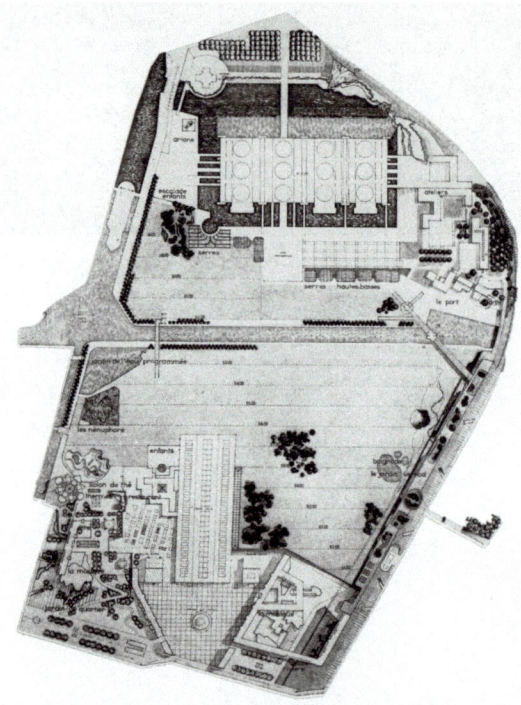
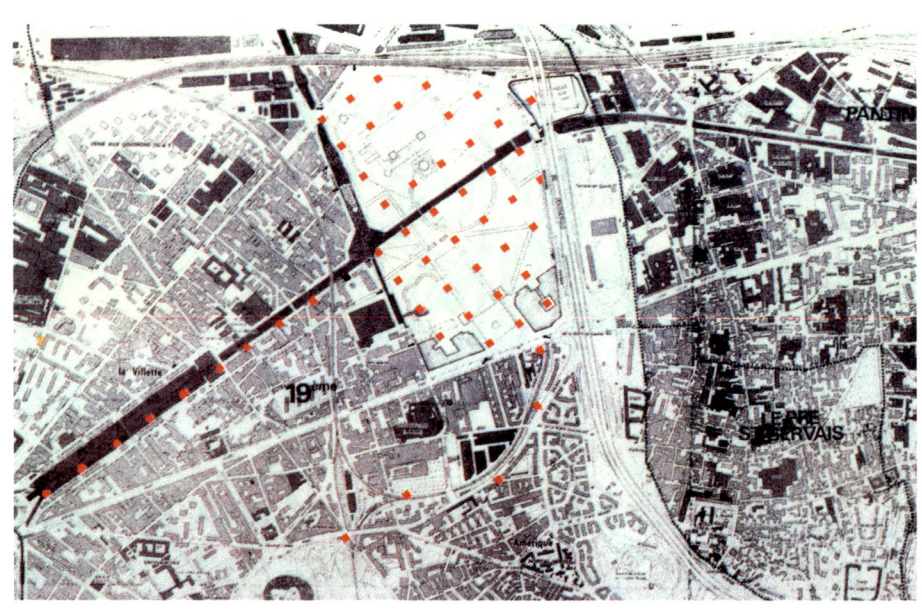

104
Leberecht Migge, public garden in the Fuhlsbüttel district of Hamburg, ca. 1910
105
Château de Fontainebleau, plan of the garden by Jacques Androuet du Cerceau, ca. 1570
106, 107
Competition for the Parc de la Villette in Paris, 1982
    Bernard Lassus, third prize
    Bernard Tschumi, first prize

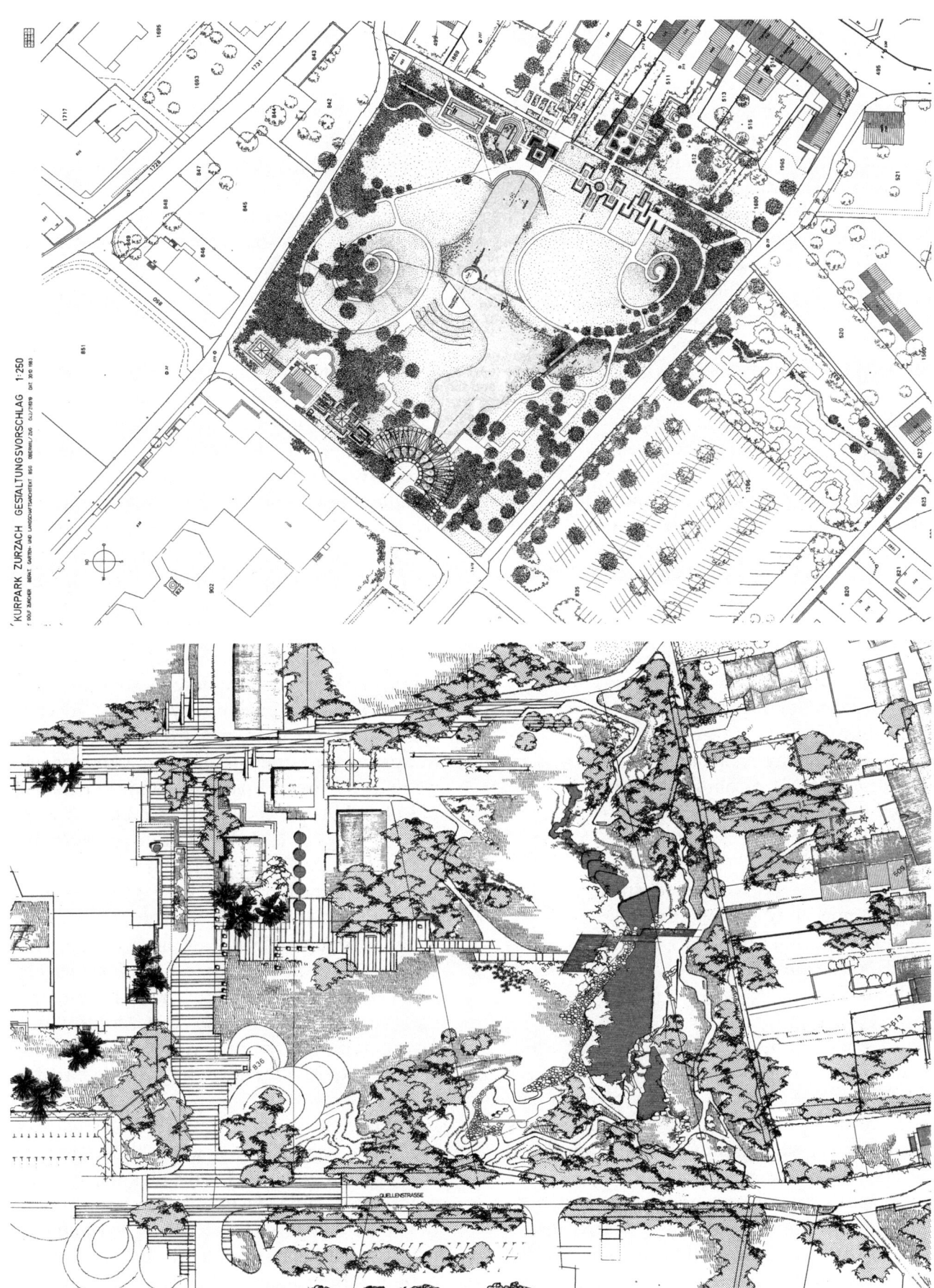

108, 109
Competition for the Zurzach Spa Gardens
Dölf Zürcher, second prize, October 30, 1983
Willi Neukom, third prize, undated (1983)

on public space and its uses by example of the prizewinning competition entries for the parks in Moabit and Frankfurt:

The city is no longer a monolithic structure. It is fractionated and dismembered a thousand times, with the medieval building next to the fully air-conditioned high-rise, the church next to the shopping center, the secondhand shop next to the high-tech boutique. Its population, too, is a colorful mix of old and young, cleric and junkie, migrant workers next to the long-established. Behaviors based on class and role have broken down. Grandma wears a Walkman, a teenager is playing Beethoven on the street, a forty-year-old is wearing tennis shoes and blue jeans, the fifteen-year-old is chicly styled.... This heterogeneity in the city and its residents demands a contemporary reaction outdoors, based on the authenticity of custom, place, and history....

La Villette by Tschumi is a built example of the theoretical concept of transparency, of the destructive, of ambiguity....

Simultaneity as a principle of the development of outdoor urban space in landscape architecture affirms the heterogeneous quality of the city and its residents. Heterogeneity, according to Lassus, is more receptive than homogeneity, provoking visuality and thus creating outdoor spaces in which society can find itself, as can the individual citizen.[178]

In this text, Kienast also evoked High Modernism and its gesture aimed at the great whole—at the utopia of a new city and a new social order—whereas postmodernism accepts the disperse, the plurality, and simultaneity.[179] Despite all of the postmodern influence, even in Kienast's smaller-scale built works this "modern" utopian quality never disappeared entirely but was instead tied either to a concern for society or to the effort to arrange the world into a pleasing whole within the boundaries of the garden.

Paradigm Shift

A landscape design based on the formal principles of collage, montage, and transparency that responds to a historical urban-planning context and quotes from the history of garden art makes it clear that a paradigm shift had occurred between the 1960s and the 1980s. Like the Kassel School, where Kienast had studied, the Texas Rangers also developed their method as a critical reaction to architectural education based on the principle of the master. The latter was advocated in the United States after the Second World War by the Bauhaus school, which argued for a kind of *tabula rasa* architecture. "Avant-garde" meant constantly having to produce something new. Drawing on theories from the history of architecture was rejected. In the basic course at ETH Zurich, which had been influenced by the Bauhaus, students were required to preserve their childlike innocence, unburdened by precursors, concentrating only on solving the problems themselves, because, it was said, every problem bore its solution within it.[180] The Texas Rangers, by contrast, focused their efforts on "abstracting from the abundance of existing works the insights or methods which, when freed from the particular and the personal of isolated cases, become transferable and available,"[181] as Bernhard Hoesli wrote in 1968 in his foreword to the German translation of *Transparency*. Theoretical principles were derived from empirical results—that is what Hoesli regarded as the fundamental value of this programmatic treatise.[182] At the same time, both *Transparency* and *Collage City* represent "a morphological approach that holds the exact description of a phenomenon as the necessary and indispensable prerequisite for any insight, understanding or knowledge."[183]

Here the goals of the vegetation studies of the Kassel School, as a new method to provide a scientific foundation for the planning of open space, are combined with the search for form of the Texas Rangers. A morphological perspective and a systematic method were core points of both approaches, and Hoesli went so far as to compare the form-organization of transparency with Carl Linnaeus's achievements in the field of botany.[184] Kienast achieved something similar in synsociology when introducing the sigma leap. Another interesting commonality is that Hoesli wanted to use the methods of transparency and collage to improve the legibility of a place by means of a design intervention.

Hence, there are strong parallels between these two attempts to reform education and practice. The Texas Rangers and the Kassel School rejected the purely intuitive approaches of the master architects. Phytosociological studies—and, in some cases, the social sciences as well—were just as focused on observation and systematization as were the studies of finding form by means of transparency. Legibility was an important goal for both. Nevertheless, in the context of landscape architecture, this critique led to iconoclasm and in the context of architecture to a flood of images, as the publication *Collage City* impressively demonstrated. In the 1990s, this flood of images spilled over into European landscape architecture, clearly as an excessive reaction to many years of hostility toward creating strong images through design. The presentation plans by the firm of Kienast Vogt Partner for the Moabiter Werder in Berlin and for the Günthersburgpark in Frankfurt are an eloquent first example of this. As part of this extreme counterattack, many professional and scholarly achievements of the reform efforts of the 1960s and 1970s were thrown overboard, and images were used above all for one purpose: to captivate.

Kienast brought together the methods of phytosociology and of transparency in his works of the 1980s, profiting from both and from their insights, which enabled him to create works that attempted, in an increasingly complicated world, to offer orientation.[185]

One's Own Garden as a Field for Experimentation

At no other site did Dieter Kienast spend as many years working as he did in his family's garden in Zurich. There, in that freely accessible space, he could try out different types of design. The constantly changing garden is thus more evidence of the developments in Kienast's work.

The Kienasts' private garden is located on Thujastrasse in the Wollishofen district of Zurich, around 4 kilometers south of the center of the city, in a residential neighborhood southeast of the autobahn to Chur, from which the neighborhood is shielded by the Muggenbühl chain of hills. It is about 400 meters from the left bank of Lake Zurich, with a heavily traveled railroad line and two major arteries passing in between. The garden grounds extend around two multifamily homes and are subdivided into many small spaces (110–112). While the entrances to the buildings are on the level of Thujastrasse, the garden is about 4 meters lower and is accessed via a ramp to the east and a spiral staircase between the two buildings. The garden can be entered directly from the lower floor of each building.

From the Nursery
to the Private Garden

The history of the Kienasts' Garden can be divided into three phases. Until 1977 the property was operated by Dieter Kienast's parents as a nursery, with cold frames, planting beds, hotbeds, a greenhouse, and sheds. The five-member family lived nearby, in the easternmost of the two buildings, which was the only one at the time, with an adjoining store on Thujastrasse for the sale of plants and flowers. The Manegg Cemetery, located at the end of the street, provided the nursery and florist with a lot of walk-in customers. The planting beds and greenhouse could be reached directly from the residential and commercial building. Traces of professional gardening activity, such as the old greenhouse, a cold frame, and a cistern, could still be found in 2016, when the ensemble was sold to new owners but Erika Kienast-Lüder was still living there (113, p. 264).

When the nursery was closed in 1977, a multifamily house was built on the property, and Dieter Kienast and his family moved into the lower floor. The rest of the lot was converted into a nearly natural garden with a wet biotope with many species, a fire pit, a sand playground, a vegetable and herb garden, and small seating areas (116–119). The goal of the redesign that Kienast planned in 1976 was "an optimal and intense use of the extremely limited open space."[186] From the time the firm of Stöckli + Kienast expanded to become Stöckli, Kienast & Koeppel in 1987, which was the same year that Günther Vogt (b. 1957) joined the firm, Kienast worked exclusively on the lower floor of the original house, in the rooms of the former nursery, with direct access to the garden (126). Erika Kienast-Lüder collaborated on his projects from 1980 onward. On Thujastrasse, professional and private life, work and leisure time, transitioned seamlessly.

The third phase of the design of the garden began in the mid-1980s, marked by a plan from 1985 and another plan from 1991; the latter was presumably drawn as documentation of the already executed changes in preparation for the first exhibition of the firm's work in 1992.[187] Both document the profound transformation of the garden at this time: from a wet biotope to a ceramic mosaic surrounded by topiary plants, from a nearly natural garden to a composition of gravel areas, pruned hedges and trees, and shrub beds framed by boxwood borders (116–124). Nevertheless, in each phase traces of the older state were incorporated into the design, so that the garden became, as Kienast put it, "part of the built and organic life history of its authors."[188]

In 1985, the following items from the redesign completed in 1979 were retained: the large vegetable, herb, and flower garden on the southwest boundary, the composting area to the west, several erratic boulders as "traces," the steps that led from the "wildest" part of the garden to the seating area in front of the Kienast family's apartment, the seating area itself, the greenhouse, and the cold frames and hotbeds (110, 111). The planting and the formal differentiation of the individual spaces changed, though there were certainly still connections to the earlier design. For example, the "wildest" area of the garden, in terms of its vegetation, located to the west of the second residence, was still densely populated in the design of 1985. But the wet biotope was replaced by a ceramic mosaic, surrounded by several trees trained in the shapes of mythical animals. The reason was simply a natural one, as Kienast explained in 1986:

> Our intensely used terrace was bounded by the most lush, wildest section of the garden, where border nitrophytes thrived alongside weather shrubs and trees from the nursery. Amid this wilderness, at the lowest point in the garden, gleamed a blue ceramic square. Grouped around it were ten (still small) mythical creatures. Previously, a dreamy, beautifully evolved wet biotope with many species had been located here. Until the ducks who arrived had, in no time at all, transformed the pond—this epitome of naturalness—into a slightly foul-smelling, bright green cloaca.[189]

Although it retained the practice, common in garden history, of dividing the space into functional areas for everyday use—sunny areas with shrubs or potted plants and shady areas with closely set trees and wild growth—the change from 1979 to 1985 was fundamental (120–124). The conversion took place in stages, beginning in 1983 without a design plan, and was carried out by three of Kienast's students from the Interkantonales Technikum Rapperswil, who at first worked on it during their school breaks: Daniel Ganz (b. 1961), Christoph Gasser, and Günther Vogt.[190] They began the redesign by installing a trapezoidal pool northeast of the garden seating and in front of the second residential building. Duckweed was placed in the pool, as is also indicated by a bright green surface on the plan from 1985 (111). At the same time, the wet biotope in the southwestern part of the garden was drained, and the new garden space prepared, in the center of which was placed a ceramic mosaic that Erika and Dieter Kienast had designed together (121). During the next stage, the playground located between the old building and the new one underwent a change: instead of a sandbox,

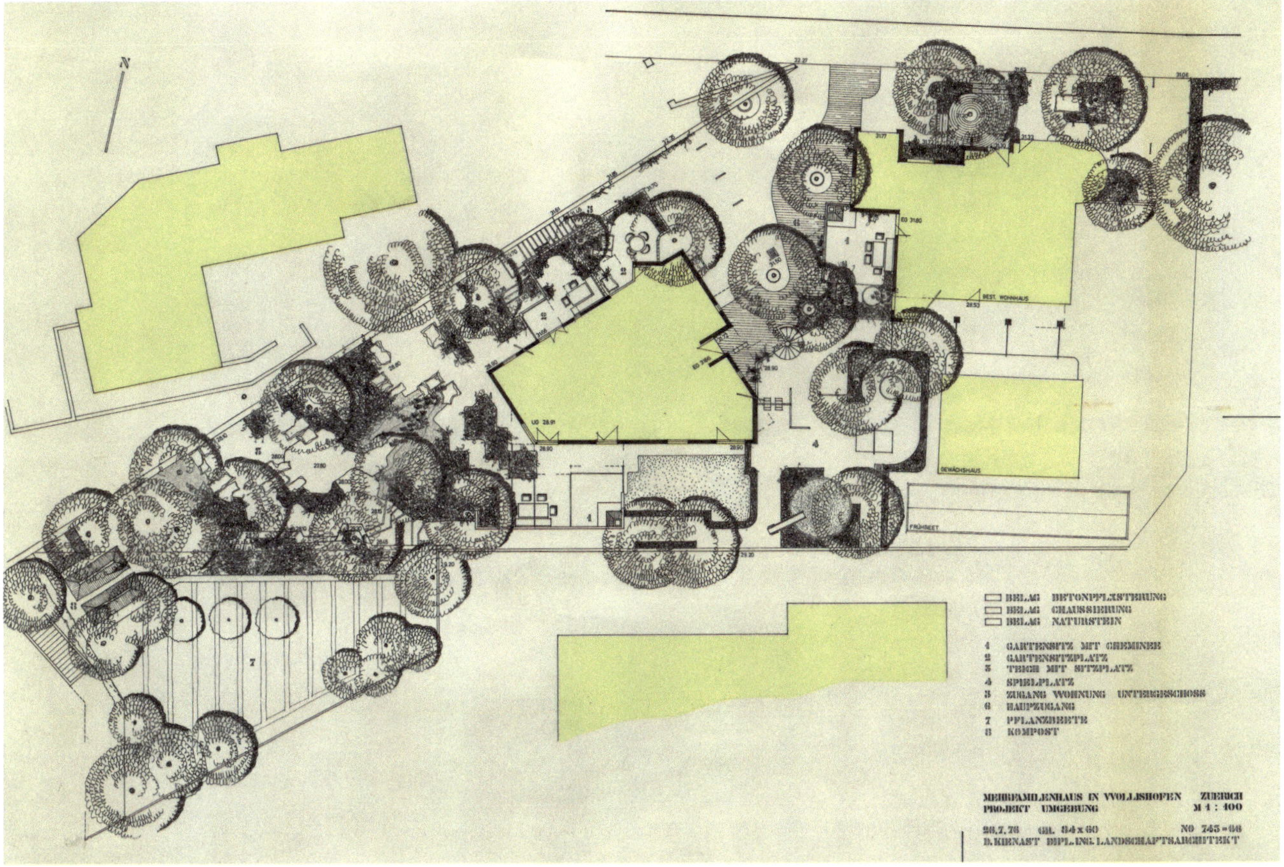

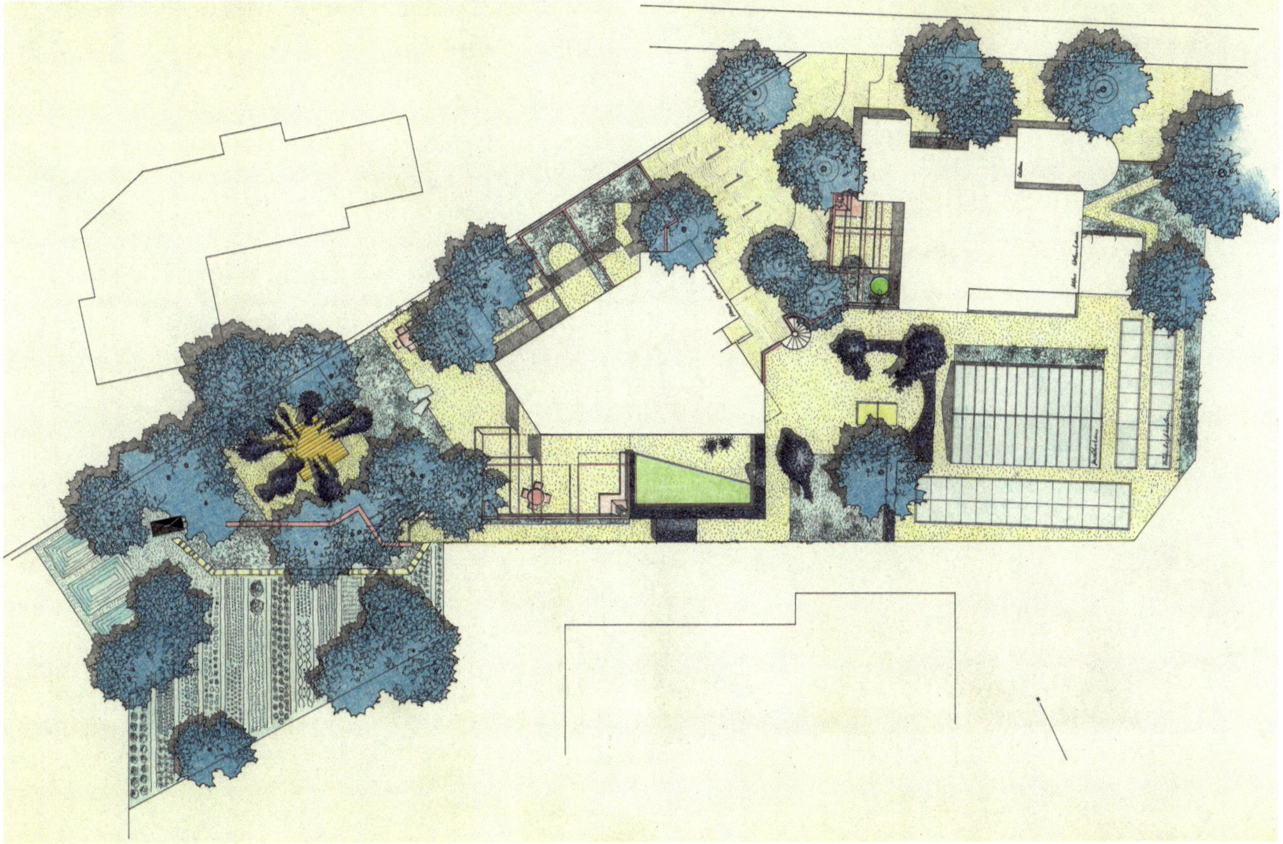

110, 111
The Kienasts' Garden in Zurich

Site plan no. 743-66, scale: 1:100, H 60 × W 84 cm, ink and pencil on tracing paper, colored film, July 26, 1976, Dieter Kienast, landscape architect

Site plan, scale: 1:100, H 60 × W 103 cm, copy of plan on lightweight cardboard, ink, graphite, and colored pencils, August 1985, signed "CM," Dieter Kienast, landscape architect

UMGEBUNGSPLAN FÜR ZWEI MFH IN WOLLISHOFEN DIETER KIENAST LANDSCHAFTSARCHI

112
The Kienasts' Garden in Zurich, site plan, scale: 1:100, H 84 × W 90 cm, copy of plan on lightweight cardboard, ink, graphite, and colored pencils, March 1991, Dieter Kienast, landscape architect

113
Greenhouse of the former nursery. The cold frames were removed in the late 1980s as part of the redesign of the garden and replaced by boxwood beds and small boxwood trees. Kienast experimented with the shapes of the trimmed boxwoods and the panels of the parterres, using, among other things, shards of glass.

114
View of hornbeams trimmed in birdlike forms (mythical animals). Photos: Christian Vogt, ca. 1991

115
View from the bridge separating the recess yard of the school for children with cerebral palsy. The view spans across the pond, which features nearly natural vegetation, toward the fenced-in Kienasts' Garden. Photo: Christian Vogt, ca. 1991

a field was left open, and it was soon overgrown with ruderal vegetation. The hedge and two trees were pruned differently, so that the latter were gradually transformed into other mythical creatures. A third "animal" was "raised" on the western corner of the new grassy field. Kienast owned a stepladder 7 meters long, on which he often stood smoking a pipe and trimming the trees and hedges, until it proved too short to control the shape of the "upward-striving animals."[191] Because of their volume (114), the actual size of the grounds is illusionistically increased, especially because they are standing in the part of the garden where a spiral staircase leads up to the front yard, located 4 meters up on the level of Thujastrasse, linking the entrances to the two residential buildings. From there one can look down on the garden and the mythical beasts rising out of it. The garden thus expands upward as well.

In his instructions for how the garden was to be redesigned, Kienast employed his newly developed method of contrasting and juxtaposing: "Built to grown; bright, sunny areas to shady, cool ones; hard coverings to soft; ruderal vegetation to potted plants; freely growing trees to trimmed trees; artificiality to naturalness."[192] By employing such strategies of composition and effect, which were intended to compel the sensory experience of the site, he was working in parallel with his other projects of this period, such as the Brühlwiese Municipal Park in Wettingen and the Zurzach Spa Gardens. The differentiated spatial organization of sequential areas that was already established in the plan from 1976 was now reinforced by individual interventions: though Kienast terraced the spaces consecutively, here and there he kept longer axes open, so that prospects alternate with close sequences of space—a method frequently employed in garden architecture to deceive the viewer into seeing the space extend beyond its actual size.[193] Kienast divided the spaces with geometrically trimmed hedges, hornbeam by preference, which were formed into real "gates" between the two parts of the garden. Apart from the "wild part of the garden" in the back, all of the ground was covered with gray gravel. The uniform ground also makes the garden look larger than it is, and it offers a monochrome backdrop against which the green plants look particularly beautiful. That has been the special quality of the Kienasts' Garden since the mid-1980s: only green plants grow there. To the extent they bloom at all, their flowers are, almost without exception, white—the garden is presented in the colors green, gray, and white.[194]

The most radical differences between 1985 and 1991 concern the northeastern and southwestern areas. In the 1985–86 season, a ramp with an adjoining stairway was added that leads down from the street level to the garden (112). Next to the ramp, at the height of the entrance to the office, four apple trees were planted, and a rectangular red mosaic designed by Erika and Dieter Kienast was placed at their center (125). The spot for the nursery's two cold frames was taken over by shrub beds framed by boxwood. Other changes followed: the vegetable and herb garden in the south end of the lot was on a site that Kienast's father had to cede to the City of Zurich for a school in the mid-1960s.[195] He was compensated financially for it and permitted to continue using the land on a lease basis. In 1986, the kitchen garden was removed and the grounds handed over to a school on Mutschellenstrasse for children with cerebral palsy. The contract for designing the school's new grounds was granted to Dieter Kienast, who converted the former vegetable garden into a recess yard and a pond with nearly natural vegetation. He separated it from his own garden by a fence that enabled each side to see the other property (115). At the suggestion of Günther Vogt, two metal bars were placed as a footbridge and connection to the now much-reduced compost area. The old wall that separates the lower lying "wild garden" from the upper lying footbridge, both running parallel to the new fence of the property, was retained: ferns were planted between the fence and the footbridge and spherically trimmed boxwoods between the footbridge and the wall (p. 262). A bench from the 1970s in a niche in the wall was allowed to weather, and today, overgrown with moss, it reveals the long history of the garden.

On the path to the compost area, a small, rectangular clearing was opened up in the southwest, on which a smaller, square field stands out (112): here the extreme mountain climber Kienast tried to grow edelweiss—a venture that never had much luck.[196] This "alpine garden" was bounded on two sides by an iron bar—another experiment in this garden with multiple references to Kienast's own biography, which were presented not without self-irony.

The outward change to the garden was associated with a transformation of its functions: In the days of the nursery, its purpose was business. From the end of the 1970s to the mid-1980s, the children of several of the tenants' families played in the new gardens; the large vegetable, herb, and flower garden served as a shared "kitchen or tenant garden."[197] Following the redesign and the opening of a branch of Stöckli, Kienast & Koeppel in 1987, the garden was "inhabited" almost exclusively by the Kienast family and the employees of the office. Its embellishment went hand in hand with it becoming increasingly private, though the Kienast family never really separated the private and the professional, and interested visitors—above all students—have always been welcome.

Changes in Kienast's Concept of Use and His Turn to Aesthetic Experience

The history of his own garden, its design and use, mirrors Dieter Kienast's personal development. In it, he expressed aesthetically his changing interests, elaborated his formal language, and explored the use of plants. His private garden was a field of experimentation, full of references to the impressions, discoveries, and experiences that Kienast made on his travels or while reading books on the garden in his extensive private library, which was also kept in the house on Thujastrasse. This included the Spanish-influenced mosaics and the topiary mythical creatures, which were probably inspired by the private garden of the Swedish landscape architect Sven-Ingvar Andersson (1927–2007) (114, 127, 128). In his private garden, Kienast experimented with such motifs and explored in miniature what he would then apply on a larger scale, such as the mosaics in the Zurzach Spa Gardens or the fields of mythical beasts and other topiary in the Moabiter Werder in Berlin (129) and in the Fasanengarten (Pheasant Garden) at the Internationale Gartenschau in 2000 (International Horticultural Show 2000) in Styria.[198]

The graphic design of Kienast's plans and his search for a way to present his work in drawings for presentation plans can also be traced in the production of plans for his own garden: the naturalistic topography of 1976, delicately drawn in ink and pencil, was followed in 1985 by a plan in the same style that was, however, colored in garish, antinaturalistic colors, with the gravel in yellow, the trees in blue, the kitchen garden and lawn in bright blue, and the topiary animals in dark blue. Finally, the plan of 1991 features only the colors gray, blue, and black, and is hatched; the material qualities of the separate areas are supposed to be conveyed by the graphic technique and the use of colored and graphite pencils of varying thickness and hardness (110–112).

Dieter Kienast's explorations in terms of program, form, spatial organization, use of plants and materials, and techniques of presentation are reflected in the genesis of his garden. He himself viewed the ruptures in his own evolution as the reason why, as he grew older, he developed a penchant for design that oscillated between opposites, as he emphasized in an interview in 1996:

> We cultivate discontinuity. Discontinuity also occurs in my own life and this is what provides stimulation. This can be strenuous sometimes, but it leads to the discovery of new horizons of experience and widening of sensory fields. Outside space must be a place perceptible by the senses. These experiences are only possible at the fractures between the poles. In other words, designing is always a process of oscillation between opposites.[199]

The Kienasts' Garden illustrates in an impressive way how Dieter Kienast's concept of use was transformed as a result of changes in his own living environment. He saw his earlier works as just as appropriate to everyday life as his later ones, but his idea of what a person needs in order to cope with daily life had changed. Emancipatory appropriation by means of participatory processes was replaced with the idea of the garden as place of "sympathy for and involvement with," where one's one creative potential is stimulated by sensory experiences. The garden should sharpen our awareness and arouse our senses, as Kienast summed it up in "Die Sehnsucht nach dem Paradies" (Longing for Paradise) in 1990, in which he observed that the ownership of a garden is defined not primarily by the size of the plot of land but by the extent of the engagement with it.[200] For Kienast, the latter included not only physical work in the garden but also emotional involvement.

Participatory appropriation or individual involvement—both attitudes have in common a utopian impulse and a subject-orientation. For Kienast, the users were always the focus, but the way he believed they can be made happy by a garden changed. This change was also expressed in the language of his programmatic texts. In 1981, he wrote about the landscape architect's work

116–124
Impressions of the Kienasts'
Garden in the late 1970s and in
May 1990. Photos: Dieter Kienast

125
In front of the entrance to the office, four apple trees were planted around a red, rectangular mosaic in 1985. Seen in the background are the new shrub beds framed by boxwood hedges.
Photo: Dieter Kienast, May 1991

126
The Zurich office of the firm of Stöckli, Kienast & Koeppel.
Photo: Christian Vogt, ca. 1991

127, 128
Sven-Ingvar Andersson in his private garden in Södra Sandby in Sweden, pruning his hawthorn sculptures. He called them "hens" and wanted them to look like they were walking through the garden.

129
Detailed plan "Spiel und Sport" (Games and Sports) for the park on the Moabiter Werder in Berlin: on the promenade along the Spree River, small fields of pruned hornbeam, yew, and boxwood were planned. Site plan with section and elevation (photo collage), scale: 1:200, H 147 × W 70.8 cm, copy of plan on lightweight cardboard, graphite and colored pencils, undated, firm of Stöckli, Kienast & Koeppel

in the tension between the poles of the contemporary dictates of aesthetics versus the dictates of nature:

The human being in the tension between garden and "nature" has to be the central theme of our engagement. That is why the question of the dictates of aesthetics versus the dictates of nature has to be asked under the premise of usefulness. . . . Against the backdrop of the social, economic, and ecological stresses on the residents, the *Aussenhaus* [literally "out-of-the-house," a coinage of Inge Meta Hülbusch to refer to open space] becomes the necessary compensating factor for the soon unbearable physical and psychological living conditions. The *Aussenhaus* can and must in this sense provide help in the banal coping with everyday life, and that makes demands of us planners that go beyond finding a form or a claim to naturalness.[201]

Kienast's design for his own garden in 1976 was developed based on these premises. In 1995, in a volume by the Bund Schweizer Landschaftsarchitekten (Association of Swiss Landscape Architects) on "good gardens" in the Canton of Zurich, he published a brief text on his own garden, by then redesigned, in which he struck a very different tone:

In the middle of the city, a sequence of small garden spaces—also a part of the built and organic life history of its authors. A garden as a playground for the landscape architect, where the separation of work, home, and free time has become obsolete. A homage to plants, from the nettle to the linden, the water hyacinth and the lemon, topiaries and the boxwood bed, the rose and the campion. A sensuous garden in the city, with a slight whiff of exhaust and jasmine, the chirping of birds and screaming of children, crunching steps on the gravel in the garden and—silently—tomcat Baldrian's hunting for a warm lunch.[202]

In both passages cited, using the garden and escaping banal everyday life play a central role. In 1981, however, there is a higher principle than "finding form" and the "need for naturalness," which can be identified as the concept defined in Kassel as the "use value of open space." In 1995, the surroundings of his own home—which in 1979 Kienast had described as "living greenery for multifamily homes in Wollishofen"[203]—became the "grown life history" and "playground for the landscape architect," which purely in terms of language already points to his now complete evolution into a designer using postmodern methods.

When characterizing the garden in 1995, he listed the plants growing there, from which anyone can picture it or imagine smelling it: nettle, water hyacinth, lemon, rose, campion . . . This mix of "feared plants," such as the nettle, which many garden owners would prefer to see banished, and with which probably every child has had painful experiences, and other, sweet-smelling plants; the reference to the smell of exhaust, which locates the private garden in an urban space; and the final image of the tomcat threatening the chirping birds—all of them bring reality into the private paradise and thus undermine it in an ironic way.

His own garden with its urban-dwelling nature becomes a space of yearning. It mirrors in miniature something that between the late 1960s and the early 1990s influenced the philosophical debate over the aesthetics of nature that had been rekindled—the theoretical approach that the experience of losing the possibility of seeing nature as an entity is compensated for by searching for nature in experience and its manifestations in aesthetic moments.[204] Ultimately, an ethics of nature is derived from this aesthetic experience of nature, requiring that we take care of our own environment, which in turn relates the aesthetic experience of nature back to our everyday actions.[205] Kienast's modified understanding of the use aspects of his gardens points in this direction as well. The aesthetic experience of nature is orchestrated by his design, in particular through the experience of contrasts.

The seed for this change was already planted in 1976 when Kienast added to the alternative proposal for Grün 80 in Basel, which he had developed with Toni Raymann, a famous quotation from Bertolt Brecht,[206] with which he also concluded his "Zehn Thesen zur Landschaftsarchitektur" (Ten Theses on Landscape Architecture), written in 1992 and revised in 1998.[207] Kienast chose a passage from *Geschichten vom Herrn Keuner* (Stories of Mr. Keuner), in which Mr. Keuner, asked about his relationship to nature, replies: "Now and then I would like to see a couple of trees when I step out of the house."²⁰⁸ Keuner justified this by saying that "thanks to their different appearance, according to the time of day and the season, they attain such a special degree of reality." Also—and this is the other crucial passage in relation to the changed "use value" of an open space for Kienast—it confuses him that in cities we "always see only commodities, houses and railways, which would be empty and pointless if they were uninhabited and unused":

In our peculiar social order, after all, human beings, too, are counted among such commodities, and so, at least to me, since I am not a joiner, there is something reassuringly self-sufficient about trees, something that is indifferent to me, and I hope that, even to the joiner, there is something about them that cannot be exploited.²⁰⁹

It is revealing that Kienast, in the context of his training in Kassel, would consistently call on Bertolt Brecht of all people as his chief witness for the aesthetic experience of nature that human beings need, and that he would retain this reference, as if he wanted to preserve the sociopolitical motivation of his oeuvre even in his turn to the aesthetic in his later works.

The Kienasts' Garden in Zurich, 1976–91. Photos: Georg Aerni, 2010

Page 262
Natural stone wall and moss-covered bench, which has been standing there since the 1970s, and boxwoods trimmed in spheres behind it

Page 264
View over the cold frames of the former nursery, now planted with peonies, toward the ivy-covered residential building. Trimmed hornbeams grow beneath the linden tree. To the right, the greenhouse

Mourning and the Experience of Nature: Consolation through Beauty as a Way of Coping with Everyday Life

In his final creative phase, Dieter Kienast understood the consolation that the beauty of a site designed with nature can give someone as one important way of coping with everyday life. This is reinforced when such sites enter into a dialogue with the landscape. In the case of places associated with religion, such as churches and cemeteries, there has been a long tradition of relating the religious site and the landscape. Cemeteries were one central design task for Kienast. In the text "Über den Umgang mit dem Friedhof" (On Dealing with Cemeteries) published in 1990, he stated that cemetery designs are "almost the only undisputed domain of our profession."[210] In it, he criticized the current overburdening of cemeteries with values and uses that distract from what Kienast saw as the true requirement when designing cemetery grounds:
In our secularized society, the cemetery is the site of the task of mourning, where we draw hope and certainty that life can remain worth living even after the death of someone we love, and where we are led to our own natural end in peaceful company.[211]

A cemetery today, he argued, is a place not so much for the dead as for the living, where they can come to terms with the experience of loss. Many cemeteries, however, no longer allow them to do so. The reasons for that, according to Kienast, lie, first, in society's encouragement to repress sickness and death from everyday life and collective consciousness and, second, in the discovery of cemeteries, once again socially conditioned, as compensation areas for urban ecology, as important habitats for plants and animals, and as nearby recreation areas for city dwellers.

The latter development, he argued, had a direct effect on the design of cemeteries and on the plants permitted there. The symbolic content of the design and the plants had become highly reduced or lost entirely. The repression of death and the demand that the cemetery be a nearby recreation area found expression, he argued, in parklike sites in which the symbols of graves were increasingly disappearing and where, in the extreme case, badminton is played on the meadow of a potter's field. He claimed that the expectation that the cemetery serve as an ecological compensatory area had led to a prohibition on roses and thujas in cemetery plantings. The arguments against these important plants in the history of the ritual of death—the rose as the flower of love[212] and the thuja as the tree of life—were said to be, in the case of the rose, that it supposedly required high maintenance and spraying and, in the case of the thuja, that it was an exotic plant and therefore not "ecologically correct."[213] The latter argument was also used to replace walls, which form a clear demarcation between the area of the dead and that of the city or countryside around them, with ecologically "more valuable" mixed hedges. But such design measures always sacrificed the cemetery's quality as a place of the "living cult of the dead," as Kienast put it.
> But where are the survivors left, with their mourning and their pain over the dead, with their fear of dying themselves? If the cemetery is no longer suitable for the task of mourning, then it has surely lost its original meaning and has become obsolete. We do not need cemeteries to dispose of dead bodies in a hygienically safe fashion.[214]

The Fürstenwald Cemetery in Chur is one of the projects in which Kienast formulated a response to the dilemma of cemeteries today. The work was a collaboration with the architects Urs Zinsli and Franz Erhard of Chur in the firms of Stöckli, Kienast & Koeppel and of Kienast Vogt Partner, with the participation of Günther Vogt, Erika Kienast-Lüder, and Peter Hüsler (130). The cemetery was a completely new site on a plot that the city had made available: outside of Chur, in a forest on a slope down to the Rhine Valley, opposite Chur's "local mountain," the Calanda Massif. The design emerged victorious from a competition organized in 1992 to 1993.

The terrain for the cemetery is surrounded by forests, meadows, and farmland. The rural

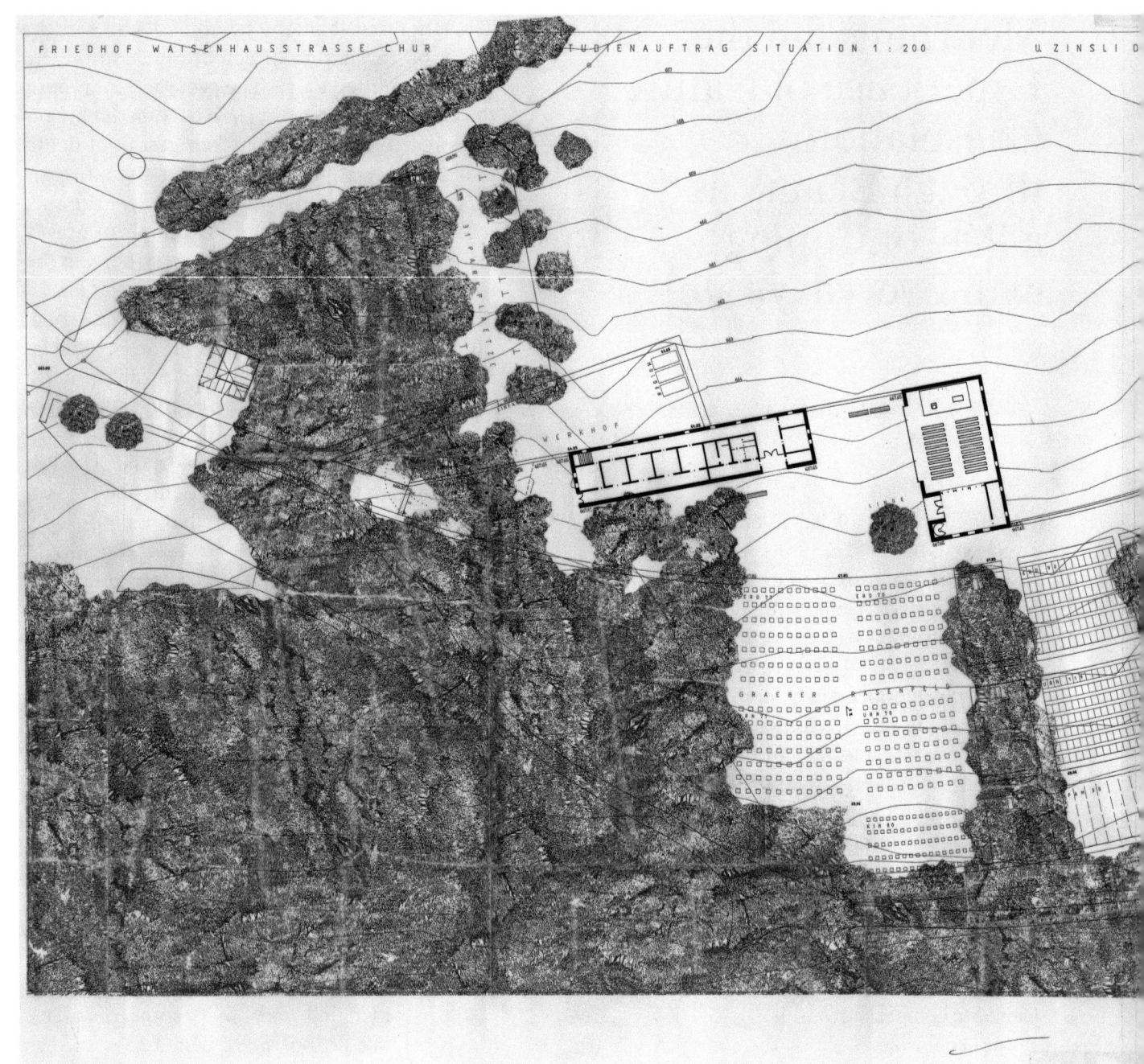

130
Fürstenwald Cemetery in Chur, study, site plan, scale: 1:200, H 78 × W 176 cm, copy of plan with collage on tracing paper, undated, firm of Stöckli, Kienast & Koeppel (with the Chur-based architects Urs Zinsli and Franz Erhard)

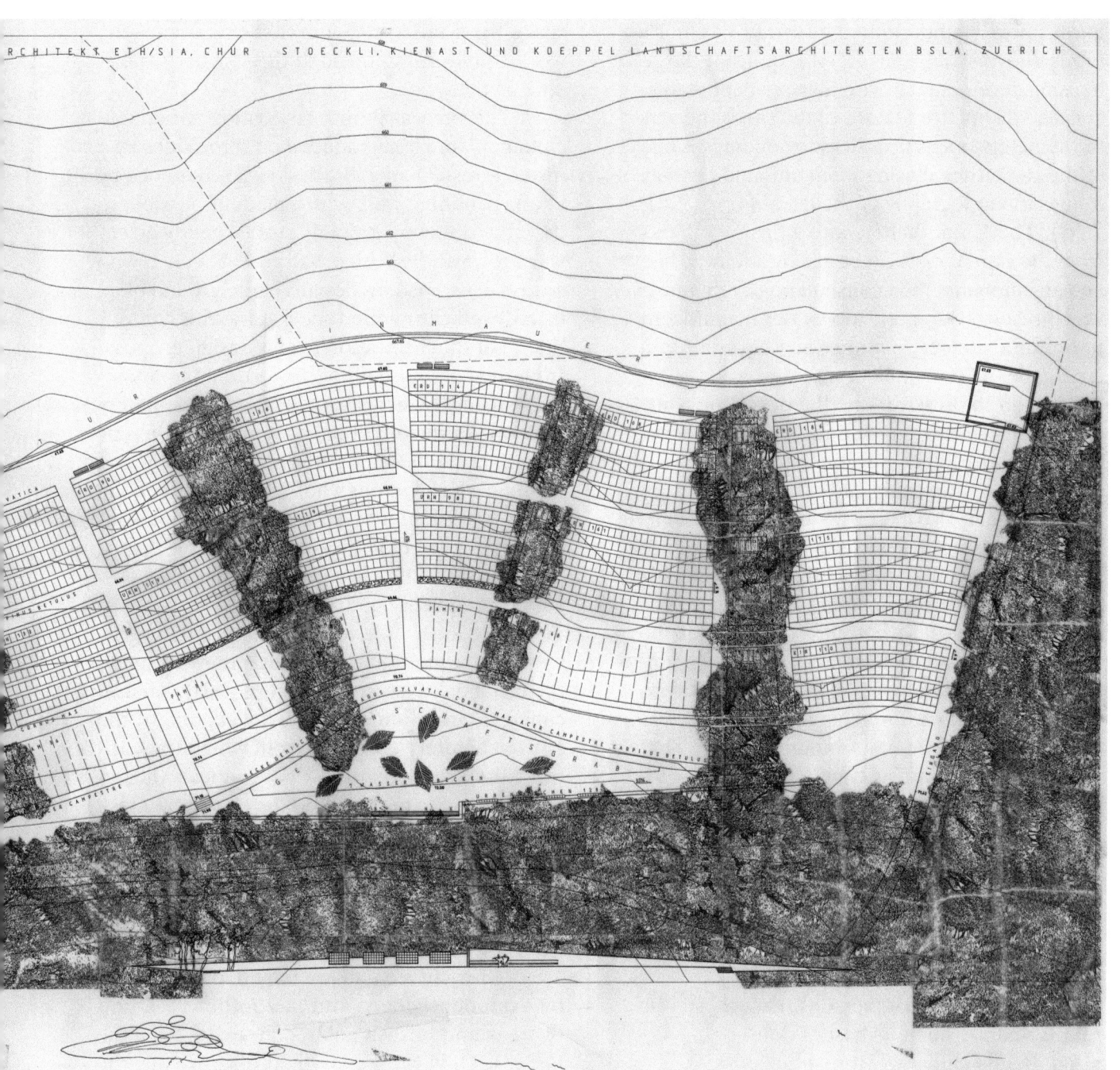

atmosphere is undermined by a view of Chur's industrial district and the surrounding large housing developments with high residential towers. The distinctive feature of the site is the way the cemetery opens up to its surroundings, while at the same time closing them out, and the way it hugs the topography of the terrain (pp. 270, 276). The Rhine Valley, into which one looks from the cemetery, is cut across by a railway line and an autobahn, their constant noise remaining very present. Behind it looms the magnificent panorama of the Calanda and the mountain landscape of Graubünden.

In an impressive way, the design not only enables the cemetery grounds to express what Kienast understands it to be, namely the place of the task of mourning and of the living cult of the dead, but it also illustrates a quality that characterizes many of his projects in the first half of the 1990s: they all permit a "Romantic" experience of nature that is tied to an urban experience of nature. Kienast's awareness of nature as the "other" plays as much a role here as the significance he attributed to orchestrating aesthetic experiences of nature. From his youth, Kienast was a passionate extreme mountain climber. That was the second familiarity with nature, after his grappling with the "nature of the city," that determined his oeuvre. The meeting of these two ways of seeing—the quotidian-urban and the Alpine-sublime—is one key to his ambiguous work.
We have different souls in us, and the Romantic side of mine is very pronounced. I like to go climbing and enjoy the grandeur of the Alps. Isn't that Romanticism at its best? I don't mind being called a Romantic as there's also the other side of me which calls my Romantic side into question. I enjoy going into the mountains, and after three days I find myself longing for the stink of exhaust fumes. I was climbing in the Mont Blanc area not long ago and just as the trip got particularly difficult, I had an incredible longing to be back in the hustle and bustle, eating a Big Mac, sitting somewhere warm and comfortable, watching TV. . . . I live in this duality of Romanticism on the one hand and the relationship to everyday realty and rationalism on the other.[215]

The Fürstenwald Cemetery contains both these poles. At the same time, the design grants the space the symbolism again that encourages the work of mourning through the sensory experience of the space.

The site was fit into an existing "woodland chamber" and is surrounded on three sides by a mixed forest. On the fourth side, it is bounded by a long retaining wall, which made it possible to have the cemetery grounds terrace gently downward toward the Rhine Valley.[216] Kienast, who always developed his designs in the ground plan, usually indicating the topography with contour lines, got the inspiration for his design proposal from the contour lines of the terrain: the retaining wall and the hedges subdividing the cemetery took up the rhythm of the contour lines, thereby accentuating the topography of the terrain and forming a harmonious whole. The gently curving wall was to be built of natural stone. For reasons of cost,[217] it was built of concrete, and the change in material resulted in a new form: The wall now cuts through the landscape with several sharp bends. It looks like an echo of the light gray bands of stone that cut through the forests on the Calanda Massif opposite (p. 280). Kienast called the retaining wall the "backbone of the entire cemetery complex, from which the various structures . . . ensue."[218] It connects the entrance, the viewing hall, and the chapel by Zinsli and Erhard in the southwest with the observation platform conceived by Kienast Vogt Partner in the northwest corner of the cemetery.

The adjoining fields with the graves follow the contour lines of the new topography. They are slightly curved and are bounded in one case by extended hedges and in another by a belt of trees and bushes planted at an angle to the hedges. Only the grassy field of graves to the south, near the entrance, is not interrupted by longitudinal hedges (130, p. 274). The species of trees are the same as the ones growing in the adjacent mixed forest, so that the forest seems to "trickle" into the cemetery. This again reinforces the connection of the grounds to their surroundings. For the belt of shrubs, lilacs that bloom and smell nice in the spring were chosen. From west to east, boxwood, hornbeam, privet, and maple determine the character of the elongated hedges. A common grave and a wall of urns are located in the upper section of the cemetery (p. 278). This part is like a gallery, and it is framed by a magnificent mixed hedge composed of the four species of the hedges that consequentially frame the different terraces below. The mixed hedge passes through various shades of green as it grows (p. 276). The common grave is designed

Fürstenwald Cemetery in Chur, 1992–96. Photos: Georg Aerni, 2008 and 2015

Page 270
Funeral chapel and viewing room by the architects Urs Zinsli and Franz Erhard crown the Rhine Valley. The cemetery, which has been fit into a "woodland chamber," extends to the north and to the east from here and is bordered to the west, facing the valley, by a retaining wall.

Page 272
The retaining wall of the cemetery cuts through the agrarian landscape with several sharp bends. For Kienast, it was the "backbone of the entire cemetery complex." It curves up slightly and is covered with broad slabs of concrete with aggregate, into which green granite slabs have been inserted. This pattern enlivens the long band, while the concrete is increasingly covered up by small, orange lichen—a patina in keeping with Kienast's aesthetic. To the east of that, elongated hedges roughly following the contour lines of the slope and a belt of trees and bushes across them divide the fields with the graves.

Page 274
View over the grassy field, with graves marked by square ashlar, toward the viewing room, the forecourt with four lindens, and the funeral chapel

Page 276
The hedges that articulate the fields with the graves lengthwise take up the course of the contour lines. The biggest one, a mixed hedge that combines the four species of the other hedges, shields the wall of urns and the common grave in front of it.

Page 278
The wall with niches for urns is made of reinforced concrete clad with natural stones from the town of Sils. It is interrupted by strictly geometrical, rectangular fields of gray tomb slabs in front of the urn niches and by abstract tree forms of gray concrete worked into the wall. The skeletons of artificial trees reduced to trunk and branches are completed by the tops of real trees that loom up out of the forest behind the wall of urns. For Kienast, the wreaths and flower arrangements placed in front of the wall of urns were the expression of a "lively cult for the dead."

Page 282
"Gradually, thoughts of the dead mingle with our experience of the immediacy of landscape and nature." Thus Kienast described the experience of arriving at the viewing platform after visiting the cemetery. The sensory perception of the presence of landscape becomes a memento mori. It can offer consolation in the experience of losing a loved one. But the contemplation of nature is a contemporary one, owing to the noise of the heavily trafficked and populated Rhine Valley. Kienast thought of and experienced nature and landscape in terms of this polarity.

as a grassy field with a very narrow, long pool of water, which is lined by a band of stone slabs. Originally, Kienast had planned several pools based on the form and structure of fallen leaves from a beech tree.

The wall with niches for urns on the mountain end, situated on the eastern side of the grounds, continues the dialogue between cemetery and landscape: it is made of reinforced concrete finished with Sils stones, taken from the mountain chain that rises directly above it. The natural stone wall is interrupted in an alternating rhythm by strictly geometrical, rectangular fields of gray tomb slabs in front of the urn niches and abstract tree forms of gray concrete worked into the wall. The skeletons of artificial trees reduced to trunk and branches are "completed" by the tops of real trees that loom up out of the forest behind the wall of urns. This structural relationship is especially strong in the autumn, when the colors of the Sils stones and the leaves of the forest correspond.

Another structural interweaving results between the ossified branches of the imitation trees and the straight beam of the cross that divides the fields of the urn niches. References to the landscape and to Christian symbolism characterize this area of the cemetery. The "living cult of the dead" is expressed in the many flower arrangements and wreaths placed in front of the wall of urns by family members. Kienast planned a low socle of concrete in front of the retaining wall for this purpose, so that flower arrangements can be placed somewhat higher, separated from the broad path that divides the urn wall from the common grave.

The Fürstenwald Cemetery is an ideal illustration of how Kienast united sensory, social, and ecological aspects in his works, how he sharpened human perception and wanted to place designing with nature in a topographical and symbolic context. Because the cemetery is located far outside the city, the residents of Chur often travel the last stretch of the journey on foot, despite the parking lot right at the entrance[219]—a remnant of the traditional ritual of the funeral procession. The mourners enter the cemetery through the gate in the southwest, its surface design introducing the theme of "building with nature": the pine needles scattered in the concrete shutters have left clear impressions behind on the artificial stone, "traces" of the enormous Scots pines that loom upward in the forest behind it. A geometrically trimmed mixed hedge follows the path to the viewing room and chapel. On the generous

282

plaza between the cubic buildings, Kienast planted four lindens within the square (p. 274). The gravel area around it is demarcated from the rest of the plaza by a narrow border of paving stones. This motif, unobtrusive within the expanse of the plaza, is a minimalist quotation of the Garden of Paradise. Near the railing guarding the plaza on the upper edge of the retaining wall, between the viewing hall and the chapel, stands a bell tower. Behind it looms the mountain landscape on the other side of the Rhine Valley, its outlines standing out strikingly in the backlighting when the sun sets.

The plaza, as a forecourt for the viewing hall and chapel, takes on a particular atmosphere thanks to this design. A grassy field next to it extends eastward toward the slope, with graves marked by square ashlars placed on the grass at regular intervals. Toward the north, the plaza leads into the access path along the retaining wall. The wall and its sharp bends draw the eyes of visitors into the depths, as the Rhine Valley and the Calanda Massif open up before them. With the City of Chur behind them, visitors appreciate only the view into the countryside. The paving is bitumen with finely ground green gravel rolled into it. The retaining wall curves slightly up and is covered with broad slabs of concrete with aggregate, into which green granite slabs have been inserted. This pattern enlivens the long band, while the concrete is increasingly covered up by small, orange lichen—a patina in keeping with Kienast's aesthetic (p. 272). Visitors can turn off the access path at any time into the fields of graves located toward the slope and go up to the common grave and the wall of urns. The upper zone of the wall peeks out from behind the enormous mixed hedge and is visible from anywhere in the cemetery. The combination of five longitudinal hedges and the tombstones, most of which are gray, produces an image rich with tension against the backdrop of the forest and mountains, because the tombstones stick up behind the hedges just like the mountain peaks above the treetops. This results in a perception of the topography being ordered by layers of green and gray, continuing into the landscape behind the cemetery. This setting of clear boundaries and the simultaneous dissolving of them by the context is a formal aspect that also recurs on the wall of urns, with its abstract tree shapes and the real treetops towering behind it.

The cemetery does, however, have formal weaknesses in the definition of its outer boundaries. In this respect, Kienast's design does not satisfy the demands made in his two texts on the cemetery: on both ends, the wall with urn niches terminates unmediated at the embankment. As a result, the eye is directed at a fence behind it as the real boundary. In the resolution on the project as a whole, however, the town council of Chur already planned to close these gaps with a later extension of the wall of urn niches as part of an expansion of the cemetery.[220] The town council believed it was important that "The view into the site and the broad landscape [be] ensured."[221] Perhaps this requirement explains why a fence draws the boundary in the north and south as well, where it is hidden by a mixture of planted trees and shrubs.

The western boundary, by contrast, is clearly formulated by the retaining wall, and the observation pavilion becomes a vantage point that attracts visitors. The path in the far north that leads from the wall of urns down the slope toward the pavilion takes a respectful turn before the oak on the terrain (p. 280); at this point, the path becomes a bridge in order to cross the oak-filled hollow and continue on its way. This "winking" motif is disturbed again by a railing on the opposite side, at the end of which the path feeds into a plaza that seems overly large in relation to the filigreed structure of the observation platform. The pavilion, designed as a cubic skeleton open on all sides, recalls the sculptures of the American artist Sol LeWitt. Its practical function is limited: the visitors who sit down on the bench on its front edge in order to enjoy the view of the Calanda Massif are not protected from the weather. The form open to the sky is best understood as a symbol and hence unrelated to its utility.

In one description of the cemetery, Kienast alludes to the path from the common grave and wall of urns to the viewing platform, where he ends his imaginary tour (p. 280):

> The path takes us along the edge of the wood to a flight of steps which provides a link to the Fürstenwald woodland trails. Passing the burial plots that are not to be developed further until the second stage, we arrive at the lookout platform at the end of the mountain path. Adjacent to a corner of the wood and two old oaks, the simple concrete structure marks a special place: the hall of rest, chapel, and burial plots have already receded into the distance. Gradually, thoughts

of the dead mingle with our experience of the immediacy of landscape and nature.[222]

In the end, it is again nature that remains—and the sensory perception of the presence of a landscape as a *memento mori*. It, too, can offer consolation in the experience of losing a loved one.

Yet at that precise point, at the end of the path, when viewing the Calanda looming on the other side of the valley, the presence of the city and the business of daily life slips into our contemplation of nature: the rumbling of the autobahn cutting through the Rhine Valley fills the ears of the viewers, as does the clattering of the constantly passing trains. In Kienast's work, as a rule, the contemplation of nature is mixed up with the noises and smells of the city, both in his specific works and in his imagination. The extreme mountain climber's experience of the landscape is combined with full awareness of the stark urbanization of Switzerland. The concept of nature on which Kienast's works are based is manifested in this polarity.

1   Dieter Kienast, "Bemerkungen zum wohnungsnahen Freiraum" (Remarks on Free Spaces Close to Habitations), *anthos* 18, no. 4 (1979): 2–9. In that issue, it is immediately followed by a text in which Kienast described the grounds of his own home. See Dieter Kienast, "Wohngrün zu Mehrfamilienhäusern in Wollishofen, Zürich" (Private Verdure of Apartment Blocks at Wollishofen, Zurich), *anthos* 18, no. 4 (1979): 10–13.
2   Kienast, "Bemerkungen zum wohnungsnahen Freiraum", 2.
3   One important foundation for the ideas expressed in this text is the publication Inge Meta Hülbusch, *Innenhaus und Aussenhaus: Umbauter und sozialer Raum,* Schriftenreihe OE Architektur, Stadt- und Landschaftsplanung 1, no. 33, 2nd ed. (Kassel: Gesamthochschule Kassel, 1981).
4   Hartmut Lüdtke, *Freizeit in der Industriegesellschaft: Emanzipation oder Anpassung?* (Opladen: Leske, 1972); quoted in Kienast, "Bemerkungen zum wohnungsnahen Freiraum," 2.
5   Kienast, "Bemerkungen zum wohnungsnahen Freiraum," 4.
6   Ibid.
7   Ibid.
8   In his 1972 documentary film *Die grünen Kinder* (The Green Children), Kurt Glor sketches everyday life in the Sunnebüel housing project in Volketswil, which had been praised as being very child-friendly. Yet the evident lack of imagination of the architecture, green spaces, and playgrounds provoked heated discussions about the construction of such large developments. The development in Volketswil was also known as "Göhnerswil," after its builder, the general contractor Ernst Göhner. See http://www.s5-stadt.ch/fileadmin/editor/S5_Flyer_Qbus_DEF_100419.pdf. In 1972, an authors' collective in the Division of Architecture of ETH Zurich published a critical study of Ernst Göhner AG and its developments: Heini Bachmann et al., *Göhnerswil: Wohnungsbau im Kapitalismus; Eine Untersuchung über die Bedingungen und Auswirkungen der privatwirtschaftlichen Wohnungsproduktion am Beispiel der Vorstadtsiedlung Sunnebüel in Volketswil bei Zürich und dem Generalunternehmung Ernst Göhner AG* (Zurich: Verlagsgenossenschaft, 1972). For a retrospective on the qualities and shortcomings of these housing developments, see Fabian Furter and Patrick Schöck, eds., *Göhner Wohnen: Wachstumseuphorie und Plattenbau,* exh. cat. (Baden: Hier und Jetzt, 2013).
9   Kienast, "Bemerkungen zum wohnungsnahen Freiraum," 5. Kienast admitted that the opportunities for open-space planning to improve this situation were limited and therefore argued for developing alternative models to large developments over the long term by working with politicians, architects, urban planners, and sociologists—in that order.
10 Ibid., 8.
11 Ibid., 6.
12 Ibid., 6–7.
13 Ibid., 2.
14 On this, see the first two theses of his "Zehn Thesen zur Landschaftsarchitektur." In the first thesis, Kienast wrote that the work of the firm "is a search for the Nature of the City, whose color is not solely green but also gray." For Kienast, the "nature of the city" meant not only "the tree, hedge, lawn; but equally the water-permeable hard surface, broad square, rigid street-gutter line, high wall …." In the second thesis, he wrote that he was interested in "the city and its inhabitants," whereby the city is "no longer a monolithic entity; instead it is dismembered and fragmented into thousands of parts." The heterogeneous mix of city dwellers necessitated "up-to-date actions and reactions in outdoor spaces that rejects a uniform greening of the city." The theses have been newly translated for this volume; see appendix, 403–5. Cf. Dieter Kienast, "Zehn Thesen zur Landschaftsarchitektur," in *Dieter Kienast: Die Poetik des Gartens; Über Ordnung und Chaos in der Landschaftsarchitektur,* ed. Professur für Landschaftsarchitektur ETH Zürich (Basel: Birkhäuser, 2002), 207–10, here 207.
15 Kienast, "Bemerkungen zum wohnungsnahen Freiraum," 2.
16 Ibid., 9.
17 Ilse Helbich, "Von unbetretbaren Gärten," in Anette Freytag and Wolfgang Kos, "Neue Parkideen in Europa: Zwischen Arkadien und Restfläche," radio broadcast, *Diagonal: Radio für Zeitgenossen,* Österreichischer Rundfunk (ORF), Österreich 1 (Ö1), first broadcast on October 10, 1998.
18 See Peter Stöckli, "Am Ende der Strasse—ein Nachruf auf Dieter Kienast," *anthos* 38, no. 1 (1999): 58–59, here 58.
19 Bernd Schubert in conversation with the author, May 16, 2008. Schubert is a former professor of landscape planning at the Interkantonales Technikum Rapperswil, now the Hochschule für Technik Rapperswil, where Kienast was a professor of garden architecture from 1981 to 1991.
20 In a lecture at the Rapperswiler Tag conference on the subject of housing, Stefan Rotzler stated, with reference to Alain de Botton, *The Architecture of Happiness* (London: Penguin, 2006), on May 16, 2008: "As a firm, we argue that every development should have a recognizable, specific atmosphere. It can be defined by the type of plants (trees, shrubs, grasses), by colors, by scents, and by particular materials. In any case, it is about recognizability and concision. And about how they can be conveyed and communicated. And for that reason, not least, about homeland." http://www.rapperswilertag.ch/r08/downloads08/07_RT08_Rotzler.pdf (link no longer active).
21 On these numbers, see "Gemeinde: Niederhasli, Bevölkerungsbestand," Statistisches Amt, Kanton Zürich, https://statistik.zh.ch/internet/justiz_inneres/statistik/de/daten/gemeindeportraet_kanton_zuerich.html#a-content.
22 www.niederhasli.ch.
23 Dieter Kienast, untitled and undated typescript, p. 1, in the file "Niederhasli-Huebwiesen/Lindenstrasse," gta Archives (NSL Archive) / ETH Zurich (Dieter Kienast Bequest).
24 The sequence is easily reconstructed based on the archival material. See the file "Niederhasli/

Huebwiesen-Lindenstrasse," gta Archives (NSL Archive) / ETH Zurich (Dieter Kienast Bequest).

25 The envelope in which the photographs are kept in the Archiv SKK Landschaftsarchitekten, Wettingen (box of object 138) is postmarked February 8, 1972, and is addressed to Peter Paul Stöckli. The sender is the garden architect Josef A. Seleger in Zurich, not the garden architect who was actually responsible, Georges Boesch.

26 The cooperative was founded in 1946 to provide work for the "craftsmen returning from military service" and now manages more than a thousand apartments in various communities in the Zurich metropolitan area. See www.bgm.ch.

27 MOBAG declared bankruptcy in June 2006. See www.mobag.ch.

28 On most of the project plans drawn by Kienast, Peter Paul Stöckli was listed as a freelancer for Georges Boesch. See the documents and plans in the Archiv SKK Landschaftsarchitekten, Wettingen (box 138) and gta Archives (NSL Archive) / ETH Zurich (Dieter Kienast Bequest: "Niederhasli/Huebwiesen-Lindenstrasse" file).

29 Because the architects' plans in the Archiv SKK Landschaftsarchitekten, Wettingen, are not complete, this number is the sum of the known number of apartment units per block and an estimate based on blocks with a similar floor plan.

30 The floor plan for the residential blocks, dated January 20, 1972, has a date stamp for receipt on February 7; the envelope in which the photographs documenting the current state of the terain were sent to Stöckli is postmarked February 8, 1972. Finally, the plan for the preliminary project is dated March 16, 1972.

31 According to Peter Paul Stöckli, the presentation plan was produced especially for this publication. Illustrated in Toni Raymann, "Privat nutzbarer Freiraum im Geschosswohnungsbau" (Open Spaces for Private Use in Apartment Housing Developments), *anthos* 22, no. 2 (1983): 2–26, here 9.

32 The photographs are in the Archiv SKK Landschaftsarchitekten, Wettingen, in an envelope postmarked April 8, 1975. It is, however, doubtful that this is the original envelope, because the sender is a company for mechanical and automotive engineering in Niederweningen.

33 Kienast, untitled and undated typescript, p. 1.

34 The explanatory notes in the text read: "Entrance zone with relaxation areas & site together for all ages; meetings, usually pedestrians, etc.; provisions for adults + children; benches, play equipment, roller-skating area; parking lots during the day—possibilities for use. Parking lots with trees—to shade from sun, block noise, dust, etc."

35 See Kienast, untitled and undated typescript, p. 2.

36 The fifth sheet in the set is dedicated exclusively to the Huebwiesen section of the development, but it is difficult to say whether it was produced before Sheet 4 or after. One of the hills for playing is marked with contour lines; it is the same hill as the one in the transitional area between the two sections of the development that first appears on Sketch 2. The individual colored areas are otherwise undifferentiated. It could have been produced after Sketch 2 or after Sketch 3 or 4. It is referred to as Sketch 5 here.

37 Kienast, untitled and undated typescript, p. 1.

38 The project was published in the journal *anthos* in 1968: "Siedlung Nussdorf, Wien" (Nussdorf Vienna Development), *anthos* 7, no. 3 (1968): 33–36.

39 See Udo Weilacher, *Visionary Gardens: Modern Landscapes by Ernst Cramer,* trans. Michael Robinson (Basel: Birkhäuser, 2001), 213.

40 See "Atrium-Siedlung Reinach/Basel" (Atrium Development Reinach/Basel), *anthos* 7, no. 3 (1968): 37–40.

41 In the same issue of *anthos* ("Albertslund-Syd," *anthos* 7, no. 3 [1968]: 30–31), a large development in Denmark was presented: the Albertslund development, built near Copenhagen for 7,000 residents. In addition to neighborhoods of row houses and multistory apartment buildings (both without private gardens), it included a large section of single-family homes with small garden seating areas of 7.5 by 7.5 meters per unit.

42 Raymann, "Privat nutzbarer Freiraum," 2–10.

43 Kienast, untitled and undated typescript, p. 3. In the label "hornbeam and oak forest," one hears Kienast the phtyosociologist speaking. This is the typical forest of the German Mittelgebirge and the climax of natural succession processes. See Heinz Ellenberg, *Vegetation Mitteleuropas mit den Alpen in ökologischer Sicht,* 2nd ed. (Stuttgart: Ulmer, 1978). Kienast regretted that the deciduous trees died in several places (he believed that they had been sprayed to death); a "thoughtful caretaker" replaced them with firs.

44 Kienast, untitled and undated typescript, p. 3.

45 The same procedure was followed for the two parts of the Huebwiesen section on the other side of Langackerstrasse and Lindenstrasse.

46 Kienast, "Bemerkungen zum wohnungsnahen Freiraum," 8.

47 Kienast, untitled and undated typescript, p. 2.

48 Ibid., p. 3.

49 This is the case, for example, with the L. Garden in Riehen of 1995 (visit in March 2008). On the L. Garden, see *Kienast: Gärten / Gardens* (Basel: Birkhäuser, 1997), 152–59.

50 See the sixth of the "Zehn Thesen zur Landschaftsarchitektur," in *Kienast,* "Zehn Thesen zur Landschaftsarchitektur," 208; see also appendix, 403.

51 I am grateful to Thom Roelly for a discussion of this subject.

52 See the editor's introduction to the special issue on the topic of "ground configurations and earth world," of *anthos* 7, no. 1 (1968): 21. Several of the essays also mention the task of working with excavated soil because it was too expensive to remove it.

53 Christian Stern, "Bedeutung und Entwicklung der Freiraumgestaltung im industriellen Wohnungsbau" (The Significance and Evolution of the Design of Open Spaces in Industrial Dwelling Construction), *anthos* 7, no. 3 (1968): 16–22, here 17. In the sentence that follows, he also mentioned the economic motivation of a rapid design of the grounds: "For the first rental or the purchase by the general contractor, immediate

54 "Wohn-, Alters-, und Siedlungsheim, Dübendorf/Zürich" (Old Age and Residential Development in Dübendorf/Zurich) [Willi Neukom], *anthos* 7, no. 1 (1968): 11–13, and Weilacher, *Visionary Gardens,* 210–12.
55 Weilacher, *Visionary Gardens,* 212–15, and Claudia Moll and Axel Simon, *Eduard Neuenschwander: Architekt und Umweltgestalter* (Zurich: gta Verlag, 2009), 109–10.
56 Weilacher, *Visionary Gardens,* 210. Cramer covered his earth modulations with lawn, but also with concrete paving.
57 Moll and Simon, *Eduard Neuenschwander,* 109.
58 I am only comparing published examples, either in the Swiss journal *anthos* or in relevant monographs.
59 Stern, "Bedeutung und Entwicklung," 22.
60 This is also true of the grounds designed by Christian Stern for the developments in Volketswil that Kienast mentions as a negative example in his text on open space near housing developments Kienast, "Bemerkungen zum wohnungsnahen Freiraum."
61 Embedding grassy areas into a well-modeled topography also played a crucial role in Cramer's designs of the grounds for housing developments and in Kienast's project for Niederhasli.
62 For the grounds of the Heuried housing development in Zurich, Ernst Cramer created a kind of hall of plane trees to provide privacy for playing children; their tops were gradually trimmed to form a continuous roof that the gazes of adults from the apartment buildings could not penetrate. See Weilacher, *Visionary Gardens,* 213–14.
63 See Ursula Seleger, "C. Th. Sörensens Herausforderung" (C. Th. Sörensen's Challenge), *anthos* 12, no. 4 (1973): 39–40.
64 The Swiss child protection organization Pro Juventute advocated playground concepts of this sort at the time. See, for example, the conference "Spielplatzgestaltung: Kinderspielplätze und Freiflächen im Siedlungsbereich," November 7–8, 1972, organized by Pro Juventute, the Bund Schweizer Gartengestalter, and the Verband Schweizer Gärtnermeister. Christian Stern had also planned a construction playground as part of his design of the grounds of the Adlikon-Regensdorf housing development. See Stern, "Bedeutung und Entwicklung," 21.
65 Dieter Kienast, "'Ein Spiel ist der Weg der Kinder zur Erkenntnis der Welt, in der sie leben und die zu verändern sie berufen sind' (Maxim Gorki)," in Professur für Landschaftsarchitektur ETH Zürich, *Dieter Kienast,* 47–52. This text was originally published without a title in 1981. The essay refers above all to Hülbusch, *Innenhaus und Aussenhaus.*
66 Kienast, "Ein Spiel ist der Weg der Kinder," 50.
67 For Burckhardt, "no-man's-lands" are "forgotten" places: remnants between functional places. Kienast proposed integrating them into the site, which in fact contradicts their definition.
68 Kienast, "Ein Spiel ist der Weg der Kinder," 49.
69 Ibid., 50.
70 Marc Schwarz and Annemarie Bucher, *D. K.: Eine Spurensuche,* DVD, Zurich, 2008, a video documentation commissioned by the chair of Christophe Girot, ETH Zurich.
71 The biology teacher Urs Schwarz was active in the Swiss city of Solothurn. That is why in Switzerland the natural garden movement is often referred to as the "Solothurn School." In 1981, the editor in chief of *Garten + Landschaft,* Eike Schmidt, stated that to his knowledge the term "Naturgarten" (natural garden) had come into common usage via a book with the title *Der Naturgarten* by Urs Schwarz, published in 1980. Schmidt recognized Schwarz as the patron saint of the modern natural garden movement but its "spiritual father" as the Dutchman Louis Le Roy, whose book *Natuur uitschakelen—natuur inschakelen* was published in 1973. Eduard Neuenschwander saw himself as the third cofounder of the modern natural garden movement. See Eike Schmidt, "Der Naturgarten: Ein neuer Weg?" (The Natural Garden: A New Direction?), *Garten + Landschaft* 91, no. 11 (1981): 877–84, here 877, and Moll and Simon, *Eduard Neuenschwander,* 179 and nn. 1, 4, and 24.
72 Rainer Zulauf in an interview (in Swiss-German) in Schwarz and Bucher, *D. K.: Eine Spurensuche.*
73 Peter Paul Stöckli in an interview in Schwarz and Bucher, *D. K.* Producing spaces by means of topography is one of the foundations of design work outdoors. The first issue of the journal *anthos* in 1968 was entirely dedicated to this subject. It also includes articles on the earth pyramids of Ernst Cramer and Hermann von Pückler-Muskau, as well as other examples of landscape architecture using earth formations, including models of terrain design using quarry stones, which may have inspired Kienast when designing his Dry Grasland Biotope in Brüglingen. Kienast subscribed to the journal and later purchased all of the back issues, had them bound, and integrated them into his library.
74 Jürg Altherr in conversation with the author, August 18, 2009. Altherr was Kienast's colleague at the Interkantonales Technikum Rapperswil. He felt that Kienast and his Dry Grassland Biotope could counter somewhat the "stodgy stuff," as Altherr described the majority of the gardens for Grün 80. For him, the Dry Grassland Biotope provided the subject of ecology with "a precise place in the debate over landscape design."
75 Dieter Kienast, "Ohne Leitbild," *Garten + Landschaft* 96, no. 11 (1986): 34–38, here 36.
76 Ibid.
77 Ibid., 34.
78 "Antrag des Gemeinderats an den Einwohnerrat betreffend Kreditbegehren von Fr. 112'000.– für die Umgestaltung des Brühlplatzes," January 8, 1976, Archiv SKK Landschaftsarchitekten, Wettingen. This document, which describes the history of the design and use of the Brühlwiese, was sent to Stöckli by Zulauf.
79 The residents' council was formed in the mid-1960s and first met in 1966. See http://www.wettingen.ch/de/portrait/geschichte/.
80 "Antrag des Gemeinderats" (see note 78).

81 "Gemeinde Wettingen: Einige Bemerkungen zum Antrag des Gemeinderats an den Einwohnerrat für die Umgestaltung des Brühlplatzes," handwritten date added later: "21.1.76," Archiv SKK Landschaftsarchitekten, Wettingen.
82 Ibid.
83 The member of the residents' council responsible was Guido Probst. See Gemeindekanzlei Wettingen to Guido Probst, February 12, 1976, with an excerpt of the minutes from a residents' council meeting, Archiv SKK Landschaftsarchitekten, Wettingen.
84 Peter Paul Stöckli in conversation with the author, January 27, 2012.
85 According to Peter Paul Stöckli's agenda and appointment calendar for 1979, the first consulting meeting was held in Baden on January 10, 1979, after which a proposal was sent to the Wettingen town council. Archiv SKK, Wettingen.
86 Ibid.
87 Peter Paul Stöckli, "Gemeinde Wettingen, Stadtpark Brühlwiese: Vorprojekt, Bericht, Kostenvoranschlag," September 26, 1979, Archiv SKK Landschaftsarchitekten, Wettingen.
88 Stöckli, agenda and appointment calendar for 1979 (see note 85).
89 Ibid.
90 The design was developed together with Peter Paul Stöckli, but Kienast was responsible for its implementation. It was also up to Stöckli to push the project through with the town council and the architects, which would not have been possible without great diplomacy and negotiating skills, because the project was regarded as very radical at the time. The many meetings and inspections are recorded in Stöckli's appointment calendars for 1979 to 1983.
91 Stöckli, "Gemeinde Wettingen" (see note 87).
92 "Ein Gesinnungswandel in Wettingen?," *Badener Tagblatt,* October 20, 1979: 30.
93 Stöckli, agenda and appointment calendar for 1979 (see note 85).
94 Kienast, undated notes (ca. 1982) on the proposal for the Stadtpark Brühlwiese in Wettingen, Archiv SKK Landschaftsarchitekten, Wettingen.
95 The approach is well documented using just a few examples in *Fred Eicher, Landschaftsarchitekt: Schulthess Gartenpreis 2004* (Zurich: Schweizer Heimatschutz, 2004).
96 Stöckli, "Gemeinde Wettingen" (see note 87).
97 Kienast, undated notes.
98 In the end, the underground garage was not built and soil was brought in to construct the sledding hills. Peter Paul Stöckli in conversation with the author, January 27, 2012.
99 Stöckli + Kienast Landschaftsarchitekten, "Gemeinde Wettingen, Stadtpark Brühlwiese. Vorprojekt. Teiletappe 2a. Kostenvoranschlag," March 8, 1982, Archiv SKK Landschaftsarchitekten, Wettingen.
100 The plan had the same title and the same number: 316-3. The only difference is the removal of the dry plaza open to the sky.
101 The plan is numbered 316-7. The only earlier plans that survived are no. 316-3 (preliminary project) and no. 316-4 (sections of the preliminary project) of 1979. Plan nos. 316-5 and 316-6 are missing. They may have made it clear whether the change to the new, more radical forms happened gradually.
102 Wettingen town council, building permit application no. 53: Stöckli + Kienast Landschaftsarchitekten, "Pyramidale Hügel bis 5 m Höhe" (submitted on June 18, 1982, approved on August 5, 1982), with plan nos. 316-8 and 316-9 as enclosures, Archiv SKK Landschaftsarchitekten, Wettingen.
103 Dieter Kienast, "Kultivierung der Brüche," in *Zwischen Landschaftsarchitektur und Land Art,* ed. Udo Weilacher (Basel: Birkhäuser, 1996), 137–56, here 144. Quote newly translated from the original German version, "Kultivierung der Brüche," in *Zwischen Landschaftsarchitektur und Land Art* (Basel: Birkhäuser, 1996).
104 See "Bepflanzungsplan für den Haupteingang," plan no. 316-26, scale: 1:100, H 67 × W 126 cm, March 16, 1983, rev. April 8, 1983, signed "VO," Archiv SKK Landschaftsarchitekten, Wettingen.
105 Minutes of the Parkanlage Brühl building committee meeting, April 5, 1983, Archiv SKK Landschaftsarchitekten, Wettingen. The building committee approved the proposal for the park, with the proviso that the mixed forest be supplemented by evergreen conifers.
106 Kienast, "Ohne Leitbild," 36.
107 The best-known example of such a "juxtaposition" in 1982 was the competition entry by the Swiss architect Bernard Tschumi for the Parc de la Villette in Paris.
108 Kienast, "Ohne Leitbild," 36.
109 See "Bepflanzungsplan für den Haupteingang." The plan called for *Clematis* "Rubens," *Parthenocissus quinquefolia,* and *Lonicera x tellmanniana* on twenty-one trellises.
110 Dieter Kienast, according to Arthur Rüegg, in a conversation with him about late Baroque phenomena; see Arthur Rüegg, speech at the memorial for Dieter Kienast at ETH Zurich, typescript, January 23, 1999, Archiv Arthur Rüegg.
111 Marc Schwarz, *In Praise of Sensuousness,* DVD, in *Dieter Kienast: In Praise of Sensuousness,* exh. cat. (Zurich: gta Verlag, 1999).
112 Kienast, undated notes.
113 See, for example, Dieter Kienast, "Von der Notwendigkeit künstlerischer Innovation und ihrem Verhältnis zum Massengeschmack der Landschaftsarchitektur" (lecture in November 1991), in *Choreographie des öffentlichen Raumes,* ed. Jürgen Wenzel (Berlin: self-published, 1994), 52–64, here 52.
114 See also the detail plan for the tree grille, plan no. 316-12, 1:10, July 28, 1983. Each tree grille was composed of two parts of 240 by 125 centimeters, which together were roughly square, to which was attached, where the tree stands, a triangle, again composed of two parts, that ran the entire length of 2.5 meters. A square standing on one point and aligned with the tip of the triangle was left open for the tree. The grilles under steel trellises at the southern entrance have two triangular parts that form a square; in the middle, a small square is left open for the plants grown there, and they were also framed with paving stones.

115 Stöckli + Kienast Landschaftsarchitekten, "Gedanken zur Gestaltung des neuen Brühlparks," [1984], copy of a newspaper article with no source indicated, Archiv SKK Landschaftsarchitekten, Wettingen, box 316-4. There is a corresponding typescript dated June 15, 1984. The program for the opening ceremony is printed next to the article. The opening speech was given by Peter Paul Stöckli.
116 Invoice from Peter Paul Stöckli, probably for a meeting of the town council, manuscript, Archiv SKK Landschaftsarchitekten, Wettingen, box 316-4.
117 Kienast, "Ohne Leitbild," 36.
118 Stöckli + Kienast, "Gedanken zur Gestaltung des neuen Brühlparks" (see note 115).
119 Minutes of the Parkanlage Brühl building committee.
120 Excerpt from the minutes of the Wettingen town council, April 14, 1983, Archiv SKK Landschaftsarchitekten, Wettingen.
121 See the plan for planting the children's playground, plan no. 316-27, March 23, 1983, rev. March 26 and April 8, 1983, Archiv SKK Landschaftsarchitekten, Wettingen.
122 Minutes of the Parkanlage Brühl building committee (see note 105).
123 Kienast, "Ohne Leitbild," 37.
124 Wettingen town council to Peter Paul Stöckli, October 11, 1982, Archiv SKK Landschaftsarchitekten, Wettingen.
125 Stöckli + Kienast to the Wettingen town council, November 10, 1982, written by Kienast (a handwritten draft also survives), and resolution of the town council on November 18, 1982, excerpt from the minutes, Archiv SKK Landschaftsarchitekten, Wettingen.
126 Kienast, "Cultivating Discontinuity," 143.
127 Quoted in Heinz Ohff, *Der grüne Fürst: Das abenteuerliche Leben des Hermann Pückler-Muskau* (Munich: Piper, 1991), 296.
128 Kienast, "Cultivating Discontinuity," 142.
129 Weilacher, *Visionary Gardens,* 117.
130 Ibid., 109.
131 On the concept for G 59, the various design stances, and the understanding of nature on which they were based, see Annemarie Bucher, "Naturen ausstellen: Schweizerische Gartenbauausstellungen zwischen Kunst und Ökologie" (DSc diss., ETH Zurich 2008).
132 Weilacher, *Visionary Gardens,* 259.
133 Kienast, "Cultivating Discontinuity," 144.
134 Ibid.
135 It is worth noting in this context that the presentation plan for the Brühlwiese Municipal Park that was produced later for an exhibition in 1991 is very similar to Cramer's presentation plan for the Garten des Poeten in terms of color and style, as if Kienast had wanted to become part of a tradition. See figs. 62 and 76 in this volume.
136 Kienast, "Vom Gestaltungsdiktat zum Naturdiktat — oder: Gärten gegen Menschen," *Landschaft + Stadt* 13, no. 3 (1981): 120–28.
137 Theodor W. Adorno, "Functionalism Today," trans. Jane O. Newman and John H. Smith, *Oppositions* 17 (1979): 31–41, here 37. Quoted (in German) in Kienast, "Vom Gestaltungsdiktat zum Naturdiktat," 120. In 1991, Kienast also used this quotation as the epigraph to his competition entry for the Günthersburgpark in Frankfurt am Main.
138 See Louis G. Le Roy, *Natuur uitschakelen, natuur inschakelen* (Deventer: Ankh-Hermes, 1973). Kienast had encountered Le Roy as visiting lecturer in Kassel (information from Thom Roelly, April 6, 2011).
139 In the final chapter of his dissertation, Kienast credited Schwarz's idea of the natural garden by initiating a critique of the stereotype of the Swiss garden. See Dieter Kienast, *Die spontane Vegetation der Stadt Kassel in Abhängigkeit von bau- und stadtstrukturellen Quartierstypen,* Urbs et Regio: Kasseler Schriften zur Geografie und Planung 10 (Kassel: Gesamthochschul-Bibliothek, 1978), 392.
140 Quoted in Bucher, "Naturen ausstellen," 212.
141 Ibid.
142 Among the earliest uses of the term "Naturgarten" is a chapter in Willi Lange's *Die Gartengestaltung der Neuzeit,* published in 1907, its point of reference being William Robinson's *The Wild Garden* of 1870. The concept seems to have "come into common usage" with the title of Schwarz's book, as the editor in chief of the journal *Garten + Landschaft* wrote in 1981: Schmidt, "Der Naturgarten." On the natural garden debate and its Swiss context, see Bucher, "Naturen ausstellen," chap. 5, 185–246, and Moll and Simon, "Naturgarten: 'Lebensraum ist Biotop,'" in *Eduard Neuenschwander,* 175–87. On the ideas behind natural gardens, see Anja Löbbecke, "Über Naturgärten: Eine Ideengeschichte und kritische Retrospektive sowie zu ihrer Bedeutung für die heutige Landschaftsarchitektur" (PhD diss., Technical University of Munich, 2013), https://mediatum.ub.tum.de/node?id=1111136.
143 Kienast, "Vom Gestaltungsdiktat zum Naturdiktat," 123.
144 Ibid., 125–26.
145 Ibid.
146 It should be noted here that Kienast's tone in the text from 1981 is didactic to the point of impertinence. The oversized pyramids also have something of a demonstrative, didactic gesture. Thanks to their convincing integration into the rural-urban ensemble of the park, however, this feature vanishes in the interplay of the various design elements.
147 Wettingen town council to Stöckli + Kienast, October 11, 1982, and excerpt from the minutes of the town council meeting of November 18, 1982, Archiv SKK Landschaftsarchitekten, Wettingen.
148 Kienast himself was in retrospect dissatisfied with the children's play area he had designed because in his view it "doesn't work very well." See Kienast, "Cultivating Discontinuity," 143. But this statement does not refer explicitly to the earth pyramids, but rather to the playground behind them.
149 See *Garten der Kunst: Österreichischer Skulpturenpark / Art Garden: Scupture Park Austria,* ed. Österreichischer Skulpturenpark Privatstiftung (Ostfildern: Hatje Cantz, 2006), and www.museum-joanneum.at/skulpturenpark.
150 Arthur Rüegg (a former colleague of Hoesli and Emeritus Professor of Architecture and Construction at ETH Zurich) in conversation with the author, October 8, 2010.

151 "Ergebnisse des Wettbewerbs für die Grün 80: Zweite Schweizerische Ausstellung für Garten- und Landschaftsbau in Basel" (Results of the Competition for 'Grün 80'—2nd Swiss Exhibition of Garden and Landscape Design 1980 in Basle), *anthos* 16, no. 3 (1977): 32–37.
152 Arthur Rüegg in conversation with the author, October 8, 2010.
153 The text was first published as "Transparency: Literal and Phenomenal," *Perspecta* 8 (1963): 45–54. The first German translation with a short commentary by Bernhard Hoesli was published in 1968 as volume 4 of the *Schriftenreihe des Instituts für Geschichte und Theorie der Architektur an der Eidgenössischen Technischen Hochschule Zürich* and volume 1 of the *Le Corbusier Studien* (Basel: Birkhäuser, 1968). Kienast had the second edition of this book from 1974 in his library. He highlighted many passages in blue. His copy still exists in the Bibliothek Dieter Kienast at ETH Zurich, which the family donated in recognition of the research of the present author in 2016.
154 Colin Rowe and Fred Koetter, *Collage City* (Cambridge, MA: MIT Press, 1978). The first German translation was published in 1984 as volume 27 of the *Schriftenreihe des Instituts für Geschichte und Theorie der Architektur an der Eidgenössischen Technischen Hochschule Zürich* (Basel: Birkhäuser, 1984).
155 On *Transparency,* see Kienast, "Von der Notwendigkeit künstlerischer Innovation," and the eighth thesis in Kienast, "Zehn Thesen zur Landschaftsarchitektur"; see also appendix, 404. On *Collage City,* see Dieter Kienast, "Longing for Paradise," in *Dieter Kienast* (Basel: Birkhäuser, 2004): 84–91; Kienast, "Von der Notwendigkeit künstlerischer Innovation"; Kienast, "Stadtlandschaft," *Zolltexte* 5, no. 2 (1995): 43–46.
156 Bernhard Hoesli, "Addendum: Transparent Form-Organization as an Instrument of Design," trans. Jori Walker, in Colin Rowe and Robert Slutzky, *Transparency* (Basel: Birkhäuser, 1997), 84–99, here 86.
157 Ibid., 86.
158 Ibid., 87.
159 Ibid., 86 (emphasis original).
160 Rowe and Slutzky, *Transparency,* 22–23.
161 Bernhard Hoesli, "Commentary," trans. Jori Walker, in Rowe and Slutzky, *Transparency,* 57–83, here 61. This definition is also offered by Rudolf Manz, a lecturer at ETH Zurich, in his publication on his work with a video on spatial analysis in architecture: Rudolf Manz, *Vito de Onto: Raumerfinder und -erforscher: … seine Freunde nennen ihn Video* (Zurich: gta Verlag, 1996), 24. On Rudolf Manz's significance to Dieter Kienast, see pages 384–89 in the present volume.
162 Hoesli, "Transparent Form-Organization," 115.
163 Kienast was appointed lecturer of landscape design at ETH Zurich in 1985. In 1982, Hoesli wrote his addendum intended for a new English edition of the Transparency essay and in 1984 he edited a German edition of *Collage City.* Both writings and their methods played a crucial role in Hoesli's teachings of design. Kienast knew Hoesli's projects and plans, not least because Hoesli had been involved with Grün 80 and hence Kienast was familiar with the discourse at ETH Zurich even before being hired there. His copy of the second edition of *Transparenz,* the 1974 German translation of "Transparency," is full of Kienast's highlighting of paragraphs he was especially interested in.
164 See Dieter Kienast, "Gedanken zum Projekt im 1. Rang: Stöckli & Kienast, Wettingen" (Project with 1st prize, Stöckli & Kienast, Wettingen), 26–27, in "Ideenwettbewerb Kurpark Bad Zurzach AG" (Competition of Ideas for Zurzach Bath Spa Park), *anthos* 23, no. 4 (1984): 24–35. The landscape architects who had been invited to the competition were Wolf Hunziker, Willi and Tobias Neukom, Stöckli + Kienast, Bernd Wengmann, and Dölf Zürcher.
165 See the summary of the competition announcement, ibid., 24–25.
166 Ibid., 26.
167 In this context, Jürg Altherr's remarks in conversation with the author, August 18, 2010, are extremely revealing: at the Interkantonales Technikum Rapperswil, where Altherr and Kienast both taught, the approach was to create a formal framework for design experiments with the rampant growth of vegetation. They worked in the tension between the natural garden ideal of Urs Schwarz and Ernst Cramer's pure geometry, with the idea of making the most of both extremes. In terms of methodology, the students were taught not to be impressed by the breadth of themes in a proposed task and not to come up with a solution that met all the demands, because that would always result in average solutions. Instead, they should think from the extremes and from the margins. "Everything good comes from the margins," was Peter Erni's slogan. It was necessary, Altherr said, to go as far to the left and right as possible and only then develop something. Moreover, he said they taught the students that they should analyze the site set for the task and at the same time continue to work on whatever happened to be occupying them personally— very single-mindedly, without being distracted in any way. This parallel development of an idea had to be pushed until it flared up, as if in a physical process. The crucial point was allowing the friction from following two paths to stay long enough for a "flash" to occur. Kienast's didactic background as described by Altherr sheds light on the pendulum movements evident in the design process for the Zurzach Spa Gardens and why the design drawing at the end was assembled from several motifs but was nevertheless new.
168 On the history and significance of the tartan grid in architecture since Andrea Palladio, see www.andrew.cmu.edu/course/48-747/subFrames/lecture/PalladianGrammar.pdf.
169 Kienast, "Ohne Leitbild," 38.
170 Ibid.
171 I am grateful to Fabienne Kienast Weber for this information. Colorful mosaics were also placed in the Kienasts' private garden in Zurich in reaction to this trip.
172 I am grateful to Kienast-Lüder for this information.

173 Kienast, "Ohne Leitbild," 38.
174 Rowe and Koetter, *Collage City,* 175. Kienast referred to this in Kienast, "Die Sehnsucht nach dem Paradies"; Kienast, "Von der Notwendigkeit künstlerischer Innovation"; Kienast, "Stadtlandschaft."
175 Rowe and Koetter, *Collage City,* 181.
176 A selection of the 471 entries submitted were published in the report on the competition: Marianne Barzilay, Catherine Hayward, and Lucette Lombard-Valentino, *L'Invention du parc: Parc de la Villette, Paris; Concours international / International Competition, 1982–1983* (Paris: Graphite, 1984). Kienast also participated in the competition, together with Stefan Rotzler, but their entry was not published.
177 The Günthersburgpark was never realized because the city government of Frankfurt changed. The project for the Moabiter Werder could only be realized in part as a result of the German government moving from Bonn to Berlin, so that parts of the land planned for it had to be built with larger housing developments and another sizable part of it became the garden for the new office of the German chancellor. The promenade as it looks today is largely the work of Günther Vogt and was completed by the firm of Vogt Landschaftsarchitekten in 2002.
178 Kienast, "Von der Notwendigkeit künstlerischer Innovation," 57–58.
179 Ibid., 58.
180 Werner Seligman, "The Texas Years and the Beginning at ETH Zurich, 1956–1961," in Jürgen Jansen et al., *Architektur lehren: Bernhard Hoesli an der Architekturabteilung der ETH Zürich / Teaching Architecture: Bernhard Hoesli at the Department of Architecture at the ETH Zurich* (Zurich: gta Verlag, 1989), 7–16.
181 Bernhard Hoesli, untitled foreword to the text, trans. Jori Walker, in Rowe and Slutzky, *Transparency,* 7–8.
182 Ibid., 8.
183 Hoesli, "Transparent Form-Organization," 85.
184 Ibid.
185 See, among other things, the third thesis in Kienast, "Zehn Thesen zur Landschaftsarchitektur," 207; see also appendix, 403.
186 Kienast, "Wohngrün zu Mehrfamilienhäusern," 10. It must have been decided to close the nursery and build a new multifamily residence already by the mid-1970s, because the first plan for the new grounds, dated September 15, 1974, replaced the nursery grounds with a playground, several seating areas, sunbeds, lawns, and a mini-landscape of quarry stones with a small pond. The header of the plan lists Kienast's father, Heinrich, as the builder-owner and Peter Paul Stöckli is listed as a "consulting architect"; Dieter Kienast drew this first plan; see plan no. 743-59, preliminary project, scale: 1:100, H 46.5 × W 83 cm, ink and felt-tip pen on tracing paper, September 15, 1974, signed "KIE, firm of Peter Stöckli," gta Archives (NSL Archive) / ETH Zurich (Dieter Kienast Bequest). The plan of 1976 was based on this one.
187 The exhibition *Dieter Kienast: Zwischen Arkadien und Restfläche,* in collaboration with Günther Vogt, with photographs by Christian Vogt, Architekturgalerie Luzern, May 31 to June 28, 1992, with accompanying catalog.
188 Dieter Kienast, "Garten Kienast, Zürich," in *Gute Gärten: Gestaltete Freiräume in der Region Zürich* (Zurich: Bund Schweizer Landschaftsarchitekten und Landschaftsarchitektinnen [BSLA], Regionalgruppe Zürich, 1995).
189 Kienast, "Ohne Leitbild," 36.
190 Ganz is now owner of Ganz Landschaftsarchitekten in Zurich; since 1986, Gasser has been the owner of Gartenkulturen, a gardening company that has since then also been responsible for maintaining the Kienasts' Garden; Vogt joined the firm of Stöckli, Kienast & Koeppel in 1987, then in 1995 cofounded the firm Kienast Vogt Partner with Dieter and Erika Kienast; since 2000 he has been the owner of Vogt Landschaftsarchitekten in Zurich and has since opened offices in London, Berlin, and Paris. For information on the various stages of the redesign, I am grateful to Daniel Ganz and Christoph Gasser for conversations on May 31 and June 1, 2015.
191 Christoph Gasser in conversation with the author, June 1, 2015.
192 Kienast, "Ohne Leitbild," 35.
193 See the analysis of the garden of the Palais Stoclet by Josef Hoffmann (1905–1911) in Anette Freytag, "Josef Hoffmann's Unknown Masterpiece: The Garden of Stoclet House in Brussels, 1905–1911," *Studies in the History of Gardens and Designed Landscapes* 30, no. 4 (2010): 337–372, here 341–48 and 356–361.
194 This too is another—certainly unintended—commonality with Hoffmann's garden of Palais Stoclet; see ibid., 349. The uniform gravel to "enlarge" gardens optically has a much longer tradition.
195 I am grateful to Erika Kienast-Lüder for this information.
196 Christoph Gasser in conversation with the author, June 1, 2015.
197 The requirement that every multifamily home should have a large kitchen garden for its tenants, which Kienast had made part of the program for the grounds of housing developments as early as 1975 in his thesis, was also mentioned in "Bemerkungen zum wohnungsnahen Freiraum" of 1979, as well as those providing private seating areas in the garden and alternative play areas for children; the text was illustrated with photographs of his own garden.
198 The Fasanengarten employed another element that had been tried out first in the Kienasts' Garden: beds framed with boxwood with a base made of shards of colorful glass, with the boxwood trimmed as topiary that rises up out of the beds. In the Kienasts' Garden in 1991, this kind of bed replaced the earlier shrub beds bordered with boxwood. I am grateful to Erika Kienast-Lüder for this information.
199 Kienast, "Cultivating Discontinuity," 152.
200 Dieter Kienast, "Longing for Paradise," trans. Bruce Almberg and Katje Steiner, in *Dieter Kienast*, 84–90.
201 Kienast, "Vom Gestaltungsdiktat zum Naturdiktat," 120.

202 Kienast, "Garten Kienast," 37. In the catalog of his first exhibition in 1992, *Dieter Kienast: Zwischen Arkadien und Restfläche,* Kienast reprinted a text by the Germanist and Slavist German Ritz that presents his private garden; he also included it in the monograph *Kienast: Gärten / Gardens.* The passage quoted is clearly influenced by the content and style of this text. Ritz compared the Kienasts' Garden to a "growing text" whose "mixture of styles" refers to "the history of the garden's life"; for Ritz, Kienast's private garden is, like the autobiographic, a dynamic form that remains open: the autobiographic "is not afraid of statements—it touches on the cliché, kitsch; it quotes what is foreign and repeats the individual; it lives from every-day-life, the garbage dump of the different and unique." See German Ritz, "Entering the Autobiographical Garden," in *Kienast: Gärten / Gardens,* 18–21, here 18, 20.

203 Kienast, "Wohngrün zu Mehrfamilienhäusern," 10.

204 On this, see Horst Rittel, "Systematik des Planens," *Werk* 54, no. 8 (1967): 505–8; Karl Heinz Bohrer, "Nach der Natur: Ansichten einer Moderne jenseits der Utopie," in *Über Politik und Ästhetik* (Munich: Hanser, 1988), 209–29; Gernot Böhme, *Für eine ökologische Naturästhetik* (Frankfurt am Main: Suhrkamp, 1989); Hartmut Böhme, *Natur und Subjekt* (Frankfurt am Main: Suhrkamp, 1988).

205 See especially Martin Seel, *Eine Ästhetik der Natur* (Frankfurt am Main: Suhrkamp, 1991).

206 See Dieter Kienast and Toni Raymann, "Grün 80; oder, Warum lieb' ich alles was so grün ist? Alternatives Konzept zur Durchführung der Gartenschau 1980 in Basel," typescript, Archiv SKK Landschaftsarchitekten, Wettingen, 1976, 7.

207 See Professur für Landschaftsarchitektur ETH Zürich, *Dieter Kienast,* 210.

208 Bertolt Brecht, "Mr. K and Nature," in *Stories of Mr. Keuner,* trans. Martin Chalmers (San Francisco: City Lights, 2001), 17.

209 Ibid.

210 Dieter Kienast, "Über den Umgang mit dem Friedhof" (On Dealing with Cemeteries), *anthos* 29, no. 4 (1990): 10–14, here 10.

211 Dieter Kienast, "Vom Kirchhof zum Erholungsraum" (1992), in Professur für Landschaftsarchitektur ETH Zürich, *Dieter Kienast,* 117–23, here 123.

212 Kienast began his project description for the competition by quoting a poem by Rainer Maria Rilke that Rilke had carved into his own tombstone: "Rose, o pure contradiction / love / to be nobody's sleep / under so many / eyelids." See Dieter Kienast, "Fürstenwald Cemetery in Chur," trans. Bruce Almberg and Katje Steiner, in *Dieter Kienast,* 116–19, here 116.

213 Kienast, "Vom Kirchhof," 121.

214 Ibid., 120.

215 Kienast, "Cultivating Discontinuity," 150.

216 The former steep terrain would not have been suitable for people in wheelchairs. See Kienast, "Fürstenwald Cemetery in Chur," 117.

217 For information on the architectural history of the cemetery, I am grateful to Alex Jost of the parks department of the City of Chur, conversation on November 27, 2008.

218 Kienast, "Fürstenwald Cemetery in Chur," 117.

219 I am grateful to the landscape architect Martina Voser of Chur for this information.

220 On March 8, 2007, the town council decided to extend the wall of urns as part of a second phase of construction; see Stadt Chur, Geschäftsnummer 328.02 / http://www.chur.ch/dl.php/de/4cd7d07fc2cc6/GRB_Friedhof_Furstenwald.pdf (link no longer active). It was to be done according to the master plan from 1994 (Stadt Chur, *Botschaft* 17 [1994]), and specifically in keeping with the original design: "The expanded design will also serve to reinforce the terrain facing the Fürstenwald, and a notch in the terrain offers spaces for mourners and other visitors to the cemetery." Stadt Chur, *Botschaft* 9 (2007): 5, http://www.chur.ch/dl.php/de/4c62725a69247/ALV_B_Friedhof_Fuerstenwald_09_07.pdf (link no longer active). As part of these measures, additional grave fields were conceived following the design of the first phase of construction.

221 Stadt Chur, *Botschaft* 9 (2007): 5. The expansion of the wall of urn niches was carried out in 2012 by the Municipal Planning Authority of the City of Chur, Parks Department based on a design proposal by the landscape architect Rita Illien (firm of Müller Illien Landschaftsarchitekten, until 2008 managing director at Vogt Landschaftsarchitekten); see http://www.chur.ch/dl.php/de/507c05c644eff/Erweiterung_Urnennischenwand_A3.pdf (link no longer active).

222 Kienast, "Fürstenwald Cemetery in Chur," 119.

III
Drawing and Perceiving

Media of Representation

In the autumn of 1977, the students of the Organisationseinheit Architektur, Stadt-, und Landschaftsplanung (Organizational Unit of Architecture and Urban and Landscape Planning) at the Gesamthochschule Kassel took up pickaxes and shovels in a spontaneous action to break up the areas in front of Building K 10 that had been paved and thus sealed off by order of the university building department during the summer vacation. They replaced them with water-permeable gravel paving on which they planted ash trees (131). This design corresponded to the "ecological planning of open space" that was being taught in Kassel. Previously, the doctoral student Dieter Kienast had insisted that, despite all the spontaneity, there had to be a plan for the new site.[1] In view of an action of civil disobedience, several of his fellow students smiled at this and interpreted it as a typically Swiss will to order.[2] Kienast's initiative to draw up a plan makes it clear, however, that he was conscious of the design consequences of planting trees, considered the future effect of these actions, and hence believed a plan was necessary.[3]

Those involved in the action of planting ash trees commented in a subsequent report in *Der Monolith,* the student newspaper of the aforementioned Organisationseinheit, that they regarded it as "completely untenable" that "the requirements of developing and preparing open space are discussed and pursued in seminars and projects while before our front door others muddle on lagging behind all the established advances in science and practice."[4] Kienast's plan was published in the *Monolith* report together with a photograph of the planting action (131). A caricature of Karl Heinrich Hülbusch smoking a pipe appears in the middle of the article. The professor of landscape planning and Kienast's dissertation adviser was quoted as saying: "This [using pickaxes and shovels to break up the ground] is a great example of how science can be plausibly pursued!" The planning action was an implementation of the call in Kassel to "make laypersons and citizens respected experts in their own world of action and living."[5] Kienast's final plan shows a clear will to order and design the open space in front of the university that was recaptured on an autumn weekend (132): thirteen ash trees were planted in four groups in front of Building K 10. They mediate between the building, which is at a slight angle to the layout of the city, the sidewalk, and the straight street. On the plan, Kienast also drew Building K 9 on the opposite side of the street and the open area in front of it, both of which are parallel to the street: here groups of bushes are planted along the building, and the sidewalk consists of concrete slabs; between the sidewalk and the street are small lawns.

In the redesigned open space in front of K 10, both the pedestrians on the sidewalk and the employees and students exiting the building are much better shielded from traffic on the street by the trees, which also provide shade. In the areas were the ash trees were planted in two stages, the students replaced the concrete slabs of the sidewalk with gravel. The parking spaces were basalt natural stone paving. Spontaneous vegetation could grow in the cracks between the paving stones.

The final plan was based on a collage. Kienast began by drawing the various areas with ink and colored pencil: the building, the new open space, and the existing open space. The two buildings are indicated by stripes, and the gravel areas and concrete slabs by a system of dots and grid networks, respectively. In an irregular style, the loops, dots, and small strokes depict "living" materials such as trees, lawns, and bushes. The volume of the bushes and treetops is suggested. The original designs were taped together to form a plan, reproduced, and then the copy was colored by hand. In the process, Kienast emphasized the volume of the treetops by increasing the density of the structure of the leaves in places using a green pen and added shadows drawn with black felt-tip pen. The "dead" materials such as concrete slabs, gravel areas, and natural stone paving are indicated with colored pencil in various shades of orange. A key explains the symbols of the individual materials and zones. Precise coordinates are indicated for the location of each newly planted tree. The recapture

PARTIZIPATION: GRÜN IN DIE NORDSTADT

Aufbauend auf dem Erkenntnisstand verschiedener Disziplinen diskutieren wir sehr eifrig die Anforderungen an die Freiraumplanung. Wir sind der Meinung, daß der Ausbau einer Hochschule und die konkrete Planung eines Standortes, an dem ein wissenschaftlicher Ausbildungsgang für Architekten, Stadtplaner, Landschafts- und Freiraumplaner besteht, von diesen professionell Betroffenen mitgetragen und beeinflußt werden sollte. Das staatliche Hochschulbauamt sah sich nicht in der Lage, die von den Planungsbetroffenen begründeten und konzipierten Freiraumplanungen weiter zu entwickeln und in kontinuierlichem Gespräch zu realisieren. Verallgemeinernd für die gesamte Baupolitik des Hochschulbauamtes steht die letzte Blitzaktion, die aus der Unkenntnis von Ausbauzielen, fehlender Einschätzung der bestehenden Situation und dem ungebrochenen Zwang zur Aktivität, zu unsinnigen Resultaten führen muß. Beim ehemaligen Verwaltungsgebäude der Henschelei (K10), in dem die OE untergebracht ist, hat das Hochschulbauamt aus heiterem Himmel ein sogenanntes Provisorium geschaffen, indem es die im Sommer erstellten Schotterflächen wieder auskoffern und durch einen 40 cm hohen Basaltschotterunterbau mit 15 cm hoher Teerdecke hat ersetzen lassen. – Ein sehr dauerhaftes Provisorium! Angesichts dieser Willkür einer bürokratischen Instanz sahen wir uns zur demonstrativen Selbsthilfe gezwungen. Es erschien uns völlig unhaltbar, daß in Seminaren und Projekten die Anforderungen an Freiraumentwicklung und -Bereitstellung diskutiert und weitergetrieben werden, während vor unserer Haustüre wie eh und je hinter jedem wissenschaftlichen und praktischen Stand zurückbleibend, weiter gewurstelt wird, und jeder mögliche Ansatz zur Realisierung durch Ignoranz vereitelt und ins Gegenteil verkehrt wird. Angesichts dieser schizophrenen Situation von Kenntnis und Situation haben wir dem "dauerhaften Provisorium" ein "praktisches Provisorium" entgegengesetzt. An dieses Exempel knüpfen wir Forderungen für die weiteren Maßnahmen und Gestaltungen. Die vorhandenen von uns entwickelten Planungsansätze, Rahmenkonzepte und typischen Beispielen, werden über den jetzigen Stand weiter entwickelt werden müssen; denn angesichts der gemachten Erfahrung und der fehlenden Freiraumkonzeption überhaupt (zu sagen: "Wir haben nichts gegen Grün" ersetzt noch nicht die Planung), die durch spontanes Stückwerk von Fall zu Fall "ersetzt" wird, sehen wir uns genötigt auch weiterhin alternativ zu arbeiten. Unsere erste Aktion stellt somit nur den Anfang dar, unsere Bedürfnisse und Kenntnisse nicht nur als Willensbekundungen sondern auch in "handgreiflicher" Form deutlich zu machen, mit denen wir auch in Zukunft auf die Entwicklung des Wohn- und Arbeitsfeldes einwirken werden.

DAT IS' DOCH 'MAL'N PRIMA BEISPIEL, WIE MAN PLAUSIBEL WISSENSCHAFT MACHEN KANN!

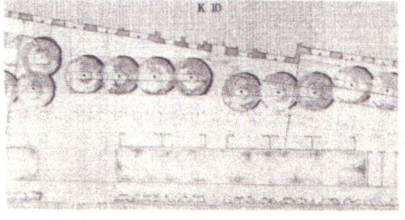

10 Eschen Pflanzaktion am 21.10.77

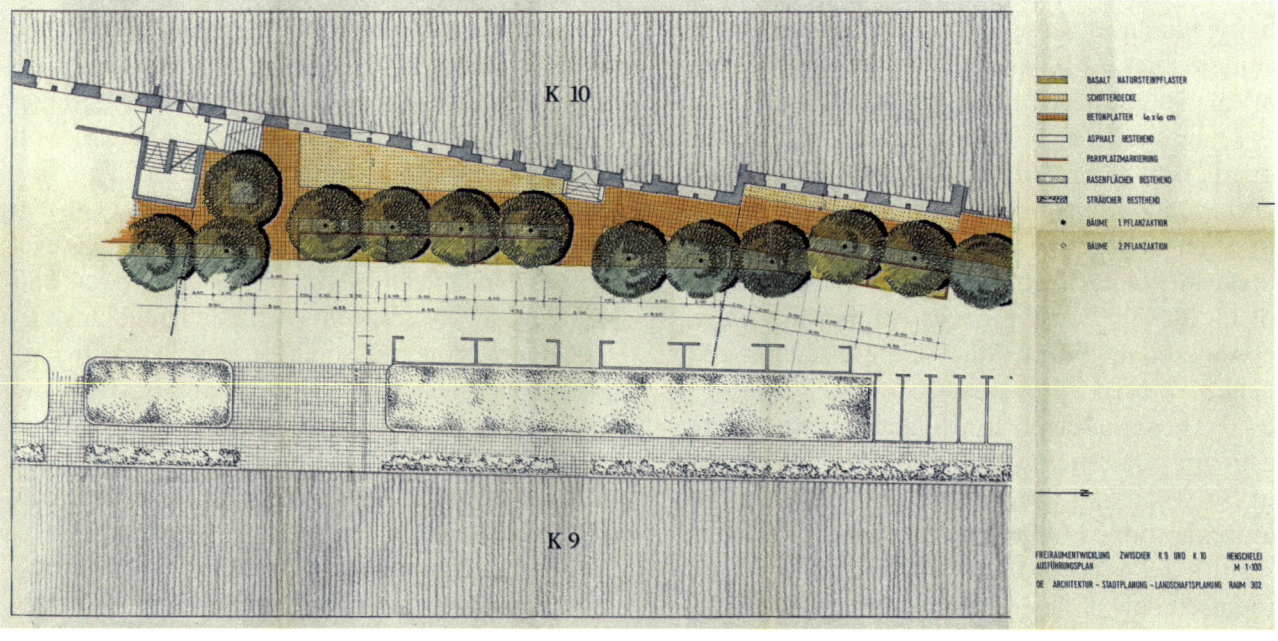

131
Article in the student magazine *Der Monolith* on the action to plant ash trees by the Organizational Unit of Architecture and Urban and Landscape Planning at the Gesamthochschule Kassel, October 21, 1977, with a plan by Dieter Kienast

132
Dieter Kienast, "Open space development between K 9 and K 10 Henschelei," final plan, scale: 1:100, H 47.5 × W 93 cm, undated (1977), Organizational Unit of Architecture and Urban and Landscape Planning

of the open space in front of the university in Kassel occurred spontaneously but not without a plan; the plan it followed was rather detailed and subtly rendered.

Kienast's Language of Form

The planting action in Kassel demonstrates that the drawing of plans was already of crucial importance to Dieter Kienast at a time when design and planning in landscape architecture were heavily dominated by social-science research, climatological studies, and phytosociological surveys. Kienast always thought of himself as a gardener, a craftsman, and an artist in one. The intermeshing of these three professions resulted in an approach to drafting and designing spaces in which a personal style he developed over the years became visible thanks to a personal language and grammar, a store of symbols and forms, which condensed into a "vocabulary for the landscape." Closely connected here are thus the perception of a space sharpened by close observation, the drawn plan for its redesign, the rendering of Kienast's ideas in order to communicate them to the clients and the public, and the experience of the place he has altered. Kienast found his inspiration for coming to terms with the phenomenological effect of his design and the topological dimension of his work in the art, architecture, and literature he encountered in Switzerland after returning from Kassel in 1979.

Kienast's work was influenced by his early training in drawing. The way he organizes space and form, as well as the way he designs individual elements and works out details, demonstrates that perceiving, drawing, and reflecting on the formal representation of intellectual concepts are key to his work. This quality is directly connected to his central concern: making an environment "more legible" by means of design interventions. Over two decades, Kienast developed a design vocabulary to enhance order and legibility that he applied ever more rigorously over the course of the 1990s. In no small measure, that led to the criticism that he was a postmodern "player with little forms"[6] who "ruined" a whole generation of landscape architects,[7] because so many of his colleagues drew inspiration from this formal vocabulary and continue to use it today, albeit with one crucial difference: Kienast employed his vocabulary in differentiated ways, always reacting very precisely to the given circumstances and anchoring it in the terrain. On the one hand, his interventions are tied to a specific place and operate topologically in that they have an effect on the terrain and at the same time create new connections between a lot and its surroundings. On the other hand, they operate in a phenomenological sense, because Kienast's intention was obvious: he wanted his design to change how people perceive a place and thus to affect their experience. His works move within the area of tension between abstraction and presence. A general vocabulary defined gradually during the genesis of the work has an effect in a specific space as a result of the perception of the people who use that space, because it makes it possible for them to experience this space in a new way.

Engaging with a chosen place in the process of drawing sketches and plans was, however, just one aspect of achieving this goal. Kienast's deep engagement with the art and literature of the 1970s and 1980s and with the strategies that these artists utilized to achieve certain effects also contributed to his specific approach to drafting, representing, and realizing his ideas. The effect of this was felt, first, in the choice of materials and forms for his designs and, second, in the various media he employed to represent them. Kienast's central concern was making a place "legible" and "activating" its qualities. This is true both of the projects themselves and of the way they are represented, both during the phase of their conception and after their realization. This is evident in the artful presentation plans, which Kienast initially drew by hand, based on which he developed a descriptive language used both in the collages of the late 1980s and the 1990s and in the first software-based plans; and also in the photographs of Christian Vogt, who was brought to Dieter Kienast's attention by Günther Vogt, and who from 1992 onward documented the built sites not in the classical sense but rather used them as inspiration to pursue a certain strand in his own artistic works. In the various media and methods of representation, we once again sense the aforementioned tension between abstraction and presence.

The style in which Dieter Kienast wrote and spoke also developed in parallel with this process. Thanks to their clarity and directness, his essays remain popular with readers of all age groups even today. All of the characteristic aspects of his work—the way he drafted, drew, collaged, planted, built, wrote, and spoke and the way Christian Vogt captured in photographs what had been realized—led to a distinctive significance and unmistakable recognizability. The latter was surely welcome if not perhaps even calculated. Anyone who knows his formal language and his modes of representation will immediately recognize "a Kienast"—that is to say, can "read" it. This is true of the built works and of the drawn or collaged plans, but also of the monographs of his works, consistently designed with black-and-white photographs in just two formats, which ultimately became the trademark of the firm Kienast Vogt Partner, which was founded in 1995.

To describe these characteristics better, it helps to return to terms from linguistics, thus bringing us to the core of the terminology of postmodern thinking and creating: it is about questions of signature, grammar, and context;[8] it is about the tensions between presence and abstraction,[9] between the general and the concrete, between so-called inscription, effectiveness, and perception. Nevertheless, it would be too simplistic to categorize Kienast as the "classic example of the postmodern landscape architect." Although he clearly and sometimes ostentatiously employed postmodern methods, he remained true to modernism in his belief in the transformational power of his open spaces. He had acquired this absolute and sometimes utopian ambition in Kassel. In this way, he evades such categorizations, and his position within the context of cultural history remains polyvalent.

One proof of this is the way the presentations of his work by means of plan, text, exhibition, photography, and video changed. Kienast is thus revealed to have been aware not only of form but also of its relativity, its effectiveness, and its individual perception. By means of his methods of representation, he tried to capture both processes of perception and material qualities and in that way cut deep to the very "essence" of things. He wanted to open up paths for the users of his grounds to have a corresponding aesthetic experience. The fact that this revealed an unmistakable personal style in the manner of a trademark was a side effect. For diversity was ultimately more crucial than uniformity, as is underscored by the different media of representation: plans, writings, photographs, videos, and exhibitions.

From Zoning to Designing

The drawing of plans had a fundamental importance for Dieter Kienast all his life. The plan for the spontaneous action in Kassel to plant ash trees is an early example of this. However, Kienast initially concentrated less on drawing and instead pursued the methods of analysis and planning that dominated the landscape architecture of the 1970s: social science research, climatological data, and vegetation studies.

For the master plan to improve the quality of open space in the Nordstadt district of Kassel, which Dieter Kienast and Thom Roelly had worked out in 1975 as part of their thesis project, the authors did not reveal any design ambitions in the presentation of their proposals for new infrastructure and maintenance measures (16). The analyses accompanying the plan, by contrast, filled more than a hundred pages. The two plans that were included as an accordion-fold appendix to his phytosociological dissertation of 1978 consist of an assemblage of zones with various dot and hatching patterns over the plan of the city. In one case, the zones indicate the various types of neighborhood in the city; in the other, the plant communities found there (17, 18). The signatures of the individual zones are explained in the legends. A similar logic of presentation is also found in the sketches and design drawings for the grounds of the housing complex in Niederhasli of 1972 (51–54). On the first sheet, Kienast indicated roughly in felt-tip pen the areas for various functions around the residences. Here, the zoning is indicated by different colors. Once again, a legend explains the assignment of the zones. In the design drawing that was developed from this first sketch (54), the individual zones are nuanced by means of drawing, but the assignment of functions to areas remained the same.

The roughness of the drawing and the choice of felt-tip pen has perplexed many connoisseurs of Kienast's later work,[10] for his drawing skills were, by this time, long since evident.

Kienast had acquired the fundamentals before entering the Hochschule der bildenden Künste in Kassel during two internships with the Swiss landscape architects Albert Zulauf and Fred Eicher. In Eicher's design for a garden, which he planned in 1969, the year Kienast began his internship with him, for the Siemer House in Gosam in the Wachau, designed by Ernst Hiesmayr, one can see how the landscape architect marked the terrain with contour lines and developed the design from the topography (133, 134). Apart from the interventions in the topography, Eicher's most important design elements are the trees, with which he created his spatial concept for the grounds, often supported by walls. Eicher liked to use so-called impoverished materials such as concrete. He integrated into his grounds geometric pools and small fountains made of concrete that look like modern sculptures. Several areas are clearly articulated by means of slight differences in height or changes in materials, which can be seen particularly well from the example of the Eichbühl Cemetery in the Altstetten district of Zurich (1966) (135). Kienast picked up on these design elements and this use of materials again in the early 1980s (136).

A comparison of Eicher's plans for the Siemer Garden of 1969 or for the New Botanical Garden in Zurich (1972–77) (137) with Kienast's presentation plans for the grounds of the Niederhasli housing colony of 1972 (138) or the competition plan submitted together with Peter Paul Stöckli in 1977 for the "Land und Wasser" (Land and Water) sector of Grün 80 (Green 80) in Basel (140) reveals several parallels, even apart from the use of trees, in the form of representation. Kienast's forms are, however, much more organic than Eicher's. When Kienast began to study with Günther Grzimek at the Hochschule für bildende Künste in Kassel in 1970, it is only a slight overstatement to say that the category of "form" in landscape architecture only had a place as organic form. Grzimek's most important project of this period, the grounds of the Olympic Stadium in Munich in 1972 (139), impressively testifies to this. This plan was clearly a model for Stöckli and Kienast's competition plan for Grün 80: the two projects are similar both in terms of the topographical subdivision of the grounds and in the course of the paths and the distribution and shape of the bodies of water.

When Kienast returned to Switzerland for good in 1979, architects there were wrestling with new forms for presenting landscaping projects that were intended to replace the organic forms then common.[11] Under the influence of such new plans, the classic example of which is the competition plan submitted for the "Land und Wasser" sector of Grün 80 by the architects Bernhard Hoesli, Arthur Rüegg, and Arnold Amsler and the landscape architect Ruedi Siebrecht (141), Kienast had found his way to a completely new language of drawing within five years. In the plans he drew shortly after that competition, like those for the Dry Grasland Biotope of Grün 80 in 1978 (25, 26), the strong influence of Swiss landscape architects and their drawing style is obvious: the layout of the design of the Dry Grasland Biotope is very reminiscent of Fred Eicher's projects. Kienast was already parting ways somewhat with the organic forms he had once used much more frequently. The drawing style, as far as the topography and the significance of trees as elements to articulate the space are concerned, also points to Eicher. Details such as stones, bushes, and trees on the site plan, the views, and the sections of the Dry Grasland Biotope are, in turn, related to the drawing technique of Willi Neukom (80). Kienast did, however, draw much more subtly than Neukom, which can be seen, for example, in the differentiation of the branches of the trees.

Later, in 1983, the firm of Stöckli + Kienast won the competition for the Zurzach Spa Gardens with a plan drawn by Kienast (142). The plan from 1983 reveals both in the layout and in the individual elements that Kienast had been studying the plans of the Bürogemeinschaft Hoesli, Rüegg, Amsler und Siebrecht. Kienast was expanding his drawing abilities during this period. Since returning to Switzerland, he had run through various phases of experimentation. Before studying in Kassel, he had mastered the craft of gardening; after returning from Kassel, he clearly wanted to perfect the craft of drawing. He impressed many Swiss architects with these abilities.[12]

In the late 1970s, many architects were extremely frustrated that many of their colleagues, especially in landscape architecture, were poor at drawing. Kienast, too, began to cultivate his drawing talent at that time. Previously, he had worked entirely in the spirit of the planning avant-garde of the time, employing the method of zoning based on use functions in accordance with the age groups of the uses, differences in temperature, urban structures, and plant societies. In the landscape of the German university system, that was the current and politically correct method of analysis and presentation. In the drawing of detailed, artful plans, which came into vogue in the 1980s, Kienast was able to establish something distinctive on many different levels in the field of Swiss landscape architecture. The highlights of his hand-drawn plans include two sheets he produced after completing the redesign of the M.-M. Garden in Erlenbach in 1989 (143). The redesign coincided with the renovation of the house by the ARCOOP firm of Ueli Marbach and Arthur Rüegg. The latter had also been involved in the competition for the "Land und Wasser" sector of Grün 80 (141) and the reformulation of the styles of representing landscapes that had clearly impressed Kienast.

The two sheets show, first, the site plan of the house and garden[13] and, second, the interlocked longitudinal and cross-sections of the grounds (143). The drawing is on beige cardstock, its materiality also integrated into the drawing: as volume for the treetops on the site plan and in the cross-sections, and as volume for the terrain and house in the cross-sections. These areas with no drawing at all are outlined in pencil and, like the treetops depicted in the cross-section, slightly differentiated. The "void" becomes a "mass" in the drawing. The details of the garden—slab-covered paths and small plazas, hedges, flower beds, a large water basis and a small pool, spheres of boxwood, lawns, and steps cut into the terrain—are subtly distinguished in pencil. Kienast was trying to illustrate the materiality of the objects as well. On both sheets, the drawing proper is framed by light hatching—a mixture of pencil and blue crayon—as if by a delicate cloud that surrounds the garden and at the same time provides a colorful background into which the parts of the garden seem to be engraved. This intertwining of figure and ground, in combination with integrating the cardstock into the rendering itself and the interlocking of sections, contributes to the artfulness of the plans. Everything seems light and playful and yet is precise down to the last details. Looking at the plans triggers joyful curiosity, and they illustrate a sensuality that is also found in the garden design they illustrate.

It is not just the plans for gardens that have this sort of effect, but also the plans for such important competition entries as the design for the Moabiter Werder in Berlin of 1990–91 (83), the expansion and redesign of the Günthersburgpark in Frankfurt am Main of 1991–92 (84), and the design of the open spaces for the grounds of Expo 2000 in Hanover of 1992 (171–174). The consequence was first prizes. By winning these competitions, Kienast's work was noticed first in Germany and Austria, and ultimately internationally as well. The attitude and the searching communicated by his plan drawings made Kienast a popular partner of young Swiss architectural firms, which in the 1980s and 1990s were attracting attention even outside of Switzerland and Europe, such as Diener & Diener, Herzog & de Meuron, Burkhalter und Sumi, and Meili Peter. Like Kienast, they were all from German-speaking Switzerland and born between 1950 and 1960, and despite differences in their careers they shared "certain experiences, images of life—symbols of the world of commodities,"[14] as the architectural historian Martin Steinmann expressed it in an essay about the question of what linked the aforementioned architects. Using the example of buildings from the early 1980s, he shows how their common experiences led them to a similar *recherche architecturale,* in which they explored questions of form, space, and construction and the central theme of the use of materials. This *recherche architecturale* was comparable to Kienast's search, since the phenomenological and topological dimension of their buildings is likewise of primary importance to this generation of architects, that is, the experience of architecture by those who live in and look at it and the local reference of buildings, as well as the consequences for construction, form, and materials. Steinmann sees the results of this *recherche* in a new sensual presence of architecture causing immediate experiences in people—separated from the efforts of architecture to span a horizon of meaning that goes beyond the present of the building and its experienceable qualities. That is why Steinmann called it "*presemiotic* experience"[15] and, in order to describe the transformation in architecture in German-speaking Switzerland, why he cites the architectural historian Bruno Reichlin: "A grammar of materials gradually developed in place of a grammar of symbols."[16]

This "grammar of materials" preoccupied Dieter Kienast as well, and in connection with his plans he was particularly moved by the question of how a highly differentiated rendering of the design elements in accordance with their materiality could be achieved. The challenge was to express on the plan the composition of the materials used and to distinguish different elements

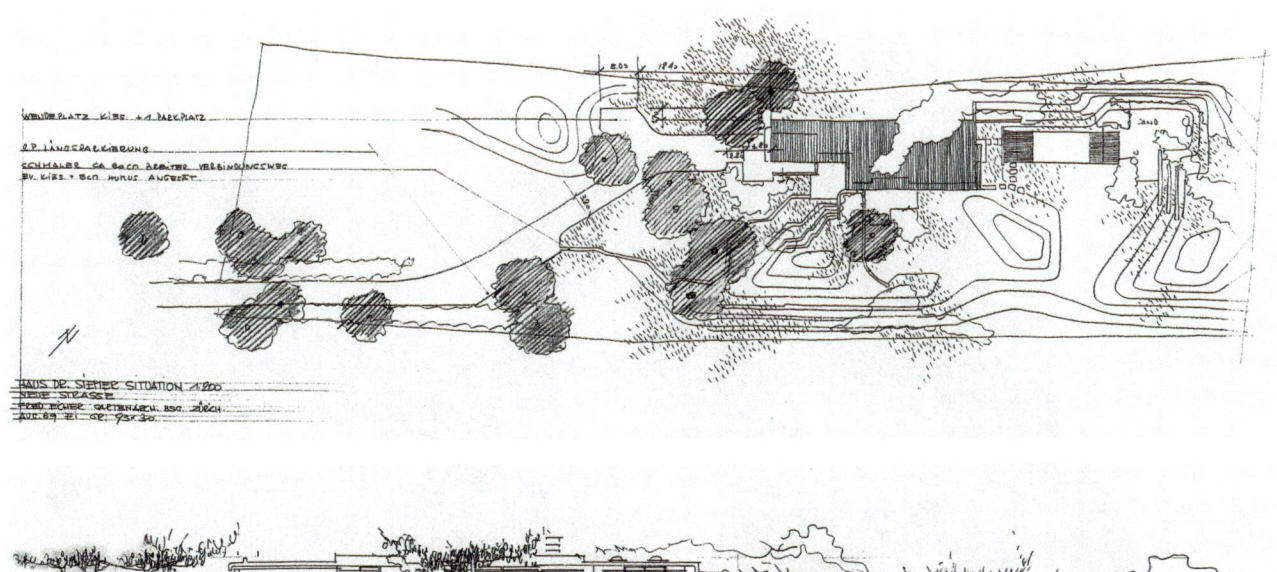
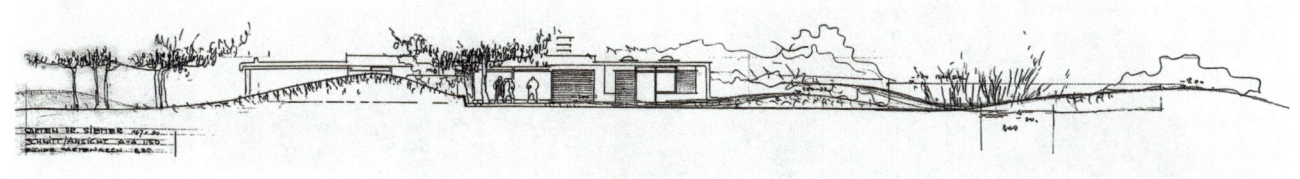

133, 134
Fred Eicher, Siemer Garden in Gossam in the Wachau, Austria, 1969, site plan, scale: 1:200, as well as section and view
135
Fred Eicher, Eichbühl Cemetery in Zurich, 1963–66. Square in front of the chapel designed by Ernst Studer
136
Recess yard of the École cantonale de langue française in Bern, 1983–91

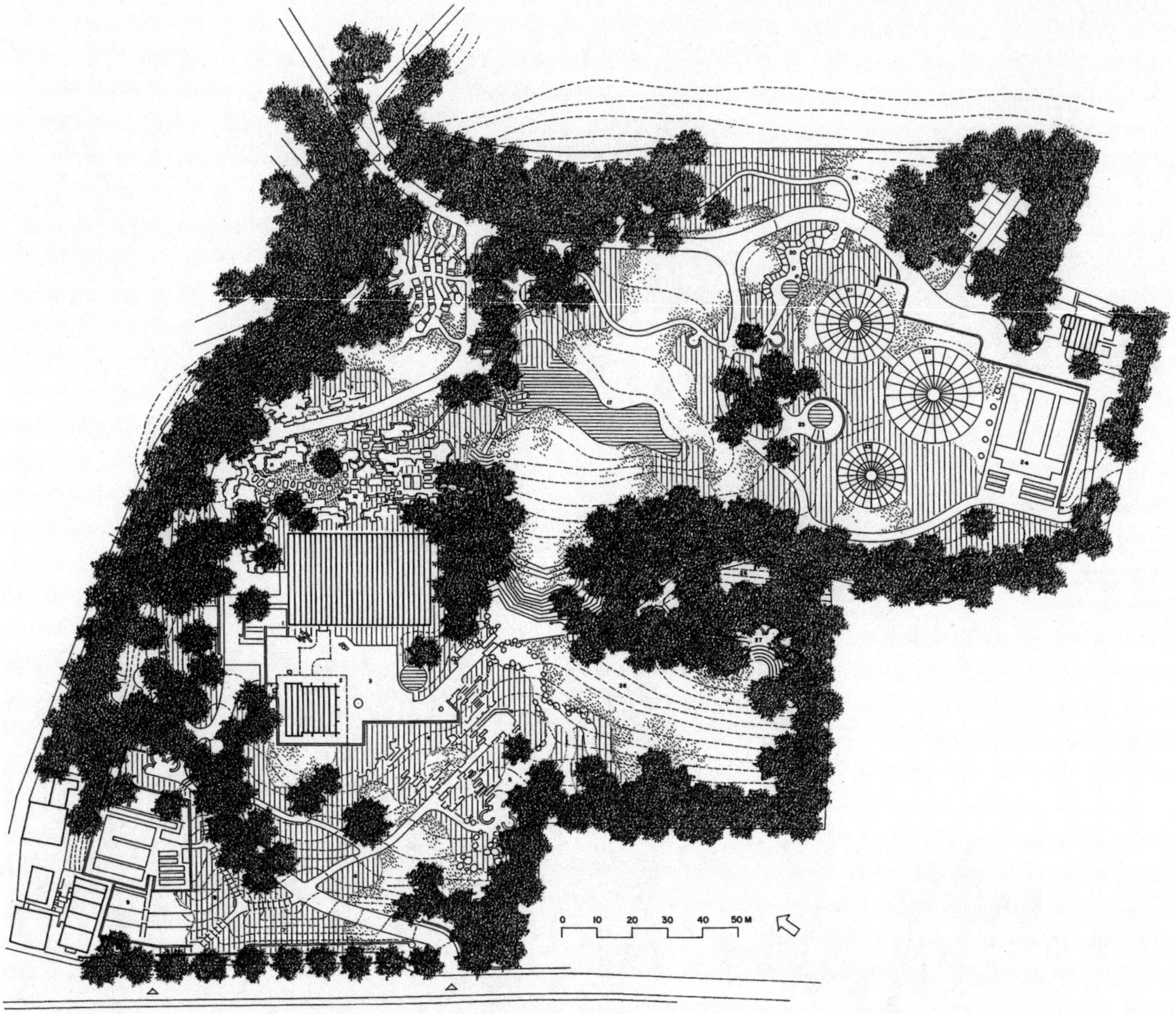

Legend

- ☐ WEGE + PLÄTZE
- ☐ BÄUME
- ☐ WASSER
- ☐ GARTEN FÜR ERDGESCHOSSWOHNUNG
- ☐ SANDSPIELPLATZ MIT WASSER
- ☐ SPIELGERÄTE
- ☐ SITZPLÄTZE MIT PERGOLA + FEUERSTELLE
- ☐ RASENSPIELFELD
- ☐ BESPIELBARER RASEN
- ☐ BOCCIA
- ☐ TISCHTENNIS
- ☐ MAL- SPIEL- + KLETTERMAUERN

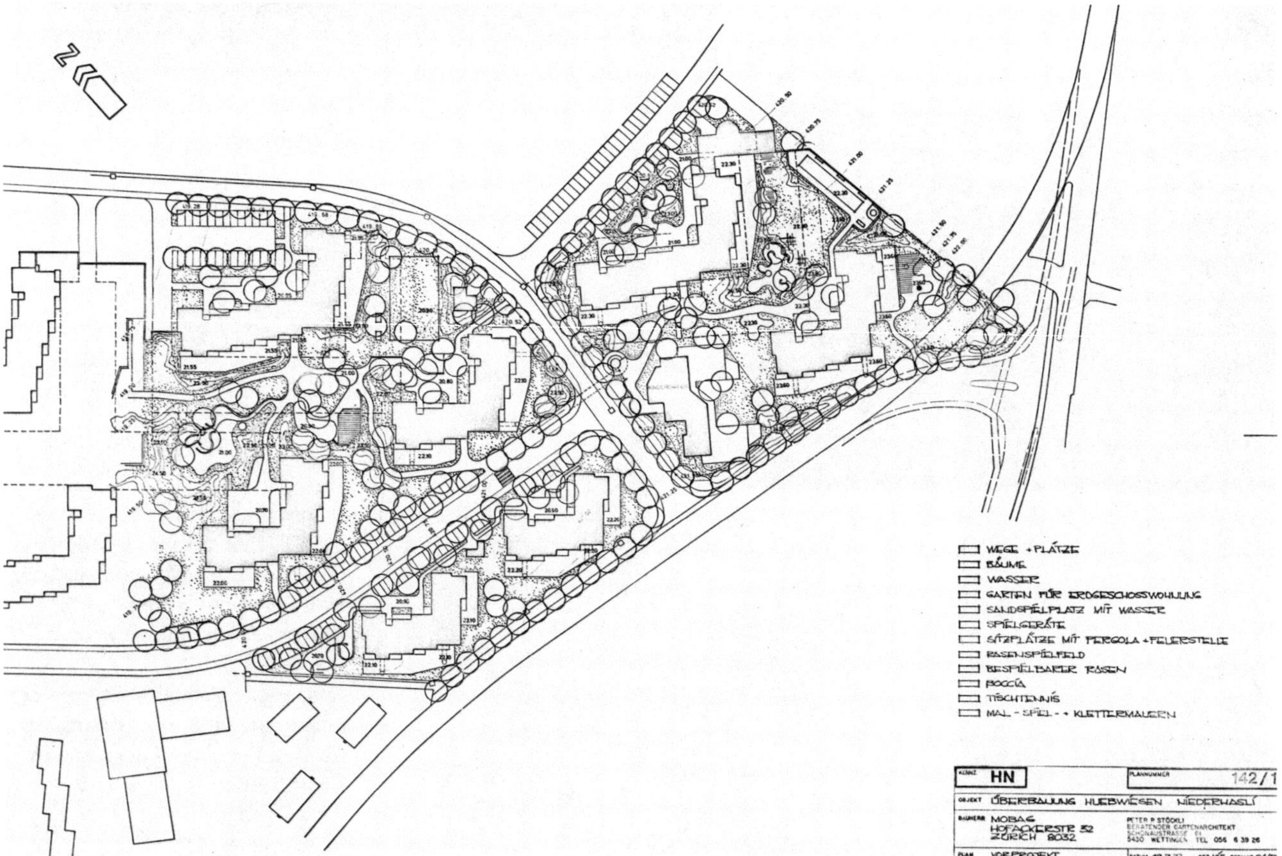

KUNDE	HN	PLANNUMMER	142/1
OBJEKT	ÜBERBAUUNG HUEBWIESEN NIEDERHASLI		
BAUHERR	MOBAG HOFACKERSTR 52 ZÜRICH 8032	PETER P. STÖCKLI BERATENDER GARTENARCHITEKT SCHÖNAUSTRASSE 61 5430 WETTINGEN TEL 056 6 39 26	
PLAN	VORPROJEKT	DATUM 19.9.72 REV.GEZ 19.12.1972	
MASSTAB	1:500		

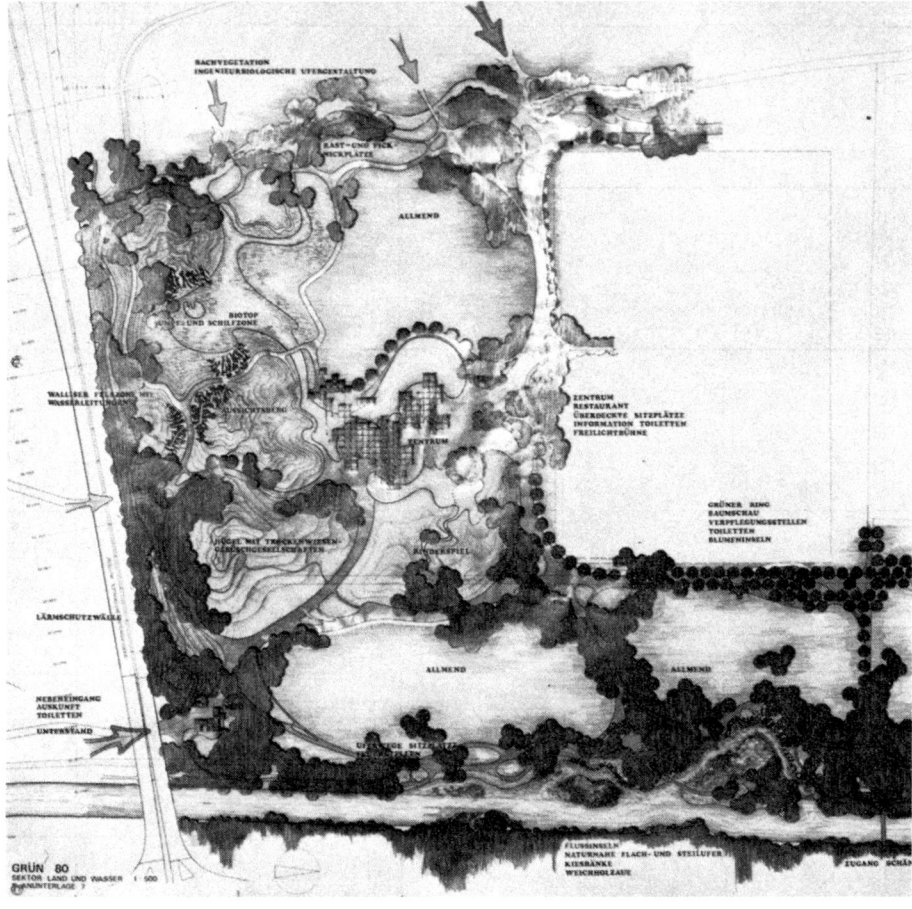

137
Fred Eicher, New Botanical Garden in Zurich, 1972–77, site plan

138
Grounds of the Huebwiesen and Lindenstrasse building project in Niederhasli, preliminary plan no. 142-1, scale: 1:500, copy, July 19, 1972, signed "Kie," firm of Peter Paul Stöckli

139
Günther Grzimek, Olympic grounds and park in Munich, 1968–72, site plan, 1969 (detail)

140
Grün 80 in Basel, "Land und Wasser" (Land and Water) sector, competition plan, scale: 1:500, undated (1977), firm of Peter Paul Stöckli, project author: Dieter Kienast

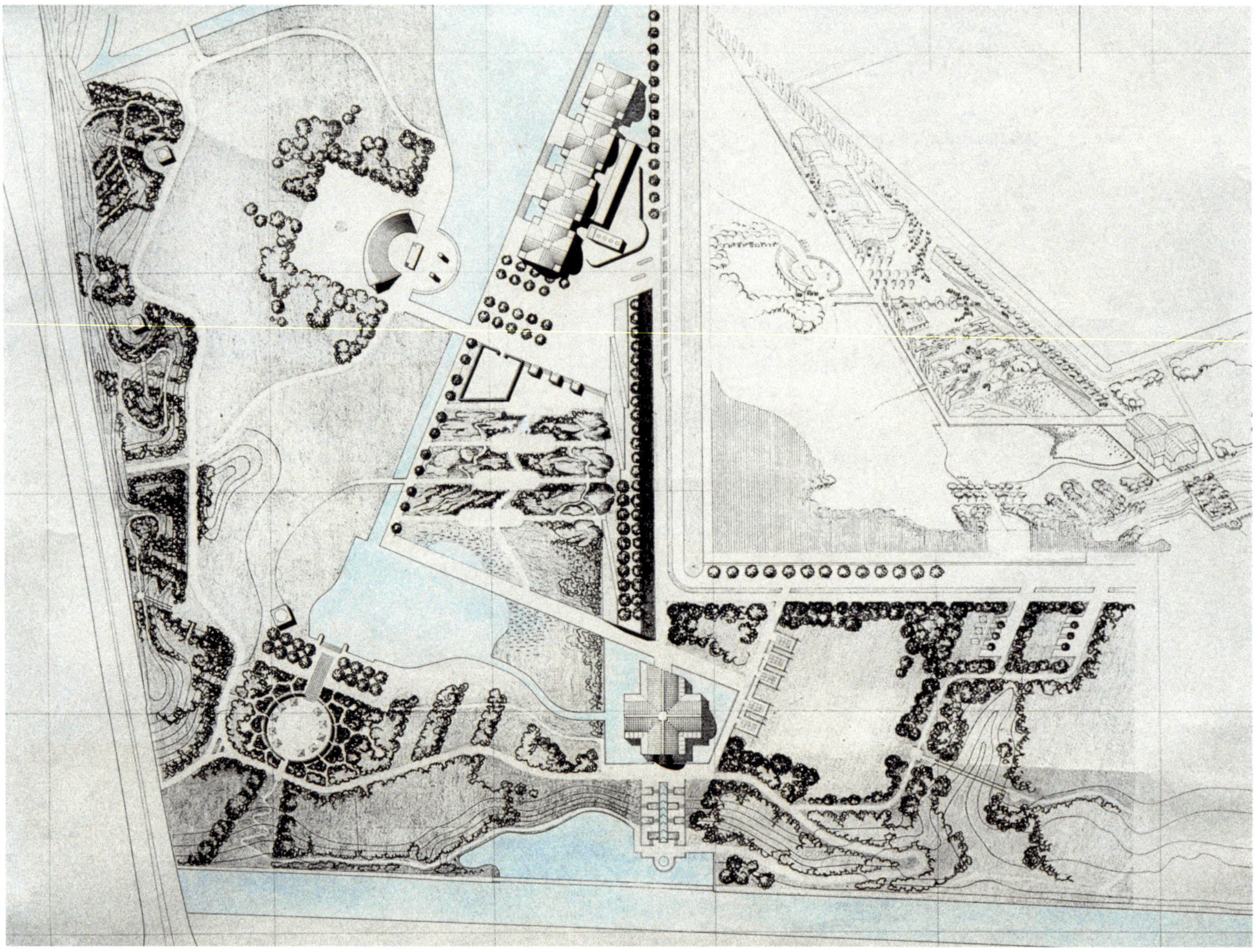

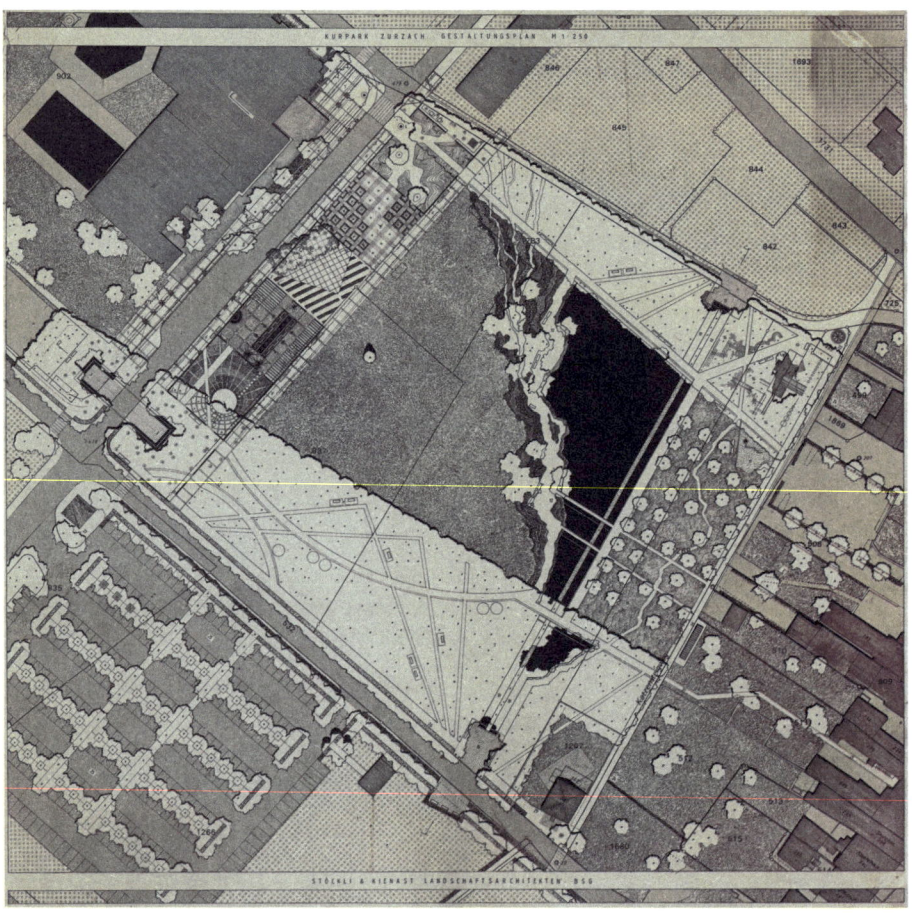

141
Bernhard Hoesli, Arthur Rüegg, Arnold Amsler (architecture), and Ruedi Siebrecht (landscape design), Grün 80 in Basel, "Land und Wasser" (Land and Water) sector, competition project, 1977 (detail)

142
Design plan for Zurzach Spa Gardens, scale: 1:250, H 100 × W 100 cm, copy of plan, undated, firm of Stöckli + Kienast

143
Dieter Kienast, M.-M. Garden in Erlenbach, longitudinal and cross section, plan no. 1005-12, 1:100, H 50.5 × W 84 cm, pencil, colored pencil on card stock, July 27, 1989, signed "Kie," firm of Stöckli, Kienast & Koeppel

GARTEN KATHARINA · TULLIO MEDICI · MALL ERLENBACH

such as hedges, shrubs, concrete walls, glasshouses, gravel, and undergrowth in a way that the materials from the plan can be experienced by the senses. Over the course of the 1980s and 1990s, Kienast developed, to that end, his own vocabulary of different signatures with their own grammar. The media of rendering plans changed in the process: drawing by hand was followed by collage with plastic film, which was followed in turn by CAAD planning and a software-based rendering technique.[17] The plans made using the new technology almost always included elements made using earlier techniques: drawings were integrated into the collages, and drawings and collages into software-based plans. But until the end of his life, Kienast did all his preliminary designs exclusively in sketches drawn by hand.[18]

In order to make landscape legible, Kienast developed in parallel with his rendering vocabulary a formal vocabulary for his interventions in the landscape that reenacts, accentuates, or counteracts the given topology. This happened as part of a "search for forms," and that he pursued in parallel with movements in literature, art, and architecture since the late 1970s. His work, which creates spaces and stimulates consciousness, embodies the topological and phenomenological dimension of postmodernism. The emancipation of the subject was always central to this when Kienast said of his era:

What is meant is not . . . the populist style of postmodernism, which has been impermissibly reduced to architecture and specifically to neoclassicism, with the truncated column, the pediment, and the oriel; what is meant is the postmodernism of the likes of Lyotard or Welsch, of Handke and Kundera, Nouvel, Herzog & de Meuron.[19]

"This Thing We Call Art": Dieter Kienast and the Phenomenology of Postmodernism—Art, Architecture, Literature

When Dieter Kienast returned to Switzerland after completing his dissertation in 1979, he encountered there a cultural atmosphere that contrasted starkly with that of Kassel. The progressive impetus of 1968 had given way ten years later to a certain "melancholy of failure,"[20] which favored a new interiority. "The hierarchy of structures proved to be more resistant than the idealism of the changers," Christel Sauer observed in retrospect, who with her husband, Urs Raussmüller, assembled a collection of art for the Migros-Genossenschaft in Zurich in the 1970s and then in the 1980s built up her own collection in the Hallen für Neue Kunst in Schaffhausen.[21] The collecting activities of Sauer and Raussmüller resulted in an intense exchange with artists who were having a profound influence on the architectural scene at the time, including Carl Andre, Joseph Beuys, Dan Flavin, Jannis Kounellis, Sol LeWitt, Richard Long, Mario Merz, Bruce Nauman, and Robert Ryman. From their perspective, the "process of changing consciousness"—which was still being pursued with great enthusiasm in the 1960s—was disappointing with respect to the range of its sociopolitical goals.[22] At the same time, several art movements of those years, such as Conceptual Art, were mutating into a "bloodless, didactic tendency that largely robbed art of its sensuous aspect."[23] The works of art that Sauer and Raussmüller had been adding to their Raussmüller Collection since 1973, and had made accessible to the public in 1984, sharpen sensory perception and provoke in viewers a new experience of space and materiality that introduces an act of making present (*Vergegenwärtigung*) that stimulates and modifies self-perception.

Urs Raussmüller described this process using the example of *Cuts,* a work by the conceptual artist Carl Andre (b. 1935) (144). First exhibited in Los Angeles in 1967, *Cuts* was integrated into the Raussmüller Collection in 2000. Rausmüller's account of it stands paradigmatically for the effect of an art to which Kienast had a close connection,[24] probably because he had wanted to initiate similar effects in his gardens.

"Cuts" is a phenomenon that encompasses much more than the physical appearance of the sculpture as such. It is not just an issue of the number and form of the concrete building blocks spread out on the floor according to a principle. It is also an issue of the air space above the capstones, the architectural space, and the entire surroundings.

And above all: it's about me—and in a way that challenges all of my senses. Here I perceive with my entire body. The thing moves me, and with every change in my perspective I experience a new constellation of the entire situation. I—and I mean *every* viewer—am an equal part of the phenomenon.... I experience perception as a feeling of being. This is where I place the greatness of this artistic endeavor.[25]

In lieu of a comprehensive claim to validity, there was now a new emphasis on experience: the art with which Kienast was coming to terms after his period in Kassel no longer claimed to be contributing enduring insights, instead focusing on the here and now of perception. Carl Andre summed it up in the formulation: "The perceptive viewer ... who's got some kind of stimulus from the work is also doing art work."[26]

Art was thus reacting to a loss of scientific and ideological certainties that went hand in hand with a fundamental change in the environment: Cities were growing, the countryside was being overdeveloped. In the Western industrialized nations, recessions and urban problems were resulting in neglected marginal areas, whereas in the economic centers speculation and pragmatism were giving rise to faceless cities and agglomerations. The economic conditions in Switzerland were better than in the rest of Europe and in the United States, but urban sprawl in the allegedly idyllic was particularly striking there. As Kienast observed of the new conditions:

The traditional city–countryside rivalry has disintegrated, the boundaries have blurred. We assume that it is impossible either to dismantle the city or to rejuvenate the countryside. Nevertheless, the legibility, the perceptibility of the world is rooted in the principle of dissimilarity. Considering this synchrony of city and countryside, the coming task is to stop the further erosion of the inner boundaries and splits. They have to become sensuously perceptible again.[27]

The paradigm shift in art with its consequences for the development of Swiss architecture and—through Kienast's influence—of Swiss landscape architecture also resulted from the experience of the tangibility of nature having been lost: specialization in the natural sciences had led to an unmanageable quantity of data, making it clear that nature as an entity could no longer be grasped by reason. This loss of a larger whole in which human beings felt protected was compensated by new insights generated through the aesthetic experience of nature. What counted was a sequence of moments of aesthetic stimulus.[28]

Against the backdrop of this new paradigm, Kienast made the switch to being a designer. The systematic methods of phytosociology with which he had previously wanted to initiate design processes in landscape architecture no longer had priority. His new striving for a formal language that would make landscape "legible" corresponded to the phenomenological spirit of postmodernism, in which artists, writers, and architects were seeking ever more subtle forms of aesthetic experience. Their different methods can be characterized as what Viktor Shklovsky referred to as "defamiliarization" or "deautomatization" in the quotations in the first chapter. By perplexing everyday perception, they make it possible to see the world in a new way: "And so, what we call art exists in order to give back the sensations of life, in order to make us feel things, in order to make a stone stony."[29]

Whereas the works of art that Kienast encountered in Kassel had been above all sensors of the social processes in which the perspectives of their creators manifested themselves, his artistic environment in Switzerland was characterized by works that intended to stimulate and modify the viewer's perspective.

Joseph Beuys: Social Sculpture and Material Tension

The initiation of social and societal processes and the sensing of oneself through the sensory experience of (the materiality of) art are two artistic poles that Joseph Beuys combined in his concept of "social sculpture." For Beuys, who had been a strong presence in Kassel and was also represented in Swiss collections, art was not a discipline but rather the absolute equivalent of creativity:

His now famous and frequently misconstrued claim "everyone is an artist" means that each individual possesses a creative capacity which, as society's veritable asset, is to be transferred from a state of dormancy to a process of free development.[30]

Creativity thus became the "liberating way to individuality."[31] Beuys saw it as the only path to achieve this and regarded art as the only valid model of a behavior directed at change. Charged with this significance, creativity and art become the existential prerequisite for the necessary process of changing the vast and inflexible entity that is society. Beuys described this process as "social sculpture." According to his idealistic view, each individual should be able, through free and creative behaviour, to act as a "sculptor of the social organism."[32]

Consequently, as Christel Sauer explains, Beuys was convinced that "the social order of the future will be formed according to the laws of art." Using the example of Beuys's work and conviction, it is possible to understand once again why Kienast's strong interest in the effects of his design interventions of the 1980s and 1990s did not signify—as the representatives of the Kassel School claimed—a complete turn away from his pronounced sociopolitical and ecological commitment of the 1970s. Rather, a transformation was taking place in which the emancipatory claim of the subject remained the focus—with the desired consequences for the development of a society that Beuys had sketched out.

Also of fundamental significance for Kienast and the Swiss-German architects of the 1980s were such central motifs for Joseph Beuys's art as the use of materials and the introduction of "material tension," which he often produced by combining organic and inorganic materials,[33] or the initiation of a relationship between the experiencing subject and the artwork that triggers his experience. Private collections like that of Christel Sauer and Urs Raussmüller had an important function in disseminating this art in Switzerland. For them, the acquisition of Beuys's *Das Kapital Raum, 1970–1977* (The Capital Room, 1970–1977), an installation with slate sabs, a concert grand piano, a film projector, loudspeakers, washtubs, and more, was the occasion to find new accommodations for the collection they had begun in Zurich (145). In the giant building of a disused worsted spinning mill on the bank of the Rhine in Schaffhausen, they found the optimal site for the enormous objects in their collection. The Hallen für Neue Kunst opened in 1984 and were open to visitors until 2014.[34] The installation of *Das Kapital Raum* was supervised by Beuys personally. He had conceived it with a planned permanent exhibition in the Raussmüller Collection in mind and had first shown it in 1980 at the Venice Biennale or, in Urs Raussmüller's words, did "a kind of test run with the 'innards' of the monument."[35] Christel Sauer observes that it is difficult to convey "Beuys after Beuys" in exhibitions: "The rare talent with which he presented things, the way he positioned them in relation to one another, charging spaces with meaning and tension, cannot be reconstructed."[36]

Lending meaning to things by positioning them in relation to one another and in space is a concern that the husband-and-wife collectors pursued for all of the works included in the Hallen für Neue Kunst. One can see parallels to Kienast's topological interventions when designing existing places, which he undertook with phenomenological motive force. They are interventions that sharpen the viewers' perception and are intended to open up for them a new view of their surroundings.

144
Carl Andre, *Cuts,* 1967. Installation at the Hallen für Neue Kunst, Schaffhausen

145
Joseph Beuys, *Das Kapital Raum 1970–1977,* 1984. Installation at the Hallen für Neue Kunst, Schaffhausen; in the background, the adjoining room with two igloos by Mario Merz

146, 147
Mario Merz, *Architettura fondata dal tempo—Architettura sfondata dal tempo,* 1977, and *Chiaro oscuro,* 1979/1984. Installation at the Hallen für Neue Kunst, Schaffhausen. Photos: Fabio Fabbrini/Raussmüller Collection

148
Sol LeWitt, *Three-Part Variations on Three Different Kinds of Cubes 3 3 2,* 1967/1974. Installation in Central Park, New York

149
Sol LeWitt, *Black Styrofoam on Black Cube and White Styrofoam on White Cube,* 1994. Installation at the Hallen für Neue Kunst, Schaffhausen. Photo: Fabio Fabbrini/Raussmüller Collection

150
Sol LeWitt, *Cube,* 1984. Model of the cube shown in the exhibition *Skulptur im 20. Jahrhundert* (Sculpture in the Twentieth Century) in the Merian Park, Basel

151
Herzog & de Meuron, Dominus Winery in Yountville, Napa Valley, California, 1996–98. Photo: Timothy Hursley

Making Present by Means of Material and Form and the Relationship between the Work of Art and the Viewer

Sharpening perception and opening perspectives were also intended by the works of the Italian Arte Povera artist Mario Merz (1925–2003), which were placed in the room next to Beuys's *Kapital*. Over the years, Merz installed four igloos there. Since 1968 he had been building these "nomads' houses" wherever he found himself.[37] The height and diameter of their form were based on the human scale, in order to offer people a symbolic home. The materials used for the igloos varied. In Schaffhausen, the choice of the material resulted from the position of the igloos in the room and hence to the landscape in front of the windows: on the side from which one had a free view of the Rhine some distance away, there rose a filigreed igloo of shattered panes of glass held by a wire frame. The airy mood in this part of the room was supported by the igloo whose panes of glass were regularly cleaned.[38]

Another image was revealed by the igloo on the opposite site, which faced the old town and its narrow streets (146), where the neighboring buildings and their tapering roofscapes loomed immediately beyond the windows. The igloo in front of that was much more compact than the one on the side with the river. Merz had collected its components in the surroundings of the Hallen: stone slabs, chain-link fencing, a wine bottle, and panes of glass covered with a layer of dust were mounted on the igloo's metal stand. A rectangular frame that seemed to grow out of the igloo held other panes of glass close together. This igloo's appearance made the city in front of the windows seem present to the viewers: the artifacts transplanted from the city to the exhibition space were removed from their functional context, so that the viewers' concentration is directed onto the material, the surface composition, the haptic qualities. The work also thrives on the light refractions in the shattered panes; the sun shining on them reveals the layer of dust. They were intended to sharpen the viewers' senses and heighten their perceptions, so that they could establish new relationships to the context. In the best case, they saw with new eyes not only the materials of the igloo but also the city beyond the windows from which the materials came.[39]

Another igloo at the Hallen reached out even further to connect to the surroundings: it was covered with grapevine branches from the neighboring vineyards (147). In the dried vines were glowing numbers from the Fibonacci sequence: 1, 1, 2, 3, 5, 8, 13, 21, and so on, in which each member is the sum of the two previous members. The Fibonacci sequence is connected to the proportional system of the Golden Section. Merz had enjoyed an education not only in philosophy but also in medicine and the sciences. The combination of decaying grapevines from Schaffhausen, the glowing numbers of the Fibonacci sequences, and the igloos built according to human dimensions reveals the whole horizon of meaning related to the question of the laws of nature and the human urge to explore them.

The igloo epitomizes the tension between the powerful insight of the 1970s and 1980s that nature as an entity can no longer be grasped by reason and the attempt to instead ensure authentic knowledge of the world, in a piecemeal way, by means of subjective experience. The drying grapevines, which were regularly replaced, also illustrated the process of the decay of nature and its cycle of birth and death. In addition to its presence, a temporality was inscribed in the object, and from its poetry of the quotidian it opened up a field of symbolic meaning that the other igloos in Schaffhausen were not able to evoke in the same way.

In their pointed symbolism, the works by European artists exhibited in the Hallen für Neue Kunst differed fundamentally from the contemporaneous works by North American artists. They shared a search for means to achieve a new degree of reality in art. The "elementary foundation for art's impact is its credibility,"[40] as Christel Sauer sums it up, and in the 1960s artists increasingly saw this credibility as dependent on the degree of reality in their art:

> *Pop Art* had taken the decisive step to a shared vocabulary of art and society. Carrying on from this, artists such as Andre, Flavin, Judd and LeWitt brought the apparent image of art up to date and radically excluded any fictive, symbolic or illusionist effects. They employed modern technical materials and industrial methods of production and reduced the artwork to the primary aspects of its determination as a real object: to volumes and material. In avoiding any manipulation going beyond the

elementary construction of simple geometrical forms (the reason for the classification *Minimal Art*), they gave the artwork a neutrality previously unknown. They concentrated perception on the thing itself, or rather on the relation between the artwork and viewer that opened up in perception.[41]

Frank Stella (b. 1936) introduced this development in 1958 with paintings showing nothing but black brushstrokes and his statement "What you see is what you see!" In the process, he defined an art with meaning and aesthetics grounded exclusively in the precise employment of its actual components—as opposed to the symbolic values of European art that Sauer emphasizes. This liberation of art from conventional expectations opened up a large scope for artists, since they no longer had to produce any additional meaning. Every material was possible, and every production method. Even ideas could be works of art. For Sauer, the characteristic feature of this era was the incredible energy that was produced by the diverse creative proposals, and that "cannot be explained in any other way than in terms of faith in the motivating power of art":

The 60s were borne to a high degree by the feeling of social and political responsibility that was apparent in the striving for equal rights and in the struggles against elitist and hierarchic structures at all levels, including art. From the resistance against a satiated, bourgeois attitude about what to expect, artworks arose in this period that literally obstruct their consumerist appropriation. The artists gave their works a spatial extension through which they demonstratively *occupied* space. . . . In the rebellion against musty institutions and out of the infinite need to allow individual insights and social goals to intermingle, they displaced their room for action into the wide open spaced, of nature *(Land Art)* or into the expanses of imagination of their viewers, who were addressed using all means possible.[42]

Christel Sauer's precise explanations are so fruitful for an understanding of Dieter Kienast's work because they reveal the connection among the seemingly so different artists about whom Kienast was enthusiastic.[43] It lies in the commonality of the discussion of phenomenological, topological, and sociopolitical questions and the links between them.

Presumably Kienast broadened the palette of his materials during his final creative years, not least because he was influenced by art and architecture. At the beginning of his career, he had manifested his love for "impoverished" materials such as concrete and steel, and in the 1990s walls of natural stone prominently entered his work.[44] He experimented with natural materials such as rammed earth, which he employed to build walls in several gardens long before it became one of the favorite materials of Swiss architects.[45] Although the symbolic and the use of recurring forms were central to Kienast's work, he almost always avoided falling into two postmodernist traps: formalism and trite visuality. He did so by anchoring his works in the place and the terrain; and, thanks to his great knowledge of plants, by integrating into his works the quality of plants as a process; and by coming to terms with the needs of the users and the "use value" of his grounds. To the end of his life, searching for clues, classification, producing order and systems of reference—all without completely disavowing an anarchic impetus—remained essential motifs of his work. The much-overused concept of expanding consciousness played a crucial role in this.

Expanding Consciousness
and Critique of Urbanism

Artists as different as Mario Merz and Sol LeWitt (1928–2007), who were close personal friends, wanted their works to seem consciousness-expanding.[46] As contemporaries, they were linked by that effort. Arte Povera, the movement with which Merz's works have been associated, conveys its program by its name. Artists worked with found, "poor" materials, which had no claim to value or durability.[47] For Christel Sauer, artists such as Mario Merz and Jannis Kounellis (b. 1936), and the suggestive images their works produce, appeal directly to the viewers' imagination:

> They built a "new reality" (Merz) from pieces of nature which, as the embodiment of an artistic formative process, stimulated creativity.

The symbolic force of the images and the poetry of the staging were conceived as a means for provoking emotions and insights through an impact on the senses. Kounellis described himself as a "political artist" who quotes history to generate a sensibility for the present.[48]

On a larger scale, such works also seemed like social criticism of advancing urbanization. In the crisis of urbanism, art became the sensorium of topological processes and realities. Consequently, Robert Smithson (1938–1973) and Gordon Matta-Clark (1943–1978) shifted their focus away from the celebrated surfaces of Pop Art to fallow fields and slums, junk and garbage.[49]

Sol LeWitt, by contrast, reflected on the gleaming surfaces of functionalist skyscrapers. His sculptures correspond to basic geometric forms and are shaped according to objective laws. Smithson saw them as dystopia: "LeWitt's show has helped to neutralize the myth of progress,"[50] as an affirmation of a rational efficiency that had replaced the modern faith in progress.

By his own account, Sol LeWitt's works are closer to architecture than to sculpture: "I have always called my three-dimensional work 'structures,' because my thinking derives from the history of architecture rather than of sculpture."[51] At the same time, they can be interpreted as commentary on the crisis experienced by architecture in the 1970s. When LeWitt liberated his transparent structures from their functional purpose, he made their inherent logic legible: the installation of the work *Three-Part Variations on Three Different Kinds of Cubes 3 3 2* (1967/1974) in Central Park in New York, with the skyline in the background (148), made it clear that this work of art created a distance from the economic hierarchy of the silhouette of New York City by focusing on the consciousness of its viewers. Pure form has a progressive potential.

But LeWitt's works also help viewers reflect on their own perception. In his earlier grid structures, LeWitt had made use of modular arrangements to evoke in the viewer's head a perception of the overall form. This emphasizes the thinking aspect of perception. One such work is in the Raussmüller Collection. Moreover, LeWitt is also represented in it by two room-high cubes of equal size that are made of the same material (149). One cube is black, the other white. The difference in color causes the white cube to be perceived as larger, and from close up the black cube is perceived as warmer. Such works of art served to remind the design disciplines of architecture and landscape architecture of the knowledge that colors and forms trigger subjective perceptions and feelings.

Dieter Kienast came into direct contact with Sol LeWitt's works and those of related artists in 1984 when, in his role as technical director of the Botanical Garden in Basel-Brüglingen, a position he held from 1981 to 1985, he was partly responsible for the exhibition *Skulptur im 20. Jahrhundert* (Sculpture in the Twentieth Century). The exhibition was organized by the Christoph Merian Foundation and the Ernst Beyeler Collection.[52] Kienast integrated the sand-lime brick cube by Sol LeWitt (150) shown there, measuring 5 by 5 by 5 meters, into an assignment he gave to his students at the Interkantonales Technikum Rapperswil, where he was professor of landscape architecture.[53] The task was to integrate this work of art into a redesign of the Kollerwiese in the Wiedikon district of Zurich into a multiuse park. Kienast's interest in this particular object was based on its form and its surface composition and its effect on the senses. In the catalog to the exhibition in 1984, the cube is said to radiate many meanings:

> Viewers experience how a formal concept produces form through both series and permutation. Despite the analytical stance, however, LeWitt's artistic imagination is expressed fully; one is astonished by the sensory richness and the diversity of the production of the so soberly programmed "concept machine." . . . The stone's horizontal position emphasizes the rest, the effect of gravity. The sculpture looms over us like a bulwark, giant and anonymous, Kaaba-like, but radiantly white, restrained and yet spectacular.[54]

In the architecture of the time, too, one of the topical issues revolved around the composition of surfaces. For the architect Jacques Herzog (b. 1950), dealing with surfaces had, as a rule, consequences for dealing with space.[55] In several buildings by the firm Herzog & de Meuron, the shell of the building penetrates the interior as well. Herzog and his partner, Pierre de Meuron (b. 1950), belonged to a new generation of Swiss-German architects who were just as influenced by artists such as LeWitt, Merz, and Kounellis as Dieter Kienast was. Their shared concern was lending architecture a specific quality by means of a sensory experience transported via the material.

Architecture

In the buildings Herzog & de Meuron had been designing since the mid-1990s, such as the Ricola Europe Packing and Distribution Building in Brunstatt (41), the Dominus Winery in the Napa Valley (151), and the studio for the artist Rémy Zaugg in Pfastatt (152), natural phenomena became part of the development process. At the Dominus Winery (1996–98), the walls are made of wire forms filled with gravel of various sizes: the walls are heavy, but they still allow light into the interior of the building. In an interview in 1988, Jacques Herzog explained that with this building he had been interested in the direct effect of the architecture on the body—the physical experience of the coolness in the interior protected by the stone walls when entering from the vineyard in California's Napa Valley[56]—as well as in the phenomenological experience of architecture when approaching it: The vineyard, he said, could barely be discovered in the hilly landscape at first, yet as one approached it, one became that much more aware of its monumentality, but as soon as one was inside it, it broke down "literally into fragments, into individual stones, and into the space between the stones, which are reflected in the light shining on the floor and walls."[57] The point was, he continued, to penetrate the architecture to a point at which human beings can perceive themselves with all of their senses in the very experience of the architecture:[58]
Our architecture aims to explore all the visual, acoustic, tactile, and olfactory experiences that we can have as human beings. It should be possible to understand that and to explain to others.... The interaction and grappling with the time in which we live is unavoidable in that process.[59]

Dieter Kienast also expressed how important it was for him that his works convey the "socially determined" quality of the time[60] and, by doing so, was reacting to current social, cultural, and ecological events. In the 1990s, this included the emerging domination of electronic media over everyday life, which gave Herzog & de Meuron crucial impetus to address the sensory experience of human beings as its opposite pole:

> We are interested in these sensory elements in architecture in part because they have become so sparse in our everyday experience.... This approach of working with the senses is radical. As a city dweller, your sensory perception is reduced to the visual, because the electronic media dominate your entire life. Expanding sensory perception in architecture suddenly puts you in a quite different position. For that reason, nature is important to us—even if it is a domesticated nature.[61]

Herzog & de Meuron were experimenting in the mid-1990s, as with the Dominus Winery in Napa Valley, with the experience of warmth radiating from the building or, as with the Ricola Europe Packing and Distribution Building, the Rudin House in Leymen, or the Studio Rémy Zaugg, with rainwater running down the facade and the associated traces and colonies of algae, lichens, and mosses. In the Studio Zaugg (1995–96) (152), they also added a rainwater basin visible through a skylight, where the pollution of the rain by the emissions of nearby industrial companies could be seen. These were "processes of consciousness raising" in the sense of a critique of urbanism like those that LeWitt, Merz, Matta-Clark, and Smithson had initiated in their works.

In addition to reflections on the consequences of urbanization and on stimulating the senses by consciously integrating natural phenomena, there was a third aspect of the art and architecture of the time that is relevant: the possibility of making temporal processes and processes of change such as growth, decay, and erosion visible. One impulse that set the direction for the European context came from the fountain designed by Meret Oppenheim (1913–1985) for Waisenhausplatz in Bern, which was dedicated in 1983 (153). The artist selected for it a looming stone column and mounted on it a metal gutter that spiraled down and was perforated in places to cause a cascade, over which the fountain has been spraying water since its inception. Oppenheim was counting on weather and gradual colonization by algae, lichens, mosses, and grasses—as is found again in several works by Herzog & de Meuron and Kienast in

314

152
Herzog & de Meuron, Studio Rémy Zaugg in Pfastatt, France, 1995–96. Photo: Georg Aerni, 2000

153
Meret Oppenheim, fountain on Waisenhausplatz in Bern, 1983. Photo: Georg Aerni, 2012

the 1990s. The strict form of the stone column was intended to counter the urban nature growing over it, making the work of art a link between architecture and urban nature, which has in the meanwhile happened.[62]

This emphatic contrasting of natural and artificial was essential to Kienast's work; in the late 1990s, Herzog & de Meuron pointedly tried to overcome it:
Bipolar models never work. . . . Playing artificiality and naturalness against each other leads to nothing. The only thing that is interesting is the question of how specifically you are in a place. Does it take on a quality that wasn't there before?[63]

Consequently, for their Ricola Marketing Building (1998) in Laufen, Herzog & de Meuron used as supporting elements artificial vines with naturalizing ivy leaves in place of real ivy, as a way of realizing their idea of letting ivy grow over the roof of the building.[64] The exterior grounds were designed by the firm of Kienast Vogt Partner under the direction of Günther Vogt; Dieter Kienast was already very ill at the time. It is idle to speculate about what direction Kienast would have taken, whether he would have abandoned the contrast of natural and artificial or continued to pursue the path of making artificiality legible. Vogt, in any case, departed from that trail and, in works such as the staged natural landscape over the underground parking garage of the Novartis Campus in Basel (2006–8), also "overcame" the bipolar model. From the later 1990s, Herzog & de Meuron regularly collaborated with the botanist and artist Patrick Blanc (b. 1953), whose *murs végétaux* are nearly the antithesis of the walls covered with spontaneous vegetation by Kienast Vogt Partner (pp. 126, 128). His attitude and practice contradict everything that was essential for Kienast and Vogt at the time, such as the collecting and use of rainwater, depicting natural processes, and initiating ecological awareness.[65]

In the words of Martin Steinmann once again, it was the "search for a form beyond the sign; beyond the other,"[66] which gave Swiss-German architecture new impulses in the early 1980s. A search for a *degré zéro* at which architecture achieves a new contemporaneity. A search that is found not only in the architecture and art of this period, but also in its literature.

When Mere Words Can Stand for Things: Peter Handke's Search for Forms

It is surely no coincidence that Dieter Kienast was particularly engaged with the work of the writer Peter Handke (b. 1942). Handke, too, found his way to a phenomenological aesthetic in the late 1970s.[67] It was first implemented in the prose text *Langsame Heimkehr* (1979, translated as "The Long Way Around," 1985). Sorger, its protagonist, who is exploring the untamed vegetation of Alaska, "was imbued with the search for forms, the desire to differentiate and describe them, and not only out of doors ('in the field'), where this is often tormenting but sometimes gratifying and at its best triumphant activity was his profession."[68]

As a trained geologist, Handke's hero is searching for a topological perspective and form of expression that transcend the modes of discovering the world found in science and ordinary language. Comparable to Handke's hero, after completing his dissertation, Kienast abandoned a traditional career in natural sciences. Instead, in his practice as landscape architect, he also concentrates on establishing a topological form of expression. His skepticism about claims of universally valid solutions had motivated his intense studies in the fields of phytosociology and planning theory, just as the beginning of Handke's career as a writer was marked by a general skepticism about language (*Kaspar,* 1968, translated in 1969) and vehement critique of the cultural establishment (*Publikumsbeschimpfung,* 1966, translated as *Offending the Audience,* 1971). Accordingly, in the late 1970s both were employing their respective aesthetic means without abandoning the distanced stance from which their creative developments had started. The quotation from Goethe that Handke noted while working on *Langsame Heimkehr* should be understood in that sense:

> The so-called classic soil is another matter. If we do not approach it fancifully but consider this soil in its reality as it presents itself to our senses, it still appears as the stage upon which the greatest events were enacted and decided. I have always looked at landscape with the eye of a geologist and a topographer, and suppressed my imagination and emotions

in order to preserve my faculty for clear and unbiased observation.[69]

As an author, Handke took his lead from Goethe's "eye of a geologist and a topographer," which sharpens topological perception.[70] Correspondingly, impression and knowledge interlock in the landscape descriptions in Handke's *Lehre der Sainte-Victoire* (1980, translated as "The Lesson of Mont Sainte-Victoire," 1985), the continuation of *Langsame Heimkehr*:

But even before setting foot on European soil, the geologist had turned himself back into me, and since then I had been living in Berlin. . . . Until then, moreover, it had never struck me that Berlin is situated in a broad glacial valley (previously the fact would hardly have interested me); the houses still seemed to have been scattered at random over a steppe-like plain. Then I discovered that a few streets away was one of the few spots in the city where the receding ice had formed a discernable slope. There lay the Matthäus Graveyard, the top of which, rising high above the surrounding country, is said to be the highest point in the Schöneberg district. (The artificial rubble mountains thrown up after the war don't count.)[71]

Nature and culture, vegetation and history, were just as much pairs of opposites as they were for Kienast. Both of them were interested precisely in the experience of boundaries and rifts. Probably for that reason, Kienast integrated the final passage from Handke's *Lehre der Sainte-Victoire* into his exhibition *Zwischen Arkadien und Restfläche* (Between Arcadia and Vacant Lot) in 1992. Handke's description of the Morzger Forest appeared as a ribbon of text that the visitors could walk past **(196–199)**. Movement in space is a central motif in Handke's work, whereby his observations and descriptions focus especially on the margins of urban space. Walking in the city, out into the periphery, crisscrossing and passing nature, is a frequently recurring motif—from *Als das Wünschen noch geholfen hat* (1974, translated as *Nonsense and Happiness*, 1976) by way of *Nachmittag eines Schriftstellers* (1987, translated as *The Afternoon of a Writer*, 1989) to *Der Große Fall* (2011, translated as *The Great Fall*, 2018). In moments of successful contemplation, Handke's narrator experiences what Viktor Shklovsky described as making "the stone stony":

> In spite of the traffic, I had a feeling of stillness, just as the day before, in the midst of the Paris noise, I had felt stillness in the street where we had once lived. I had thought of taking someone with me; now I was glad to be alone. I was walking on "the road." In the shady roadside ditch I saw "the brook." I stood on "the stone bridge." Here were the cracks in the rock. There, bordering a side road, were the pines; large, at the end of the road, the black-and-white of a magpie.[72]

At such moments, Handke finds a "legibility" and "experienceability" of landscape that corresponds to his poetic ideal: "At the end of the story, it has to be achieved that mere words can stand for things."[73] In his euphoric, epiphanic tendencies, Handke's formalism goes beyond Shklovsky's aesthetics of influence. Stone, street, and bridge are no longer things, just as the author ceases to be a person: "In order to write, I must first myself be a form: that is, a form has to approach the form."[74] The aesthetic experience touches on the mystical here.

It is known that Kienast would first spend hours in the given landscape before designing a garden or other work, in order to observe all the conditions very precisely before finding his forms at the drafting table.[75]

Form and Landscape

Becoming one with the landscape through quiet contemplation or by wandering through it in order to do justice to the landscape by means of design interventions—other landscape architects, too, are characterized by such an approach, such as Georges Descombes (b. 1939) and Günther Vogt, seeing themselves as crucially influenced by Peter Handke's literature.

The French-Swiss architect Descombes used a Handke quotation as the epigraph to his design of a section of the "Weg der Schweiz" (Swiss Path) hiking trail: "Quelque chose commença qui était déjà là" (Something that was already there began).[76] Descombes regarded the quotation as both a manifesto for his design

154
Grounds of Tate Modern in London, 1995–2001, Kienast Vogt Partner / Vogt Landschaftsarchitekten. Photo: Christian Vogt, ca. 1998

interventions in a section of the hiking trail and as a motto for his way of working in general, because it summed up for him his approach to landscape and territory.[77] The brief sentence expresses his core concern: the uncovering of something that is already present but is only manifested by the form, and this form results from a design that at the same time exposes one's one ideological attitude, one's own view of a place and a landscape.[78] Descombes roamed for weeks through the area around the 3.5-kilometer-long section of the trail between Morschach and Brunnen, becoming familiar with the countryside and its residents, with the traces of events in natural history and cultivation, in order to—working with the artist Carmen Perrin (b. 1953) and with the participation of Richard Long (b. 1945) and Max Neuhaus (1939–2009)—make the traces of natural and cultural history in the landscape more visible.

Descombes took a similar approach as part of an interdisciplinary team of landscape architects, engineers, hydrologists, biologists, and environmental technologists for the prizewinning renaturation of the small Aire River near Geneva.[79] He acquired an immense amount of scientific knowledge and saw his task as the "architect in the landscape" in giving the natural scientists a sensory understanding of the landscape.[80] He regarded it as necessary to move in the river and the surroundings, to confront the forces of water, rather than "planning" on his drawing board the river liberated from its canal. Only such an approach would make it possible to uncover the historical development of and cultural imprint on the landscape: "Landscapes themselves perform as living structures. One goal of the project was to position the river in a way that it becomes itself an agent in design."[81] On the other hand, he saw his task as creating images that do justice to the requirements and possibilities of a landscape and that at the same time evoke memories. For him, memories are linked with the design technique of montage, with making traces and layers valuable: the Aire River was supposed to find a natural path again for a stretch of 4 kilometers, but it is accompanied by terraces, new rest stops, and observation platforms, and the channel of the removed canal is transformed into a new open space.

Günther Vogt connects his interest in Peter Handke's literature to the same question of sensory experience in a world dominated by digital media that Jacques Herzog had discussed:
Our daily life is influenced by digitalization: computer, internet, television, cell phones. When the digital world becomes so all-embracing the banality of everyday life—which can be very beautiful and very surprising—is sometimes lost. By this I mean the simple, genuine experiences: weather, landscape, ugly urban landscapes, beautiful city landscapes.[82]

For Vogt, digital dominance destroys both attention to surroundings and the formation of a critical distance from what is seen: "Quite literally what is lacking is emptiness and quiet, and for me finding these is very important." The writer Peter Handke "once put it beautifully: he said something along the lines of that you don't obtain quiet and emptiness by being silent but by giving silence and emptiness a form."[83] This pausing can perhaps best be seen in Vogt's work in urban space, when the open spaces are standing there calmly and matter-of-factly and transition into an urban space that they make more legible. Yet only the attentive viewer consciously recognizes this, whether in the outdoor grounds for Tate Modern in London (begun with Dieter Kienast, 1995–2001, continued 2005) (154), Münsterplatz in Konstanz (2005–2006), the courtyards and parking lot of the Justizzentrum Aachen (2004–2007), the forecourt of the Festspielhaus Bregenz (2005–2006), or the Elisabethenanlage in Basel (2006–2007).[84] These "placeholders" for a coming-to-rest open up the possibility of bringing about the attentiveness that Vogt regards as the foundation for a critical observation of one's surroundings. The phenomenological, topological, and sociopolitical impulses come together again here. Such interventions that transitioned into foregone conclusions were explored in works that Vogt and Kienast developed in the 1990s.

A Vocabulary for the Landscape I: Shaping the Terrain

The effort to design his gardens and grounds to make it easier to experience the city and the countryside was as important to Dieter Kienast as it was to establish forms of representation for his plans that reveal the materiality of the built works. In both areas, Kienast developed over the years a differentiated vocabulary that he applied in both the terrain and the plan.

>   Writing versus
>   Vocabulary: The E. Garden
>   in Uitikon-Waldegg

"The transformation of a modest garden of a single-family home into spacious grounds is the unfinished story of the E. Garden, which we are tackling step by step together with the residents,"[85] Kienast wrote in 1995 on a garden created in two stages, in 1989 and in 1993, which stretches out on a slope in Uitikon-Waldegg on the back of Zurich's Uetliberg (155, 156). The garden had become more than a "meaningful pleasant view from the living room": it was a "place of daily work and relaxation, of expectation and fulfillment, a place of memory, present, and future."

These solemn words seem to suit this garden perfectly; large letters on its southern edge read: "Et in Arcadia ego" (I Too in Arcadia). This phrase in concrete also has a quite prosaic purpose: it protects those strolling in the garden from crashing down the forest slope, which descends 50 meters from here, and thus unintentionally falling out of Arcadia. This balustrade has been made famous by the photographs of Christian Vogt (157), which have been published multiple times, in which the words are seen rising up over a lawn covered with autumn foliage, lending the garden an aura of melancholy and transience.[86]

Kienast's repeated use of monumental phrases cast in concrete for gardens and public grounds probably contributed to his reception as the *homme de lettres* of the Swiss landscape architecture scene.[87] The garden historian Brigitte Wormbs sees multiple layers of meaning in the balustrade:

> "Et in Arcadia ego"—set in concrete letters as a balustrade onto the threshold between the tamed "nature" of the garden and the wildness of the steep wooded slope—at the same time offers a concrete support in and a spiritual enrapture from the material world.[88]

For Wormbs, Kienast's experiments with writing are images in which the writing is a carrier of thought, which lies behind a sustained concrete literalness. Kienast found inspiration for this in the artistic works of Jenny Holzer (b. 1950) and Ian Hamilton Finlay (1925–2006), the latter of whom had applied five pairs of stone tablets to trees in the landscape garden next to the Botanical Garden (158), in the sculpture exhibition at Merian Park in Basel, of which Kienast was technical director at the time. The Latin name of the tree in question was engraved on a rectangular stone tablet; on the tree next to it were mounted oval stone tablets with the names of famous pairs of lovers who had carved their names in tree bark, including Shakespeare's Rosalind and Orlando and Rousseau's Julie and Saint-Preux.[89] Kienast assessed the work of art he had experienced in Basel:

> With this modest installation, Finlay weaves literature into the art of the garden, the reference to place into natural science, poetry into physical activity, creating a complex web of mutual relationships and dependencies. Finlay's work becomes the icon of contemporary art that no one understands. The tablets were small; they did no more damage to the tree than a carefully attached birdhouse, but they were not understood because they negate the basic ambition of

155
Extension to the E. Garden in Uitikon-Waldegg, views, sections, details: pool and balustrade "Et in Arcadia ego" (I Too in Arcadia), forest path with bench and handrail (1:20/1:1), pergola (1:20). H 100 × W 70 cm, mixed media (copy of plan, collage, drawing) on card stock, hand colored, June 22, 1993, firm of Stöckli, Kienast & Koeppel

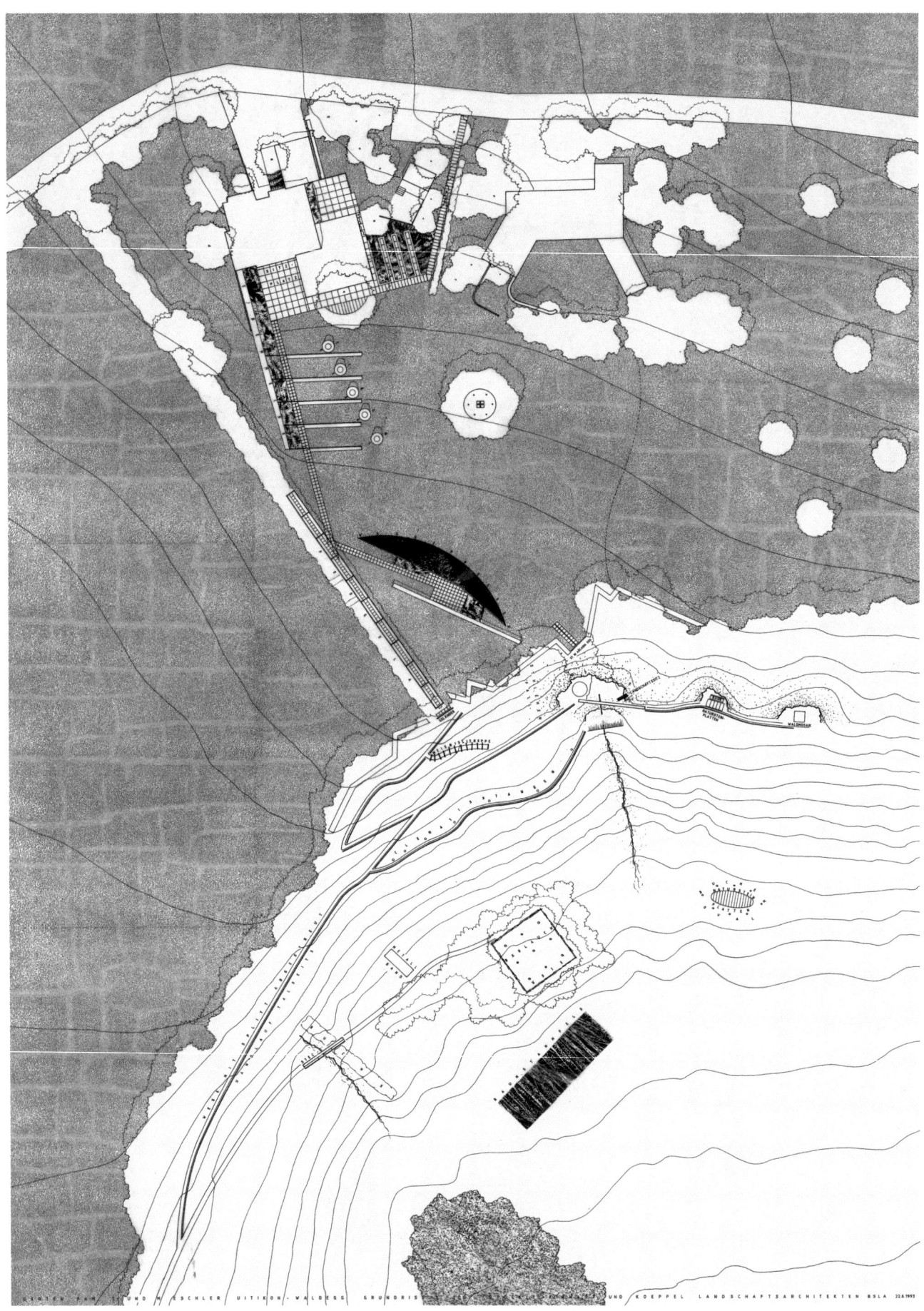

156
Extension to the E. Garden in Uitikon-Waldegg, site plan, scale: 1:200, H 112 × W 73.5 cm, copy of plan, June 22, 1993, firm of Stöckli, Kienast & Koeppel

157
E. Garden, pool and balustrade.
Photo: Christian Vogt, ca. 1996

158
Ian Hamilton Finlay, *Names for Trees*, 1984. Installation in the exhibition *Skulptur im 20. Jahrhundert* (Sculpture in the Twentieth Century) in the Merian Park, Basel

accepted art: they are neither inherently beautiful, nor do they make their surroundings beautiful. The work requires time and commitment—both luxuries that we want to provide only to a limited extent.[90]

These sentences speak to Kienast's topological interest in creating relationships and interlinked settings both in the specific space and in the space of the imagination, as well as to his desire to heighten human attentiveness and the willingness to take time and get involved in an experience. One means by which Kienast tried to establish such multidimensional relations is working with writing. On the viewer platform of the M. Garden in Ulm, the words "ogni pensiero vola" (every thought flies) are formed into a balustrade from which a spectacular view of the city and its landmark opens up: the cathedral with the highest church tower in Europe. In combination with the view, these words, which are extremely famous in garden circles, take on a new meaning. They are from the Mannerist garden in Bomarzo (1564–70), near Viterbo in Italy, where they are inscribed above the gaping maw of an eerie stone garden sculpture.[91] In the M. Garden, the words literally support the roaming gaze and roaming thoughts of the viewer.[92]

Kienast did not, however, choose the saying only because of the view gliding over the city as a reference to "flying thoughts": Bomarzo is a *bosco,* a forest. The *bosco* is an important element in the Renaissance art of the garden. Even in gardens that otherwise featured geometric design, there was always a *bosco* as a symbol of wild nature as opposed to the garden created by the human hand. The viewing platform in the M. Garden is also located in such a *bosco,* in the wildest part of the grounds, whose surfaces near the house are dominated by gravel and concrete. After crossing the dense *bosco,* one is surprised to arrive at the platform with the incredible panorama and find the saying from Bomarzo on the balustrade. This considered integration of garden history demonstrates that Kienast's texts are not one-dimensional, formal games but are topologically situated.

However much interest or even philosophical and scholarly debate Kienast's use of writing in landscape architecture has sparked,[93] it has been equally met with skepticism from art historians.[94] They often regard the written balustrades as superficial, rather banal attempts of the landscape architect to try to look artistic. The tenor of the criticism is that Kienast's work with writing has tended rather to harm his reputation. This overlooks the fact that Kienast's interest in art and literature and his search for forms that sharpen perception have found expression in just a few experiments of this kind. In terms of his grammar for the landscape, the writing in the E. Garden is more of a footnote that took on the importance attributed to it in the current discourse on Kienast's work only as a result of Christian Vogt's photographs and Kienast's later description of the project.[95] The design qualities of the garden in Uitikon-Waldegg that stand out are different ones. The E. Garden is an example of Kienast's formal vocabulary that understands, accentuates, or counteracts the existing topography in order to lend the garden and the surrounding landscape a new look by giving it form. The crucial factors in such an effect is the arrangement of the elements and their spatial relationships to one another; in other words, their topology.

The Garden:
Topography and Topology

Located on the side of the Uetliberg that is opposite Lake Zurich, beneath a ridge but still high above the bottle of the valley, from the highest points in the E. Garden an impressive panorama of several chains of hills leading up to the Alps opens up (p. 332). The garden stretches from the client's residence and the neighboring house of his daughter down the slope to the south to a striking edge of the grounds, behind which a forest looms (p. 334). A hedge of whitethorn more than a 100 meters long terminates the lot on the west in the direction of the valley. The forest is owned by the E. family, as is the meadow of fruit trees and the adjacent pasture east of the garden that move up the slope to the north. Much like the layout of the classic villas of antiquity and the Renaissance, the E. Garden is thus a "third nature" associated with art integrated into an agricultural context ("second nature") with a border on the forest beyond and a view of the Alps (symbols of "wilderness" or "first nature").[96] At the edge of the slope and the southern border of the garden, the dense forest is broken up by a natural open strip: a stream has eaten its way into the ground of the landscape and formed a small mountain gulch (156). Beyond the edge of the grounds, the slope drops around 50 meters into this ravine, and so the strip gives the garden a natural viewing platform (160), from which, on clear days, the impressive Alpine mountain massif of the Eiger, the Mönch, and the Jungfrau and their snow-covered peaks can be seen.

By means of succinct formal interventions from 1989 onward, Kienast restructured the individual spaces of the E. Garden and changed their relationship to the surroundings to make the landscape visible in a new way using the design vocabulary applied by him. Just how attentively he developed his forms from the existing situation and the needs of its users is indicated by the client's description of how the preliminary project in 1989 came about: when Mr. and Ms. E. asked whether Kienast would design their garden for them, he lay down in the meadow for several hours and allowed the place and its surroundings to affect him.[97] In an extended conversation, he then inquired into the wishes of Mr. and Ms. E. regarding plants and their maintenance. After a few days, he showed them a first sketch, and the first stage of the redesign was begun. The clients had only a single objection: they declined to have the word "differentia" cast in concrete for use as a balustrade on the edge of the ravine, both to prevent falls and to catch the eye.[98] By placing the word "differentia" at this critical spot in the garden, Kienast wanted to draw attention to "the different natural forms of garden and wilderness, the hidden change in altitude or the change from the former to the new garden design."[99] He liked to use the word "different" to describe his design method and did so often.[100]

Kienast's true design vocabulary for the landscape is revealed in the striking interventions in the topography. He had the originally steep slope turned into a gradually descending one,[101] so that from the deepest point of the garden in the south it opens up like a funnel or a slow-rising semicircle of an ancient theater toward the two residences and the meadow with fruit trees to the north (pp. 336, 338). At the deepest part of the slope, where on his first visit to the grounds he had found a damp area and the corresponding vegetation, Kienast placed a pool.[102] Its orientation and form fit both the new design of the terrain and the rhythm of the contour lines. Shaped like a segment of a shallow arch on the side of the slope, the pool has a stick-straight edge on the other side, which folds slightly outward in its second third.

Behind the pool and shifted slightly, Kienast placed a mixed hedge as another "backbone" of this deepest point, where, to return to the image of the theater, the stage would be (p. 334). This hedge as background resulted from a refinement of the preliminary project, where the hedge was still intersected, and thus counteracted, by four trees planted in a diagonal line. The hedge serves like a stage backdrop behind a seating area in the garden, which is framed by a blooming bush and a bed of shrubs. The mixed hedge stands at an angle to the long whitethorn hedge that terminated the garden to the west. The mixed hedge and the pool form the imaginary base for the terrain that opens up above, giving the garden an interior spatiality.[103]

The unusual form of the pool is due, first, to its location within the grounds and, second, to technical necessities.[104] The curve transitions to the slope, and the angle deftly turns to the viewing platform with its balustrade of words and to the meadow with fruit trees that rises to the side of the buildings. As a result, the angled pool becomes not only an anchoring point on the grounds but also a spatial hinge.

On the side facing the slope, the pool is framed by a steel plate and facing the seating area with concrete; the folded diagonals of its edge here faintly recall a shoreline. This already suggests that the pool, whose form is highly artificial, is perceived in the terrain itself as a natural necessity—an experience that Sacha Fahrni and Christian Gut, who had studied architecture at ETH Zurich, also had when entering the garden:

> On closer inspection on site, one observes with astonishment that the form and the placement are not arbitrary at all but could only derive from the specific qualities of the place.... The rays of the sun dance on its surface. The sky, the tops of the forest trees, the whole landscape is reflected in the water. The garden is present in its entirety here, even though in fact only part of it can be seen.[105]

The slope's edge facing the forest is marked by a geometrically pruned hedge of field maple, which was planted partially in a zigzag and partially in a straight line (156). The guard rail in the form of the phrase "Et in Arcadia Ego" written above the gulch is the same height as the hedge and is integrated into it (p. 334). Behind it loom the mighty treetops of the forest. As with the pool, an obviously artificial form has been developed out of its natural quality and at the same time embedded into shaped nature resulting in the overlapping of naturalness and artificiality that Kienast so liked. At the same time, the striking forms of the hedge reinforce the natural qualities of the forest.

The shaping of the section of the garden slope directly in front of the house is also significant in the sense of a vocabulary for the landscape: an existing broad, slab-covered terrace is followed by a cascade of lawn terraces with steps lined by small, geometrically pruned boxwood hedges, and with conically pruned yews on the eastern ends (p. 338). On the western end, the terraces meet at an angle with the yew hedge that runs up the slope, its direction determined by the position of the house. This hedge, which continues the western wall of the house, also frames the sun terrace in front of the house like a low wall. Connecting directly to the latter, stretching between the yew hedge and a joining path that accompanies the descending lawn terraces, is a bed of shrubs originally planted by Nicole Newmark[106] and refined and maintained over the years by Ms. E. On the lowest grassy step, the hedge and the bed end, while the path continues straight and flat until it meets the whitethorn hedge that borders the property. There the path takes a bend toward the pool and leads to the seating area in the garden.[107]

Where the path meets the whitethorn hedge, a pergola was added in 1993, during the second stage of the garden redesign, running parallel to the hedge, and in the meanwhile it is overgrown with white roses (p. 340). When walking in it, one passes the branch leading to the pool and arrives in the very front at the edge of the slope, where the field maple hedge separates the garden from the forest. A small "window" was cut into the whitethorn hedge (155), which surprises us with a view of the hilly landscape to the west. At the end of the pergola, a bench for the man of the house has been placed near the edge of the slope, facing the open landscape and offering an unrestricted view over the fields to the chains of hills.

The Garden in the Vicinity of the House

When sitting on the large terrace in front of the house, the yew hedge on the western side blocks the view of the surrounding landscape. That is somewhat strange, in light of the spectacular view one has here at the highest part of the garden. Kienast deliberately chose this height for the hedge so that the landscape is not seen when sitting; instead, the eye is directed to family and friends dining at the table and to the garden. As soon as one stands up, however, the landscape appears behind the hedge. This playing with hiding and discovering has a long tradition in garden history.[108] Hiding certain structures and blocking views that suddenly appear when the viewers change their own position in space is supposed to increase the pleasure of discovery and rediscovery.

To make this interplay possible, Kienast had the existing whitethorn hedge, which was more than a 100 meters long, massively cut back on the western edge of the garden to restore the view of the first chains of hills. Only then did he add the yew hedge. The layering of chains of hills that is part of the surrounding landscape is repeated by layering the two hedges. In this way, Kienast's interventions create a new visual relationship between the space of the garden and its surroundings.

Around the house, in the cracks between the concrete slabs of the sun terrace and seating area, and in the cracks for slabs put down for the connecting paths, thyme and mint were sown;[109] on the east side, an herb and flower garden was planted. Halfway up the slope, between the residences and the pool, a circle of six tall beeches with spherically pruned tops has been placed, symbolizing the number of grandchildren and the community of the E. family (159).[110] Kienast also removed all of the existing plants on the grounds that in the meanwhile blocked the view of the surrounding landscape, like a dense blackberry bush in front of the viewing platform in the south or the shrubs that separated the garden from the meadow with fruit trees and the adjacent pasture in the east.[111] Very much in the historical tradition of the art of the garden, Kienast examined the existing garden in Uitikon-Waldegg for its "capabilities," searching for the *genius loci,* as this process was called in the context of the English landscape garden,[112] and redesigned it accordingly.

E. Garden in Uitikon-Waldegg, 1989 and 1993. Photos: Georg Aerni, 2012

Page 332
View from the sun terrace of the residence, across the lawn terraces of the garden, into the surrounding hilly landscape to the southwest. On clear nights, the view extends to the Bernese Alps.

Page 334
At the lowest point of the gently descending slope, there is a pool of water in the form of a segmental arch, immediately adjacent to which is a seating area and a bed with bushes with white flowers and shrubs. A tall mixed hedge offers a stage-like scenic "backdrop." The forest looms beyond the edge of the slope.

Page 336
View from the pool toward the meadow with fruit trees and the adjacent pasture to the northeast

Page 338
The slope in front of the residence split into five grass-covered terraces, terminated by a slender boxwood hedge and row of conically pruned yews. On the west side, a stepped path and a bed of shrubs follow the terraces and a yew hedge provides a backdrop. The sun terrace in front of the house forms the upper termination.

Page 340
A steel pergola overgrown with roses leads along the whitethorn hedge to the forest; wandering through it, one passes a "window" cut out of the hedge that provides a view of the hilly landscape to the west.

Page 342
A blue-painted handrail runs like Ariadne's thread through the dense forest at the base of the garden.

159
E. Garden in Uitikon-Waldegg. The small grove of six tall beaches symbolizes the number of grandchildren and the cohesion of the family. Photo: Christian Vogt, ca. 1996

160
On clear days, from the observation point at the balustrade one can see the snow-covered peaks of the Eiger, Mönch, and Jungfrau mountains. Beyond the edge of the grounds, the slope drops 50 meters down into a gorge that forms a natural corridor through the forest. Photo: Georg Aerni, 2012

Phenomenology: Et in Arcadia Ego

Over the course of the years, during which a friendship grew between the garden's owners and the landscape architect, on "numerous garden tours" they discussed "possible and necessary design work," "pored over literature," "on our many excursions both at home and abroad, . . . collected new impressions and brought back many plants and planted them," and also discovered together the text for the balustrade: "Et in Arcadia ego."[113]

This coincided with the second stage of the garden redesign in 1993—the motto in concrete letters appears on the plan for it (156)—when Mr. and Ms. E. wanted to develop the forest in front of the garden. For the steep path that Mr. E. had built from the garden into the slope of the gulch, Kienast designed a red bannister intended to guide like Ariadne's thread along the new path through the dense forest (155). At the client's request, the bannister was painted not red but blue (p. 342). The placement of the bannister has an impressive effect: it is formed in such a way that it really does snake through the forest like a thread. At the same time, it is a very dynamic element. In the steep parts of the narrow, constantly winding path, it seems almost to plunge into the depths of the gulch. It brings out the gradient of the forest; the presence of the blue bannister also makes its thicket seem much wilder than in the places where the handrail is interrupted. The experience of contrast of which Viktor Shklovsky, cited above, speaks in the context of how art functions can be followed here very well.

Further interventions in the forest included installing benches, emphasizing the stream that emerges from the slope and runs into the gulch, the planting of a rectangular field of wild garlic, and a "beech oval," for which an oval plaza of natural stone slabs was to be placed around the beech. These interventions seem like using the art of the garden to domesticate the wilderness.

"Et in Arcadia ego," the new motto for the balustrade, alludes to Virgil and his *Eclogues,* in which Arcadia is the place of desire and a "spiritual landscape," as well as to two paintings by the French Baroque painter Nicolas Poussin.[114] In Poussin, it points to a rift in the idyll: "Et in Arcadia ego" is the inscription on a sarcophagus placed in a landscape, a subtle indication that death exists even in the happy land of Arcadia.[115] Even if Kienast placed his use of the phrase in the tradition of German idealism, thus merely drawing attention to the happiness of living in the peaceful land of pastoral Arcadia,[116] its placement in the E. Garden nevertheless has the effect of refracting its meaning: the reference to the presence of death in the garden, as a guardrail cast in concrete that can be seen from afar in front of a slope 50 meters deep, does not lack a certain irony in the accuracy of expressing what it is doing—in purely functional terms. As a balustrade, "Et in Arcadia ego" is not an autonomous monument but a design element integrated into a functional context. The way it is made points to this as well: the words are spanned between two crossbars that support them, and yet at the same time they stand out against them. The impression is clear and unambiguous: it is a balustrade with a handrail and not an autonomous sculpture (160).

The question "monument or balustrade?" has a specific effect on the question of care and maintenance: a few years ago, Ms. E. wanted to know whether she should have the lichens and moss that had grown over the letters cleaned off completely, since it was the opinion of people visiting the garden that it was a sculpture, or whether it should be allowed to weather.[117] Because the motto should be seen as a balustrade and not as a sculpture, after a visit to the site Ms. E. was advised not to spoil the natural process of weathering by sandblasting it to clean it.[118] Because it was in good condition, there was no reason to change its present character; on the contrary, nature's reconquering processes were always an integral component of Kienast's work and visible aging in the wake of natural processes was of essential importance to him. The space of associations that the words open up is reinforced by the natural erosion processes seen when looking at the nostalgic Swiss landscape of the Eiger, the Mönch, and the Jungfrau.

In the E. Garden, Kienast employed design methods that run through his entire oeuvre as a recurring vocabulary and always come together in new ways for a different "narrative": a pronounced modeling of the terrain, pools framed with steel plate, geometrically pruned hedges, bushes with white flowers and beds of shrubs, boundaries drawn with walls or balustrades, artful topiaries, consistent contrasts of natural and artificial elements, and much more. Adapting them to a specific place in order to make it legible again links Kienast to the artistic trends of his day. Like Sol LeWitt's minimalist structures and Peter Handke's *Lehre der Sainte-Victoire,* his spaces modify the viewers' perception and contribute to the phenomenological efforts of postmodernism.

A Vocabulary for the Landscape II: Drawing and Representation as Processes for Understanding Topography, Form, and Material

How can the qualities of the materials used when designing a garden be represented on a plan? How can a representation differentiate between individual design elements? These were questions that clearly preoccupied Dieter Kienast intensely. He developed a refined vocabulary for representing different forms and signatures with a special grammar. The media of rendering plans—drawing, collage, CAAD—changed over the years; they transitioned, were mixed up. It was only drafting proper that was always and exclusively done by drawing by hand.

Apparently, Kienast worked with four means of representation, which can almost always be assigned, chronologically as well, to certain phases of designing and documentation: the hand-drawn sketch, the drawing of a draft[119] (sometimes freehand, sometimes with a straightedge and stencils), the technical execution drawing, and the presentation plan. Except in the case of competition entries, the last of the four was generally made after a project was completed, very often by Kienast personally, and it served as documentation.

Kienast often gave these plans made afterward to the clients in question, who frequently displayed them in the vestibule or salon of the house, next to paintings, sculptures, and other art objects. The gesture of drawing such artful plans afterward and giving them to the clients as a farewell gift underscores Kienast's self-image as a craftsman and garden artist. Moreover, the impression conveyed by these detailed and truly lovingly and carefully designed plans is that he was concerned not least about reflectively meditating on the work created. With knowledge of the inevitably occurring processes of change and transience resulting from the dynamic forces of nature and the interventions of the users in the realized grounds, the presentation plan preserves for eternity an ideal or memorial image of the work.

Artistic quality as an alternative image to the horticultural is also emphasized by the use of color: after a phase of experimenting with an antinaturalistic palette in the mid-1980s (111),[120] Kienast ultimately turned to limiting his hand-drawn presentation plans to shades of gray, blue, and black. He placed particular value on using only ink and graphite and colored pencils. Felt-tip pens were banned from the office from then on. This process of abstraction in the rendering ran parallel to a process of abstraction in the design itself, in which Kienast employed his vocabulary of forms and materials in increasingly reduced ways. Only in the mid-1990s did he again expand his palette of forms, materials, and plants. It would, however, be wrong to reduce Kienast's presentation in plans to the purely visual. Rather, the effort to translate the various elements of the garden and its materiality into drawing triggered a process of reflection in Kienast that is comparable to the much-discussed "artistic research."[121]

After redesigning his own garden on Thujastrasse in Zurich, Kienast produced a presentation plan measuring 84 by 90 centimeters in 1991 (112). For other projects, the firm—and for competitions, the entire family—would help to produce this artful visual material.[122] Kienast drew the plan of his private garden with black ink on eggshell cardstock. The ground plans of the adjacent buildings, of the street north of the garden, and of the boundaries of the garden itself and the buildings, walls, steps, pergolas, and so on, were drawn with straightedge and stencils. The outlines of the trees, the topiaries, and the large stones, as well as the pools of water and playground of the school for children with cerebral palsy on Mutschellenstrasse, adjacent to the private garden, were rendered freehand by Kienast. The individual areas were differentiated by a system of hatching in different colors, style, and intensity.

To mark the surroundings, Kienast used a long straightedge that stretched across the entire width of the paper and was fastened to a rail and moved along it. He advanced it millimeter by millimeter and drew one fine blue line after the other. He left out only the garden located in the center and the treetops that extend out beyond its edges. That task alone must have taken many hours, if not days. To differentiate the area further, he drew hatching freehand over this linear structure in order to distinguish areas such as the walkway, the street, and the adjacent buildings. This resulted in the garden area having a composition of blue surfaces of varying intensity, whereby the lines and hatching are so delicate that the eggshell ground shines through everywhere, lending the drawing an uncommon legerity.

The separate parts of the garden itself are depicted in extreme detail (161, 162). Kienast was working here with the colors black, gray, and blue, and with the drawing implements pen and ink, and with graphite and colored pencils. Only the red rectangular mosaic under the square of apple trees has real color in his plan, and so it stands out against the dominant gray-blue. Once again, Kienast combined areas of hatching of varied intensity and always allowed the support below to shine through, as if the point of a garden were to create a topography over an existing terrain. Only the treetops rising high above the ground and the ground plans of the residences standing on the property were not designed by Kienast. They remained completely free: their color comes from the pure, bright cardstock. They are the elements furthest from the ground. The cast shadows of the trees and buildings were applied with dense hatching in pencil, adding an unusual tension to the depiction: on the one hand, the trees and buildings seem two-dimensional because of the top view; on the other hand, the cast shadows suggest plasticity. Their identity with the support likewise has the effect of an ambiguous image: now they seem to stick to the ground, now to float high above it.

Other elements closer to the ground, such as the topiaries, hedges, two large stones, and the foliage and trellises, were drawn with varied hatching and also have shadows. The imaginary light comes from the south; only in very sporadic places are the cast shadows indeed slightly varied, placed in such a way that the forms stand out better from the ground. There is an overarching logic in the presentation, such as the combination of top view and three-dimensional formation, a central light source, as well as an internal logic that results from the relationship of the elements to one another, with nuances that deviate from the overarching logic.

It becomes particularly striking how substantial it was for Kienast to reproduce by means of drawing the materiality of the elements of the garden in as highly differentiated a way as possible: to do so, he used primarily pencils with

161, 162
The Kienasts' Garden in Zurich, presentation plan, 1991, ink, graphite and colored pencils on card stock (details)

Eastern front garden with greenhouse, beds, and parterres, large topiary trees, and a red ceramic mosaic surrounded by four apple trees

Western rear garden with ceramic mosaic and topiary and the adjacent garden of the school for children with cerebral palsy, with its pond and platform

leads of different hardness and thickness. This gives each element, each area, its own signature. The drawing style is thus connected to the drawn area, which is particularly well illustrated by the eastern part of the garden (161): framed by a hedge that runs along the edge of the garden and surrounded by a patch of gravel, it has the greenhouse, a cold frame with its glass top removed, three parterres framed by boxwoods, and the hedge-lined niche with the animal figures. To the south, it connects to a small, rectangular lawn with a freestanding tree. The patch of gravel that is the foundation for this part of the garden is drawn with light hatching on the presentation plan, using a very soft, thick graphite pencil. These structures are probably the result of frottage, a technique that the painter and draftsman Max Ernst discovered for art. Kienast probably placed the cardstock on a wall with a delicate grain and rubbed it with pencil to transfer the pattern.

The beds, the parterres, and the lawn are much darker and probably rendered with a pencil of a different thickness, yet the cardstock still shines through. Possibly these areas were hatched while resting on the same surface used for the patch of gravel. By contrast, a much harder pencil was used for the cast shadows. They are denser than the areas with gravel or plants. The glass roof of the greenhouse is drawn with a pencil that was presumably even harder, applied with only slight pressure. Here the structure of the cardstock itself is evident, which suggests that it was placed on a different surface when drawn. As a result, the visual effect is completely different: the gray areas that represent the glass roof of the greenhouse look smooth and have a diaphanous quality, while the ones with coarser grain, which indicate gravel and plants, have a stronger materiality in a sense because of that grain.

The hedges and topiaries have a completely different signature: on top of the ground applied with blue pencil, many small and tiny irregular squiggles have been drawn with a hard, very delicate graphite pencil, imitating the leaf structure of the hedges and the pruned trees. The topiaries are accentuated even more strongly with blue pencil. The pools of water in the garden are drawn with jet-black ink, with small patches condensed into groups left open (162). The cardstock that shines through in these areas suggests the glimmering of reflections of the sun on the rippling surface of the water.

Despite all abstraction of the antinaturalistic palette of the rendering, the plan reveals a clear connection between the given topography, the materiality present, and the form and texture of how they are drawn. Kienast was clearly experimenting here and appropriating, by means of artistic activity, knowledge about the possibilities of rendering materiality and topography.[123] For example, in certain places he integrated the cardstock on which he drew the plan into the drawing itself, making its structure visible through the hatching. In addition, he changed the surface under the cardstock in order to use its specific qualities to depict different materials. In order to formulate the structure of hedges and topiaries, he developed his own drawing signature. The abstract and the concrete, the physical and the seemingly dematerialized, dovetail in these plans.

Is Landscape Architecture Art? Craft as Link between Gardening and Artistic Activity

The "gray plans," as these hand-drawn presentation plans dating between 1985 and 1992 are called, became Dieter Kienast's trademark. The first one was the project plan for the Zurzach Spa Gardens in 1985 (100). The first-prize-winning competition entries for the Moabiter Werder in Berlin (1990–91, two-stage competition) (83) and the Günthersburgpark in Frankfurt am Main (1991–92, two-stage competition) (84) combined hand-drawn gray or gray-and-blue plans with a collage technique of integrated photographs and recurring figures. In addition to anonymous people, one finds the easily recognizable silhouettes of Charlie Chaplin, James Dean, and Alfred Hitchcock, as well as Giulietta Masina in her role as Gelsomina in Federico Fellini's *La Strada*. These figures were later collected in a samples file, together with slides recording situations in traffic and everyday urban life; its contents were then used for presentation plans.

These successful competition entries are all by the firm of Stöckli, Kienast & Koeppel, renamed after the landscape architect Hans-Dietmar Koeppel had been added as partner in 1987.[124] Günther Vogt joined on April 1, 1987, and soon became Dieter Kienast's most important employee and later his partner.[125] Because Koeppel took over the tasks of landscape planning and environmental design,[126] Kienast was able to focus more on artistically oriented design from here on out, and to concentrate entirely on residential gardens, on competitions for parks, cemeteries, and outdoor grounds for companies. Previously, he had also worked on tasks such as the grounds and greenery for the Herteren landfill and the renaturation of the Eichrüteli gravel pit in Mülligen.[127]

Autodidacticism had been the philosophy of the company at the time, as Peter Paul Stöckli judged in retrospect in a conversation.[128] He added that after founding their partnership in 1979 he and Kienast had taken on many assignments for which they had not known when bidding how they would manage them. Stöckli is one of many of Kienast's colleagues who emphasised that he always regarded autodidacticism and the craft aspects as absolutely crucial: get started, work, and learn more in the process.[129] Stöckli's statement confirms that Kienast was acquiring new knowledge in the process of drawing and pursuing a kind of "drawing research," whether in design sketches (88–99) or presentation plans.

Kienast's former colleagues at the Interkantonales Technikum Rapperswil, like the botanist Katharina König Urmi (1935–2008), confirm that when teaching, too, Kienast placed great emphasis on the craft skills of the students.[130] "We are not artists, we are artisans," he would preach repeatedly, she indicates, and if one of the students was less talented or less inspired artistically, Kienast always respected that, as long as he or she was good at the craft. Despite this emphasis on the practical when teaching, it is impossible to overlook that Kienast regarded his own work as close to the production of art. His renderings for plans are imbued with his interests in art and in the crafts in equal measure, whereas in his experimental search for appropriate ways to translate the elements and materials in the garden into a drawing, Kienast's interest in gardening is also brought to bear. In a lecture in 1991, Kienast asked a question of his audience: "Ist Landschaftsarchitektur Kunst?" (Is Landscape Architecture Art?). A eulogy to the craft of gardening followed:

The history of garden art teaches us that there was a time when it was called the highest of all the arts. But in 1990 Lucius Burckhardt declared concisely that designing gardens is garden art. I prefer to stick to the quotidian term "craft," to which I grant my undivided respect. It contains not only the practicing of the profession in the moment, but also centuries of tradition and innovation. Craft is the foundation in which history and contemporaneity are equally embedded. Our task is grounded in the cultural knowledge of many generations. It consists simply of rediscovering this knowledge and interpreting and updating it. Knowledge means familiarity with cultural history and sociology, as well as designing, familiarity with plants, and working with our neighboring disciplines. The things that are important for the further development of landscape architecture are ability in the crafts, innovation, and creativity.[131]

His rendering techniques demonstrate Kienast's craft skills not only in drawing but also in collage. This technique introduced motifs from popular culture into his plans. Popular culture was a defining source of inspiration from the late 1980s onward. Kienast was enthusiastic about films, pop music, poetry, and comics. For example, in the B. Garden in Gockhausen (1992–95) (163), which he designed under the motto "Monsieur Hulot in the Garden" for a client who had trained as an ornamental ironworker and had long worked in the iron business, Kienast designed a garden with many wrought-iron motifs on a small plot. This included a bench, a small garden house, two pools in the shape of a beech leaf, and several fanned-out iron plates that both open up to the street and provide privacy. Kienast placed this playful semantic overloading in the context of the garden that was the site of Jacques Tati's legendary film *Mon Oncle* (1958).[132] In the sixth of his "Zehn Thesen zur Landschaftsarchitektur" (Ten Theses on Landscape Architecture), Kienast formulated explicitly why popular culture was so important to him: it helped him to place his garden art in the context of a "socially determined" quality.

Grappling with the events of the here-and-now requires the inclusion of the wider cultural environment, engaging with film and video, philosophy and literature, music and advertising. We listen to Bach and Schönberg, but similarly Laurie Anderson and Phil Glass. We immerse ourselves in Sol LeWitt and Walter De Maria, Christo and Carl Andre. We find receptions of the themes of nature and the garden not only in Goethe's *Elective Affinities* or Stifter's *Indian Summer* but also in Bloch's *Alienations,* Handke's *The Lesson of Mont Sainte-Victoire,* or Sennett's *The Conscience of the Eye.* In *Mon Oncle,* Jacques Tati leads us from his lovingly tended roof garden through urban wastelands to the bizarrely designed garden of his brother-in-law, while in *The Draughtsman's Contract* Peter Greenaway imparts a lesson in the art of the garden and its social determination.[133]

The collages, which gradually replaced the hand-drawn presentation drawings, are also marked by his interest in craft, art, and gardening. The replacement of hand-drawn plans with collages had two central causes. First, the company was always trying to keep up with the latest design and rendering techniques—it was one of the first landscape architecture firms in Switzerland to work with CAAD.[134] Second, the production of plans was growing as the number of commissions increased, so that Kienast continued to sketch by hand, but the hand-drawn presentation plans gradually disappeared, because he simply no longer had the time and the demands were too great. Many presentation plans were reproduced from the collages and then hand-colored by employees of the firm or members of the Kienast family.

The Relating of Drawing, Collage, and Software-Based Presentation Techniques

When switching from drawing to collage, Kienast continued his earlier search for a hand-drawn translation of materials and surfaces into textures and signatures, but using a new technique. His visual language ranged from drawing to collage and software-based rendering techniques, but the various techniques were also mixed.

In the Zurich office of Stöckli, Kienast & Koeppel, several binders with patterns for the textures of plants, as well as the surface structures of various materials, were compiled (164).[135] Certain textures were used several times, others found or produced for a particular object. In one of these sample folders, for example, there is a photograph of fallen pine needles. It is reasonable to assume that they were especially arranged this way and then photographed. To make a model for branches lying on top of one another, the copies of individual leaves were cut out and then layered and pasted (167). There are various sample sheets with representations of lawns: in addition to sheets of Mecanorma and Letraset, there are also models that were copied until their pattern became slightly irregular, which is how Kienast preferred to represent lawns. Several models also have strokes or dots made by hand before they were copied. Most of the models for structures of trees and leaves are from a book nearly four hundred pages long: *L'architettura degli alberi* (The Architecture of Trees) by Cesare Leonardi and Franca Stagi.[136] Photographs of existing gardens also found their way into the sample books, such as of the carefully raked gravel of the famous Ryōan-ji Garden in Kyoto, which served to illustrate areas of gravel or sand. CAAD was also developed in parallel with collage. Once again, the knowledge derived from an earlier technique—in this case, collage—was applied here. Despite the possibilities of CAAD programs, collage continued to be used regularly for competition plans until Kienast's passing.

During the eight years when Kienast was experimenting with collage, in many cases he preferred the textures he had developed earlier in his search for adequate presentation in drawing, for example, for lawns, water, and solitary trees. Until Mecanorma and Letraset sheets began to be used regularly, lawns were painstakingly stroked and dotted by hand. The sheets were included in the sample book for collages and used for several plans, but Kienast must have been dissatisfied with their effect, because he very quickly returned to stroking and dotting by hand to obtain a better structure. If one compares the depiction of lawns on the partially hand-drawn project plan for the Zurzach Spa Gardens of 1985 (165) and on the collaged competition plan for the Internationale Gartenschau 2000 (International Horticultural Show 2000) in Styria of 1997 (166), it is clear that Kienast ultimately chose pictorial models that were close to the structures drawn by hand. The same is true of pools of water, the ideal presentation of which Kienast had clearly found in the plan for the garden on Thujastrasse, where the cardstock shining through a lake of black ink represents the reflections of lights (162). He tried to achieve this effect in collage by using certain details of images of water (166).[137] Finally, with trees it is

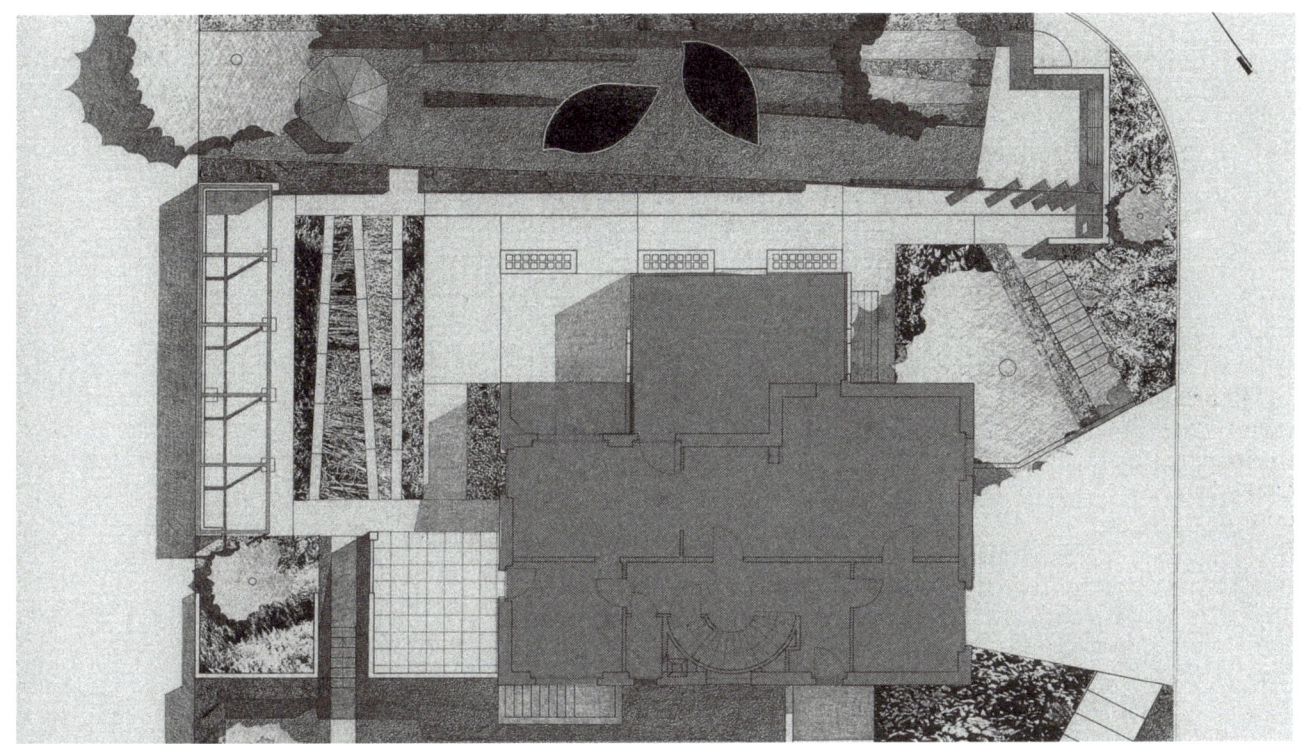

163
B. Garden in Gockhausen,
project plan, 1994, scale: 1:50,
ink, film, and colored pencil
on copy of plan, firm of Stöckli,
Kienast & Koeppel
164
Pattern books with textures, firms
of Stöckli, Kienast & Koeppel
and Kienast Vogt Partner

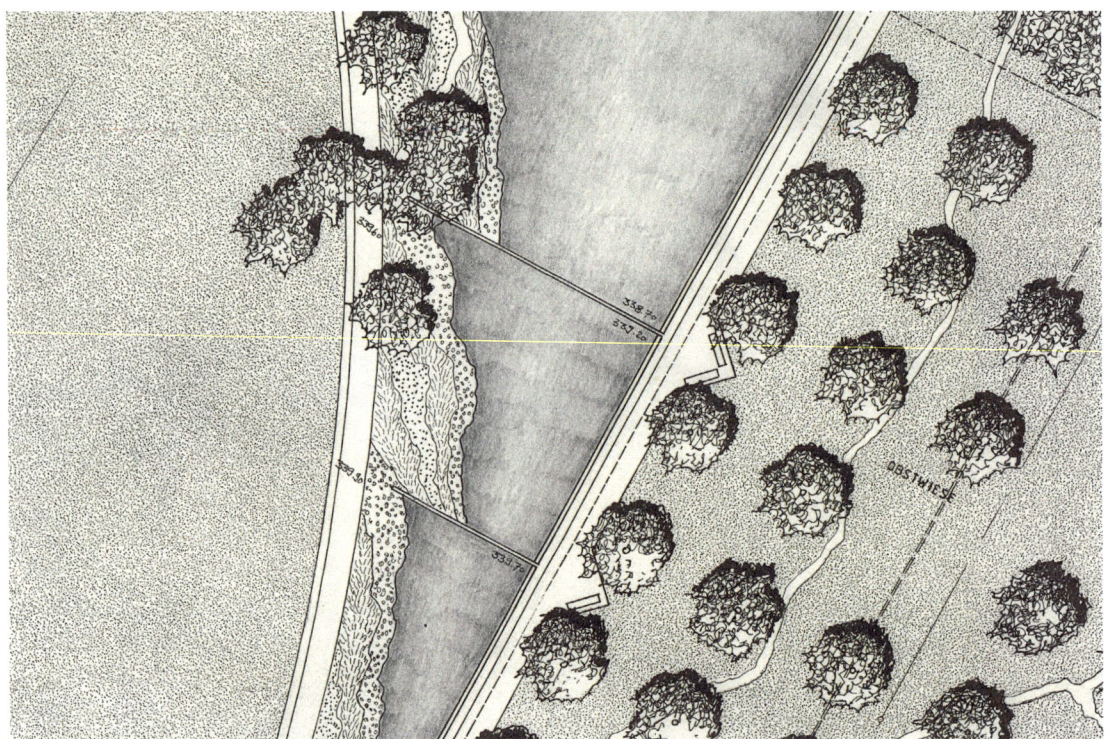

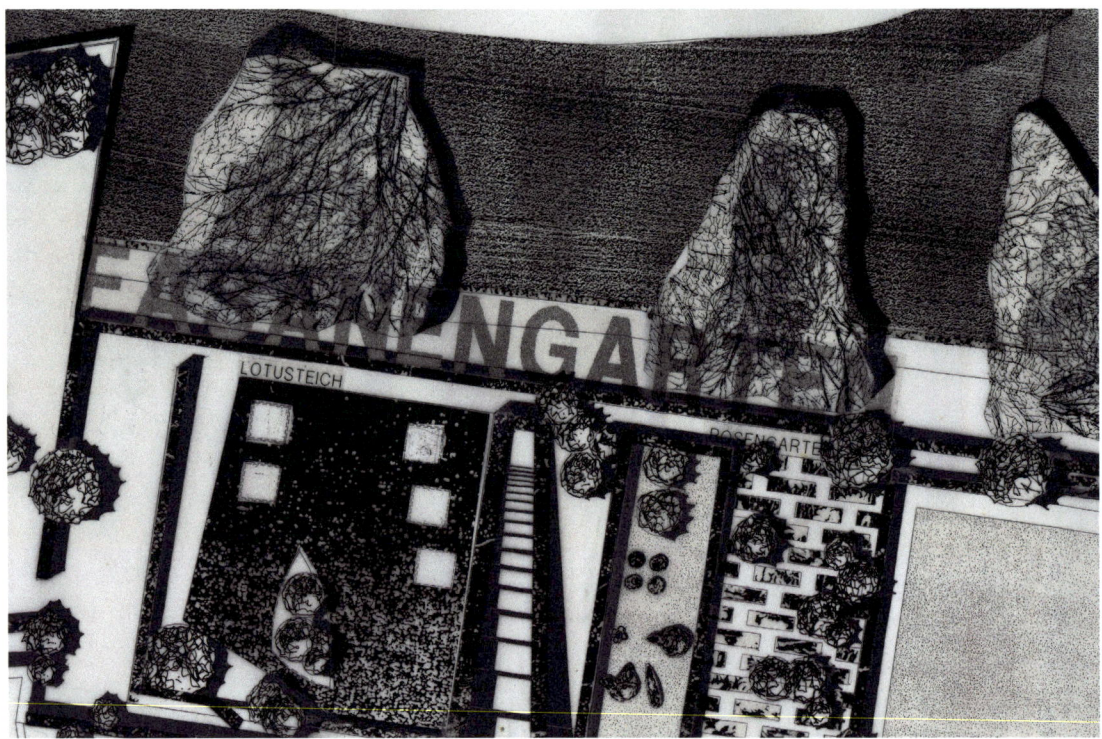

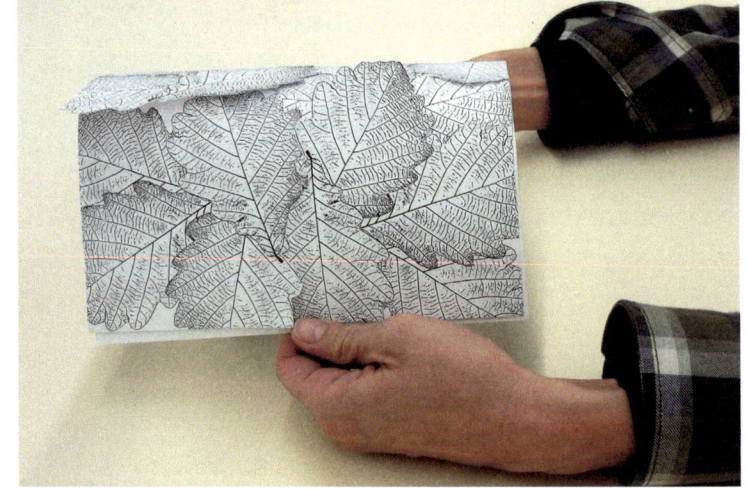

165
Zurzach Spa Gardens, project plan, 1985, copy of plan on card stock, ink, and graphite pencil (detail)

166
Internationale Gartenschau 2000 (International Horticultural Show 2000) in Styria, model for the competition plan, 1997, film, ink, and graphite pencil on tracing paper (detail)

167
Sample from a pattern book: overlapping copies of oak leaves

necessary to distinguish between solitary trees, groups of trees, and forests. Whereas areas of forest are represented by collaged views of trees or the texture of their needles, solitary trees were drawn by hand in ink in both 1985 and 1997—though the squiggles of ink representing the treetops on the plan of 1997 were placed on a circle, which is how trees were represented on early CAAD plans.

The presentation plans submitted to the invited competition for the Internationale Gartenschau (IGS) 2000 in Graz exemplify Kienast's continued effort to translate materiality into a texture or signature, but now using the technique of collage. Patterns were prepared for these plans, copied, and then supplemented with color elements or details added in Photoshop to create the presentation plan proper. The pattern for the site plan[138] shows this very clearly (168). Various areas of the garden show are depicted on a sheet of tracing paper measuring 130 to 200 centimeters: Ackergarten (Arable Garden), Blumengarten (Flower Garden), Berggarten (Mountain Garden), and Fasanengarten (Pheasant Garden) as a collage on the reproduction of a plan drawn with CAAD. All of the solitary trees were drawn by hand with ink over the CAAD tree circles and supplied with shadows on the back of the tracing paper. Once again, Kienast integrated the materiality of the support into the representation, since when seen from the front the shadows appear somewhat diminished by the texture of the paper. The incorporation of the paper also supports greater nuancing of the shades of gray. Kienast used the same approach for the topiaries in the Pheasant Garden and vineyards on the edge of the Arable Garden.

To represent other areas of the grounds, textures from the sample files were copied and pasted to thin sheets of tracing paper (168–170): the grassy areas of the gardens and surroundings are represented by delicately dotted surfaces; the lawn on the earth pyramids of the Mountain Garden by photographs of rougher lawn structures; the forested areas in the Mountain Garden, where there are many conifers, by photographs of fallen pine needles; the groups of deciduous trees by photographs of overlapping, bare treetops; the banks of the Schwarzl Lakes by a reproduction of the gravel areas of the Ryōanji-Garden; the bodies of water in the Mountain Garden and the Pheasant Garden by photographs of surfaces of water with reflected light, and so on. The literalness and hence legibility of the collages, and of the realized grounds, take on concrete physical form in one element: in the Mountain Garden, Kienast constructed a "landscape reader," an earth pyramid that unfolds like a book and has a text; it is indicated on the plan by a pasted-on newspaper clipping (170).[139]

The greatest diversity of structures is offered by the display area called the "arable garden," which consisted of a number of fields with different sorts of grains and vegetables (169). Rectangular clippings from a number of patterns have been pasted on here, including various textures of leaves and soil surfaces.

On the one hand, the overall impression is of a very graphic image. On the other hand, the photographic textures and the way they are combined—for example, the various triangular lawn surfaces to illustrate the earth pyramids in the Mountain Garden—powerfully stimulates the viewers' sculptural imagination. It is remarkable how much was invested in the legibility of the plan, which is again very clear from the example of the earth pyramids: a slight gap was left between the sides of the pyramids where the tracing paper can be seen, thus emphasizing the form of the pyramids and the structure of their angles.

Just as the finished grounds are permeated with the intention to make the place legible through the design and the use of vegetation, so are the plans: legibility was a central concern, but not only in order to make a space comprehensible from the plan—the techniques of representation go far beyond that. The plan is intended to give viewers a feeling of the materiality and the atmospheric quality of the planned grounds. The Letraset-and-collage technique had the advantage that the plans could be copied often during the various phases without losing the different nuances of gray and the structures in black and white, since the copiers of the time were set to capture and reproduce such information. The loss in the case of colored models was much greater.[140] Color accents were therefore added only directly to the finished presentation plan—either, as with the plan for IGS 2000, with color film that was pasted on at the end[141] or by hand coloring.

The vocabulary for the representation of landscape was also adopted on the software-based CAAD plans, for example the presentation plans for the outdoor grounds of Expo 2000 in Hanover produced between 1995 and 1998. Before the CAAD plans—their details presumably created using programs such as CorelDRAW, QuarkXPress, and Photoshop—there were several experimental phases in the

168
Internationale Gartenschau 2000 (International Horticultural Show 2000) in Styria, model for the competition plan, 1997, scale: 1:500, H 135 × W 200 cm, film, ink, graphite pencil on tracing paper, firm of Kienast Vogt Partner

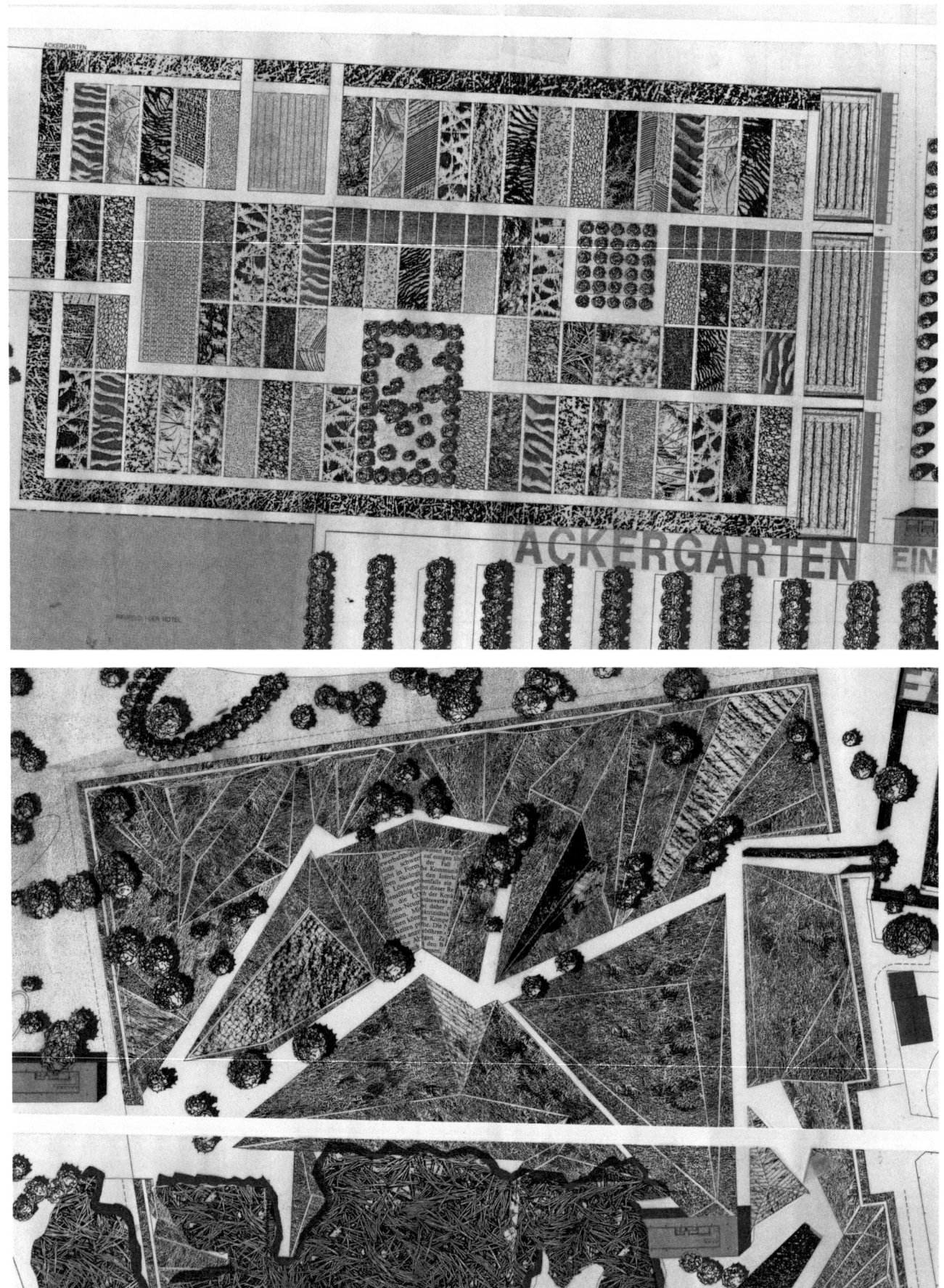

169, 170
Internationale Gartenschau 2000 (International Horticultural Show 2000) in Styria, model for the competition plan, 1997, film, ink, graphite pencil on tracing paper (details)

Ackergarten (Arable Garden) with fields of vegetables and grains

Berggarten (Mountain Garden) with earth pyramids and "landscape reader"

planning production for this project using a mixed technique of collage and computer-drawn plans. In the software-based master plan (171), apart from the planning of the grounds with the structuring grids of trees, the areas designed by Kienast can be seen: the Allee der Vereinigten Bäume (Boulevard of United Trees) in the south, the Parkwelle (Park Wave) in the west, the Expo-See (Expo Lake) next to it under the Expo-Dach (Expo Roof) by the Swiss architects Herzog & de Meuron, and the Erdgarten (Earth Garden) in the east. On the software-based plans, too, Kienast made use of a collection of surface structures and textures in order to retain the visual language he had developed (172). In the case of the Parkwelle, the variations in the density of the texture of the lawn reveal where the wave valleys and the wave mountains lie. Despite the top view of the topography, the depiction reveals how it had been developed as a means of representation on hand-drawn plans. The solitary trees are all indicated by the same image; each of them is given a shadow. When smaller and larger trees overlap, the shadows are arranged the same way, just like with the collages, on which one tree was pasted over the other.

For the rendering of the Earth Garden, the enlarged structure of a leaf served as the ground. The individual spaces of the garden have been arranged over it in various geometric forms. The hedges of decorative fruit trees that Kienast planned to use to demarcate it are—unusual for his plans—drawn in as black circles and bars. The pools, beds of shrubs, and collections of ferns found, among other places, in the garden spaces are each marked by a texture, much like the different fields of the Arable Garden of IGS 2000 (169). The water surfaces on the CAAD plans have been simulated by an apparent reflection of blue sky and white clouds.

Kienast developed a visual language that, despite different techniques, employs a common vocabulary in which the textures used to depict certain elements are similar in all techniques. The textures were employed "literally," and the various elements composed, following the logic of a grammar. The plans are often mixed forms combining different techniques of representation, which also influenced one another in their development. The connection between drawing, collage, and computer rendering is also easily recognizable in the sections and perspectives: if one compares the presentation plans completed around 1995 for Expo 2000 (173) to the presentation plans for the Günthersburgpark in Frankfurt am Main of 1991 (84), the phases in the development between drawing, collage, and CAAD and their interconnections are clear.

The competition plan of 1991 was largely drawn by hand on gray cardstock, supplemented by photographs of the existing terrain and by figurines, photographs of tree structures and of a bamboo grove from the sample books. Written explanations of the concept for the park, the master plan, two sections, and four perspectives are combined on the cardstock. An elongated section of the terrain runs vertically through the center of the presentation from bottom to top. It deftly links the left side, featuring the concept and the cross-section of the terrain, and the right side, showing the perspectives of individual areas of the park. From the section of the terrain, it becomes clear that Kienast had the idea to introduce a very gentle, flowing "wave of lawn" into the park. The first-prize-winning proposal for the Günthersburgpark was never realized, but Kienast did use the idea of the wave of lawn later in a very prominent way for the Expo 2000 in Hanover (172).

In the case of the perspectives produced around 1995 for the outdoor grounds of the Expo (174), the relationship between drawing and collage is precisely reversed: various silhouettes for depicting trees were copied from the sample books onto tracing paper. The tree branches were cut out and later added in ink by hand. That made it possible to produce an effect of depth by allowing the branches to grow smaller in the distance without having to reduce the structure of the treetops. The figurines were also taken from the firm's sample books.

In the case of the CAAD perspectives (173), this technique of representation, which had been developed for collage, was adopted and translated into the new medium. Between the tree silhouettes and the figurines, there are computer-drawn elements such as green cones and green walls, representing the lawn cone and the hedge-enclosed garden areas of the Earth Garden. The cross-sections feature the same tree silhouettes as did the perspectives and collage of 1995. The edges have a vegetal structure; a blue sky with small white clouds is the backdrop for all of the images.

These comparisons reveal a logic that runs through all types of plans. The changing media influence the method of depiction only a little. The sole exception is the sky, probably created with QuarkXPress or Photoshop. The horizon

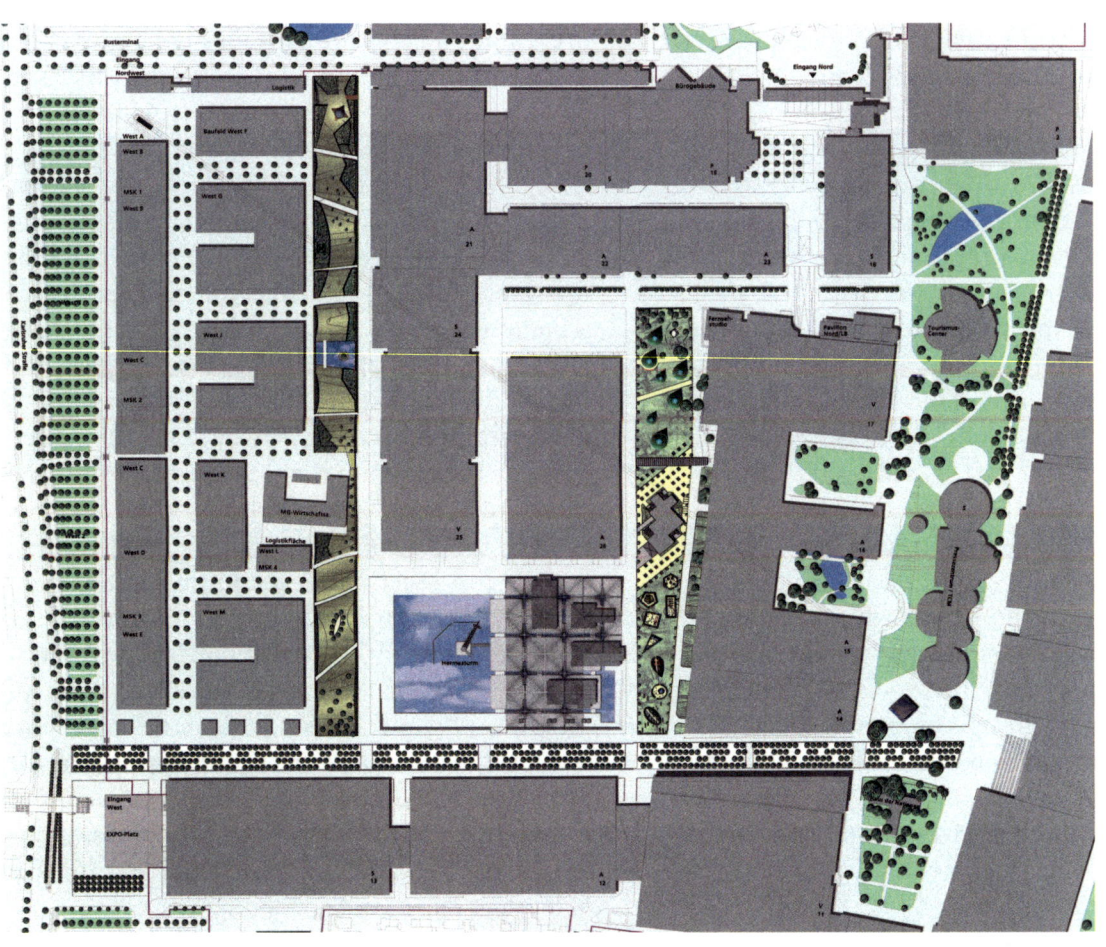

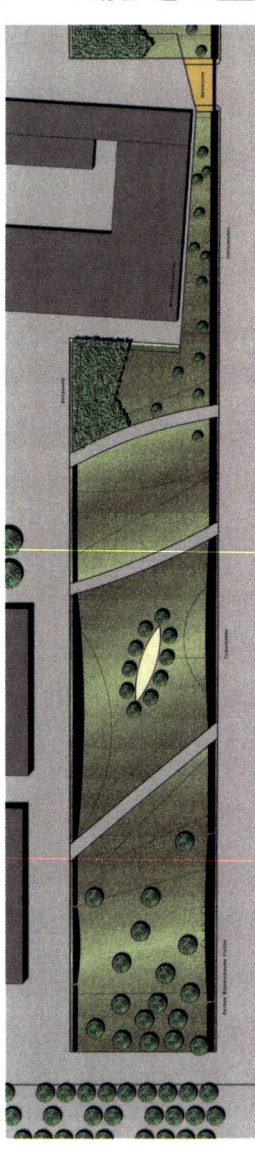

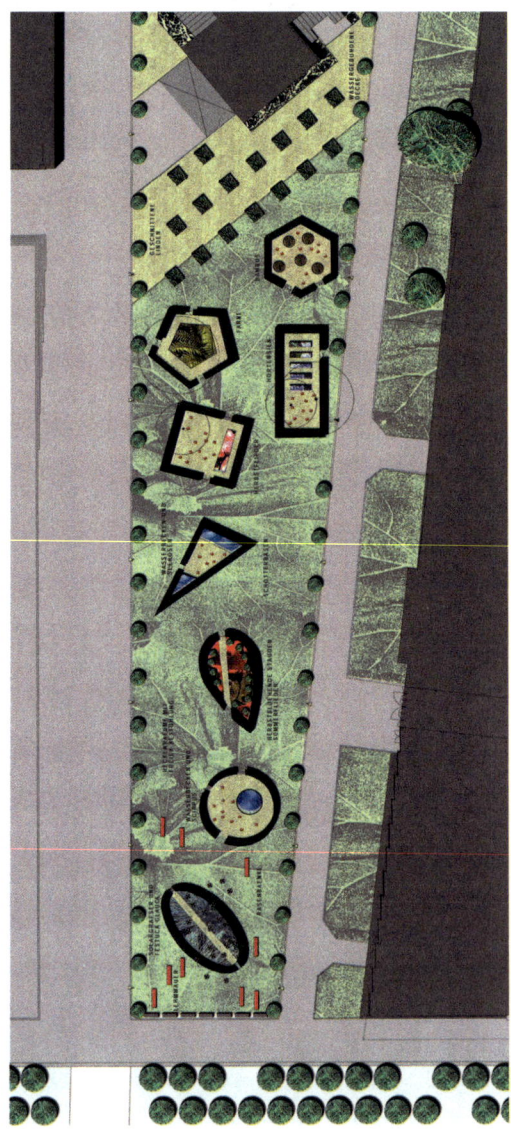

171–174
Expo 2000 in Hanover
　Master plan for the exterior grounds with the Allee der Vereinigten Bäume (Boulevard of United Trees) in the south, the Parkwelle (Park Wave) in the west, and the Erdgarten (Earth Garden) in the east, CAAD, ca. 1995, firm of Kienast Vogt Partner
　Details of Parkwelle (Park Wave) and Erdgarten (right)
　Sections and perspectives (details)
　Perspective "Paradise is just where you are," ca. 1995

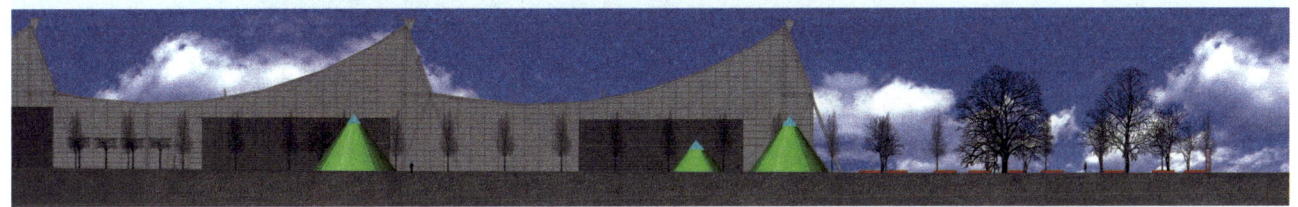

SCHNITT C-C M 1:

PERSPEKTIVE HECKENRÄUME   PERSPEKTIVE RASENKEGE

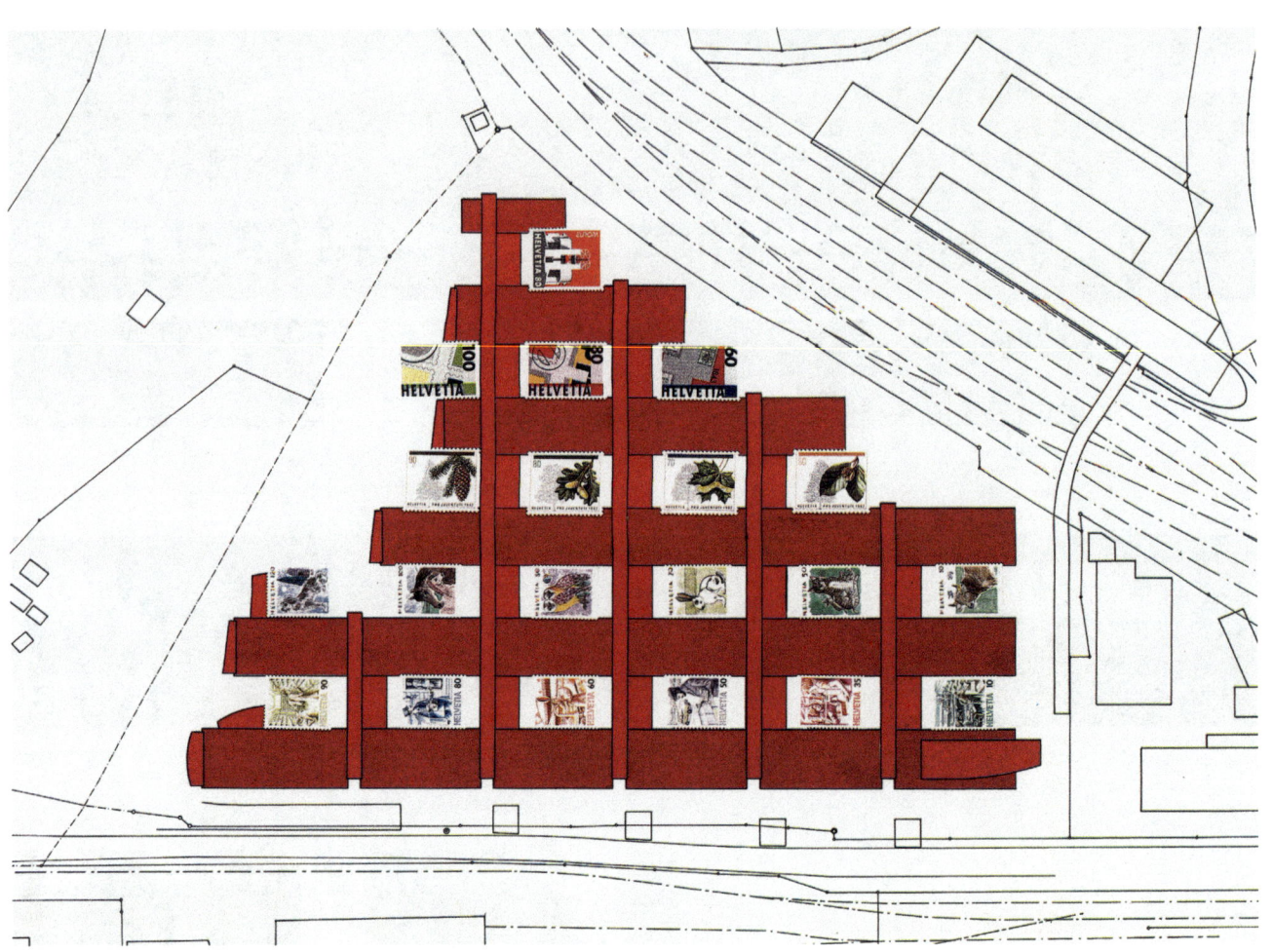

175
Garden courtyards and grounds
for the PTT/Swisscom Administration Building in Worblaufen,
competition plan, site plan
with ground plan, scale: 1:500,
H 60 × W 83.5 cm, postage stamps
on copy of plan, June 1993,
firm of Stöckli, Kienast & Koeppel

line is set very low; the cloud formations are symmetrically mirrored several times and are quite spectacular. It is the sky as it can be actually experienced in the landscape around Hanover: the broad plane of the northern German lowlands that begins here makes the horizon line seem low, and the sky is characterized by the forms of cirrocumulus clouds that are typical of this region. Even in these seemingly artificial images with a strange mirroring of cloud formations, there is a clear connection to the immediately perceivable reality of the existing landscape depicted on the plans.

Kienast rigorously pursued his effort to retain the language of pictorial translation of the material qualities he had discovered in a triad of gardening, craft, and artistic interests. One special case in the production of plans was the competition plan presented to the client in 1993 for the courtyard gardens of the PTT/Swisscom Administration Building (until 1998: the Post-, Telefon- und Telegrafenbetriebe, or PTT) in the Worblaufen area of Bern (175). The building by Christian Indermühle Architekten is divided into longitudinal and latitudinal bars; with its building-block logic, it is fitted into the triangular lot and makes as much use of it as possible. The associated open space is divided into twenty rectangular courtyard gardens. Kienast and Günther Vogt developed a concept for their design that would run through the possibilities of certain natural themes and forms. The competition plan uses minimal means to convey this impressively and very precisely: each courtyard in the schematic red ground plan has a stamp pasted on it, a reference to the client, PTT, which was responsible for delivering the mail. On closer inspection, it becomes clear that each horizontal row of stamps belongs to a series that, as Arthur Rüegg put it in a description of the PTT/Swisscom project, forms "families."[142]

In the lowest row of the ground plan, which tapers toward the top, various postal professions are represented; the next has animals, then trees and their leaves, then variations on graphic depictions of the Swiss cross. At the very top, in the last courtyard, is a stamp from the postal service's so-called Europe series with modern buildings in Switzerland, in this case a single-family home by Mario Botta. The stamps overlap somewhat with the edges of the building, which for Rüegg results in a "dual readability" of the concept: on the one hand, they "furnish" the given areas, in a sense; on the other hand, they appear "as an autonomous pattern on a red base. They act both as filling, as background and as the principal element, the foreground":[143]

> Minimal means are used to outline the idea, without making it necessary to reveal details at such an early stage. Each layer of space to be traversed from the baseline is assigned to an immediately recognisable theme, varied from court to court, so that clearly identifiable places are created. The individual courts still have to be designed, but—as the preliminary projects also dating from 1993 show—they can, thanks to the precise idea, readily be given concrete form.[144]

The first conceptual presentation based on "families" of stamps that belong together was followed by detailing and working out the individual courtyard in the design in that, analogously to the families of stamps, one idea (for example, the use of a wide variety of pools) runs through a series of courtyards. The connection to art, such as the phenomenon of series in contemporary art, and to garden art, such as the allusion to the parterres and bosquets of the Renaissance and Baroque, is evident in this work.

Drawing and Literalness/
Abundance and Models:
Kienast/Vogt

The implementation of the courtyard gardens for PTT/Swisscom was completed in 1998. The courtyards drew from the rich diversity of garden culture for the selection of plants as well as for the formal idiom, whereas otherwise Kienast was concerned, especially in his work in urban contexts, with a very reduced use of plants in the spirit of a new legibility of the city, using plants like words, in a sense, to refer to something (*Iris pseudacorus* for "Here it is damp," *Allium ursinum* for "Here is the undergrowth of a row of trees," and so on). In the courtyard gardens for PTT/Swisscom, the influence of Günther Vogt can clearly be felt: the works from the final years of their collaboration are marked by a garden art that falls back on the entire abundance of the plant world. This includes, for example, the artificial rain forest landscape for the Masoala-Halle (Masoala Rain Forest Hall) in Zurich's zoo (1994–2003). Kienast and Vogt worked out the layout in collaboration with Walter Vetsch; Vogt was responsible for the selection of plants and the implementation. This drawing on lavish abundance was developed further by Vogt in the early years of the firm Vogt Landschaftsarchitekten, which he founded in 2000. One example is the grounds for the headquarters of the Helvetia Patria insurance company in St. Gallen, conceived under the name Blumenberge (Flower Mountains) between 2001 and 2004, for which Vogt's firm was awarded the Schulthess Gartenpreis in 2010.[145] The site plan indicates the topography, the paths, and the spatial structure established by trees. However, the crucial information to be able to imagine the project with all its subtlety and complexity lies in the planting schema and monthly tables. They are sorted according to color aspects and illustrate the distribution of the individual groups of plants in an area of 4 to 25 meters and their flowering seasons. The groups of plants selected were subdivided into "Solitary Herbaceous Plants," "Main Plants," "Ground Cover," "Bedding Plants," "Bulb Plants," and "Filler." The associated month tables show the flowering seasons in each case and their changing colors over the course of the year. This makes it clear that landscape architecture, in comparison to architectural design, has a much more complex interplay of space and time, since the construction material is the plant. In the design of the grounds of Helvetia Patria, the plants are distributed spatially according to their structure, color, growth, and flowering season, and they transform this space in their dynamic from month to month. Because of the cycle of the plants, certain colors and forms appear and disappear again in different places according to the season. Growth, flowering, dying, and care constantly change the structures as well as the spatial and atmospheric appearance of the grounds.[146]

Comprehensive knowledge of plants and great power of imagination as to how the complex dynamic processes will be expressed in a design are necessary for such a work to succeed. One does not find in Günther Vogt the penchant for drawing that Kienast developed over the course of the 1980s and that became typical of his work. Vogt continued to develop his own style after Kienast died and has visibly changed it several times since as well, depending on the needs of a place, a program, or a client. Nevertheless, he believes that there is "still a unifying thread" that connects him to Kienast even today:

> Dieter Kienast and I both had tremendous knowledge about plants, but in extremely different areas. His PhD was in urban vegetation. I also know quite a bit about this field of potentially natural vegetation, but my interest was more in cultivated plants. After Dieter Kienast's death, no one in the office other than me possessed this excellent knowledge of plants. To this day, I am trying to maintain this knowledge and this quality of plant use that has always been the hallmark of this office.[147]

Although the quality of the draftsman that one observes in Kienast's work is an essential feature of his approach, his works were not that of a sole individual. Günther Vogt has described their collaboration as follows:

> Dieter Kienast had a passion for drafting, while I loved to discuss content and was interested in the craftsmanship aspect. During our partnership, we discussed each and every project with one another. The time came when we would hardly have been able to work without the other. This collaboration became a real obsession and it represented a complementarity that had not existed in the beginning, but which evolved over time.... When Dieter Kienast and I spoke together toward the end of our cooperation, not everyone understood what we were saying; we had developed a kind of secret language.[148]

Dieter Kienast described his cooperation with Vogt in a similar way:

> We understand each other spontaneously without a lot of words. We discuss and also fight with each other, pencil in hand. After talking, however, as a rule the deed will follow: drawing. That is often my part, but the fundamental concept we always, whenever possible, develop together. Our projects do not result from working alone.[149]

Drawing was Kienast's task. He was the draftsman, and the one who designed and searched by drawing. The penchant for drawing runs like a trail through his oeuvre and, like the use of plants, reveals a *manière* that was also subject to different processes of reinterpretation over time. Kienast's use of drawing—not only to document and present his works but also to draft them—is the crucial aspect of his final designs. He almost never developed the ideas for his own designs using models.[150] For his students in Rapperswil, there was a large sandbox for experiments.[151] Moreover, along with the sculptor Jürg Altherr and the architectural theorist Peter Erni, Kienast encouraged students to work with models.[152] In the firm of Kienast Vogt Partner, however, models were built, if at all, only after the plans had been drawn.[153] They served primarily to communicate the ideas to the clients.

This method was fundamentally different from that of Günther Vogt, who uses models almost exclusively to develop design ideas at Vogt Landschaftsarchitekten. In this three-story office in Zurich, one floor is dedicated to the model workshop. There are also intense studies of materials, field studies with long walks, and an image database developed especially for the office in which all visualizations are stored. "Making is thinking," writes Alice Foxley, the former head of research in Vogt's office:

> We design using physical models and stimulate the walker's perspective. We revise them time and again, trying out design solutions and techniques. The simultaneous representation and reality of models influences the way we think about landscape architecture.[154]

In recent years, Günther Vogt has made an essential contribution to once again making the appropriateness of materials and plants for their location and the concentration on the sensory experience while walking central themes in landscape architecture. On these issues, the attitudes and values of Vogt and Kienast were in agreement. In a series of exhibitions, including the installations *The Mediated Motion* developed with the artist Ólafur Elíasson for the Kunsthaus Bregenz in 2001, Vogt tried to make it possible to experience the diversity and complexity of nature intellectually and physically.

> Olafur Eliasson originally wanted to bring a park—or the idea of a park, a landscape or a garden—into the museum . . . . In the very imposing architectural spaces created by Peter Zumthor for the Kunsthaus Bregenz, I found it difficult to build a real model, in the sense of a landscape or a park. So we decided to work only with phenomena: with water and aquatic plants floating on it, with rammed earth that is not horizontal but slightly inclined, with fog. These are very simple phenomena that everyone knows—but only from the real landscape and not in the context of a museum. Even I was astounded at how well it worked: that simply by changing the context you can suddenly be so impressed by a simple phenomenon.[155]

By means of reduction in the exhibition space, of the isolation of a specific phenomenon, a focus on perception resulted that produced an atmospheric design of the landscape,[156] as Vogt attempted in other projects as well, such as the small forest of birches in the inner courtyard of the Hotel Greulich (2000–2003), the rain forest pastiche in the Masoala-Halle in Zurich's zoo, the home of FIFA (2003–2006), and the artificial landscape of the Novartis Campus in Basel. Vogt sometimes calls his finished projects "models." On the one hand, this is because he wants to preserve an openness even in projects, a quality that distinguishes the physical working model, because anyone who looks at it brings his or her own idea and interpretation of the space to be realized.[157] The "lack of precision" and partial abstraction of the model suggest and promise things, without becoming explicit; this interpretive freedom, the promise of a "more," should, in his view, also remain open in the realized works.[158] On the other hand, in the realized project a leap in scale occurs with which Vogt wants to poetically implement a kind of condensed reality of a landscape. When looking at the birch grove in the Hotel Greulich or at the landscape on the Novartis Campus, people are confused: Are they in reality now or in a model? Vogt sees this confusion as "ideal."[159] By doing so, he is picking up on the old garden traditions of playful imitation and densification and making a clear paradigm shift from "legibility" to "landscape as a cabinet of curiosities":[160]

> We have to rely a bit on people's real experience of landscape. Although I have to admit that I do increasingly see a problem in the fact that many people read our projects as nature because they are no longer able to compare them with the real landscape. But I think if you look closely, you can see that our landscapes are man-made,

and you also see how they're made. In the Novartis Campus, for example, in the karst landscape or in the gravel banks in the park that don't drop off but stand there like a wall, like a built wall, you can see already that it is "only" a reconstructed, model landscape. I consider this model character of the landscape important, because the landscape architecture of the late twentieth century in particular really struggled with the question of what new content might be available for nature in the city. We operate to a great extent with images in this connection . . . . And if we can no longer rely on people having had a real, live experience with landscapes and in landscapes, then things become difficult.[161]

Kienast's Vocabulary of Planning and Designing

The sketch and the drawing were the media in which Dieter Kienast expressed himself.[162] His imagination enabled him to visualize his designs spatially when drawing.[163] The landscape architect Henri Bava (b. 1957), who succeeded Kienast as professor at the Institut für Landschaft und Garten (Institute of Landscape and Gardens) of the Universität Karlsruhe (TH) in 1997, believes that Kienast, when drawing, imagined himself in the space he was putting to paper and was able to visualize it from the inside out.[164]

This is also confirmed by the landscape planner Thomas Proksch (b. 1960), who took over, for Kienast Vogt Partner, the construction management of the Mountain Garden for the Internationale Gartenschau 2000 in Styria. Proksch received drawings from Kienast in which he had already noted all the altitudes and slopes of the earth pyramids, which together form a landscape that looks folded.[165] According to Proksch, the dimensions were implemented precisely as Kienast had indicated on the first working plan and never needed to be corrected. Kienast often marked the contour lines of the terrain on his drawings to make them easier to understand spatially (156). For the Fürstenwald Cemetery in Chur (130), they even became the immediate source of inspiration for the design: the hedges that subdivide the cemetery into steps adopt the rhythm of the contour lines. They accentuate the topography in this way and cause the designed landscape to fit in organically with the existing one.

One example of how a graphically developed design idea can reshape an existing landscape by incorporating a very personal reservoir of landscape images is the design for the Kronsberg in Hanover. In 1992, as part of the planning for Expo 2000, the State Capital of Hanover and the State of Lower Saxony announced an international urban and landscape planning competition by invitation, in order to solicit and present ideas for a structural proposal for the area around the Kronsberg near the Hanover trade fair grounds.[166] After a workshop with five firms had been organized to design the landscape architecture for this new district of the city in 1994, the firm Kienast Vogt Partner was commissioned in 1995 to design the open spaces of the Hanover trade fair site where the Expo 2000 would be held and to plan the landscape for the Kronsberg.

The Kronsberg is a long hill with a broad ridge southeast of Hanover near the grounds for the Expo 2000; it is the highest elevation point of the surrounding flatlands. Because the Kronsberg is an important local recreation site for the city's residents, the City of Hanover wanted to avoid its overdevelopment at any cost and instead decided to redesign a landscape that had until then been used for agriculture.[167] On the basis of a proposal presented by Dieter Kienast at the workshop in 1994 (176), the recently founded firm of Kienast Vogt Partner worked out a master plan for the area on the western slope of the Kronsberg, where a large housing development was to be built under the motto of "ecological housing construction."

The point of departure for the design was a kilometers-long boulevard that would form the backbone of the planning and also serve as a threshold against the expansion of the settlement.[168] The choice of trees also made the boulevard a symbolic threshold between the city and the countryside: the row of trees facing the cities is composed of lindens, as a reference to the famous historic linden boulevard along the large urban planning axis between Hanover's center and the Georgengarten, near Herrenhausen. Cherry trees were planted on the side facing the countryside. This choice of trees once again supported Kienast's desire to use design interventions to make a landscape "more legible."

Several "ribbons of park" at right angles to it simultaneously supplement the boulevard and break it up. They intervene in the housing colony and serve as open spaces for recreation and playgrounds for children. Today, the urban planning terminations of these open aisles are formed by viewing platforms facing the city that are surrounded by oak trees on the side facing the Kronsberg. In the design by the firm of Kienast Vogt Partner,[169] which was only partially realized, the ribbons of park extended well into the agrarian landscape. Beyond the boulevard, and interrupted by these aisles, commons were planned and next to them a forest that was supposed to be made denser by reforestation and would have run along the ridge of the Kronsberg.

As is clear from the master plan and several sketches by Kienast (179), artificial hills of raised earth were meant to rise out of the ribbons of park at the peak of the Kronsberg as landmarks and vantage points. Three were realized. A wreath of trees marks each of the accessible hilltops. When standing on one of the hills, the other raised areas can be perceived in the distance, including the slag heaps of the nearby potash works (177). The designed placement of the trees on the hills triggers an effect of recognition: One feels a connection to the other hilltops and is inspired to climb them, too. The hilltops create a visual link over and beyond the designed landscape, while the straight boulevard physically holds it together. Its effect of depth encourages strollers and joggers to move forward and offers them protection from heat and rain.

The marking of hilltops by solitary trees is widespread in the cultivated landscape of Switzerland (178). In some areas, trees are often planted on otherwise bare hilltops to celebrate the birth of a child. It is impossible to say whether this transfer of a familiar image from Kienast's homeland to the landscape expanses of Lower Saxony was unconscious or deliberate. The design of the Kronsberg shows that Kienast retained, even when working on a large scale, both the graphic character of his designs and the object-based design with a landscape vocabulary adapted to the surroundings.[170]

Just how Kienast established a recurring form in design as vocabulary can be illustrated by the mimetic translation and reinterpretation of leaf forms, a motif that occurred often between 1992 and 1995. It first appeared in the revised competition plan for the Günthersburgpark in Frankfurt am Main of 1992[171] and is found in different variations, such as the overview plan for the grounds of the Waldhaus Psychiatric Clinic in Chur of 1993 (180), which shows the site with a ground plan. In addition to a grove in the form of a pointed leaf,[172] in one of the courtyard gardens for the structures housing beds, the viewer's eye is struck by the leaf motives loosely distributed on the access road to the main building and the lawns on its sides. They are oversized for the scale of the drawing and scattered as if they had fallen from the surrounding trees and were left lying randomly on the street and lawns.[173] These leaf forms were realized as yew massifs planted in leaf shapes in the soil, demarcated by steel plates (183). These volumes are an intelligent trick for spatially articulating and opening up the asphalt streets, which have to remain free of all obstacles because they function as access roads for emergency vehicles. There is enough room for a vehicle, and in an emergency it would also be easy to drive over the small yew massifs.

For the Sp. Garden in Egnach on Lake Constance (1991–93), Kienast had the idea of creating large decorative beds in the form of beech leaves (185). Once again, leaves that had fallen on the ground were a source of inspiration for this form:

> During our first walk through the garden, we noticed beech leaves on the lawn which formed a contrast to the rich green of the lawn itself. We have taken up this theme in the design of the bush beds. The beech leaves were enlarged 50 times, interpreted as bush beds and are placed without orientation on the square and the lawn.[174]

In the B. Garden in Gockhausen, two small pools in the shape of beech leaves were placed in the lawn (163); in the courtyard garden of Swiss Re on Gotthardstrasse in Zurich, Kienast used tree grilles in the form of the leaves of the katsura trees standing in these grilles (p. 148). In the courtyard gardens of the PTT/Swisscom Administration Building in Worblaufen, an oversized metal water lily leaf decorates a pool in which real water lilies float (184).

These motifs always have a reference to the tension between the natural and the artificial; and they represent a way of running through all the possible designs of natural forms, as is so often found in Kienast's projects. The mimetic and essentially anecdotal use of these leaf forms originates from his interest in imitating and

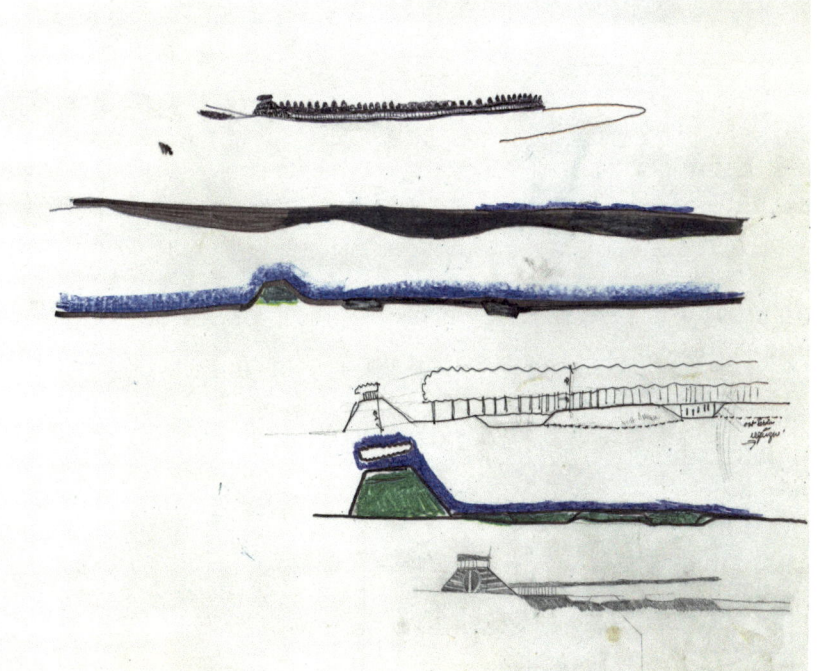

176
Landscape design for the Kronsberg in Hanover, master plan "Gesamtkonzept" (Overall Concept), scale: 1:2000, H 100 × W 200 cm, copy of plan on tracing paper, film, ink, graphite and colored pencils, sketches by Dieter Kienast pasted on, 1994, firm of Stöckli, Kienast & Koeppel

177
View, looking east, of a Kronsberg hilltop with a ring of trees, surrounded by other artificial hills, such as slag heaps of the potash work, 2008

178
Hilly landscape in the Hirzel region, Canton of Schwyz. Photo: Fritz Maurer

179
Dieter Kienast, landscape design for the Kronsberg in Hanover, sketch of a hill with ring of trees and forest, ink and chalk on sketching paper, undated

defamiliarizing observed reality in his designs, and it strikingly emphasizes once again Kienast's continual formation of a "literal" design vocabulary for landscape. The medium of drawing serves here as the morphological pattern for it. The two-dimensional notation of a form is developed into a spatial volume or object for the specific garden grounds.

In the cases of the katsura leaf grilles, the metal water lily leaf, and the garden pavilion at the Waldhaus Psychiatric Clinic, where the roof and floor are leaf-shaped (181), the leaves that provided the model become objects in space. The natural form that is imitated, and at the same time defamiliarized artistically, is placed anecdotally, even poetically, in the forest like a sign Kienast writes:

A small aperture in the wood provides us with . . . a close-up view of the pavilion. The leaf-shaped roof is covered in sheet metal, its underside consists of a fan-shaped structure of light-coloured plywood. The roof is supported by a black concrete pillar and a slender steel tube, painted red. Matching the roof, the floor is made of concrete, dyed pale blue. As this pavilion is located in woodland, it is surrounded by allium ursinum, lilies of the valley, ferns and anemones. Its form and the choice of plants and materials show that the pavilion addresses the location without subordinating itself to it, it is both external to and a part of the wood.[175]

A sketch, not drawn by Kienast, shows that in addition to the idea ultimately implemented—a seating area covered by a leaf—there was also an idea to have an open corolla (182). The petals would have formed a kind of wreath around the seating area. Presumably, this variant was rejected because of the difficulty in executing it and because of the weather conditions in Chur.

The flowers-and-leaves pavilion planned for the forested area not far from the clinic's main building looks playful and fairy-tale-like. Kienast begins his own description just quoted above of the leaf-shaped pavilion with a sentence that could have been taken from a fairy tale: "A second garden pavilion lies in the midst of a dense foliage of trees. A small aperture in the wood provides us with . . . ."[176] In the next sentence, he begins a concrete description filled with technical terms. Popular culture is echoed in many of Kienast's works of the 1980s and 1990s. This is true not only of the objects conceived for the garden grounds, like the colorfully painted pavilion with leaf roof here, or the many views and cross-sections in which, as he began to apply collage technique more and more, figures from film and television are found. It was also expressed in the way Kienast wrote and spoke about his work.

364

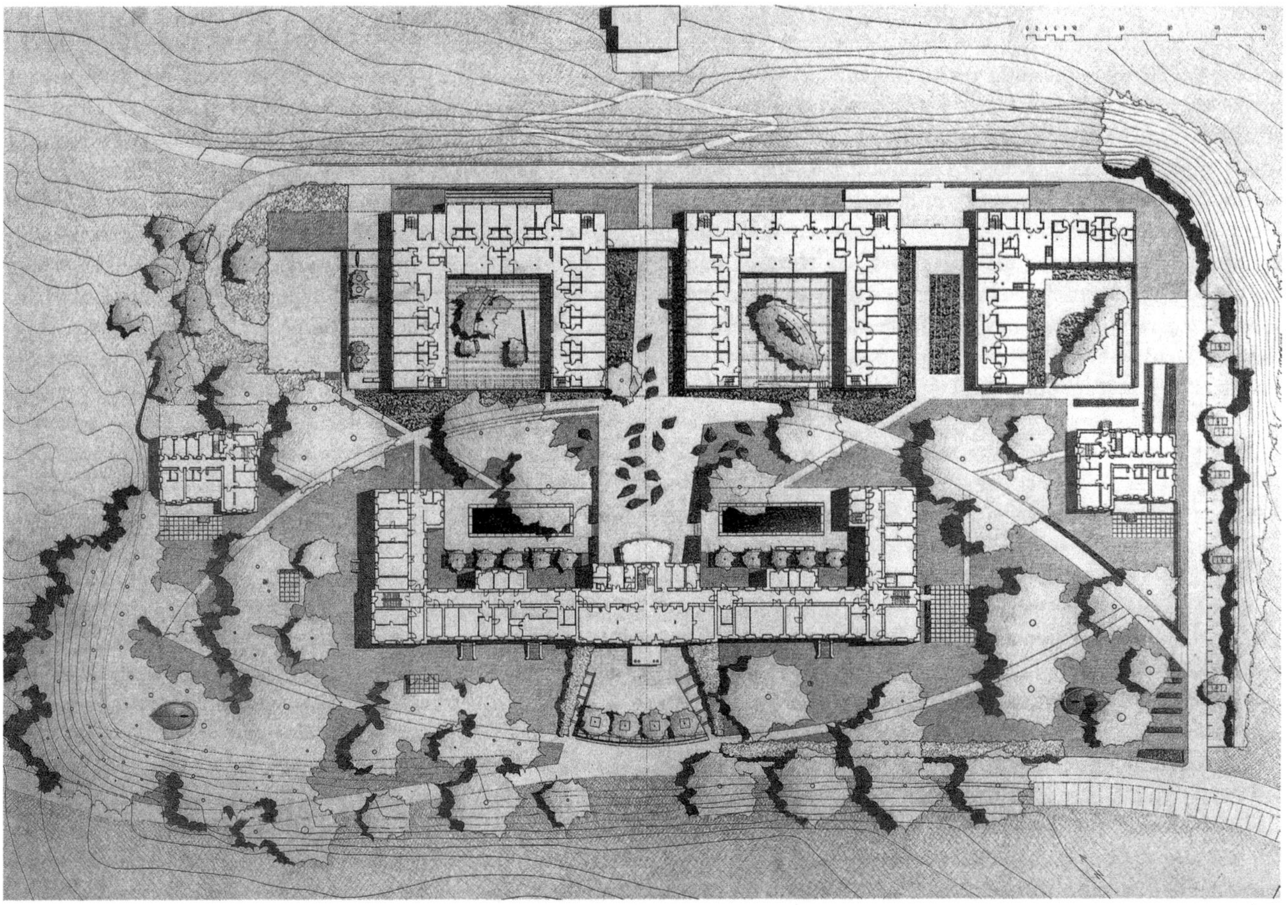

180–182
Waldhaus Psychiatric Clinic in Chur, 1987–96
    Overview plan, scale: 1:200, H 100 × W 141 cm, copy of plan mounted on a board, hand colored, 1993, signed Dieter Kienast, Günther Vogt
    Pavilion. Photo: Christian Vogt, ca. 1997 (detail)
    Sketches for the pavilion

183
Waldhaus Psychiatric Clinic
in Chur, 1987–96. Yew massifs
in the shape of leaf. Photo:
Christian Vogt, ca. 1997

184
PTT/Swisscom Administration
Building in Worblaufen,
1993–98. Garden courtyards,
metal waterlily leaf in the pool.
Photo: Christian Vogt, ca. 1998

185
Sp. Garden in Egnach,
1991–93. Shaped like leaves.
Photo: Christian Vogt, ca. 1996

Mechanisms of Representation: Texts, Photographs, Exhibitions, and Video

Over the course of his career, Dieter Kienast developed diverse media of representation and, in the process, set new standards for the discipline of landscape architecture. He was one of the first landscape architects in Switzerland to have his plans exhibited in a gallery; his collaboration with the photographer Christian Vogt demonstrated a new kind of photographic presentation of works of landscape architecture; and he was one of the first in his profession to be enthusiastic about the medium of video and its possibilities for illustrating the dynamic factors of movement and changing weather conditions that are so important in landscape architecture. But, this author of artfully drawn plans owes his enduring popularity not least to his reputation as a writer of powerfully expressive texts.

Kienast's Way of Writing and Describing: The Rhetoric, Style, and Illustration of His Texts

The numerous short essays in which Dieter Kienast expressed his thoughts on fundamental issues of garden culture are essential to any account of his work.[177] In them, he often combined programmatic positions with descriptions of his own projects. Because even today there are few writing and theorizing landscape architects in the German-speaking world, his texts are read often and with pleasure. When writing, Kienast was always concerned about language that was clear, easily understandable, and directly involved his readers. That is also true of the works on phytosociological research, with their numerous facts and scientific terms. Reinhold Tüxen, in his evaluation of Kienast's dissertation, explicitly praised its language:

> Both areas [the complete notation of syntaxonomically ordered plant societies of part of a city and, developing from that, a synsociological system of sigma societies] as well as their evaluation for planning are distinguished by a clear presentation that, remarkably, dispenses with superfluous foreign and trendy words and obtains an acuity and intelligibility for that very reason.[178]

From the 1980s onward, Kienast addressed changing themes related to the planning of open space and landscape architecture and chose the form of the essay to do so. He published in all of the important German-language journals on architecture and landscape architecture and took positions on the current discussions. This included questions of design and use of grounds around residences, the controversy over the natural garden, the environmental movement, education in landscape architecture, and the importance of gardens and open spaces for urban planning. Kienast often illustrated his essays with comics, photographs from newspapers, illustrations from books, and photographs he took himself, in which he captured everyday situations and model or off-putting projects (186, 187). He presented his own works as best practices.

Two general questions recur in ever new variations: To what extent do gardens reflect a society's relationship to nature? What design should be developed today to reflect critically on the current relationship to nature and to develop it in a "good sense"? Kienast's search for answers finds a visible parallel to the essays in the grounds he designed. In the text "Vom Gestaltungsdiktat zum Naturdiktat; oder, Gärten gegen Menschen?" (From the Dictate of Design to the Dictate of Nature; or, Gardens against People?), written in 1981 (186), Kienast wrote of this:

> Turnover and profit can, nevertheless, scarcely be realized with nearly natural garden forms in which construction and maintenance are kept within modest limits. That is why "green industry" praises the aseptic garden as the new form of the garden and simultaneously holds it up as

the incarnation of the natural. This is intended to achieve two things: first, to satisfy the human desire for more naturalness and, second, to manipulate the concept of nature to ensure long-term profit and turnover. Hover lawn mowers, electric hedge shears, verticutters, rose shears, complete fertilizers, and poisons replace the servants working in the garden and ensure the continued existence of garden culture, ultimately providing the proof that the owner of such a garden also has culture.[179]

There are still clear echoes here of the Kassel School. Over the years, Kienast's texts became increasingly literary.[180] They are written in clear language and with a scholarly ambition that is supported by numerous references to the specialized literature. With the exception of the "Zehn Thesen zur Landschaftsarchitektur" (Ten Theses on Landscape Architecture), developed in 1992 as a way of establishing his positions in the context of his appointment to the Universität Karlsruhe (TH),[181] all of his texts have footnotes.

Kienast used language to communicate to the reader the sensory experience of a garden. To do so, he used the present tense exclusively, as the most direct tense.[182] Writers usually employ the present tense as a stylistic means to heighten the intensity of a narrative and often combine it with a description from the first-person perspective. Kienast, too, was obviously interested in intensity and, beyond that, in the suggestion of authenticity and immediacy. Nevertheless, he never used the first-person singular, instead always speaking of "we." His descriptions of his projects were written at times as if he were walking through the grounds.[183] The "we" makes the reader a companion on his voyages of discovery. This form of imaginary walk occurs for the first time in "Die Sehnsucht nach dem Paradies" (translated as "Longing for Paradise") of 1990, when in conclusion Kienast described an ideal garden:

And it is on this note that we stroll through the garden, enjoy the shade provided by the mature trees, the fascinating wilderness created by elderberries, buddleiea and echium, are almost overpowered by the unbelievable scent of mock orange, delight in the patter of the water from the fountain, enjoy grapes fresh from the vine, dream as we sit under the apple tree, stride through the rose-tree trellis, sit down on a sunlit terrace and take a deep breath of fresh air.[184]

The heightened intensity of the description is reinforced by combinations such as "unbelievable scent of mock orange," "grapes fresh from the vine," "deep breath of fresh air," thus stimulating the readers' imagination to experience the garden with all of their senses: touching, smelling, tasting, seeing, hearing. By doing so, Kienast also awakens his readers' memories of their own time spent in a garden or landscape. The readers are thus drawn into the narrative and their individual desires for nature are fomented again.

Elsewhere, when explaining his conceptual ideas, the "we" stands for the collective of the firm in which the designs were produced. The first sentences of his "Zehn Thesen zur Landschaftsarchitektur," which he revised many times until 1998, are categorical and cannot be misunderstood:

Our work is a search for the Nature of the City, whose color is not solely green but also gray. . . . Our interests focus on the city and its inhabitants. . . . We assume that it is impossible either to dismantle the city or to rejuvenate the countryside. . . . We understand garden architecture as an expression of the spirit of the times.[185]

The programmatic introductory sentences of the theses are supplemented in the course of the individual paragraphs by explanations in which, at least in the short theses, no more "we" appears. The collective form of address is limited to a few sentences that, however, immediately convey to readers the manifesto-like character and the "we" feeling of the authors. In Kienast's second thesis, another feature of his texts emerges: the mixture of technical and foreign words and expressions from colloquial language:

Our interests focus on the city and its inhabitants. The city is no longer a monolithic entity; instead it is dismembered and fragmented into thousands of parts. City dwellers are a kaleidoscope-like mixture of young and old, immigrant workers and long-established locals, clergy and junkies, managers and ecofreaks. This heterogeneity demands up-to-date actions and reactions

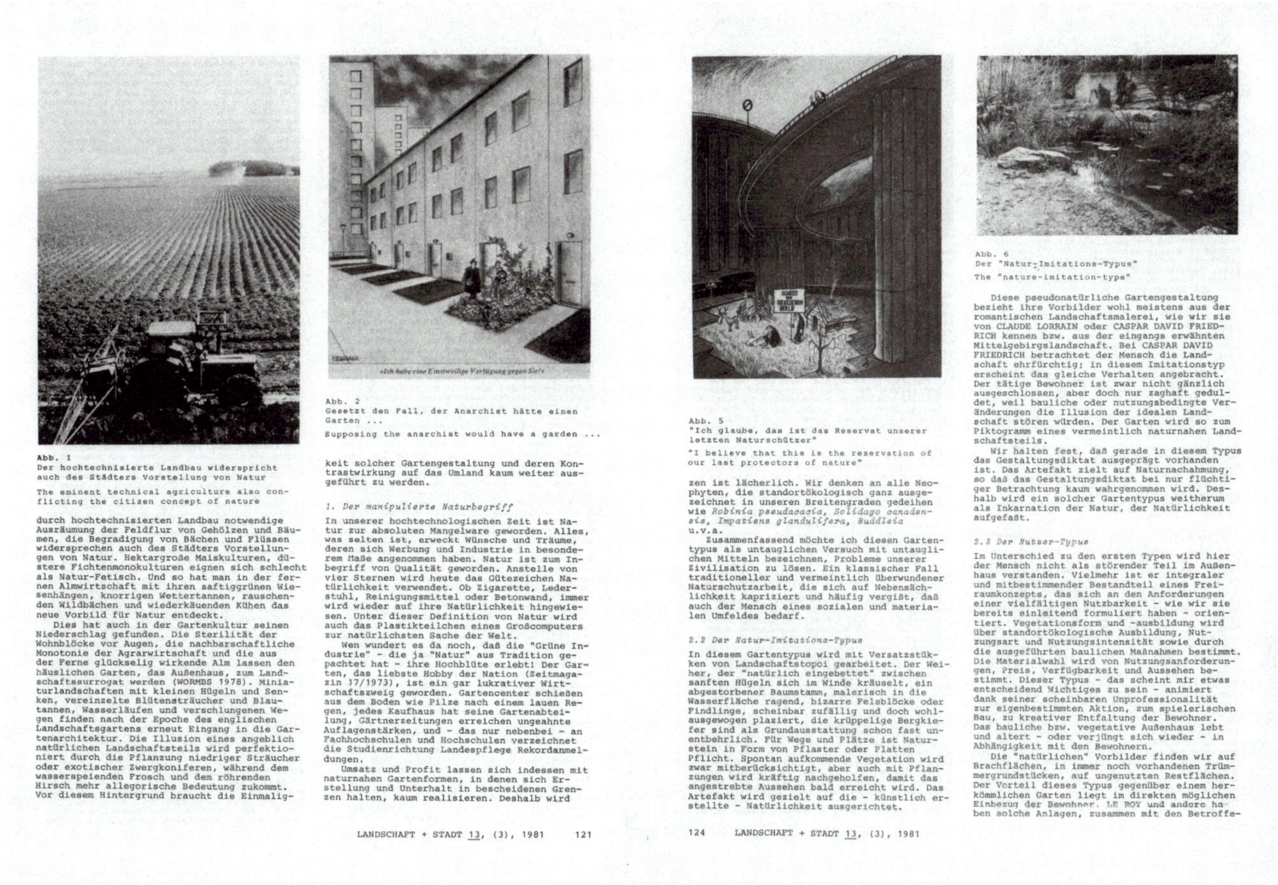

Abb. 1
Der hochtechnisierte Landbau widerspricht auch des Städters Vorstellung von Natur

The eminent technical agriculture also conflicting the citizen concept of nature

durch hochtechnisierten Landbau notwendige Ausräumung der Feldflur von Gehölzen und Bäumen, die Begradigung von Bächen und Flüssen widersprechen auch des Städters Vorstellungen von Natur. Hektargroße Maiskulturen, düstere Fichtenmonokulturen eignen sich schlecht als Natur-Fetisch. Und so hat man in der fernen Almwirtschaft mit ihren saftiggrünen Wiesenhängen, knorrigen Wettertannen, rauschenden Wildbächen und wiederkäuenden Kühen das neue Vorbild für Natur entdeckt.

Dies hat auch in der Gartenkultur seinen Niederschlag gefunden. Die Sterilität der Wohnblöcke vor Augen, die nachbarschaftliche Monotonie der Agrarwirtschaft und die aus der Ferne glückselig wirkende Alm lassen den häuslichen Garten, das Außenhaus, zum Landschaftssurrogat werden (WORMBS 1978). Miniaturlandschaften mit kleinen Hügeln und Senken, vereinzelte Blütensträucher und Blautannen, Wasserläufen und verschlungenen Wegen finden nach der Epoche des englischen Landschaftsgartens erneut Eingang in die Gartenarchitektur. Die Illusion eines angeblich natürlichen Landschaftsteils wird perfektioniert durch die Pflanzung niedriger Sträucher oder exotischer Zwergkoniferen, während dem wasserspeienden Frosch und dem röhrenden Hirsch mehr allegorische Bedeutung zukommt. Vor diesem Hintergrund braucht die Einmaligkeit solcher Gartengestaltung und deren Kontrastwirkung auf das Umland kaum weiter ausgeführt zu werden.

»Ich habe eine Einstweilige Verfügung gegen Sie!«

Abb. 2
Gesetzt den Fall, der Anarchist hätte einen Garten ...

Supposing the anarchist would have a garden ...

1. Der manipulierte Naturbegriff

In unserer hochtechnologischen Zeit ist Natur zur absoluten Mangelware geworden. Alles, was selten ist, erweckt Wünsche und Träume, deren sich Werbung und Industrie in besonderem Maße angenommen haben. Natur ist zum Inbegriff von Qualität geworden. Anstelle von vier Sternen wird heute das Gütezeichen Natürlichkeit verwendet. Ob Zigarette, Lederstuhl, Reinigungsmittel oder Betonwand, immer wird wieder auf ihre Natürlichkeit hingewiesen. Unter dieser Definition von Natur wird auch das Plastikteilchen eines Großcomputers zur natürlichsten Sache der Welt.

Wen wundert es da noch, daß die "Grüne Industrie" - die ja "Natur" aus Tradition gepachtet hat - ihre Hochblüte erlebt! Der Garten, das liebste Hobby der Nation (Zeitmagazin 17/1973), ist ein gar lukrativer Wirtschaftszweig geworden. Gartencenter schießen aus dem Boden wie Pilze nach einem lauen Regen, jedes Kaufhaus hat seine Gartenabteilung, Gärtnerzeitungen erreichen ungeahnte Auflagenstärken, und - das nur nebenbei - an Fachhochschulen und Hochschulen verzeichnet die Studienrichtung Landespflege Rekordanmeldungen.

Umsatz und Profit lassen sich indessen mit naturnahen Gartenformen, in denen sich Erstellung und Unterhalt in bescheidenen Grenzen halten, kaum realisieren. Deshalb wird

LANDSCHAFT + STADT 13, (3), 1981 121

Abb. 5
"Ich glaube, das ist das Reservat unserer letzten Naturschützer"

"I believe that this is the reservation of our last protectors of nature"

zen ist lächerlich. Wir denken an alle Neophyten, die standortökologisch ganz ausgezeichnet in unseren Breitengraden gedeihen wie *Robinia pseudacacia*, *Solidago canadensis*, *Impatiens glandulifera*, *Buddleia* u.v.a.

Zusammenfassend möchte ich diesen Gartentypus als untauglichen Versuch mit untauglichen Mitteln bezeichnen, Probleme unserer Zivilisation zu lösen. Ein klassischer Fall traditioneller und vermeintlich überwundener Naturschutzarbeit, die sich auf Nebensächlichkeit kapriziert und häufig vergißt, daß auch der Mensch eines sozialen und materialen Umfeldes bedarf.

2.2 Der Natur-Imitations-Typus

In diesem Gartentypus wird mit Versatzstükken von Landschaftstopoi gearbeitet. Der Weiher, der "natürlich eingebettet" zwischen sanften Hügeln sich im Winde kräuselt, ein abgestorbener Baumstamm, malerisch in die Wasserfläche ragend, bizarre Felsblöcke oder Findlinge, scheinbar zufällig und doch wohlausgewogen plaziert, die krüppelige Bergkiefer sind als Grundausstattung schon fast unentbehrlich. Für Wege und Plätze ist Naturstein in Form von Pflaster oder Platten Pflicht. Spontan aufkommende Vegetation wird zwar mitberücksichtigt, aber auch mit Pflanzungen wird kräftig nachgeholfen, damit das angestrebte Aussehen bald erreicht wird. Das Artefakt wird gezielt auf die - künstlich erstellte - Natürlichkeit ausgerichtet.

124 LANDSCHAFT + STADT 13, (3), 1981

Abb. 6
Der "Natur-Imitations-Typus"

The "nature-imitation-type"

Diese pseudonatürliche Gartengestaltung bezieht ihre Vorbilder wohl meistens aus der romantischen Landschaftsmalerei, wie wir sie von CLAUDE LORRAIN oder CASPAR DAVID FRIEDRICH kennen bzw. aus der eingangs erwähnten Mittelgebirgslandschaft. Bei CASPAR DAVID FRIEDRICH betrachtet der Mensch die Landschaft ehrfürchtig; in diesem Imitationstyp erscheint das gleiche Verhalten angebracht. Der tätige Bewohner ist zwar nicht gänzlich ausgeschlossen, aber doch nur zaghaft geduldet, weil bauliche oder nutzungsbedingte Veränderungen die Illusion der idealen Landschaft stören würden. Der Garten wird so zum Piktogramm eines vermeintlich naturnahen Landschaftsteils.

Wir halten fest, daß gerade in diesem Typus das Gestaltungsdiktat ausgeprägt vorhanden ist. Das Artefakt zielt auf Naturnachahmung, so daß das Gestaltungsdiktat bei nur flüchtiger Betrachtung kaum wahrgenommen wird. Deshalb wird ein solcher Gartentypus weitherum als Inkarnation der Natur, der Natürlichkeit aufgefaßt.

2.3 Der Nutzer-Typus

Im Unterschied zu den ersten Typen wird hier der Mensch nicht als störender Teil im Außenhaus verstanden. Vielmehr ist er integraler und mitbestimmender Bestandteil eines Freiraumkonzepts, das sich an den Anforderungen einer vielfältigen Nutzbarkeit - wie wir sie bereits einleitend formuliert haben - orientiert. Vegetationsform und -ausbildung wird über standortökologische Ausbildung, Nutzungsart und Nutzungsintensität sowie durch die ausgeführten baulichen Maßnahmen bestimmt. Die Materialwahl wird von Nutzungsanforderungen, Preis, Verfügbarkeit und Aussehen bestimmt. Dieser Typus - das scheint mir etwas entscheidend Wichtiges zu sein - animiert dank seiner scheinbaren Unprofessionalität zur eigenbestimmten Aktion, zum spielerischen Bau, zu kreativer Entfaltung der Bewohner. Das bauliche bzw. vegetative Außenhaus lebt und altert - oder verjüngt sich wieder - in Abhängigkeit mit den Bewohnern.

Die "natürlichen" Vorbilder finden wir auf Brachflächen, in immer noch vorhandenen Trümmergrundstücken, auf ungenutzten Restflächen. Der Vorteil dieses Typus gegenüber einem herkömmlichen Garten liegt im direkten möglichen Einbezug der Bewohner. LE ROY und andere haben solche Anlagen, zusammen mit den Betroffe-

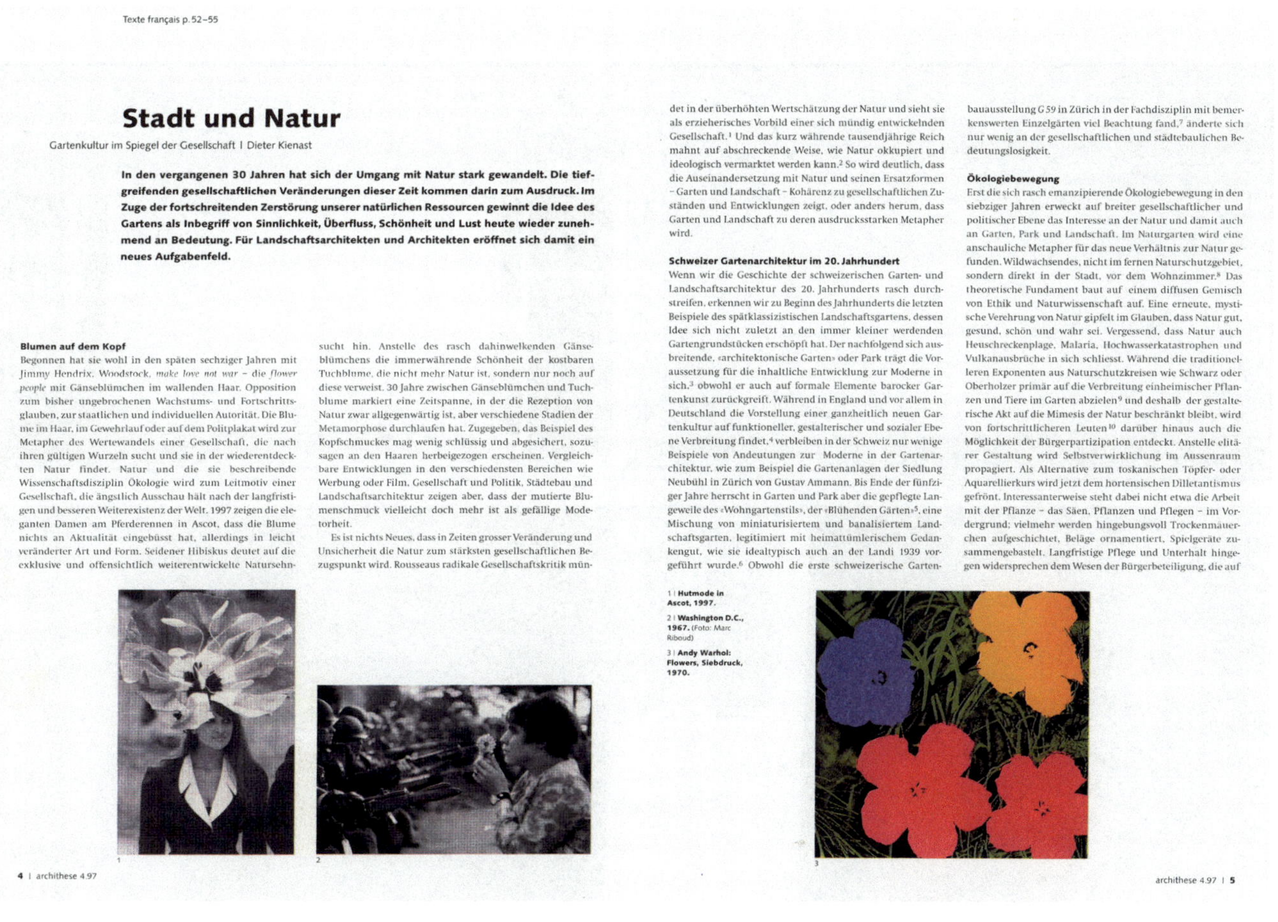

Texte français p. 52–55

Stadt und Natur

Gartenkultur im Spiegel der Gesellschaft | Dieter Kienast

In den vergangenen 30 Jahren hat sich der Umgang mit Natur stark gewandelt. Die tiefgreifenden gesellschaftlichen Veränderungen dieser Zeit kommen darin zum Ausdruck. Im Zuge der fortschreitenden Zerstörung unserer natürlichen Ressourcen gewinnt die Idee des Gartens als Inbegriff von Sinnlichkeit, Überfluss, Schönheit und Lust heute wieder zunehmend an Bedeutung. Für Landschaftsarchitekten und Architekten eröffnet sich damit ein neues Aufgabenfeld.

Blumen auf dem Kopf

Begonnen hat sie wohl in den späten sechziger Jahren mit Jimmy Hendrix, Woodstock, *make love not war* – die *flower people* mit Gänseblümchen im wallenden Haar. Opposition zum bisher ungebrochenen Wachstums- und Fortschrittsglauben, zur staatlichen und individuellen Autorität. Die Blume im Haar, im Gewehrlauf oder auf dem Politplakat wird zur Metapher des Wertewandels einer Gesellschaft, die nach ihren gültigen Wurzeln sucht und sie in der wiederentdeckten Natur findet. Natur und die sie beschreibende Wissenschaftsdisziplin Ökologie wird zum Leitmotiv einer Gesellschaft, die ängstlich Ausschau hält nach der langfristigen und besseren Weiterexistenz der Welt. 1997 zeigen die eleganten Damen am Pferderennen in Ascot, dass die Blume nichts an Aktualität eingebüsst hat, allerdings in leicht veränderter Art und Form. Seidener Hibiskus deutet auf die exklusive und offensichtlich weiterentwickelte Natursehnsucht hin. Anstelle des rasch dahinwelkenden Gänseblümchens die immerwährende Schönheit der kostbaren Tuchblume, die nicht mehr Natur ist, sondern nur noch auf diese verweist. 30 Jahre zwischen Gänseblümchen und Tuchblume markiert eine Zeitspanne, in der die Rezeption von Natur zwar allgegenwärtig ist, aber verschiedene Stadien der Metamorphose durchlaufen hat. Zugegeben, das Beispiel des Kopfschmuckes mag wenig schlüssig und abgesichert, sozusagen an den Haaren herbeigezogen erscheinen. Vergleichbare Entwicklungen in den verschiedensten Bereichen wie Werbung oder Film, Gesellschaft und Politik, Städtebau und Landschaftsarchitektur zeigen aber, dass der mutierte Blumenschmuck vielleicht doch mehr ist als gefällige Modetorheit.

Es ist nichts Neues, dass in Zeiten grosser Veränderung und Unsicherheit die Natur zum stärksten gesellschaftlichen Bezugspunkt wird. Rousseaus radikale Gesellschaftskritik mündet in der überhöhten Wertschätzung der Natur und sieht sie als erzieherisches Vorbild einer sich mündig entwickelnden Gesellschaft.[1] Und das kurz währende tausendjährige Reich mahnt auf abschreckende Weise, wie Natur okkupiert und ideologisch vermarktet werden kann.[2] So wird deutlich, dass die Auseinandersetzung mit Natur und seinen Ersatzformen – Garten und Landschaft – Kohärenz zu gesellschaftlichen Zuständen und Entwicklungen zeigt, oder anders herum, dass Garten und Landschaft zu deren ausdrucksstarken Metapher wird.

Schweizer Gartenarchitektur im 20. Jahrhundert

Wenn wir die Geschichte der schweizerischen Garten- und Landschaftsarchitektur des 20. Jahrhunderts rasch durchstreifen, erkennen wir zu Beginn des Jahrhunderts die letzten Beispiele des spätklassizistischen Landschaftsgartens, dessen Idee sich nicht zuletzt an den immer kleiner werdenden Gartengrundstücken erschöpft hat. Der nachfolgend sich ausbreitende, «architektonische Garten» oder Park trägt die Voraussetzung für die inhaltliche Entwicklung zur Moderne in sich,[3] obwohl er auch auf formale Elemente barocker Gartenkunst zurückgreift. Während in England und vor allem in Deutschland die Vorstellung einer ganzheitlich neuen Gartenkultur auf funktioneller, gestalterischer und sozialer Ebene Verbreitung findet,[4] verbleiben in der Schweiz nur wenige Beispiele von Andeutungen zur Moderne in der Gartenarchitektur, wie zum Beispiel die Gartenanlagen der Siedlung Neubühl in Zürich von Gustav Ammann. Bis Ende der fünfziger Jahre herrscht in Garten und Park aber die gepflegte Langeweile des «Wohngartenstils», der «Blühenden Gärten»[5], eine Mischung von miniaturisiertem und banalisiertem Landschaftsgarten, legitimiert mit heimattümlerischem Gedankengut, wie sie idealtypisch auch an der Landi 1939 vorgeführt wurde.[6] Obwohl die erste schweizerische Gartenbauausstellung *G 59* in Zürich in der Fachdisziplin mit bemerkenswerten Einzelgärten viel Beachtung fand,[7] änderte sich nur wenig an der gesellschaftlichen und städtebaulichen Bedeutungslosigkeit.

Ökologiebewegung

Erst die sich rasch emanzipierende Ökologiebewegung in den siebziger Jahren erweckt auf breiter gesellschaftlicher und politischer Ebene das Interesse an der Natur und damit auch an Garten, Park und Landschaft. Im Naturgarten wird eine anschauliche Metapher für das neue Verhältnis zur Natur gefunden. Wildwachsendes, nicht im fernen Naturschutzgebiet, sondern direkt in der Stadt, vor dem Wohnzimmer.[8] Das theoretische Fundament baut auf einem diffusen Gemisch von Ethik und Naturwissenschaft auf. Eine erneute, mystische Verehrung von Natur gipfelt im Glauben, dass Natur gut, gesund, schön und wahr sei. Vergessend, dass Natur auch Heuschreckenplage, Malaria, Hochwasserkatastrophen und Vulkanausbrüche in sich schliesst. Während die traditionelleren Exponenten aus Naturschutzkreisen wie Schwarz oder Oberholzer primär auf die Verbreitung einheimischer Pflanzen und Tiere im Garten abzielen[9] und deshalb der gestalterische Akt auf die Mimesis der Natur beschränkt bleibt, wird von fortschrittlicheren Leuten[10] darüber hinaus auch die Möglichkeit der Bürgerpartizipation entdeckt. Anstelle elitärer Gestaltung wird Selbstverwirklichung im Aussenraum propagiert. Als Alternative zum toskanischen Töpfer- oder Aquarellierkurs wird jetzt dem hortensischen Dilletantismus gefrönt. Interessanterweise steht dabei nicht etwa die Arbeit mit der Pflanze – das Säen, Pflanzen und Pflegen – im Vordergrund; vielmehr werden hingebungsvoll Trockenmäuerchen aufgeschichtet, Beläge ornamentiert, Spielgeräte zusammengebastelt. Langfristige Pflege und Unterhalt hingegen widersprechen dem Wesen der Bürgerbeteiligung, die auf

1 | **Hutmode in Ascot, 1997.**

2 | **Washington D.C., 1967.** (Foto: Marc Riboud)

3 | **Andy Warhol: Flowers, Siebdruck, 1970.**

1

2

3

4 | archithese 4.97

archithese 4.97 | 5

186
Dieter Kienast, "Vom Gestaltungsdiktat zum Naturdiktat; oder, Gärten gegen Menschen," *Landschaft + Stadt,* no. 3 (1981): 121 and 124

187
Dieter Kienast, "Stadt und Natur: Gartenkultur im Spiegel der Gesellschaft," *archithese,* no. 4 (1997): 4 and 5

188, 189
Christian Vogt, *Deckenbilder*
(Photographs of Ceilings), 1989

190, 191
Christian Vogt, untitled panoramas, 1989

in outdoor spaces that rejects a uniform greening of the city.[186]

Flippant everyday language conveys familiarity, yet readers get the impression that an insight based on scholarship is being formulated here. It makes the text accessible to a younger and a nonacademic audience as well.

The production of familiarity is also cultivated as a method in the project descriptions, passages of which are presented in the form of joint imaginary walks. These descriptions were published in four monographs on his works published as of 1997. In them, the physical presence of the gardens and grounds suggested by the medium of language meets the black-and-white photographs of Christian Vogt, which make Kienast's work abstract again in turn.

Gray Landscapes:
Christian Vogt's Black-and-White Photographs as Parallel Artistic Production

In 1992, Dieter Kienast exhibited for the first time works produced in the firms of Stöckli + Kienast and Stöckli, Kienast & Koeppel. The Architekturgalerie Luzern, Switzerland, published a catalog with a light gray, minimalistic cover to accompany the exhibition *Dieter Kienast: Zwischen Arkadien und Restfläche* (Dieter Kienast: Between Arcadia and Vacant Lot).[187] At first glance, all one sees on the cover is the name Dieter Kienast—in small, slender letters; the subtitle is written vertically like a ribbon along the right edge. All of the photographs for the exhibition and the accompanying catalog were taken by the internationally active photographer Christian Vogt (b. 1946). From the time of this first collaboration, Kienast had all of the other publications of projects on which he had worked documented exclusively by Christian Vogt. After Kienast's death, Günther Vogt continued to work with Christian Vogt, but the monographs on his works were published with color photographs.

The exhibition in Lucerne was presumably organized because of Kienast's extraordinary planning materials. It was one of the first and few exhibitions in which the works of a contemporary landscape architect were presented in a public space.[188] At the same time, this show, with its catchy title, motivated him to produce more artistic presentation plans like those of his own garden (112) and those of the Brühlwiese Municipal Park (62). Christian Vogt's photographs are not so much documentation of the work as an artistic production with its own logic that must be read in parallel with Kienast's oeuvre. Vogt recalled that Kienast had given him full freedom when photographing the projects selected, saying: "I do my work, and you do yours."[189]

The proposal to contact the photographer, who already at the time was famous outside of Switzerland and had been awarded several prizes, came from Günther Vogt,[190] who had noticed him as a result of a solo exhibition at the Architekturmuseum Basel.[191] Among the black-and-white photographs presented there in 1990 was a series of "ceiling photographs," in which Christian Vogt had recorded the spatial qualities of ceilings in unusual detail (188, 189), along with a series of panoramas in which interiors and their doors and windows onto spaces beyond play an particular role (190, 191). Both series make it clear that Vogt had pronounced interest in a graphic pictorial composition, in spatial structures, in depicting surfaces of different materiality, and in questions of temporality and nonsimultaneity.

In the photograph of an interior with a door onto a terrace, the two worlds are sharply separated and connected (190). The person in the room is reflected both in the glass of a picture frame and on the screen of a television—it is not only about time but also about space and about a spatial location of human beings in their environment. The viewer becomes another reference point in the space that spreads out before him or her in the photograph. The panoramas cannot be taken in with one look: the format of the panorama forces the viewer's eyes to roam. Vogt contrasts the horizontal extension with vertical spatial structures that articulate the image like cuts. He found these verticals particularly interesting, like polarities in general.[192]

If one compares Christian Vogt's search to the one that preoccupied Dieter Kienast at this time—designing with contrasts; graphic pictorial compositions and their spatial qualities; representing structures, textures, and surfaces; translating various material qualities into pictures—it seems almost inevitable that the two began to work together regularly. For the historian of photography Martin Gasser, Vogt practices "photography as an independent visual medium with specific means, one whose design strategies approached art and literature."[193] Herein, too, lies a parallel to Dieter Kienast's phenomenological strategies. Urs Stahel and Martin Heller write of Vogt's panorama photographs of 1989 and 1990 in the exhibition catalog *Wichtige Bilder: Fotografie in der Schweiz* (Significant Pictures: Photography in Switzerland):

Panorama, see "'all-round view, vista, cyclorama': origin 18th century, from Greek *pan-*, "all". . . + *horama,* "to see, the seen, the appearance" . . . . According to this, the actual meaning approximates to "showing everything." Looking at everything. Looking at, not just seeing: that would be too little, also too superficial. The gaze of the lover wants to feel what is seen confirmed with all the senses, with the entire body. Christian Vogt is a lover. His photography manifests a profound tenderness toward what he sees. Every situation he captures or forms, every portrait becomes for him the potential declaration of love to the state of being.[194]

The intensity formulated here is strongly reminiscent of Kienast's impetus to grasp the world and different facets of being, to make it visible and experienceable by the senses by means of one's own formal vocabulary. Parallels can be drawn between the work of this period by Kienast and that of Christian Vogt: in how they approached their own work and how they attempted to transfer what is perceived into a picture. Indeed, both the grounds that Kienast realized and the presentation plans he produced testify to an intense perception of the place, into which the specific garden, cemetery, park or square is fit, and of the elements designed by him, their form, their material, their effect on the senses.

Christian Vogt chose just two formats to present the works that Kienast had produced for various firms over the years: the panorama, which he had also used for the photographs in his exhibition in Basel, and the square. By his own account, Vogt was particularly interested in the way the two formats are perceived differently by the viewer:[195] The square can be grasped in one look. It is a self-contained image that refers to itself. In the panorama, the eye has to roam. The image is first scanned and grasped by the eye, bit by bit. It thus takes on an aspect of temporality.

All of the photographs were taken in black-and-white. According to Vogt, hardly any other color is as difficult to capture in all of its nuances as green.[196] The black-and-white photograph is well suited to it because it permits distinctly subtle gradations and an exploration of structures and surfaces. Vogt feels that color photographs of gardens risk becoming sweet and pleasing, especially when the sun is shining. That is why he works exclusively under overcast skies. He regards many shadows in a garden as confusing for a successful photograph, because they attract attention themselves and, in his view, distract the viewer too much.

Both the formal discipline and the choice of black-and-white technique per se—or, more precisely, the photography of shades of gray—were just as unusual in the landscape photography of this period as Kienast's vocabulary.[197] Vogt's artistic work was therefore perfectly suited as a trademark depiction of Kienast's work and later the work of Günther Vogt as well.[198] The choice fits well with the production of "gray plans," which had begun already in the mid-1980s and went hand in hand with a search for new concepts of form.

Christophe Girot (b. 1957) coined the term "gray landscapes" in this context in order to point out that, in his view, Kienast wanted to counter the exclusive green with something else that would make the dialectic of natural and artificial in landscape architecture more visible:

> We have reached the point where the word "green" has become confusing and—if I may say so—obsolete. It can neither describe nor determine our practice today. Green is far too limiting, and probably the landscape on which we work would be better and more appropriately described with the word "gray." By "gray" I don't mean a leaden gray, not the unmoving, unresponsive gray that makes one think of decay and death. Rather, I mean that several of the most impressive cultural and natural landscapes that I know are best depicted in subtle shadings of gray. . . . Gray permits thousands of nuances and possibilities; it includes everything and excludes nothing. It depicts all the components of nature side by side on a common plane of representation that would otherwise remain unnoticed.[199]

The illustration of various components on a common representational plane can be demonstrated

192
Rütihof Cemetery in Baden,
1985–90. Entrance. Photo:
Christian Vogt, ca. 1991

193
Rütihof Cemetery in Baden,
1985–90. Photo: Christian Vogt,
ca. 1991

with two photographs by Vogt that were on display in the Kienast exhibition in Lucerne (192, 193). The pictorial composition of the photographs of the extension to Rütihof Cemetery in Baden is extremely graphic. Vogt often chose a close framing, which brings out well the surfaces depicted. The fence of the cemetery that thrusts from the front edge of the image deep into the pictorial space divides the panorama into two halves and guides the viewer's gaze along with it.

On one side of the wall lies a meadow with fruit trees; on the other side, the cemetery and, following the concrete wall, a gravel area. In it stand the trees recently planted along the lane. Extending from there, separated by hedges, are the grassy areas and the graves of the cemetery. The photographs clearly reproduce the individual structure: the meadow partly covered with leaves; the smooth surface of the broad concrete wall, which barely rises above the meadow, not enormous but rather placed like a ribbon into the landscape, its slight topographical differences thus revealed by the wall; the grainy, yellow gravel areas; the dark fruit trees with their delicately ramified, bare branches; the newly planted tree-lined lane, where some of the trees still have their last leaves; the gravel areas of the cemetery; the hedges; the gravestones. Everything has its materiality and its surface structure, and the photographs make them visible.

This translation of existing materiality into a pictorial structure is interesting insofar as it is also recognizable in the plans that Kienast drew by hand, such as the project plan for the Zurzach Spa Gardens of 1985 (100) and the presentation plan for his garden on Thujastrasse of 1991 (112). On the plan for his private garden, Kienast tried to transfer the material qualities such as grain, leaf structures, or smooth surfaces to the representation by drawing and hatching with pencils of various hardness and thickness and by making use of the material quality of the cardstock itself and of the different structured surfaces on which he placed the cardstock. Later he transferred to collages his search for a vocabulary to reproduce materials in highly differentiated ways. The collaboration with Christian Vogt, in whose photography these issues of representation are so essential, inspired Kienast to find the solutions he developed in 1991. His effort to find formal correspondences in plans to the qualities of the material and of space achieved a new intensity of research at that time, which suggests processes of exchange with Vogt and his work.[200]

Another characteristic feature in Christian Vogt's photographs is the black edge of the negative that frames all of them. It confirms the detail selected by the photographer. By retaining this edge, the photographer indicates to the viewers that the framing chosen when the photographer pushed the release is the same one they are seeing here and has not been altered subsequently. This method derives from photojournalism and is intended to underscore the authenticity of a photograph.[201] In the 1980s and 1990s, the black edge became a topos of *auteur* and art photography.[202] By insisting that the frame of the negative not be cropped when exhibiting or publishing his photograph, Vogt emphasizes that the photograph is meant to be thus and not otherwise and that it was not subsequently reworked or altered by him or others.

The photographs presented in the exhibition *Dieter Kienast: Zwischen Arkadien und Restfläche* and published in the catalog reveal the spectrum of Kienast's works at the time: from the private garden as individual paradise (114) to open spaces in the agglomeration (37), which reflect hard human interventions in the landscape. Such photographs belong to the tradition of the New Topographics of the 1970s, with the difference being that Christian Vogt did not set out on his own initiative to explore the "man-altered landscape," since the photographs were commissioned.[203] Both photographs illustrate again the characteristics of Vogt's photography: the interest in structures and surfaces, the translation of many of the shades of green of a garden into shades of gray, the strict, often graphic pictorial compositions, and the mix of strong abstraction of the subject represented with its simultaneous presence.

With the help of the graphics of his plans in shades of gray and blue, Kienast was searching for artistic abstraction and, as far as possible, a presence of the materials in the drawn representation; this search continued in the collages. As an author, he tried to produce presence by means of visual description. Heinz Wirz, the owner of the Architekturgalerie Luzern and the organizer of the exhibition *Dieter Kienast: Zwischen Arkadien und Restfläche,* tells of another variation on mixed media: Kienast distributed small scented candles on top of the partition walls with the black-and-white photographs and the gray plans, the scents intended to accompany and supplement the "visual trail" of a visit to the exhibition.[204] According to Wirz, Kienast sowed the seeds of several weeds at the entrance—

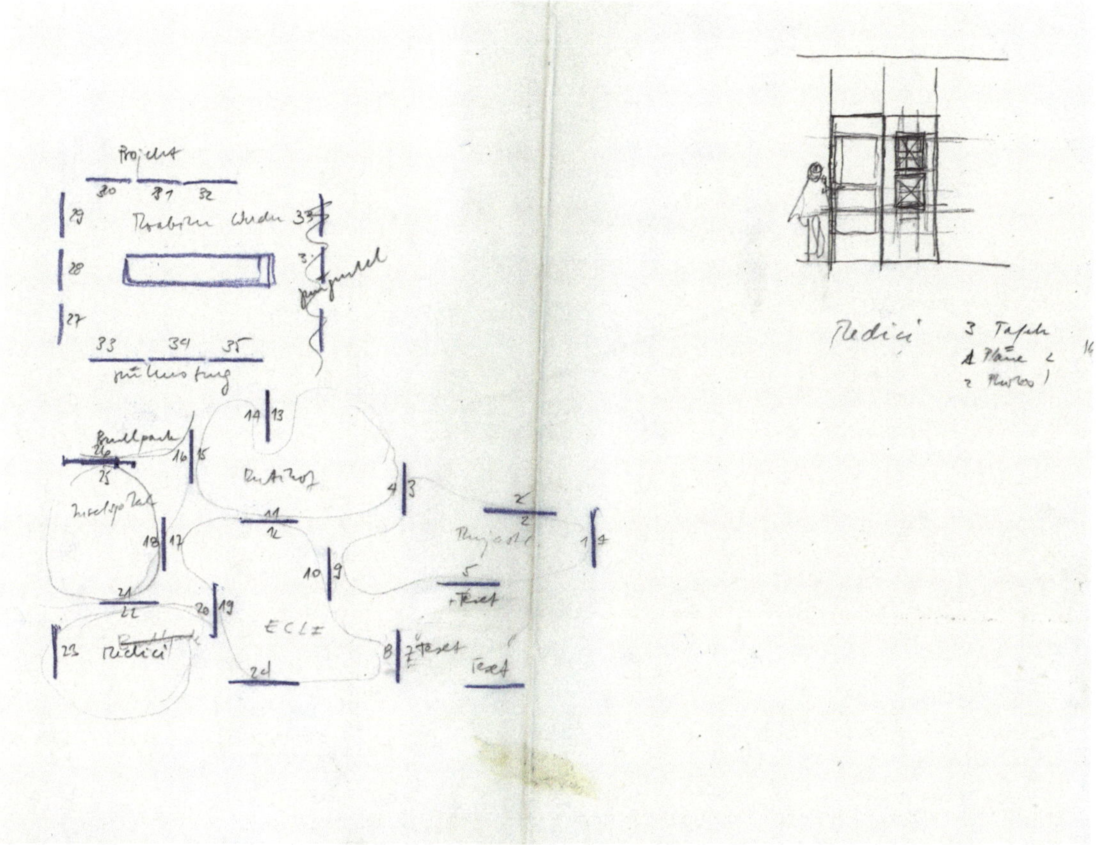

194, 195
The exhibition *Dieter Kienast: Zwischen Arkadien und Restfläche* (Between Arcadia and Vacant Lot) at the Architekturgalerie Luzern, 1992. Sketch by Dieter Kienast for the positioning of and view into the exhibition

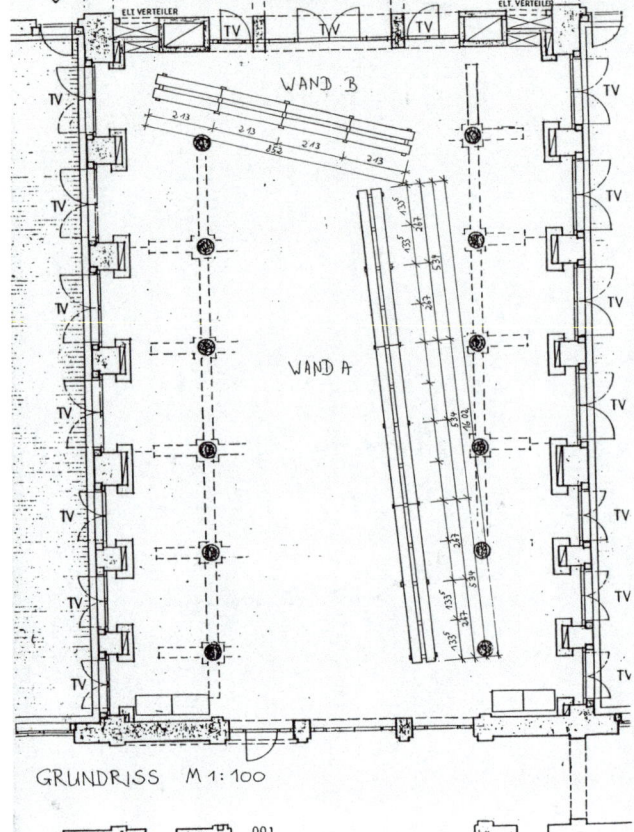

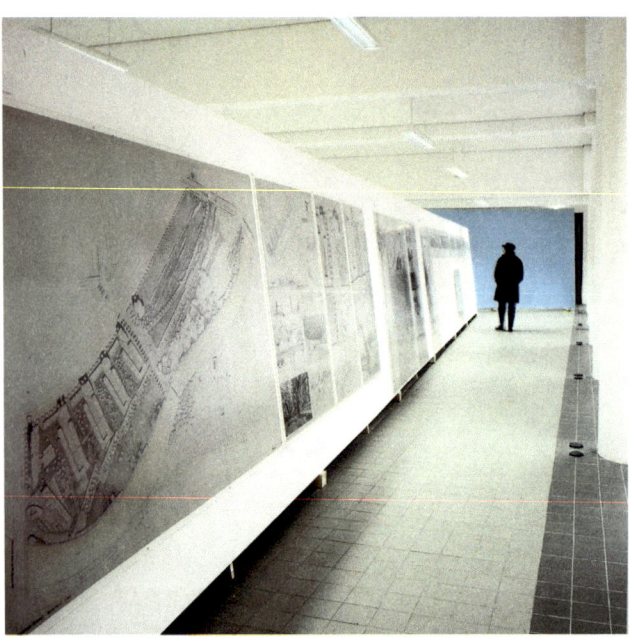

196–199
The exhibition *Dieter Kienast: Zwischen Arkadien und Restfläche* (Between Arcadia and Vacant Lot) at the Universität Karlsruhe (TH), 1993. Exhibition views and floor plan

a late trace of Kassel in Kienast's aestheticized exhibition world.

These actions open up another aspect of his work: the sensory supplement of the pictorial or the rift with the pictorial—by extending sensory perception, by the dynamic of plants, by the haptic qualities of the materials he used, or by use. The pictorial and the escape from the pictorial are always in tension in Kienast's work.[205]

On Handke's Trail:
An Exhibition as
a Stroll between
Arcadia and Vacant Lot

In Kienast's work, the pictorial is connected with movement, as it is traditionally in that of most garden architects. Consequently, he designed the structure of his Lucerne exhibition as a course, so that visiting it was like taking a stroll. The exhibition architecture seems to have been inspired by the structure of the labyrinth, which is so central to the history of the garden. As is clear from sketches in which Kienast recorded the course of movements and from slides of the installed exhibition, Kienast placed several partition walls in the room at angles that force visitors to find their way through (194, 195).

Standing prominently on a wall amid the documents on Kienast's work was the final passage from Peter Handke's *Die Lehre der Sainte-Victoire* ("The Lesson of Mont-Sainte-Victoire") (198). This essay-like text addresses the author's wrestling with capturing the landscape in language. Handke's narrator describes this difficulty using the example of his hiking around Montagne Sainte-Victoire in southern France, which was painted a number of times by Paul Cézanne.[206] The narrator sets himself the goal of doing what the painter managed—conveying reality in universal forms: his language, too, should be timeless and eternal, unfashionable and with no pragmatic function. At the end of his reflections, Handke's narrator has found a literary expression that comes close to the ideal of Cézanne. It is, however, not another, now linguistic description of Sainte-Victoire, since that would be merely a "lesson," but rather a multipage description of the Morzg Forest, near Salzburg, where Handke frequently took walks.

Visitors to Kienast's exhibition could walk along a narrow ribbon of text and read just that passage:

> In the round pond, now free of ice, the water circles almost imperceptibly. It teems with fish. Pieces of something that looks like volcanic tufa, but is actually polystyrene, are floating on the surface. At the edge of the pond, a raft hammered together from doors is rocked by a sudden wind squall as on an ocean wave. An evening shower taps lightly, kindly, on the wanderer's forehead.
>
> On the threshold between forest and village, the cobbles of the Roman road reappear. Here there's another woodpile, covered with a plastic tarp. The rectangular pile with the sawed circles is the only brightness against a darkening background. You stand there and look at it until nothing remains but colors: the forms come later. They are gun barrels pointed at the beholder, but each of them individually is aimed at something else. Exhale. Looked at in a certain way—extreme immersion and extreme attention—the interstices in the wood darken, and something starts spinning in the pile. At first it looks like a scarred piece of malachite. Then the numbers of color charts appear. Then night falls on it and then it is day again. After a while the quivering of unicellular organisms; an unknown solar system; a stone wall in Babylon. World-spanning flight, a concentration of vapor trails, and finally, a unique blaze of colors, taking in the entire woodpile, reveals the footprint of the first man. Then inhale. And away from the forest. Back to the people of the present; back to bridges and squares; back to streets and boulevards; back to stadiums and newscasts; back to bells and department stores; back to drapery and glittering gold. And the pair of eyes at home?[207]

Handke's filigreed description, which embraces all of the senses, recalls Kienast's vocabulary for the landscape. A decade apart in time, the Austrian author and the Swiss landscape architect, each in his own medium, went about establishing a connection between the individual and the experienceable world. For both, this connection resulted from the possibility of sensitive feelings and intellectual impressions in the recipient.

The year of the exhibition in Lucerne, Dieter Kienast took a position as full professor of landscape architecture at the Universität Karlsruhe (TH), where he was head of the Institut für Landschaft und Garten. One of his conditions when negotiating the position was that the exhibition would be shown at the Universität Karlsruhe as well.[208] From January 16 to 27, 1993, it could be seen in the three-nave hall of the rectangular floor plan on the ground floor of the institute at Englerstrasse 11 (196–199).

Kienast placed two partitions in the hall: a double wall 16 meters long that extended diagonally into the hall, on which the plans and photographs could be viewed on both sides; and at a right angle to it a wall 8.5 meters wide painted blue, on which, at around chest height, the quotation from Handke's *Die Lehre der Sainte-Victoire* was mounted as a continuous narrow strip.

The exhibition was choreographed as a kind of walk. The hall in which the exhibition was installed has three tall, glazed perforations with doors on each of the short ends. The central entry door on the middle axis of the hall and the right-hand door were locked and covered with white, transparent paper. The exhibition poster was prominently displayed on the center door: a composition with the minimalist dust jacket of the catalog and a poem by Hans Magnus Enzensberger.[209] The entrance to the exhibition was to the left of it. From there visitors could already see the long wall with Christian Vogt's photographs and the plans mounted beneath them. Several meters behind the entrance a curtain of strips of paper hung from the ceiling to the floor, so that the blue wall on the far end of the hall appeared only after one took a few steps.

The double partition wall, 2.15 meters tall, was composed of two plywood boards painted white and placed at a slight angle to each other to form an acute angle. The presentation plans, some of which were mounted on Plexiglas panels, and the photographs, wedged between two Plexiglas panels—always a combination of a panorama with a square photograph—were screwed to the partition wall.[210] The individual panels were arranged to create a continuous band along which visitors could walk.

When visitors arrived at the 2.7-meter-tall plywood board, which was slightly angled, painted blue, and situated at a right angle to the other wall, they saw the quotation from Peter Handke, placed at a height of 1.43 meters: a delicate white stripe with words the visitors had to decipher step by step. When they arrived at the end of the Handke text, the back of the double partition wall opened up, on which they could see depictions of the Kienast family's private garden (112) and the M.-M. Garden in Erlenbach (143), as well as the large competition plans for the Moabiter Werder in Berlin (83, 129) and the Günthersburgpark in Frankfurt am Main (84). Walking in the space and finding one's place amid the delicate structures of the photographs and presentation plans, seeking and following the Handke quotation, and physically confronting the oversized plans necessitated a constant walking forward and back and changing of perspective between close and far.

The slightly slanted, long walls, which sat on stands rather than directly on the floor, so that they seemed to float a few centimeters above it, produced a dynamic pull into the depths that animated visitors to walk quickly. By contrast, the rich structures of plans and photographs and the words of the Handke quotation could only be taken in by slowing one's gaze and pace. This staging could not be repeated in this form, because the spaces of the other institutions that were interested in taking the exhibition were not suitable for it.[211]

Abstraction and
Presence: The Monographs
on the Works

Another medium to disseminate Kienast's work in which Christian Vogt's photographs play a role were the bilingual (German and English) monographs on his works that Birkhäuser published from 1997 onward (200, 201). *Kienast: Gärten / Gardens* (1997) was published during Kienast's lifetime and edited under his direction. The plans are printed in black-and-white, in keeping with the photographs. Erika Kienast-Lüder then edited *Kienast Vogt: Aussenräume / Open Spaces* (2000) and *Kienast Vogt: Parks und Friedhöfe / Parks and Cemeteries* (2002), for which the plans were printed in color. In 2004, on the occasion of the anniversary of Birkhäuser's landscape architecture

Stadt und Natur – Gartenkultur im Spiegel der Gesellschaft

in: archithese 4 / 1997

Blumen auf dem Kopf

Begonnen hat sie wohl in den späten sechziger Jahren mit Jimmy Hendrix, Woodstock, «make love not war», mit den *flower people* mit Gänseblümchen im wallenden Haar – die Opposition zum bisher ungebrochenen Wachstums- und Fortschrittsglauben, zur staatlichen und individuellen Autorität. Die Blume im Haar, im Gewehrlauf oder auf dem Politplakat wird zur Metapher des Wertewandels einer Gesellschaft, die nach ihren gültigen Wurzeln sucht und sie in der wiederentdeckten Natur findet. Natur und die sie beschreibende Wissenschaftsdisziplin Ökologie wird zum Leitmotiv einer Gesellschaft, die ängstlich Ausschau hält nach der langfristigen und besseren Weiterexistenz der Welt. 1997 zeigen die eleganten Damen beim Pferderennen in Ascot, dass die Blume nichts an Aktualität eingebüsst hat, allerdings in leicht veränderter Art und Form. Seidener Hibiskus deutet auf die exklusive und offensichtlich weiterentwickelte Natursehnsucht hin. Anstelle des rasch dahinwelkenden Gänseblümchens die immerwährende Schönheit der kostbaren Tuchblume, die nicht mehr Natur ist, sondern nur noch auf diese verweist. 30 Jahre zwischen Gänseblümchen und Tuchblume markieren eine Zeitspanne, in der die Rezeption von Natur zwar allgegenwärtig war, aber verschiedene Stadien der Metamorphose durchlaufen hat. Zugegeben, das Beispiel des Kopfschmuckes mag wenig schlüssig und abgesichert, sozusagen an den Haaren herbeigezogen erscheinen. Vergleichbare Entwicklungen in den verschiedensten Bereichen wie Werbung oder Film, Gesellschaft und Politik, Städtebau und Landschaftsarchitektur zeigen aber, dass der mutierte Blumenschmuck vielleicht doch mehr ist als gefällige Modetorheit.

Es ist nichts Neues, dass in Zeiten grosser Veränderung und Unsicherheit die Natur zum stärksten gesellschaftlichen Bezugspunkt wird. Rousseaus radikale Gesellschaftskritik mündet in der überhöhten Wertschätzung der Natur und sieht sie als erzieherisches Vorbild einer sich mündig entwickelnden Gesellschaft.[113] Und das kurz während tausendjährige Reich mahnt auf abschreckende Weise, wie Natur okkupiert und ideologisch vermarktet

[113] Jean-Jacques Rousseau in seinen Schriften, z.B. Rousseau richtet über Jean-Jacques, oder im Roman Julie ou La nouvelle Héloïse, 1761.

Morgenstimmung in Vicenza, Seminarreise 1996; Foto: Wolfram Müller

200, 201
Monographs *Kienast: Gärten/Gardens*, 1997, and *Kienast Vogt: Aussenräume/Open Spaces*, 2000

202
The essay "Stadt und Natur" reprinted in 2002 by the Chair of Landscape Architecture at ETH Zurich in an anthology of texts, with a photograph by Wolfram Müller showing Dieter Kienast on a seminar trip to Vicenza in 1996

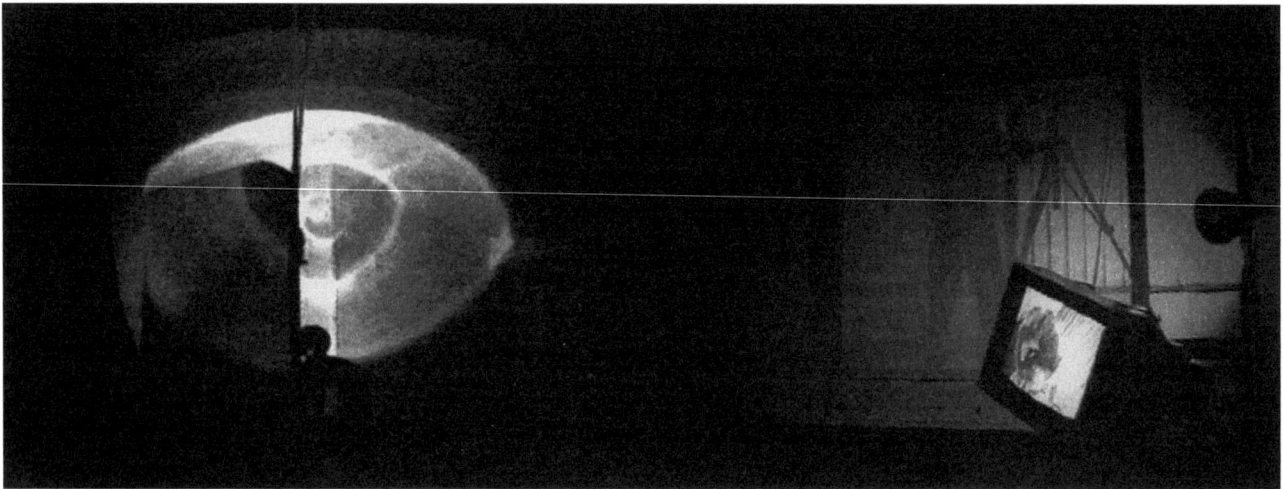

203
Sp. Garden in Egnach, presentation plan for the project, scale: 1:100, H 60 × W 120 cm, ink, film, chalk on copy of plan, 1991 and 1996 (revision), firm of Stöckli, Kienast & Koeppel

204, 205
Sp. Garden in Egnach, 1991–93. The seating areas facing Lake Constance and pathway through the reed belt to the wooden jetty. Photos: Christian Vogt, ca. 1996

206
Rudolf Manz, *Sonnenlicht-Augenlicht-Videolicht: Videobilder sind Bilder aus Licht* (Sunlight-Eyesight-Video Light: Video Images Are Pictures Made of Light), 1996

series, a best-of paperback edition compiled by Kienast-Lüder was published under the simple and signet-like title *Dieter Kienast*. It presented again selected projects from the three previous monographs, supplemented by others.[212]

Each volume begins with one or two introductory essays by guest authors, including Erik A. de Jong, Walter Prigge, Kai Vöckler, German Ritz, Robert Schäfer, and Brigitte Wormbs. Descriptions of the projects follow, most of which were written by Dieter Kienast himelf,[213] but some were written by employees of the office such as Erika Kienast-Lüder and Günther Vogt, colleagues such as Kamel Louafi and Arthur Rüegg, or former assistants such as Thomas Göbel-Groß and Udo Weilacher. The paperback edition of 2004 had additional essays by Kienast.

The series of monographs was planned soon after the exhibition in Lucerne. As part of the preparations for it, the Zurich office Stöckli, Kienast & Koeppel prepared numerous presentation plans using the collage technique. Kienast finished several of them by hand, such as the plan for the redesign in 1993 of the Sp. Garden in Egnach. The atmosphere of this garden located on Lake Constance is characterized by a broad belt of reeds. On the plan drawn by Fabienne Kienast (b. 1973), produced during her internship in her father's firm, a lot of room is left to represent the full depth of the field of reeds and the entire length of the pathway and the wooden jetty jutting out into the lake (203).[214] The lawn, the trees, the beech-leaf-shaped beds of shrubs, the flower beds, and the hedges are depicted using the familiar collage technique with different structures. The surfaces of the gravel-covered front yard and the water-filled swimming pool were marked with delicate hatching in pencil, referring in one case to gravel and in the other to water. Kienast drew the reeds himself with a soft, thick pencil. The strokes run parallel to the footbridge and are placed so that an irregular rhythm results and the strokes gather into clouds, as it were. The visual impression calls to mind the image of wind brushing through reeds and causing their stalks to wave.

The Sp. Garden was presented in the 1997 publication: under the "gray photographs" of Christian Vogt (204, 205), corresponding excerpts from Dieter Kienast's project description were placed. The combination of the description and the black-and-white photographs results in a changing mixture of abstraction and presence; text and image alternate in the role of the more strongly abstracting element. For his texts, Kienast chose the form of an imaginary walk taken with the reader. Along the way he offers objective information about the history, the categorization of natural space, and the topography of the garden, as well as the materials and plants found there and introduced in the redesign. As so often, at the end of his description he turns to the surrounding landscape to illustrate for the reader the dominant atmosphere:

> From the garden level, Lake Constance remains hidden beyond the high reeds. Its presence is manifested through the muffled sound of the waves and the unmistakable smell of the lake. A narrow pathway and wooden jetty lead through the reeds to the open water surface. Sitting on the bench on the jetty facing the land, we recognize the small silhouette of the house with the surrounding tree crowns; beneath us is the clear water and towards the lake are expansive fields of reeds. In the dim light of dusk we can just discern the remains of the distant and foreign shoreline.[215]

Kienast extremely precisely observed the landscape surrounding a garden; ultimately, these observations determined the design. On Lake Constance, he made the reed the main character of the image determining the garden: in order to hide a swimming pool constructed in front of the belt of reeds, Kienast integrated a hedge of hornbeams as partial termination for the garden. When seen from the garden, the reeds now appear to grow out of the hedge (204). From the seating areas installed under six lindens, one can meditate on this view and hear the rustle of the rees.

> The formerly uneven lawn surface becomes the horizontal base plane in its precise design above which—similar to height levels—the hedge, reeds, tree crown and sky have a significant appearance.[216]

In the monographs, the plan, text, and photograph for each project were intended to reproduce a certain existing atmosphere or the one created by the design. Depending on the object, sometimes a slight irony is heard in the narrative, or everyday experiences appear vividly to the mind's eye, such as in the description of the grounds of the Waldhaus Psychiatric Clinic in

Chur, published in the next volume, *Kienast Vogt: Aussenräume / Open Spaces* (2000). The final passage addresses the physical strain of visiting the garden, for example, tiredness and sweating as signs of physical exertion, whereby Kienast is formulating an alternative image to the deserted photographs of Christian Vogt, which are, despite the visual presence of the material, generally enraptured.

We proceed further along the circular route and on the valley side arrive at the foot of a large bastion in line with the middle of the building. The blue projection covers the wooden bench. It is hot, we are a little tired. Above us we can hear the unmistakable sounds of a garden restaurant. Somewhat overcome by the heady smell of lilac, we ascend the steps to the terrace and, relieved to find an empty table, we sit down. In the shade of the low lime trees, we enjoy an almost overwhelming view of the city, the Rhine valley and—partly obscured by the afternoon haze—the Calanda [mountain] after which the drink in front of me is named.[217]

Despite such rhetorical rifts and the colorful diversity that the collaged plans introduce to the monographs, which are otherwise illustrated exclusively with black-and-white photographs, the series of "gray monographs" has imposed a "gray aesthetic" on the reception of Kienast's work that still obscures it today.

One example of that is the collection of Kienast's essays published in 2002 by the Chair of Landscape Architecture at ETH Zurich and edited by Udo Weilacher with Thilo Folkerts (b. 1967).[218] On the one hand, this very laudable publication is eminently useful, since it brings together in one volume texts that had been scattered across many specialized journals (202). On the other hand, because the comics, photographs, and collages that originally illustrated them were omitted, their content is smoothed over, since the illustrations originally supplemented, reinforced, or counteracted the statements of the essays and conveyed their own messages (187). The essays were set in a uniform typeface in a new edition, interrupted by single photographs showing Dieter Kienast on excursions or in workshops. As a result, the content of the texts, in which Kienast expressed himself critically, amusingly, provocatively, and sometimes also sarcastically on issues of planning open spaces, is necessarily veiled by a solemn mood that strips them of some of their ambivalence and ambiguity.

Space and Video:
In Praise of Sensuousness

Dieter Kienast found his way out of the "gray aesthetic" through an interest in video. His work with this medium of the moving image led him to focus more on the question of how the components central to design in landscape architecture, such as designing a space with living materials and their phenomenological possibilities, can be conveyed. For the perception of a landscape design, weather conditions—rain, wind, snow, sun, and shade—and the changes of the day and of the seasons are elemental. The medium of video opened up new paths for presentation. The question of how to translate the qualities of materials into surface structures receded into the background, and the visualization of the spatial and temporal-dynamic aspects of a designed open space came to the fore.

During his time as a lecturer on landscape design in the Division of Architecture (since 1999: Department of Architecture) of ETH Zurich, Kienast came into contact with the work of his colleague Rudolf Manz (1932–2013). Manz had a lectureship on the recording and depicting of space using the medium of video. Kienast taught at ETH from 1985, and as part of his elective subject taught architecture students the basics of landscape architecture.[219] He also directed theses for the elective subject and supervised design diplomas, with assistance from Günther Vogt.

Rudolf Manz was working with space, the "medium of architecture."[220] He saw himself as an "inventor of space and researcher of space," and in 1996, to conclude his activity at ETH Zurich, he published his approaches and research results in a book that bore the title of his self-invented alter ego: *Vito de Onto ... seine Freunde nennen ihn Video* (Vito de Onto ... His Friends Call Him Video).[221] Under the motto "video ergo cogito," Manz presented therein the shared interests and levels of understanding between architecture and video (206). The medium of video offered him an opportunity to train and develop architectural thinking, since video is a "manipulator and inspirer of space": "Vito de Onto ... uses videomotorics to uncover spatial moods and effects and to moderate the conversation between space-shaping senses, things, and thoughts."[222] The perception of space is very closely tied to dynamics and movement:

> Viewers always experience space dynamically, that is, in the context of movement: movement of the eyes, the head, the body, and movement of countless objects and things. It is characteristic of movement that objects change their position in space and that time is necessary for that. Movement makes it possible to experience time and space in the first place, in a sense producing space and time: space and time are the parameters of architecture—and video, too, works with these parameters.[223]

Space, time, and movement are even more critical for landscape architecture than they are for architecture, since in addition to its connection to human beings and their movement, landscape architecture also works with dynamic material, and its works are totally subject to the influences of weathering. Another point of contact between Kienast and Manz was the latter's definition of "space" as a product of the physical and psychological relations of human beings and their environment. Becoming aware of this relationship requires both intellect and sensuousness. In this combination of intellect and sensuousness, the perception of the surrounding space can be expanded by analyzing the medium of the video:

> Perception includes more than taking in sense impressions; it also has a cognitive dimension. It includes searching for, finding, and giving meaning and the recognition of connections that create meaning: thinking is also part of perception.... We want to think with video—about space and architecture.[224]

This approach of understanding space as something that can be experienced through observation, personal involvement, and intellectual engagement, along with the will to study it through creative engagement, corresponds to Kienast's intention in design and representation. In all media, from drawing by way of photography to language and the staging of exhibition architecture, Kienast had a motivation similar to Manz's when it was about developing the notation and representation with which he could communicate his own experiences engaging with landscape and its influence on the design of his projects. The combination of sensory experience and intellectual intensification that Manz sought coincided with Kienast's central concern. Also, in terms of the utility and use value of an (open) space, both had similar ideas:

> Space is alive. In the process it is revealed that unpretentious spatial material, space without ambition and presumption, is particularly suited to such ideas of use. Space without aesthetic desire, raw space, raw space in an "unhewn state." That is what we turn to, that is our model.[225]

When Kienast was appointed Professor of Landscape Architecture at the Universität Karlsruhe (TH) in 1991, he saw an opportunity to get the necessary funding for his teaching in order to give new media a permanent place in it. He kept his teaching position at ETH Zurich despite his appointment in Karlsruhe, and hence maintained contact to his colleagues as well.[226] He sent his assistant in Karlsruhe, Wolfram Müller (b. 1950),[227] who had already worked for his predecessor, the Swedish landscape architect Gunnar Martinsson (1924–2012), to Zurich to have Manz advise him on the subject of video.

The experimental character of Manz's teaching and his methods were not transferred to Karlsruhe, but modern equipment was purchased on his recommendations. This included video cameras; a powerful computer; the most important multimedia programs, such as Photoshop, QuarkXPress, and more; two hard disks to store student works; a CD burner and a plotter linked to the computer; and an endoscope camera with which the students could film in their models.[228] Kienast himself tried out this technology for the Masoala Rain Forest Hall at the Zoo Zürich, on which the firm Kienast Vogt Partner was working at the time. He had constructed the model with the help of students. The photographs inside the model were taken by Wolfram Müller, who was teaching the students to shoot and edit video material in the introductory courses. Kienast was one of the first professors at the Universität Karlsruhe (TH) to introduce digitalization; after a short time, all of the plans for the design assignments were being distributed digitally,[229] and the students could submit their finished projects on a

CD-ROM. Even if Kienast himself did not switch to digital media for designing and rendering, but remained true to drawing by hand, he always wanted to have the latest equipment—whether at his chair or in the office.

When he was appointed the first Professor of Landscape Architecture at ETH Zurich, Kienast began to use video experimentally. The chair had been newly established, presumably because of the effect that Kienast's work had on the Swiss architecture scene. A decision of the school's administration on November 18, 1996, directly appointed Kienast to this position, which he officially began on April 1, 1997.[230] For the new teaching with video, he hired the architect Marc Schwarz, who had studied at ETH and, as assistant to Rudolf Manz, had moved from architecture to film by teaching himself.[231] Because Manz retired the year that Kienast took up the new position at ETH, Kienast recruited Schwarz for his chair and asked him to develop research and teaching using the medium of video. As he had previously in Karlsruhe, Kienast set aside a generous budget for it, which made it possible to work with the latest technology: several video cameras and a high-end editing table were acquired and editing areas for students were created.

In preparation for his inaugural lecture at ETH, Kienast commissioned Schwarz to film at the lake basin in Zurich. He wanted to use urban development at the lake basin to analyze how the city had grown and what phases and layers of the landscape and urban planning could be identified. According to Schwarz, he found filming landscape difficult after having filmed architecture.[232] In architecture, there are edges everywhere that offer orientation, and one can play with their shifts in the experimental analysis of space; in landscape, by contrast, nothing is static and little is vertical. Schwarz understood that it was primarily about the phenomenology of structures and minimal topographies.

To practice filming landscape, Schwarz sought out the Brühlwiese Municipal Park in Wettingen, which he regarded as a laboratory for found landscape objects: here the quotation of a little forest, there the quotation of a boulevard, at the end of it a geometric pool, lawns, hedges, and earth pyramids (pp. 200–210). In the tiniest space, the most diverse landscape elements could be found, with which Schwarz could explore recording and presenting the space of the landscape using the medium of video. Only while practicing filming the park in Wettingen did he learn that it had been designed by Dieter Kienast.[233] He then showed him his shots, and Kienast was so moved by the cinematic translation of his work that shortly thereafter he approached Schwarz with the idea of a joint exhibition.

Under the title "Lob der Sinnlichkeit" (In Praise of Sensuousness), Kienast wanted to install a presentation of his works in the main hall of the Semper building to mark the beginning of his professorship at ETH. The proposal that he presented to Schwarz for it was realized just over a year after Kienast's death, with few changes, as the exhibition *Dieter Kienast: Lob der Sinnlichkeit* (207).[234] Kienast had planned to place four rectangular "gardens" framed with green Andeer stone slabs in the central aisle of the main hall. Each would be covered with a layer of gravel in a different color. Rather than plants, the garden square would be decorated with plans and video monitors. In one of the two side aisles, Kienast wanted to place a long strip of Oriental carpets and present the works of his students between them. As it was realized, the four "gardens" were filled with white, green, gray, and black gravel, and planes and monitors were combined (208). Kienast had planned to integrate the fountain in the entrance to the hall into the first "garden." A large projection screen was hung behind the figures of the Three Graces in the fountain. The video portraits by Marc Schwarz were projected onto it (209); their subjects were gardens, parks, and cemeteries—works by the three firms Stöckli + Kienast, Stöckli, Kienast & Koeppel, and Kienast Vogt Partner. Those entering the main building of ETH were greeted by the scent of the fresh rose petals strewn on the gravel of the first garden, which were changed several times during the exhibition.

In the three other gardens, the presentation plans were spread out directly on the gravel. Kienast had wanted to protect them with unbreakable glass so that the visitors could have walked right over them (210, 211). This gesture sums up Kienast's search for a way to translate topography into the presentation: the plans themselves were lying on the "ground" of the garden, almost like microtopographies above the existing topography, as layers with history over a new layer that opens up a new meaning. The ground plans exhibited thus visually continue the topography of the exhibition floor. The perspectives and sections, in turn, are connected to the cinematic portraits of the projects, which could be seen on monitors placed at an angle in the gravel (212).

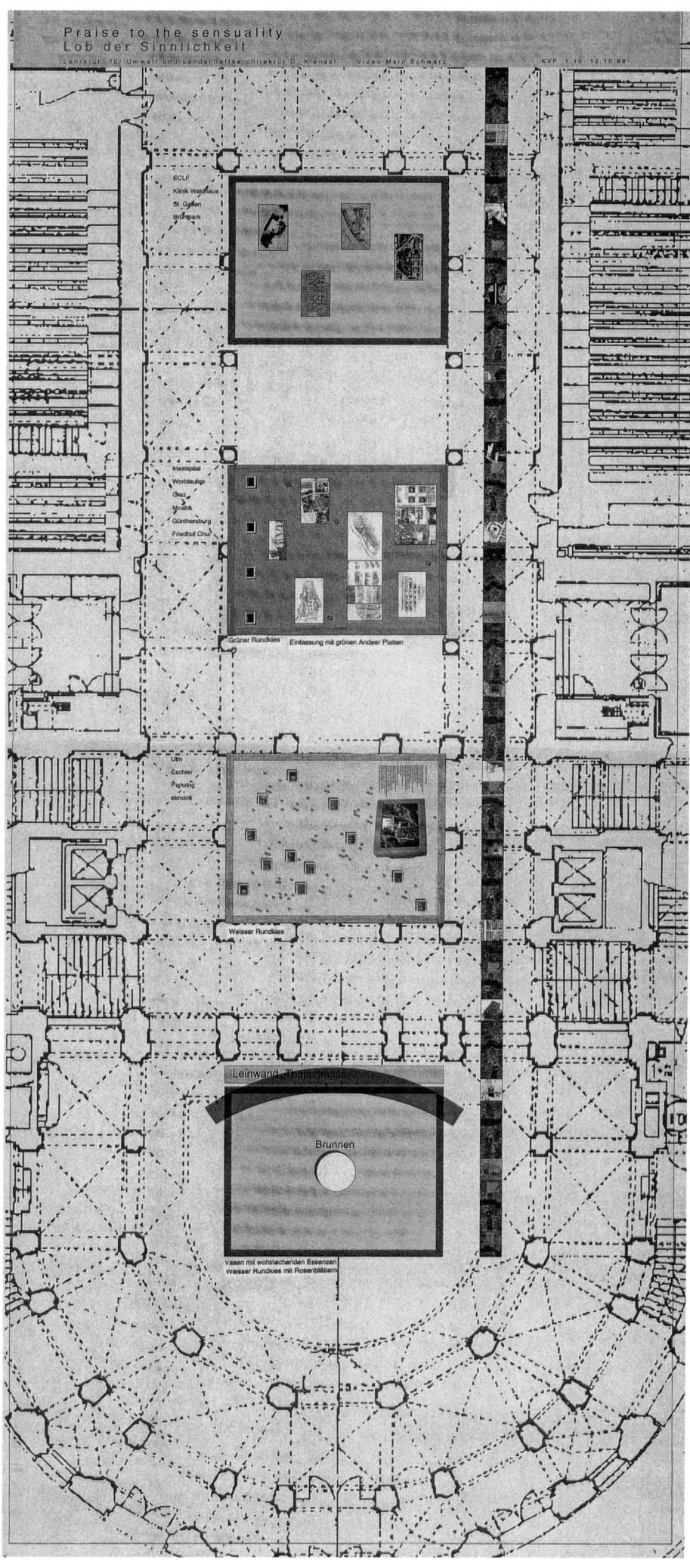

207
Exhibition *Dieter Kienast: Lob der Sinnlichkeit* (In Praise of Sensuousness), hall of the main building of ETH Zurich, floor plan of the exhibition architecture, scale: 1:10, October 12, 1998, Lehrstuhl für Umwelt und Landschaftsarchitektur D. Kienast / Büro Kienast Vogt Partner

208–211
Exhibition *Dieter Kienast: Lob der Sinnlichkeit* (In Praise of Sensuousness) in the hall of the main building of ETH Zurich, 1999. The floor plans of the projects lying on gravel continue the topography of the exhibition gardens. Videos of the projects in the exhibition where shown on monitors. The fountain in the main entrance was integrated into the exhibition architecture.

The video portraits of the grounds convey things that only the moving medium of film can make visible and audible: the rustling of a Lombardy poplar in front of a high-rise, the rippling of the surface of the water in a concrete pool, the ringing of cowbells in front of the cemetery in Chur, a park going from winter to spring when the snow has melted and the plants are budding, the change in a garden when a cloud passes overhead stripping the trees and hedges of their shadows . . . In his portraits, Marc Schwarz combines shots in which the camera remains still for as long as ten seconds, and only the leaves of a hedge move, with sequences in which he goes through the garden with the camera, and as a result the space seems to spin itself around. He makes slow dissolves from a winter image to a summer image or uses a dissolve to make autumn foliage appear on a meadow.

Several of the shots are very closely framed. One of Dieter Kienast's preferred design elements, the rows of hedges behind one another, then seem to be pressed into the images: all that can be seen is their layer, no beginning, no end. Then the camera opens up the expanse of a landscape in which grounds such as Fürstenwald Cemetery are standing. The space focused and constructed by the framing of the shot permits concentration on details or on the existing relationships between architectural elements that would escape the viewer without the pictorial space of the video. Distortions from reflection or changes in duration are perceived as a result of the intensification of the form of representation, for example the crunching of gravel under the feet of the cameraman moving through the space. Marc Schwarz's video portraits implement the teachings of Rudolf Manz.

Video is a challenge to the imagination. It evokes ideas and expectations, shapes attitudes and stances. Video is a provider of impulses, a stimulus. Video points to unexpected thoughts and perceptions, and making or watching it triggers an autonomous act of the imagination. The video impulse influences our behavior of perception and apperception—also in relation to space and architecture.

Video seeks and finds the dynamic: sun, rain, clouds, people, light, and shadows and everything that is in motion—even the course of time.[235]

In the context of his affinity to Handke's poetic effort in *Die Lehre der Sainte-Victoire,* Kienast's enthusiasm for the medium of video, which opens up a new dimension of perception, is another attempt to make it possible to experience landscape through the senses.

Marc Schwarz put his video portraits together into a film, which was pressed on DVD and included in the exhibition catalog *Dieter Kienast: In Praise of Sensuousness.*[236] In the exhibition itself, visitors walked around the individual "gardens," between the plans and monitors, alternating between visual and haptic experience, with some children literally touching the gardens and making castles from the gravel. Kienast's proposed ideas of having small containers releasing different scents and placing small bouquets of flowers on the gravel were not implemented.

By presenting works by his students from two decades of teaching—the Interkantonales Technikum Rapperswil, the Universität Karlsruhe (TH), and ETH Zurich—in the places between Oriental carpets, Kienast was referring to an old topos in the history of the garden: the linking of garden and Oriental carpet. The French philosopher Michel Foucault devoted a separate observation to it in his canonic text on heterotopias:[237] both gardens and carpets are traditionally rectangular. In the center of the garden is a fountain—very much like the one Kienast had conceived for the entrance to the exhibition—and the plants around it exemplify the vegetation of the world. The garden expresses a perfect order. This structure is repeated in the patterns of the carpets:

> If one considers that Oriental carpets were originally reproductions of gardens—"winter gardens" in the strict sense of the term—one understands the legendary value of flying carpets, of carpets crossing the world. The garden is a carpet on which the entire world ends up fulfilling its symbolic perfection, and it is at the same time a mobile garden crossing space. Was it a park or carpet, this garden that the narrator of *A Thousand and One Nights* describes? We see that all the beauties of the world are collected in this mirror.[238]

Foucault's sentences express the utopian goal that Kienast was approaching in the years before he died unexpectedly: making it possible to experience the beauty of the world in the mirror of the garden; symbolically arranging the world

212
St. Gall Municipal Park; Mountain Garden at the IGS 2000 in Styria; Ground Wave at the Allenmoos Outdoor Swimming Pool in Zurich; grounds of the Lory Hospital in Bern; Fürstenwald Cemetery in Chur; the Kienasts' Garden in Zurich. Video stills from Marc Schwarz's film *Dieter Kienast: In Praise of Sensuousness,* 1999

to perfection in a garden plan; believing in the garden as a space that can open up other spaces of the imagination. That was Kienast's utopia of the 1990s. In his finished works and in the media for representing them, he tried to approach it: through images, haptics, the senses, and language. Kienast's utopian impetus reflects not least his now most often quoted sentences, set down in his text "Die Sehnsucht nach dem Paradies" (Longing for Paradise) of 1990:

The garden is the last remaining luxury of our time as it demands something which has become rare and precious in our society: time, attention and space. It is the representative of nature, where we again use mind, knowledge, and manual skills in a careful approach to the world and its microcosm. . . .

Emphasis should be on its need for attentive care not easy care. Physical work is only one aspect, the other is dialogue, one is tempted to say sympathy for and involvement with this strange object that we call a garden. Having a garden is not primarily defined by the size of the plot of land but also by the extent of the approach to it.

Kienast's text ends with the existential demand "Do you want to have a garden?"[239]

The final question also helps to fathom the decisions we make in life—annually, weekly, daily. That is perhaps Dieter Kienast's most powerful legacy: a query that has to be posed again and again and answered anew—as a personal and social barometer.

1. Confirmed by oral testimony from Heidrun Hubenthal, Kassel, and Roman Raderschall, Meilen. On the latter, see the interview in Marc Schwarz and Annemarie Bucher, *D. K.: Eine Spurensuche,* DVD (Zurich: schwarzpictures, 2008).
2. I am grateful to Heidrun Hubenthal, Kassel, for this information.
3. In reaction to a lecture in which the author told this anecdote, Günther Vogt, a partner at Kienast's firm, said it had made sense for Kienast to have drawn that plan, because he was probably the only one in the atmosphere in Kassel at the time, which was characterized by discussions and actions, who could draw that well. The lecture and podium discussion were part of the symposium "Natur entwerfen: Zur Aktualität des Werks von Dieter Kienast (1945–1998)," December 5, 2008, at Schaulager in Basel.
4. "Partizipation: Grün in die Nordstadt," *Der Monolith: Studentische Zeitschrift der Organisationseinheit Architektur, Stadt- und Landschaftsplanung, Gesamthochschule Kassel* 12 (December 1977), title page.
5. Karl Heinrich Hülbusch, "Notizbuch der Kasseler Schule: Programmatische Anmerkungen," in *Krautern mit Unkraut oder: Gärtnerische Erfahrung mit der spontanen Vegetation,* Notizbuch der Kasseler Schule 2 (Kassel: Arbeitsgemeinschaft Freiraum und Vegetation, 1986), 158–63, here 159.
6. Reto Mehli, "Das Lei(d)tbild 'Landschaft': Zur Kritik ästhetischer Leitbilder in der Gartenarchitektur," in *Reise oder Tour?,* Notizbuch der Kasseler Schule 26 (Kassel: Arbeitsgemeinschaft Freiraum und Vegetation, 1992), 128–56.
7. Eduard Neuenschwander in conversation with Christophe Girot, Peter Märkli, Claudia Moll, and Axel Simon, October 1, 2007, sound recording, Archiv Moll/Simon.
8. These are themes of postmodern philosophers such as Jacques Derrida; see, for example, Jacques Derrida, "Signature Event Context," in *Margins of Philosophy,* trans. Alan Bass (Chicago: University of Chicago Press, 1982), 307–30.
9. See, for example, Jacques Derrida, "Différance," in *Margins of Philosophy,* 3–27.
10. From the publications on Kienast from 1997 to 2004, we are familiar only with the delicately drawn gray and blue plans and the collages and software-based plans. When the author showed the early felt-tip-pen drawings in a lecture, several participants in private conversation afterward made no secret of their displeasure about this supposed undermining of the traditional image of Kienast as the author of beautiful plans. For the lecture, see Anette Freytag, "Siedlung Lindenstrasse/Huebwiesen in Niederhasli (1972–75) von Dieter Kienast und Peter Paul Stöckli," Rapperswiler Tag, theme "Wohnen!" (Housing!), Hochschule für Technik Rapperswil, May 16, 2008, (http://www.rapperswilertag.ch/r08/downloads08/05_RT08_Freytag.pdf (link no longer active).
11. For more detail on this, see chapter II in this volume, 219–29.
12. Arthur Rüegg in conversation with the author, October 8, 2010.
13. An illustration for the ground plan can be found, for example, in *Kienast: Gärten / Gardens* (Basel: Birkhäuser, 1997), 42.
14. Martin Steinmann, "The Presence of Things: Comments on Recent Architecture in Northern Switzerland," trans. Ingrid Taylor, in *Construction, Intention, Detail: Five Projects from Five Swiss Architects / Fünf Projekte von fünf Schweizer Architekten,* ed. Mark Gilbert and Kevin Alter, 2nd ed. (Zurich: Artemis, 1994), 8–25, here 9.
15. Ibid., 11.
16. Bruno Reichlin in 1993, quoted in ibid., 12. Steinmann specified in note 4 of his text that Reichlin's semiotic studies since the 1970s had made a crucial contribution to the evolution of architecture in German-speaking Switzerland.
17. For more detail on this, see chapter II in this volume, 341–57.
18. Fabienne Kienast Weber in conversation with the author, November 11, 2010.
19. Dieter Kienast, "Von der Notwendigkeit künstlerischer Innovation und ihrem Verhältnis zum Massengeschmack der Landschaftsarchitektur," lecture in November 1991, in *Choreographie des öffentlichen Raumes,* ed. Jürgen Wenzel (Berlin: Jürgen Wenzel, 1994), 52–64, here 58.
20. Christel Sauer, "Entering Twentieth Century Art," in *Main Stations: Newman, Pollock, Beuys, Broodthaers, Klein, Warhol, LeWitt, Johns, Stella, Ryman, Kounellis, Nauman, Weiner,* ed. Urs Rausmüller and Christel Sauer, exh. cat. (Schaffhausen: Neue Kunst Bücher, 1995), 7–19, here 17.
21. Ibid.
22. Ibid.
23. Ibid.
24. See Thesis 6 of Dieter Kienast, "Zehn Thesen zur Landschaftsarchitektur," in *Dieter Kienast: Die Poetik des Gartens; Über Ordnung und Chaos in der Landschaftsarchitektur,* ed. Professur für Landschaftsarchitektur ETH Zürich (Basel: Birkhäuser, 2002), 207–10, here 208; see appendix, 403.
25. Urs Raussmüller quoted in Christel Sauer, "Carl Andre: Cuts, 1967," in *Carl Andre: Cuts,* ed. Urs Raussmüller and Christel Sauer (Basel: Raussmüller Collection, 2011), n.p.
26. Carl Andre, quoted in Maurice Berger, *Robert Morris: Minimalism and 1960s* (New York: Harper & Row, 1989), 92.
27. Kienast, "Zehn Thesen zur Landschaftsarchitektur," 207; see appendix, 403.
28. See Karl Heinz Bohrer, "Nach der Natur: Ansichten einer Moderne jenseits der Utopie" in *Über Politik und Ästhetik* (Munich: Hanser, 1988), 209–29.
29. Viktor Shklovsky, "Art as Device (1917–1919)," in *Viktor Shklovsky: A Reader,* ed. and trans. Alexandra Berlina (New York: Bloomsbury Academic, 2017), 73–96, here 80. Shklovsky's essay was first published in 1916. The Western reception of Russian formalism began in the 1960s in the context of French structuralism.
30. Christel Sauer, "Joseph Beuys," in Rausmüller and Sauer, *Main Stations,* 20–25 and 21.
31. Joseph Beuys, quoted in ibid.
32. Ibid.
33. For example, in the action *7000 Eichen* (7000 Oaks) in Kassel: one of the basalt stelae stacked

in front of the Fridericianum was removed for every tree that was planted in the city and rammed into the soil of the city next to the growing and changing trees. For Beuys, the inorganic basalt stelae were "landmarks" of the "symbolic beginning" to point to a "redesign of all of life, all of society, all of ecological space." See Marion Kittelmann, "Heils-Wahn und Zeit-Schwindel im Werk von Anna & Bernhard Blume," PhD diss., Universität Wuppertal, 2002, http://nbn-resolving.de/urn:nbn:de:hbz:468-20020428, 86. The growing trees became "elements of regeneration" and "forms of sculpture": "It is about the metamorphosis of the social body in itself to bring it to a new social order for the future." Joseph Beuys, in Richard Demarco, "Conversations with Artists: Richard Demarco Interviews Joseph Beuys, London (March 1982)," *Studio International* 195, no. 996 (September 1982): 46–47, here 46.

34 The halls were closed in 2014 in the context of a legal battle over the Beuys work, and the art collection, with the exception of that work, was moved to Basel, where Christel Sauer and Urs Raussmüller built a new exhibition site. *Das Kapital Raum* was sold to the Berlin collector Erich Marx.

35 Urs Raussmüller and Christel Sauer, "Ich mag Kunst, die nicht lügt," interview in *Du: Das Kulturmagazin* 796 (2009): 72–79, here 73.

36 Sauer, "Joseph Beuys," 25.

37 Urs Raussmüller and Christel Sauer, eds., *Mario Merz: Senza titolo* (Basel: Raussmüller Collection, 2011), n.p.

38 Ibid.

39 Günther Vogt described the experience in this way in an interview. See "'Actually, the Case Studio Is My Secret Garden': On Chance, and Chance Encounters," in Günther Vogt, *Landscape as a Cabinet of Curiosities: In Search of a Position*, ed. Rebecca Bornhauser and Thomas Kissling, trans. Jennifer Taylor and Roderick O'Donovan (Zurich: Lars Müller, 2015), 129–70, here 136–37.

40 Sauer, "Ein Einstieg in die Kunst des 20. Jahrhunderts," 15.

41 Ibid.

42 Ibid.

43 They are assigned to movements as different as Pop Art, Conceptual Art, Arte Povera, Minimal Art, and Land Art. See Thesis 6 in Kienast, "Zehn Thesen zur Landschaftsarchitektur," 208; see appendix, 403.

44 For example, the first proposal for the Fürstenwald Cemetery in Chur proposed a retaining wall of natural stone that would run along the contour line of the existing topography in order to make it visible.

45 For example, in the K. Garden on the Parkring in Zurich—see Anette Freytag, "Garden K in Zurich by Kienast Vogt Partner," in *Best Private Plots / Die besten Gärten: Internationale Beispiele zur Gartenarchitektur 2006* (St. Pölten: Land Niederösterreich, Abteilung Umweltwirtschaft und Raumordnungsförderung, 2006), 20–22—and in the K.-U. Garden in Maur. Günther Vogt, who regards contemporary art as an important stimulus for his work, used rammed-earth walls for the exterior grounds of the Novartis Campus in Basel, for example. See Alice Foxley, *Distance and Engagement: Walking, Thinking and Making Landscape; Vogt Landscape Architects* (Baden: Lars Müller, 2010).

46 Christel Sauer in conversation with the author, April 29, 2012.

47 Sauer, "Ein Einstieg in die Kunst des 20. Jahrhunderts," 17.

48 Ibid.

49 Philip Ursprung, *Die Kunst der Gegenwart: 1960 bis heute* (Munich: Beck, 2010), 39.

50 Robert Smithson, "Entropy and the New Monuments," in *The Writings of Robert Smithson*, ed. Nancy Holt (New York: New York University Press, 1979), 9–18, here 13.

51 Sol LeWitt, in "Sol LeWitt by Saul Ostrow," interview, *Bomb* (Fall 2003): 25–29, here 29, http://bombmagazine.org/article/2583/sol-lewitt.

52 *Skulptur im 20. Jahrhundert*, exh. cat., Basel (Basel: n.p., 1984).

53 Dieter Kienast and Guido Hager, *Gartenarchitektur—Freiraumgestaltung: Nachdiplomstudiengang 89/98*, Schriftenreihe Abteilung Grünplanung, Landschafts- und Gartenarchitektur ITR— Ingenieurschule Interkantonales Technikum Rapperswil 1 (Rapperswil: Abteilung Grünplanung, Garten- und Landschaftsarchitektur, 1992), 73–101, here 73.

54 *Skulptur im 20. Jahrhundert*, 168.

55 Jacques Herzog, "Mit allen Sinnen spüren," interview by Sabine Kraft and Christian Kühn, *ARCH+* 31, no. 142 (1998): 32–40, here 32.

56 Ibid., 34.

57 Ibid., 39.

58 Ibid., 34.

59 Ibid., 32.

60 Thesis 6 of Kienast, "Zehn Thesen zur Landschaftsarchitektur," 208; see appendix, 403.

61 Herzog, "Mit allen Sinnen spüren," 33–34.

62 Initially, plants did not grow at all, until the biologist and urban ecologist Klaus Ammann was brought in. He had small grates built into the metal gutters and attached various mosses on the column to aid growth, which soon had an effect; see Michele Andina, *The Oppenheim Fountain*, short documentary film, swissinfo.ch, 2012, https://www.youtube.com/watch?v=Zl69u2wyk94. In the meanwhile, the fountain is overgrown with tufa and vegetation; the rampant tufa is regularly removed for reasons of safety.

63 Herzog, "Mit allen Sinnen spüren," 34.

64 Ibid., 35–36.

65 See Anette Freytag, "Topology and Phenomenology in Landscape Architecture," in *Landschaftsarchitekturtheorie: Aktuelle Zugänge, Perspektiven und Positionen*, ed. Karsten Berr (Wiesbaden: Springer, 2018), 195–225, here 201–2. For his *murs végétaux*, Patrick Blanc uses PVC sheets into which he places plants, which then have to be changed regularly so that the green walls do not become unsightly. Drinking water is usually used to water them.

66 Steinmann, "The Presence of Things," 24.

67 I am grateful to Tarek Münch for many productive suggestions regarding Peter Handke.

68 Peter Handke, "The Long Way Around," in *Slow Homecoming*, trans. Ralph Manheim (New York: Farrar, Straus, and Giroux, 1985), 3–138, here 3.

69 Johann Wolfgang von Goethe, *Italian Journey*, trans. W. H. Auden and Elizabeth Mayer (New York: Penguin, 1970), 125.

70 Peter Handke, "The Lesson of Mont Sainte-Victoire," in *Slow Homecoming*, trans. Ralph Manheim (New York: Farrar, Straus, and Giroux, 1985), 139–212.

71 Ibid., 186–87.

72 Ibid., 159.

73 Peter Handke, *Die Geschichte des Bleistifts* (Frankfurt am Main: Suhrkamp, 1985), 320.

74 Ibid., 269.

75 Ms. E. (June 20, 2008) and Marc Schwarz (February 15, 2011) in conversation with the author. Kienast advised the video filmmaker Marc Schwarz to observe a landscape or place for hours before filming it in order to allow it to affect him.

76 The French translation of Handke is quoted in Georges Descombes, *Voie suisse: L'itinéraire genevois; De Morschach à Brunnen* (Geneva: Republic and Canton of Geneva, 1991), 7. Peter Handke, *Phantasien der Wiederholung* (Frankfurt am Main: Suhrkamp, 1983), 57. The Swiss Path was established in central Switzerland in 1991 on the occasion of the 700th anniversary of the Swiss Confederation. As national hiking route number 99, it goes along Lake Urner, which is part of Lake Lucerne. Every canton was allowed to design part of the 35-kilometer-long path. The Canton of Geneva selected Descombes to design the "Geneva stretch" from Morschach to Brunnen. Almost nothing remains today of his interventions, which were documented in the publication in 1991.

77 Descombes, *Voie suisse*, 61.

78 Georges Descombes in conversation with the author, August 5, 2012.

79 *Groupement Superpositions: Schulthess Gartenpreis 2012 / Prix Schulthess des Jardins 2012* (Zurich: Schweizer Heimatschutz, 2012).

80 Georges Descombes, "Superposition," lecture at the conference "Thinking the Contemporary Landscape: Positions and Oppositions," Hanover-Herrenhausen, June 20, 2013, http://www.video.ethz.ch/conferences/2013/ila/05_thursday/2ea2d207-bec0-44de-89bc-c503917d1ca5.html.

81 Ibid.

82 Quoted in "'Design Can Only Be Discussed between People on an Equal Footing': Designing, Debating, Cooking, and Traveling," in Bornhauser and Kissling, *Landscape as a Cabinet of Curiosities*, 91–128, here 100.

83 Ibid.

84 See also *Von Büchern und Bäumen: Vogt Landschaftsarchitekten / About Books and Trees; Vogt Landscape Architects*, exh. cat. (Basel: Architekturmuseum Basel, 2004); Kirstin Feireiss and Hans-Jürgen Commerell, ed., *Lupe und Fernglas: Miniatur und Panorama; Vogt Landschaftsarchitekten / Magnifying Glass and Binoculars; Miniature and Panorama*, exh. cat. (Berlin: Aedes, 2007); Günther Vogt, ed., *Miniature and Panorama: Vogt Landscape Architects; Projects 2000–12*, trans. Laura Bruce and Steven Lindberg (Baden: Lars Müller, 2012).

85 Quoted in Dieter Kienast, "Garten E. am Uetliberg," in *Gute Gärten: Gestaltete Freiräume in der Region Zürich* (Zurich: Bund Schweizer Landschaftsarchitekten und Landschaftsarchitektinnen [BSLA], Regionalgruppe Zürich, 1995), 31.

86 For example, in Kienast, "Where Is Arcadia?," in *Kienast: Gärten / Gardens* (Boston: Birkhäuser, 1997), 82–95, here 92–93, and Dieter Kienast, "Cultivating Discontinuity," in *Between Landscape Architecture and Land Art*, ed. Udo Weilacher (Boston: Birkhäuser, 1996), 137–56, here 155.

87 See, for example, Udo Weilacher, "Die Sinnlichkeit architektonischer Strenge: Zum Tod des Landschaftsarchitekten Dieter Kienast," *archithese* 29, no. 1 (1999): 74–75, here 74.

88 Brigitte Wormbs, "Diction and Contradiction: Ways of Reading History and Tales," in *Kienast: Gärten / Gardens*, 8–15, here 14.

89 *Skulptur im 20. Jahrhundert*, 214–15.

90 Kienast, "Von der Notwendigkeit künstlerischer Innovation," 60.

91 Illustrated in *Kienast: Gärten / Gardens*, 123. On Bomarzo, see, among others, Horst Bredekamp, *Vicino Orsini und der heilige Wald von Bomarzo: Ein Fürst als Künstler und Anarchist*, 2 vols. (Worms: Werner, 1995).

92 There are illustrations of the viewing platform in *Kienast: Gärten / Gardens*, 131–35.

93 See also Stefanie Krebs, *Zur Lesbarkeit zeitgenössischer Landschaftsarchitektur: Verbindungen zur Philosophie der Dekonstruktion*, Beiträge zur räumlichen Planung 63 (Hanover: Institut für Grünplanung, 2002).

94 See, for example, Brigitte Franzen, *Die vierte Natur: Gärten in der zeitgenössischen Kunst* (Cologne: König, 2000), 207. The present author has also encountered this skepticism again and again when giving talks.

95 Kienast, "Where Is Arcadia?" In his idea of the garden published in 1995, in only one sentence does he speak of "letter balustrade," leaving the words themselves unmentioned. Dieter Kienast, "In Praise of Ambiguity," in *Kienast: Gärten / Gardens*, 120–35, here 131.

96 The concept of "three natures," which can be traced back to the Renaissance theorists Jacobo Bonfadio and Bartolomeo Taegio, who appealed to Cicero, was reintroduced to the discussion by the historian John Dixon Hunt: Cicero called the cultivated agricultural landscape "altera natura," in order to distinguish it from the "prima natura," the wilderness untouched by human beings, in which he included the forest, the mountains, and the sea. In the Renaissance, when searching for a word for the nature of the garden, which no longer served an economic purpose, a nature associated with art, it was then called the *terza natura*. See John Dixon Hunt, "The Idea of a Garden and the Three Natures," in *Greater Perfections: The Practice of Garden Theory* (Philadelphia: University of Pennsylvania Press, 2000), 32–75, here 32–34.

97 Ms. E. in conversation with the author, June 20, 2008.
98 The phrase is noted on the plan of the preliminary project.
99 Kienast, "Where Is Arcadia?," 85. Kienast wrote the text in the first-person plural. When conceiving the project, he surely consulted with Günther Vogt, Erika Kienast-Lüder, and other members of the office.
100 See Arthur Rüegg, speech at the memorial for Dieter Kienast, ETH Zurich, typescript, Zurich, January 23, 1999, pp. 1, Archiv Arthur Rüegg.
101 The work required to move the soil was, according to Ms. E., considerable and was carried out by the Oeschger garden construction company. The contour lines on the drawing make it possible to assess the magnitude of the intervention.
102 Ms. E. in conversation with the author, June 20, 2008.
103 For Sacha Fahrni and Christian Gut, who analyzed the E. Garden in 2008–2009 as part of a diploma thesis in an elective subject at ETH Zurich, the garden is "like a shell . . . that holds people"; the "garden site" is like a "Greek theater." See Sacha Fahrni and Christian Gut, "Gartenräume von Dieter Kienast: Der Gartenraum im Verhältnis zur Landschaft," diploma thesis in an elective subject, ETH Zurich, Chair in Landscape Architecture, Prof. Christophe Girot (supervision: Anette Freytag), 2009, 34–35.
104 Ibid., 36. There was a technical, engineering reason for the side of the pool to be angled: it absorbs the water pressure.
105 Ibid., 35–36.
106 Nicole Newmark worked for several years in the firms of Stöckli, Kienast & Koeppel and Kienast Vogt Partner; her specialty was designing shrub beds and mixed borders.
107 The white stripes on the slabs of the garden paths that are visible in the photographs are special traces of use. They helped the owner, who had been visually impaired for many years, find her own way through the garden.
108 See, among others, Hermann von Pückler-Muskau, *Hints on Landscape Gardening: English Edition with the Hand-Colored Illustrations of the Atlas of 1834,* ed. John Hargraves (Basel: De Gruyter, 2014; reprint of 1st ed. of 1834).
109 Kienast, "Where Is Arcadia?," 84.
110 Ms. E. in conversation with the author, June 20, 2008.
111 Kienast, "Where Is Arcadia?," 83–84.
112 See, among others, John Dixon Hunt and Peter Wills, eds. *The Genius of the Place: The English Landscape Garden, 1620–1820* (Cambridge, MA: MIT Press, 1993). The English landscape architect was known as "Capability Brown" because he studied the "natural capabilities" of the countryside he designed.
113 Quoted in Kienast, "Where Is Arcadia?," 85.
114 Ibid. One of the paintings is in the Musée du Louvre in Paris; see ibid., 84.
115 Ibid., 85. Kienast refers to Erwin Panofsky.
116 Ibid. Kienast refers to Herder, Schiller, and Goethe.
117 Oral request to the author, September 17, 2010.
118 A visit to the garden by Albert Kirchengast and Dunja Richter of ETH Zurich and the author, September 21, 2010; Anette Freytag to Ms. E., September 24, 2010.
119 These projects were often preliminary ones presented to clients.
120 The landscape architect Andreas Tremp, who was working for Kienast during these years, reported how unusual it was at the time to draw blue trees and the like, and with what great skepticism they were met by the firm's employees. Andreas Tremp in conversation with the author, October 19, 2010.
121 See, among others, Bettina Heintz and Jörg Huber, eds., *Mit dem Auge denken: Strategien der Sichtbarmachung in wissenschaftlichen und virtuellen Welten* (Zurich: Voldemeer, 2001). I am grateful to Annemarie Bucher for discussion of this topic.
122 Fabienne Kienast Weber in conversation with the author, November 11, 2010.
123 "The work materializes knowledge; it makes it possible to understand the world by way of aesthetic experience," as Anke Haarmann put it in a lecture on artistic research. She points out that, in this approach, artistic practice is not only understood separately from the completed work but the process of creating an artwork makes the phase of investigation the center of attention. Haarmann summarizes the process of understanding, stimulated by artistic production, as follows: "Making art then means—at first programmatically and in general—researching something with the specific means of art, finding out something about the world, society, or people." See Anke Haarmann, "Artistic Research / Künstlerische Forschung," lecture at the Universität der Künste, Berlin, 2008, www.aha-projekte.de/HaarmannArtisticResearch.pdf.
124 See the documents on the registration of the company name Stöckli, Kienast & Koeppel Landschaftsarchitekten AG in the office of the company register of the Canton of Aargau on November 26, 1986. On January 1, 1987, the new partnership became official. Archiv SKK Landschaftsarchitekten, Wettingen.
125 The year 1987 brought other changes to Kienast's work: in addition to the partial separation in terms of the nature of his work from that of his firm partners Peter Paul Stöckli, who gradually began to explore the area of historical preservation of gardens, and Hans-Dietmar Koeppel, there was also a complete geographic separation. Because the expansion had made the office in Wettingen too small, from that year onward Kienast worked along with Erika Kienast-Lüder, Günther Vogt, and the other employees assigned to him on the ground floor of number 11 on Thujastrasse in Zurich—with his own garden just outside the windows. Vogt, who had previously studied with Kienast in Rapperswil, shared with Kienast and his wife, Erika, both artistic and literary interests. The intense office work of both surely reinforced the increasing artistic direction of the projects.
126 Hans-Dietmar Koeppel in conversation with the author, September 8, 2006.

127 On these and similar projects, such as the proposals to increase planting along cantonal roads, see the list of works in the present volume.
128 Peter Paul Stöckli in conversation with the author, September 1, 2006.
129 Ibid. Günther Vogt and long-standing colleagues such as Toni Raymann, Jürg Altherr, and others, made similar assessments.
130 Katharina König Urmi in conversation with the author, September 14, 2006.
131 Kienast, "Von der Notwendigkeit künstlerischer Innovation," 58–59.
132 Erika Kienast-Lüder, "Monsieur Hulot in the Garden," in *Kienast: Gärten / Gardens,* 142–43.
133 Kienast, "Zehn Thesen zur Landschaftsarchitektur," 208; see appendix, 403–4.
134 Fabienne Kienast Weber in conversation with the author, November 11, 2010. Peter Petschek, who worked in the office later, had learned to draw with CAAD in the United States and brought this knowledge to the Zurich office of Stöckli, Kienast & Koeppel in the early 1990s. The first object drawn by computer was the design for the grounds of ETH Zurich of 1992; see the overview plan of the preliminary project (scale: 1:1,000, 80 × 113 cm), June 24, 1992, gta Archives (NSL Archive) / ETH Zurich. Here, too, special attention was paid to depicting the plantings, the terrain, and the meadow with fruit trees.
135 The collection in these files has grown over the years and has been taken from different sources. From the Interkantonales Technikum Rapperswil and from the Institut für Landschaft und Garten in Karlsruhe, there are also scripts containing models for depictions in plans that Kienast had made for his teaching. I am grateful to Fabienne Kienast Weber for permitting me to examine the sample files.
136 Cesare Leonardi and Franca Stagi, *L'architettura degli alberi,* 3rd ed. (Milan: Mazotta, 1998).
137 Fabienne Kienast Weber explained that the patterns for water were supposed to be used only for certain details, namely, where the highlights on the waves are more dense or brighter. This also corresponds to the effect that Kienast had developed in the plan for the garden on Thujastrasse he drew in 1991. Fabienne Kienast Weber in conversation with the author, February 8, 2011.
138 The original of the presentation plan submitted could not be found in any of the archives. It is illustrated in: *Dieter Kienast* (Basel: Birkhäuser, 2004), 195.
139 In order to implement the "landscape reader," the letters of the text were stamped in sheets from expanded metal and mounted on the earth pyramid. Grass grows in the openings of the stamped letters, turning the letters into a passage from *Das Schilcher ABC* by the Styrian author R. P. Gruber: "The average areas of this world are malleable and therefore already distorted and deformed. The 'cultural area' is an area that forms. Actually, it is not an area at all, as an area has a passive character: it defines itself as the opposite of humans, as that which humans have distanced themselves from. In other words, area means opposition and, at the same time, it is transcended by distancing or superiority. In other words, insofar as the character of potentially being formed is always part of the concept of area, the notion of a cultural area which forms itself cannot actually exist." See *Kienast Vogt: Parks und Friedhöfe / Parks and Cemeteries* (Basel: Birkhäuser, 2002), 180–81.
140 For the discussion of software-based rendering techniques, I am grateful to James Melsom of the Landscape Visualization and Modeling Lab (LVML), Professur für Landschaftsarchitektur, Christophe Girot, ETH Zurich.
141 The most important element pasted on afterward is the photograph of a blue sky with clouds that illustrates the Schwarzl Lakes stretching out north of the grounds. The sky is thus mirrored on the surface of the water, so that viewers become more aware of the "God-like" perspective of the top view.
142 Arthur Rüegg, "Dieter Kienast: Paradise Is ...," in *Dieter Kienast: In Praise of Sensuousness,* exh. cat., with a DVD by Marc Schwarz (Zurich: gta Verlag, 1999), 12–14, here 14. See also Arthur Rüegg, "Swisscom Administrative Building, Worblaufen Bern," in *Kienast Vogt: Aussenräume / Open Spaces* (Basel: Birkhäuser, 2000), 74–79.
143 Rüegg, "Swisscom Administrative Building," 76.
144 Ibid., 77–78.
145 On this, see Vogt, *Miniature and Panorama,* 109–23, and *Vogt Landschaftsarchitekten: Radikal, präzis und poetisch; Schulthess-Gartenpreis 2010* (Zurich: Schweizer Heimatschutz, 2010). The two-page series of images intended to illustrate this work (illustrated in, among other places, Vogt, *Miniature and Panorama* [see note 84], 112, 114, and 119) had to be omitted at the request of Günther Vogt.
146 The planting was done in cooperation with Pit Altwegg, and the maintenance is the responsibility of the Eberle landscaping company in Herisau.
147 Günther Vogt, "'Anyone Can Change the Model with Just a Single Stroke of the Hand': On the Process of Design and the Design of Processes," in Bornhauser and Kissling, *Landscape as a Cabinet of Curiosities,* 171–210, here 176.
148 Ibid., 174 and 176.
149 Kienast, "Cultivating Discontinuity," 148.
150 There were exceptions as part of workshops or design workshops to which Kienast was invited. A photograph taken by Udo Weilacher in 1996 shows Kienast working on a silk model produced as part of the Internationale Städtebauliche Entwurfswerkstatt in Krefeld-Fichtenhain; see Professur für Landschaftsarchitektur ETH Zürich, *Dieter Kienast,* 146.
151 Jürg Altherr in conversation with the author, August 20, 2009.
152 Günther Vogt in conversation with the author, May 26, 2008.
153 I received oral information that confirms this from Günther Vogt, Andreas Tremp, and Fabienne Kienast Weber, among others. The Kienast Archive at ETH Zurich has no models apart from a small model of the earth pyramid in the form of a book ("landscape reader") in the mountain garden of the Internationale Gartenschau (International Horticultural Show) 2000 in Styria.

154 Foxley, *Distance and Engagement,* 28. For the illustration, see Vogt, *Miniature and Panorama,* 538 and 540.
155 Günther Vogt in "'Actually, the Case Studio Is My Secret Garden,'" 137.
156 Ibid., 137–38.
157 See "Postscript: Four Words and One Supposition," in Bornhauser and Kissling, *Landscape as a Cabinet of Curiosities,* 211–21, here 217.
158 Ibid.
159 Günther Vogt in "'Actually, the Case Studio Is My Secret Garden,'" 143.
160 *Landscape as a Cabinet of Curiosities* is the title of a book that presents a series of interviews with Günther Vogt, edited by Bornhauser and Kissling.
161 Günther Vogt in "'Actually, the Case Studio Is My Secret Garden,'" 144.
162 In sixteenth-century art theory, drawing was considered the supreme art, without which sculpture and architecture would be unimaginable. The *concetto* expressed in the drawing was understood to be the theory behind the creative practice.
163 Similar remarks have been made of Kienast's teacher Fred Eicher. The architect Ernst Hiesmayr wrote of their collaboration on a housing development in the Nussdorf section of Vienna: "He brilliantly designs the garden using only the contour lines of the geometer; his imagination is unsurpassed." Quoted in Ernst Hiesmayr, "Garten der Siedlung Nussdorf, Wien," in *Fred Eicher Landschaftsarchitekt: Schulthess Gartenpreis 2004* (Zurich: Schweizer Heimatschutz, 2004), 14–15, here 14.
164 Henri Bava in conversation with the author, May 11, 2008.
165 Thomas Proksch in conversation with the author, September 17, 2007.
166 In 1993, Michele Arnaboldi, Raffaele Cavadini, and Guido Hager (Locarno/Zurich) won first prize. For the landscape planning design, however, a workshop was organized in 1994 to which five other landscape architecture firms were invited, including Stöckli, Kienast & Koeppel. The firm Kienast Vogt Partner, founded in 1995, then got the commission.
167 For detailed information and a visit to the Kronsberg together on July 7, 2008, I am grateful to Kaspar Klaffke, who at the time of the workshop was the executive head of the landscaping department of the state capital of Hanover.
168 A precise description of the project is found in Dieter Kienast, "Landscape Design for Kronsberg in Hanover," in *Kienast Vogt: Parks und Friedhöfe / Parks and Cemeteries,* 126–31. On the background, see Heike Brenken et al. *Naturschutz, Naherholung und Landwirtschaft am Stadtrand: Integrierte Landnutzung als Konzept suburbaner Landschaftsentwicklung am Beispiel Hannover-Kronsberg; Abschlussbericht zum E + E-Vorhaben "Naturschutzorientierte Entwicklung im suburbanen Bereich am Beispiel Hannover-Kronsberg" und Zwischenbericht der wissenschaftlichen Begleituntersuchungen,* Angewandte Landschaftsökologie 57 (Münster-Hiltrup: Landwirtschaftsverlag, 2003).
169 While working out and implementing the master plan between 1994 and 1997, differences arose between the firm of Kienast Vogt Partner and the City of Hanover. Parts of the plan were never realized; the firm ultimately withdrew.
170 This method was also used for a few other works on a large scale: the Moabiter Werder in Berlin (competition in 1990–91, implemented in altered form by Vogt Landschaftsarchitekten), the grounds of Expo 2000 in Hanover (competition in 1992, implemented), lakeshore design of the Triechter, Sursee (project, 1991–95), and Tempelhof Airport in Berlin (study for new use, 1998, for which there are sketches by Kienast, reworked by Vogt Landschaftsarchitekten, not implemented).
171 On this, see the site plan submitted in 1991 for the first round of competition, and in 1992 for the second, in *Kienast Vogt: Parks und Friedhöfe / Parks and Cemeteries,* 71 and 73.
172 Kienast had a number of such groves in the shape of a leaf in projects conceived between 1991 and 1995: the first ones appear in the competition plan for the Moabiter Werder (1990–91) and the Günthersburgpark (overview plan, second round of the competition, 1992). Nine of these groves in a highly abstracted form are distributed through the exterior grounds for the Zentrum für Kunst und Medientechnologie in Karlsruhe (Center for Art and Media Karlsruhe, ZKM) (winning competition entry, 1995, implemented). The trees there are standing in roughly hip-height pedestals lined with steel sheet. The motifs are a combination of the leaf-shaped groves of trees that Kienast had used earlier and his leaf-shaped hedges and beds of shrubs distributed through the grounds. See Thomas Göbel-Groß, "The Grounds of the Center for Art and Media, Karlsruhe," in *Kienast Vogt: Aussenräume / Open Spaces,* 152–63, here 153.
173 Kienast explains in his description: "Tubs of clipped yew are scattered over the main area like fallen leaves of chestnuts." See Dieter Kienast, "Garden of the Waldhaus Psychiatric Clinic, Chur," in *Dieter Kienast,* 104–15, here 108.
174 Dieter Kienast, "Outlooks and Insights," in *Kienast: Gärten / Gardens,* 96–98, here 98.
175 Kienast, "Garden of the Waldhaus Psychiatric Clinic, Chur," 108–9.
176 Ibid., 108.
177 On this, see the bibliography in the present volume, 415–17.
178 Reinhold Tüxen, "Vorwort," in Dieter Kienast, *Die spontane Vegetation der Stadt Kassel in Abhängigkeit von bau- und stadtstrukturellen Quartierstypen,* Urbs et Regio: Kasseler Schriften zur Geografie und Planung 10 (Kassel: Gesamthochschul-Bibliothek, 1978), n.p.
179 Dieter Kienast, "Vom Gestaltungsdiktat zum Naturdiktat; oder, Gärten gegen Menschen?," *Landschaft + Stadt* 13, no. 3 (1981): 120–28, here 121–22.
180 This change in style was probably influenced by his reading of more fiction literature. According to Erika Kienast-Lüder (oral information to the author), after finishing work the couple often read together until the early morning hours in their private apartment right next to the office. Kienast-Lüder also edited many of her husband's texts, for example, when working on the

181 Information from Udo Weilacher, lecture in March 2006 at ETH Zurich. Most of the ten theses had been formulated earlier in a lecture that Kienast had given in Berlin in 1991. See Kienast, "Von der Notwendigkeit künstlerischer Innovation."

182 The now excessive use of the present tense is criticized by Ulrich Greiner, "Falsches Präsens: Warum unterwerfen sich so viele Autoren dem Diktat der Gegenwart?," *DIE ZEIT* 19 (April 30, 2009), feuilleton.

183 This stylistic means is also employed by the garden theorist Brigitte Wormbs, for example, when she presented the unrealized project for the Günthersburgpark in Frankfurt am Main—in the form of an imaginary walk—in the exhibition catalog *Dieter Kienast: Zwischen Arkadien und Restfläche,* exh. cat. (Lucerne: Architekturgalerie Luzern, 1992). Kienast was already using the immediate description of senses in texts from the late 1980s.

184 Dieter Kienast, "Longing for Paradise," in *Dieter Kienast,* 84–90, here 90.

185 Kienast, "Zehn Thesen zur Landschaftsarchitektur," 207–8; see appendix, 403.

186 Kienast, "Zehn Thesen zur Landschaftsarchitektur," 207; see appendix, 403.

187 Architekturgalerie Luzern, May 31 to June 28, 1992.

188 The fifteen exhibitions that the Architekturgalerie Luzern had presented earlier were dedicated exclusively to the work of architects, including Heinz Tesar, Peter Zumthor, Adolf Krischanitz, and Marie-Claude Bétrix. Kienast had originally planned to call the exhibition "Zwischen Restfläche und Elysium" (Between Vacant Lot and Elysium); see Dieter Kienast to Lucius Burckhardt, October 4, 1991, gta Archives (NSL Archive) / ETH Zurich (Dieter Kienast Bequest).

189 Christian Vogt in conversation with the author, June 5, 2008.

190 Günther Vogt and Christian Vogt are not related.

191 *Christian Vogt: Fotografische Arbeiten im Architekturmuseum in Basel,* exh. cat. (Basel: Architekturmuseum, 1990). They were also connected through the garden lover and shrub specialist Nicole Newmark, who knew Christian Vogt well because they were related and also because she was a friend of Kienast's and occasionally worked in his firm.

192 Christian Vogt in conversation with the author, June 5, 2008.

193 Text on the exhibition *Christian Vogt: Today I've Been You,* exh. cat. Fotostiftung Schweiz Winterthur, October 24, 2009, to February 14, 2010, www.fotostiftung.ch/index.php?id=187.

194 Urs Stahel and Martin Heller, *Wichtige Bilder: Fotografie in der Schweiz,* exh. cat. (Zurich: Der Alltag, 1990), 74.

195 Christian Vogt in conversation with the author, June 5, 2008.

196 Ibid. Günther Vogt, in conversation with the author, May 26, 2008, was of the same view.

197 The precursors who created their own aesthetic in black-and-white landscape photographs during the 1970s include Bernd and Hilla Becher, Robert Adams, Lewis Baltz, and Frank Gohlke, among others.

198 The photographs for the first exhibition of the works of Vogt Landschaftsarchitekten (Architekturmuseum Basel, November 27, 2004, to January 30, 2005), which also presented several projects begun at Kienast Vogt Partner, corresponded precisely to the method developed by Christian Vogt in 1991 for Stöckli, Kienast & Koeppel: square, panorama, black and white. See *Von Büchern und Bäumen: Vogt Landschaftsarchitekten / About Books and Trees: Vogt Landscape Architects,* exh. cat. (Basel: Architekturmuseum Basel, 2004). Not until 2006 did Christian Vogt photograph in color several works by Vogt Landschaftsarchitekten for the first monograph on its works. See Vogt, *Miniature and Panorama.*

199 Christophe Girot, "Graue Landschaften," in Professur für Landschaftsarchitektur ETH Zürich, *Dieter Kienast,* 7–10, here 7.

200 It is worth noting in this context the similarity of the look of the hornbeam topiary animals in Kienast's garden as seen in Vogt's photograph and in Kienast's drawing (see figs. 112 and 114).

201 See, for example, the work of the Swiss photojournalist Jean Mohr (1925–2018), who had been producing reportage from Asia, Africa, and the Middle East since the late 1950s and always had his photographs printed with the black edge of the negative. Revealingly, the Swiss writer Nicolas Bouvier has said of Mohr's work: "During the 30 years I have been looking at Jean Mohr's photographs, they inspire in me a feeling one experiences very rarely: unlimited confidence. I believe that the objects and the people are exactly as he shows them to me." Nicolas Bouvier, quoted in "Jean Mohr," in *Photo Suisse,* ed. Christian Eggenberger and Lars Müller, trans. Ishbel Flett and Catherine Schelbert (Baden: Lars Müller, 2004), 54–57, here 57.

202 See the Swiss examples in Stahel and Heller, *Wichtige Bilder.*

203 See *New Topographics: Photographs of Man-Altered Landscape,* exh. cat. (Rochester, NY: International Museum of Photography at George Eastman House, 1975) and Britt Salvesen et al., eds., *New Topographics: Robert Adams, Lewis Baltz, Bernd and Hilla Becher, Joe Deal, Frank Gohlke, Nicholas Nixon, John Schott, Stephen Shore, Henry Wessel Jr.,* exh. cat. (Göttingen: Steidl, 2009).

204 Heinz Wirz in conversation with the author, June 2006.

205 I am grateful to Annemarie Bucher for a discussion on this topic.

206 A drawing of Montagne Sainte-Victoire by Paul Cézanne appears on the cover of the first edition of Peter Handke's text.

207 Handke, "The Lesson of Mont Sainte-Victoire," 210–11.

208 Wolfram Müller in conversation with the author, June 12, 2008. Müller was an assistant professor at the Institut für Landschaft und Garten of the Universität Karlsruhe (TH) at the same time as both Kienast and the latter's predecessor, Gunnar Martinsson. He was responsible for the

208 installation architecture and organization of the additional venues of the Lucerne exhibition (Karlsruhe, Technische Universität München-Weihenstephan, December 7, 1993, to February 1, 1994).

209 Untitled poem by Hans Magnus Enzensberger, which begins with the lines: "Was einst baum war, stock, hecke, zaun; / unter gehn in der leeren schneeluft / diese winzigen spuren von tusche / wie ein wort auf der seite riesigem weiss: / weiss zeichnet dies geringfügig schöne geäst / in den weissen himmel sich, zartfingrig, / . . . ." (What was once tree, stump, hedge, fence; / disappears in the empty snow air / these tiny traces of ink / like a word on the page of immense white: / white draws this minimally beautiful branch / in the white sky, delicate-fingered). Quoted in *Dieter Kienast: Zwischen Arkadien und Restfläche*, 3.

210 The panels with the plans and the panes of Plexiglas with photographs are now in the gta Archives (NSL Archive) of ETH Zurich (Dieter Kienast Bequest).

211 At the Technische Universität München-Weihenstephan, and at the Seedamm Center in Zurich-Pfäffikon, the exhibition was shown in modified form. In addition to the plans and Vogt's photographs, large-format color slides of selected sites taken by Wolfram Müller were also projected.

212 It included, among other things, the competition entries for the Töölönlahti Park in Helsinki (no prize) and the Conrad Gessner Park in Zurich-Oerlikon (second prize).

213 Erika Kienast-Lüder revised many of the project descriptions for the monographs and wrote several of them herself, usually under her husband's name.

214 Kienast's daughter, Fabienne, was involved in the production of plans as part of a three-month internship in the office in 1994. Fabienne Kienast Weber in conversation with the author, November 11, 2010.

215 Kienast, "Outlooks and Insights," 98.

216 Ibid.

217 Dieter Kienast, "Garden of the Waldhaus Psychiatric Clinic, Chur," 109.

218 Professur für Landschaftsarchitektur ETH Zürich, *Dieter Kienast*, which grew out of *Textbuch Landschaftsarchitektur: Texte Dieter Kienast*, ed. Professur für Landschaftsarchitektur, ETH Zürich, Professor Christophe Girot (Zurich: Institut für Landschaftsarchitektur, ETH Zürich, 2000).

219 From 1985 to 1989, Dieter Kienast taught two semester hours, and from 1990, four. In 1990, he applied to have the elective subject "Landschaftsgestaltung" (Landscape Design) renamed "Gartenarchitektur" (Garden Architecture), but this was denied both in the department conference and in the conference of professors of the Division of Architecture. Written communication from Ursula Suter, staff of the Department of Architecture, February 21, 2011.

220 Rudolf Manz, *Vito de Onto: Raumerfinder und -erforscher; … seine Freunde nennen ihn Video* (Zurich: gta Verlag, 1996), 10.

221 The name of this alter ego is clearly derived phonetically from the words "video" and "ontology."

222 Manz, *Vito de Onto*, 10.

223 Ibid.

224 Quoted in the six-point program published in the 1996 yearbook for the Department and Division of Architecture. *Jahrbuch / Yearbook 1996. ETH Zürich, Departement und Abteilung für Architektur*, 106 (Zurich: self-published, 1996).

225 Ibid.

226 As Gerhard Schmitt, professor at ETH and former member of the school administration, emphasized in an interview, this counted very much in Kienast's favor. See Schwarz and Bucher, *D. K.: Eine Spurensuche*.

227 Kienast's team in his chair in Karlsruhe also included Thomas Göbel-Groß, Marketa Haist, Ulrike Rothe, and Udo Weilacher.

228 Wolfram Müller in conversation with the author, March 16, 2011.

229 Ibid.

230 Written communication from Ursula Suter, staff of the Department of Architecture, February 21, 2011. For this new task, Kienast hired as assistants the sculptor and architect Christoph Haerle (b. 1958) and the landscape architect Guido Hager (b. 1958), both of whom he knew from the Interkantonales Technikum Rapperswil. The landscape architect Udo Weilacher, Kienast's assistant in Karlsruhe, began his doctorate at ETH, in which he examined the archives of the garden architect Ernst Cramer, whom Kienast greatly admired. See Udo Weilacher, *Visionary Gardens: Modern Landscapes by Ernst Cramer*, trans. Michael Robinson (Basel: Birkhäuser, 2001).

231 While studying at ETH Zurich, Marc Schwarz had attended Kienast's elective course, and he chose to major in landscape design under Franz Oswald. Dieter Kienast and Günther Vogt were his advisors. Marc Schwarz in conversation with the author, February 15, 2011.

232 Ibid.

233 Udo Weilacher had informed Schwarz about this. Schwarz was already familiar with the park as "a place that one knew even as a city dweller in Zurich even though it was in the provinces." Ibid.

234 The exhibition, which was realized with the participation of Guido Hager and Christoph Haerle from the exhibition department of the Institut für Geschichte und Theorie der Architektur (Institute for the History and Theory of Architecture, gta), was on view from December 10, 1999, to January 20, 2000, in the main hall of the Semper building of ETH Zurich. Three editions of the catalog (German, English, and French) were published with a DVD by Marc Schwarz to accompany the exhibition. See *Dieter Kienast: In Praise of Sensuousness*. Then the exhibition was shown in modified form from June 30 to August 19, 2000, under the title *Dieter Kienast architecte-paysagiste: Images des jardins* at the Musée des arts décoratifs in Lausanne, and from October 23 to November 11, 2000, as *Dieter Kienast: Gärten* at the Architektur Forum Ostschweiz in St. Gallen.

235 Manz, *Vito de Onto*, 11 and 20.

236 See *Dieter Kienast: In Praise of Sensuousness*. Marc Schwarz is not entirely satisfied with his film. The technology of video and DVD was not yet

sufficiently advanced at the time to reproduce such delicate movements as rustling leaves and water surfaces in a compressed version. Schwarz is also of the view that, influenced by the pain of the loss of Dieter Kienast, the film had become an "act of mourning" and therefore missed the opportunity to covey the true intention of Kienast's works. After *D. K.: Eine Spurensuche (1. Etappe)* (2008), Schwarz is planning a third film which will combine a cinematic analysis of the projects with interviews of his contemporaries. Marc Schwarz in conversation with the author, February 15, 2011.

237 According to Foucault, heterotopias juxtapose "in one real place several spaces that are normally incompatible." Michel Foucault, "Les hétéropies," in *Die Heterotopien: Der utopische Körper // Les Hétéropies: Le corps utopique* (Frankfurt am Main: Suhrkamp, 2005), 37–52, here 44. Foucault includes theater, cinema, and garden as examples. Kienast integrated a German translation of Foucault's "Des espace autres" of 1967, which is also about heterotopias, in his *Textbuch Landschaftsarchitektur,* which was compiled by Matthew Davies for teaching at ETH Zurich. Cf. Michel Foucault, "Of Other Spaces," trans. Jay Miskowiec, *Diacritics* 16, no. 1 (Spring 1986): 22–27.

238 Foucault, "Les hétéropies," 45.

239 Kienast, "Longing for Paradise," 89–90.

Ten Theses on Landscape Architecture

As a short theoretical basis for our work at the chair, we invoke 10 theses, which I formulated for the exhibition in the Architekturgalerie Luzern in 1992 and which, in a slightly revised version, we consider continue to remain valid.

1.
Our work is a search for the Nature of the City, whose color is not solely green but also gray. The Nature of the City rests in its elements: the tree, hedge, lawn; but equally the water-permeable hard surface, broad square, rigid street-gutter line, high wall; and the unobstructed fresh-air or visual axis, the center and the periphery.

2.
Our interests focus on the city and its inhabitants. The city is no longer a monolithic entity; instead it is dismembered and fragmented into thousands of parts. City dwellers are a kaleidoscope-like mixture of young and old, immigrant workers and long-established locals, clergy and junkies, managers and eco-freaks. This heterogeneity demands up-to-date actions and reactions in outdoor spaces that reject a uniform greening of the city.

3.
The traditional city–countryside rivalry has disintegrated, the boundaries have blurred. We assume that it is impossible either to dismantle the city or to rejuvenate the countryside. Nevertheless, the legibility, the perceptibility of the world is rooted in the principle of dissimilarity. Considering this synchrony of city and countryside, the coming task is to stop the further erosion of the inner boundaries and splits. They have to become sensuously perceptible again.

4.
The city, with its outdoor spaces, cannot be planned as a whole. We trust in mosaic-like interventions in the hope that they will result in meaning and the ability to be experienced, not just of the particular place but also of the whole.

5.
We pay particular attention to the innumerable non-places produced by bureaucratic planning and design. We consider that urban-planning interventions, which equally include landscape architecture, are all the more crucial on the periphery—in those unloved leftover metropolitan spaces.

6.
We understand garden architecture as an expression of the spirit of the times. It is anchored in current social, cultural, and ecological events, which in turn can only be understood in their historical context. For us this means critically confronting the core themes in garden art, or better garden culture, which include not only the achievements of the Age of Feudalism but equally the modest gardens of the simple folk. For us, working with our co-disciplines of architecture, engineering, and the visual arts is not an unwelcome obligation but a welcome axiom. Working together generates mutual innovations. Grappling with the events of the here-and-now requires the inclusion of the wider cultural environment, engaging with film and video, philosophy and literature, music and advertising. We listen to Bach and Schönberg, but similarly Laurie Anderson and Phil Glass. We immerse ourselves in Sol LeWitt and Walter De Maria, Christo and Carl Andre. We find receptions of the themes of nature and the garden not only in Goethe's *Elective Affinities* or Stifter's *Indian Summer* but also in Bloch's *Alienations*, Handke's *The Lesson of Mont Sainte-Victoire*, or Sennett's *The Conscience of the Eye*. In *Mon Oncle*, Jacques Tati leads us from his lovingly tended roof garden through urban wastelands to the bizarrely designed garden of his brother-in-law, while in *The Draughtsman's Contract* Peter Greenaway imparts a lesson in the art of the garden and its social determination.

7.

A further basis of our landscape architecture is the reference to place. This all-too-often adulterated idea is crucial to our work because it prevents solutions from becoming arbitrary and interchangeable, giving more scope to the particular than the generic. Starting from a reading and analysis of the place—its cultural, ecological, and social state—we develop a concept that tests the viability of what exists, either keeping it completely intact, reshaping it, newly interpreting it, or indeed neglecting it. The decisive aspect in this process remains the authenticity of place, as defined by form, material, and use. This is the opposite of a timid conservational approach, which aspires more to recreate the past and which does little to help the future to again become the past. Gardens, parks, and squares should be allowed to tell their history, but they should also be able to tell new stories. They are poetic places of our past, present, and future.

8.

With the transformation of Baroque gardens into English landscape gardens in the eighteenth century, a garden form emerged in which both of their contradictory essential features are combined. Scornfully derided as an inferior potpourri of styles, recently people have now come to recognize the quality and modernity of these gardens. In their analysis of Cubist painting and architecture, Rowe & Slutzky define transparency as over-layering, multidimensionality, and mutual permeation of various structures and systems, allowing a simultaneous and ambivalent reading. To us, the principle of transparency seems ideally suited to the creation of urban outdoor spaces. It affirms the variety, the heterogeneity of the city and its inhabitants, can absorb the old and the new, evokes a pictorial quality—dialectical places in which society, but also individuals, can re-find themselves.

9.

Nature has become scarce in the city and the countryside. The natural bucolic has now become the highest quality label. Like the cultural amenities of the past, a city's nature amenities now define its location advantages. We think, considering this unprecedented social consensus, that formulating concepts for the Nature of the City is imperative. The essential thing is to rediscover the plant as an urban element, and not to consider it simply as an ecological or dendrological factor, as an architectural filling element. We should learn that there are different shades of green, that plants rustle differently in the wind, that not just flowers but also fallen leaves have a fragrance. We should include shade, take account of the impression of the bare branches in winter, express the symbolism of plants, and feel their sensuousness.

10.

It is fashionable for people to prove their progressive credentials by promoting naturally grown, native vegetation. The regulations and rules read that planting exotic trees and shrubs is forbidden. Waving the warning about imminent ecological collapse, plane trees, butterfly bush, and sweet mock orange have been declared "enemies of the garden," and instead nettles, plantains, and mugwort have become protected species. Quite rightly, the environmental movement has criticized the absurdity of the weekly ritual of spraying herbicides and the planting of ground-covering monocultures. Nonetheless, this strict rejection of cultivation, selection, and grafting is equally irresponsible because it negates centuries of manual craft and thus garden culture in itself. No one would seriously argue that the crab apple is superior to the Bern rose apple—and every now and then we all secretly love to eat potatoes! "Give the Foreigners a Chance" was Jürgen Dahl's plea in an article in *Die Zeit*, by which he meant plants, animals, and people alike. We wholeheartedly support this call. City vegetation thrives with and from its polarity: it is trimmed and grows wild, is multicolored and uniformly green, lush and bare, native and foreign. Plants are useful. They improve the climate and are a habitat for animals and people. But plants also stand for the city's promise of nature, which has a special significance for our everyday lives. Bertolt Brecht expressed it as follows: "Asked about his relationship to nature, Mr. K said: 'Now and then I would like to see a couple of trees when I step out of the house.

Particularly because, thanks to their different appearance, according to the time of day and the season, they attain such a special degree of reality. Also, in the cities, in the course of time, we become confused because we always see only commodities, houses and railways, which would be empty and pointless if they were uninhabited and unused. In our particular social order, after all, human beings, too, are counted among such commodities, and so, at least to me, since I am not a joiner, there is something reassuringly self-sufficient about trees, something that is indifferent to me, and I hope that, even to the joiner, there is something about them that cannot be exploited.' 'Why, if you want to see trees, do you not sometimes simply take a trip into the country?' he was asked. Mr. Keuner replied in astonishment: 'I said, I would like to see them *when I step out of the house.*'"*

* Bertold Brecht, "Mr. K and Nature," in *Stories of Mr. Keuner*, trans. Martin Chalmers (San Francisco: City Lights, 2001), 17.

This is translated from the last amended version from November 1998 by Professor Dieter Kienast, Chair of Landscape Architecture at ETH Zurich, as published in *Dieter Kienast: Die Poetik des Gartens; Über Ordnuung und Chaos in der Landschaftsarchitektur*, ed. Professur für Landschaftsarchitektur ETH Zürich (Basel: Birkhäuser, 2002), 207–10.

Biography

Dieter Alfred Kienast, born on October 30, 1945, in Zollikon, Canton of Zurich, was the youngest of the three children of Heinrich and Elisabeth Kienast-Sommerauer.

Grew up at his parents' nursery on Thujastrasse in Zurich, where later he also had his home, office, and the private garden he used as a field of experimentation.

Attended school in Zurich and apprenticed as a gardener with Gebrüder Hottinger. In his free time, Kienast was a passionate extreme mountain climber.

From 1970 onward, Dieter Kienast collaborated with his wife, Erika Kienast-Lüder (d. 2017).

1966–1967
Internship with Albert Zulauf (b. 1923), Landschaftsarchitekt BSG (Bund Schweizer Gartengestalter; Association of Swiss Garden Designers), in Baden.

In March 1967, Kienast meets there his future firm partner, Peter Paul Stöckli (b. 1941).

1969
Birth of daughter Nicole.

1969–1970
Internship with Fred Eicher (1927–2010), Landschaftsarchitekt BSG, in Zurich.

1970–1971
After briefly attending the Technische Universität München-Weihenstephan, where he finds the teaching "too much like secondary school," Kienast begins studying landscape architecture at the Hochschule für bildende Künste (Fine Arts College) in Kassel at the Lehrstuhl für Landschaftskultur (Chair of Landscape Culture) of Günther Grzimek (1915–1996), who is succeeded by Peter Latz (b. 1939) in 1973.

From the winter semester of 1971–72, he enrolls in the interdisciplinary study course in Architektur, Stadt- und Landschaftsplanung (Architectural, Urban, and Landscape Planning) at the newly founded Integrierte Gesamthochschule Kassel (Integrated University of Kassel) (transitional phase; this course of study officially opens in the winter semester of 1975–76).

Lucius Burckhardt (1925–2003) and Karl Heinrich Hülbusch (b. 1936) become Kienast's most influential teachers.

1972
Begins working in the office of Peter Paul Stöckli, Landschaftsarchitekt BSG, in Wettingen during summer vacations.

1973
Birth of daughter Fabienne.

1975
Receives his diploma as an engineer of landscape planning at the Gesamthochschule Kassel with a thesis, cowritten with Thom Roelly, titled "Freiraumplanung Kassel-Nordstadt unter besonderer Berücksichtigung der Vegetation und ihrer Verwertung für Analyse und Planung" (Planning of Open Spaces in the Nordstadt District of Kassel, Paying Particular Attention to the Vegetation and Its Utility for Analysis and Planning).

1978
Receives his PhD in engineering with a dissertation on phytosociology in the city, with Karl Heinrich Hülbusch as his thesis advisor, titled "Die spontane Vegetation der Stadt Kassel in Abhängigkeit von bau-und stadtstrukturellen Quartierstypen" (Spontaneous Vegetation in the City of Kassel in Relation to Types of Neighborhoods in Terms and Architectural or Urban Structure).

Appointed Landschaftsarchitekten BSG; Kienast is active in, among others things, the working group for university education.

1979
Moves into the ground-floor apartment of a newly constructed apartment building on Thujastrasse in Zurich after his parents' nursery is closed in 1977.

He takes up a position as a lecturer at the Interkantonales Technikum Rapperswil (ITR).

1980
Kienast becomes a partner in the firm Stöckli + Kienast Landschaftsarchitekten, Wettingen.

1981–1985
Technical director of the Botanical Garden in Brüglingen, Canton of Basel-Landschaft. In 1984, Kienast organizes the international exhibition *Skulptur im 20. Jahrhundert* (Sculpture in the Twentieth Century) of the Ernst Beyeler Collection in Merian Park, of which the Botanical Garden forms a part.

1981–1991
Professor of Garden Architecture at the ITR. Collaborative teaching with the sculptor Jürg Altherr (1944–2018) and the architect and theorist Peter Erni (b. 1942).

1985–1997
Lecturer for the elective course in landscape design at ETH Zurich—Eidgenössische Technische Hochschule (Swiss Federal Institute of Technology).

1986
Joins the Schweizerischer Ingenieur- und Architektenverein (Swiss Society of Engineers and Architects, SIA), Zurich Section.

1987
Founding of the firm Stöckli, Kienast & Koeppel Landschaftsarchitekten, Wettingen and Zurich, with Hans-Dietmar Koeppel (b. 1944).

Günther Vogt (b. 1957) joins the firm as an employee. Kienast had met Vogt as a student at the ITR.

Dieter Kienast, Erika Kienast-Lüder, and Günther Vogt work, along with other employees, exclusively at the Zurich office on Thujastrasse.

1990
Opening of an office in Bern.

1992–1997
Professor of Landscape Architecture and head of the Institut für Landschaft und Garten at the Universität Karlsruhe (TH), Fakultät Architektur.

1995
Founding of the firm Kienast Vogt Partner Landschaftsarchitekten, Zurich and Bern, with Erika Kienast-Lüder and Günther Vogt.

1997–1998
First Professor of Landscape Architecture at ETH Zurich, Department of Architecture.

After a brief but severe illness, Dieter Kienast dies in Zurich on December 23, 1998.

List of Works

Based on the structures of the landscape architecture firms in which Dieter Kienast was a partner, his papers are divided into three parts.

Most of the works he produced from 1972 to 1987 are in the firm of SKK Landschaftsarchitekten in Wettingen. Dieter Kienast worked on those projects from 1972 to 1986 as an employee and project leader in the landscape architecture firm of Peter Paul Stöckli. In 1980, he became a partner in the firm Stöckli + Kienast, which was expanded to Stöckli, Kienast & Koeppel when Hans-Dietmar Koeppel joined in 1987. The team around Dieter Kienast and Erika Kienast-Lüder, with Günther Vogt as head employee, had its offices on Thujastrasse in Zurich at that time, which had been the Kienasts' shared home and workplace since 1979.

A large number of the plans and the photographic documentation of the works compiled by the firm for the works from 1987 to 1995 are part of the papers that were initially stored for several years as a temporary deposit in the gta Archives of ETH Zurich before Erika Kienast-Lüder donated it to the NSL Archive, Netzwerk Stadt und Landschaft (Network City and Landscape) (part of the gta Archives) on November 21, 2013. Another focus of this part of his papers is the competition and presentation plans shown in various exhibitions, as well as the plans of very early projects, primarily private gardens, that Dieter Kienast took on as freelance work while studying in Kassel. This archive also has all of the documentation of Kienast's studies in Kassel (1970–78) and the teaching materials he compiled while working as a professor at the Interkantonales Technikum Rapperswil (1981–91) and the Institut für Landschaft und Garten of the Universität Karlsruhe (TH) (1992–97), and as a lecturer of landscape design (1985–97) and first Professor of Landscape Architecture in the Department of Architecture at ETH Zurich (1997–98).

In 1995, Dieter Kienast co-founded—with Erika Kienast-Lüder and Günther Vogt—the firm Kienast Vogt Partner in Zurich and Bern. The firm continued after Kienast's death in December 1998 until the end of 1999. In 2000, Günther Vogt established the firm Vogt Landschaftsarchitekten in Zurich with offices in Munich (2002–2009), London (since 2008), and Berlin (since 2010). Many of the projects from this final period had been begun by Kienast Vogt Partner and were continued and executed by Vogt Landschaftsarchitekten. When he founded his firm, Günther Vogt moved the relevant plans and dossiers to the new location on Stampfenbachplatz in Zurich. He also possesses the majority of the documentation of the projects from 1987 onward, including cost estimates, plans of the planting designs, settlements of accounts, working plans, drawings of details, and a series of presentation plans.

The documents at both SKK Landschaftsarchitekten in Wettingen and Vogt Landschaftsarchitekten are organized in the two places by object numbers and are available to researchers. The archive transferred to ETH Zurich has been cataloged digitally, and scholars can request its data sets from the holdings of the SKK and Vogt firms as well.

Database

Because there are so many documents and because they are stored at three locations, all of the material was recorded in a database according to archival and scholarly criteria. The database was compiled by the present author as part of her preparation of Kienast's archival materials as the basis for her dissertation.[1] It includes more than two thousand data sets and provided the basis for the catalog of the NSL Archive that is currently accessible at the gta Archives.[2] The precise storage location in the relevant archive is indicated in each data set.

List of Works

The current circumstances of who possesses the material and its very abundance made it impossible to record all of the planning material in detail. The transfer of the data sets in the database into a list of works and the checking and completion of it was done in 2011 in collaboration with the archivist Michael van der Steeg (of the Chair of Landscape Architecture, Christophe Girot, and the NSL Archive).

The dating of the works is based primarily on the available planning material. The early work was carefully processed and documented. For the objects from the mid-1980s onward, in several cases the dating was adopted from the list of works edited by Erika Kienast-Lüder in 2002 in *Kienast Vogt: Parks und Friedhöfe / Parks and Cemeteries* (Basel: Birkhäuser, 2002), 282–88.

Dieter Kienast always worked within an office structure. The works are therefore always attributed to the relevant firms and their employees.

Competition entries (comp.) were only included if they received first prize, if they received a different prize but were partially realized, or if the competition entries were published in one of the four publications on Kienast's works from 1997 to 2004. At the express wish of their owners, the private gardens realized since 1985 are described anonymously.

---

1. This work was done with constant discussions with the director of the NSL Archive, Marta Knieza 2005 to 2012, and with assistance from Siri Frech and Caroline Mathis.
2. The catalog was edited by the director of the NSL Archive, Alex Winiger.

Key to the List of Works

1972–79
  Peter Paul Stöckli
1980–86
  Stöckli + Kienast
1987–94
  Stöckli, Kienast & Koeppel
1995–99
  Kienast Vogt Partner

* Dating based on the list of works from 2002 compiled by Erika Kienast-Lüder
° Revision of the project by Stöckli, Kienast & Koeppel or SKK Landschaftsarchitekten
+ Completion or revision of the project by Kienast Vogt Partner
++ Completion or revision of the project by Vogt Landschaftsarchitekten
◊ Project began in the year indicated, but the end date of planning/implementation was not researched.

List of Works

Switzerland
Canton of Aargau
Aarau
— Aargauisches Versicherungsamt (commissioned study, proposal for the landscape architecture), 1994*
— Buchenhof Administration Building (design of the grounds), 1991*
— Casinogarten Restaurant (planning for the design of the grounds), 1994
— Gerhard Garden, 1973
— Scheibenschachen neighborhood (pilot study for planning green spaces), 1978

Auenstein-Veltheim
— Auenstein–Veltheim cantonal road (project for planting improvements), 1980

Baden
— Extension to the Kappelerhof School (design of the grounds), 1990–92
— Liebenfels Cemetery (extension), 1983◊
— Rütihof Cemetery (extension), competition entry, first prize, 1985(comp.)–90

Bad Zurzach
— Coop Genossenschaft, store (design of the grounds and roof greening), 1985, 1993–95
— Rheumatology Clinic (forecourt), 1985
— Zurzach Spa Gardens, competition entry, first prize, 1983 (comp.)–85

Brugg
— Brugg Cemetery, 1980–81
— Brugg central bypass (landscape planning), 1980–82 Döttingen/Klingnau
— Döttingen/Klingnau bypass (landscape planning), 1981°◊

Gebensdorf
— A. Garden, 1983–84

Gränichen
— Schiffländi Retirement Home (design of the grounds), 1985–86
— Village center (design of the square), 1989

Hägglingen
— Hägglingen Cemetery (extension), 1972–74
— Geissmann Garden (preliminary project), 1972

Holderbank
— Holderbank Cement und Beton (HCB) (since 2001: Holcim AG, now: HolcimLafarge Ltd.), Schümel quarry (plan for recultivation), 1978°–99

Laufenburg
— Coop Genossenschaft, store and housing for cooperative employees (design of the grounds), 1995–97

Mägenwil
— Meier Garden (redesign), 1972

Mülligen
— Eichrüteli gravel quarry (plan for gravel stripping and recultivation), 1979°◊

Neuenhof
— Nackenbachtal (landscape planning) and Brühl pond (planting design), 1978

Obersiggenthal
— Schul-/Oberdorfstrasse (landscape planning), 1984
— Sternenplatz, Nussbaumen, competition entry, first prize, 1979*(comp.)–89

Oeschgen
— Oeschgen Cemetery (landscape design), 1974

Othmarsingen
— Fischer Garden (redesign), 1978–79

Rekingen and Tegerfelden
— Holderbank Cement und Beton (HCB), Musital quarry (plan for recultivation), 1982–2002°

Rudolfstetten
— Rudolfstetten Cemetery (extension), 1985–89

Schinznach-Bad
— Schinznach Cemetery (extension), 1980–84

Schinznach-Dorf
— Degerfeld gravel quarry (plan for gravel stripping and recultivation), 1980, 1984°◊

Spreitenbach
— Längacker residential tower (design of the grounds), 1973–75

Unterehrendingen
— Schoolhouse (design of the grounds), 1989–91

Untersiggenthal
— Juchli Garden, 1984

Villmergen
— Cantonal road 252 (Bullenberg) (project for planting improvement), 1980

Walterswil
— Safenwil Retirement Home (design of the grounds), 1988*

Wettingen
— Brühlwiese Municipal Park ("Brühlpark"), 1979–84
— Garden of the W. and S. single-family houses, 1985–86
— Rüttimannn Garden (redesign), 1974
— Sankt Bernhard Retirement and Nursing Home (design of the grounds), 1977

Wohlen
— Cellpack AG (design of the grounds), 1976
— H. Garden, 1987*
— PTT (Post, Telefon- und Telegrafenbetriebe, since 1998: Post and Swisscom AG), telecommunication center (design of the grounds), 1979
— Schweizerischer Bankverein (since 1998: UBS) (design of the grounds), 1981

Wölflinswil
— Wölflinswil Cemetery (extension), 1978–84

Canton of Appenzell Innerrhoden
Appenzell
— Museum Liner (design of the grounds), 1998*

Canton of Basel-Landschaft
Augst
— Roman Gardens in Augusta Raurica (preliminary project for temporary gardens as part of Grün 80, not realized), 1979

Laufen
— Ricola AG, marketing building (design of the grounds), 1998++
— R. Residence and Studio (design of the grounds), 1994*

Münchenstein (Brüglingen)
— Grün 80: 2. Schweizerische Landesausstellung für Garten- und Landschaftsbau (Green 80: Second Swiss National Exhibition of Horticulture and Landscape Architecture), 1977–80
——— Botanical Garden, southern section (diverse interventions), 1978–79
——— Dry Grassland Biotope (now part of the Botanical Garden), 1978–80
——— "Grüne Universität" (Green University) sector, competition entry, no prize, 1977
——— "Land und Wasser" (Land and Water) sector, competition entry, no prize, 1977

Therwil
— K. Garden, 1997

Canton of Basel-Stadt
Basel
— F. Garden (project, not realized), 1994
— F. Garden (project, partially realized), 1995
— F. Hoffmann-La-Roche Pension Fund (interior green space), 1989
— Luzernerring/Bungestrasse housing development (roof garden), 1993
— Schweizerische Unfallversicherungsanstalt (The Swiss Accident Insurance Company, Suva) (garden courtyards), 1993

Riehen
— G. Garden, 1997*
— Gottesacker Riehen (cemetery) (design of the entry and grounds), 1992
— K. Garden, 1994
— L. Garden (redesign), 1995

— Vogelbach housing development (design of the grounds), 1992–94

Canton of Bern
Aarberg
— Retirement home (consultation, design of the grounds), 1987*

Bätterkinden
— Dorfmatt School Complex (design of the grounds), competition entry, first prize, 1992*

Bern
— Bärenplatz/Waisenhausplatz (design of the square), competition entry, first prize, 1990 (comp.)–99
— Betriebsgesellschaft Brunnadern (design of the grounds), 1991*
— Bümpliz Schoolhouse (design of the grounds), 1984–85
— Dental Clinic (design of the grounds), 1988*
— École cantonale de langue française (French School of Bern) (design of the grounds and playing fields), competition entry, first prize, 1983–84 (comp.), 1987–91
— Inselspital (hospital) (master plan green spaces), 1985*°
——— Pathology Department (design of the grounds), 1989*
——— New Building for Wing 2 of the Polyclinic (design of the grounds), 1990*
——— Psychiatric Polyclinic (roof design), 1990*
——— Renovation Imhoof Pavilion (design of the grounds), 1990*
——— Dormitory high-rise (path connecting the entrance), 1991*
——— Wing 1 of the Polyclinic (flat roof greening), 1991*
——— New Building for the Women's Clinic (design of the grounds), 1994*
— La Villette Office Building, Effingerstrasse, 1990*
— Lory-Spital (hospital) (design of the grounds), 1982*/1986–88
— Murifeld municipal sports complex, competition entry, first prize, 1983*(comp.)°
— Proposal for dynamic landscape planning for Bern, 1998*
— PTT (Post, Telephon, and Telegraph Company), Eigerhaus (design of the grounds), 1987
— Regional Library (planning for the design of the grounds), 1993*

Biel
— Centre PasquArt, extension and renovation (design of the grounds), 1996°

Burgdorf
— Steinhof work sides (design of the grounds), 1993*

Hilterfingen-Hünibach
— Schönegg Retirement Home (design of the grounds), 1989*

Hinterkappelen
— B. Garden, 1995*

Interlaken
— Congress Centre Kursaal Interlaken (design of the forecourt), 1996*

Ittigen
— K. Garden, 1990*

Köniz
— Landorf rural boarding school (design of the grounds), 1989

Lyss
— Lyss Retirement Home (design of the grounds), 1984–89

Muri
— Gümligen Castle Park (proposal to redesign the Castle Park), 1990°

Spiez
— Lakeshore, competition entry, no prize, 1995

Wengen
— Bahnhofplatz (redesign), 1990*

Wiedlisbach
— Rebstock Restaurant (design of the grounds), 1978–79

Worblaufen
— Swisscom AG (until 1998: Post-, Telefon- und Telegrafenbetriebe, PTT), administration building (design of the grounds and garden courtyards), competition entry, first prize, 1993 (comp.)–98

Zollikofen
— Häberlimatte (design of the grounds), 1996*

Canton of Freiburg
Murten
— SBB-Ausbildungszentrum (Swiss Federal Railroad Training Center) Löwenberg (design of the grounds), 1994–95

Canton of Geneva
La Capite
— Garden for the Sauvin/Mahler Houses (project), 1996*

410

Canton of Glarus
Glarus
— "Altersgerechte Wohnungen" (senior citizens apartments) (design of the grounds), 1995–96

Canton of Graubünden
Chur
— Bündner Kunstmuseum (Grison's Museum of Fine Arts) (planting design for existing grounds), 1990–91
— Fürstenwald Cemetery, competition entry, first prize, 1992/93 (comp.)–96
— In den Lachen neighborhood (proposal for the design of the grounds), 1997
— Kantonsbibliothek/Staatsarchiv and Regierungsplatz (design of the grounds), 1993–95
— Waldhaus Psychiatric Clinic (design of the grounds, several phases), 1987–96
— Z. Garden (redesign), 1993

Scuol
— Quadras Spa and Swimming Center (design of the grounds), 1988*, 1991*
— Two Village Squares (preliminary project for the design of the squares), 1991

Canton of Lucerne
Lucerne
— Grand Hotel National (front yard), 1996

Meggen
— F.-K. Garden, 1996–97

Sursee
— Nature conversation plan and master plan for the development of green spaces for the city of Sursee (continued consulting), 1993–95
— Sankt Urbanhof, 1998*
— Triechter (design of the lakeshore of Sempachersee), 1991–95

Canton of Nidwalden
Stansstad
— STEINAG Rozloch AG, quarry (plan for siliceous limestone stripping and recultivation), 1975–86

Canton of Schaffhausen
Schaffhausen
— F. Garden (project plan), 1993

Canton of Schwyz
Altendorf
— Lindt & Sprüngli AG (design of the grounds), 1981–84

Reichenburg
— Burg School Complex and Retirement Home (design of the grounds), 1996*

Schwyz
— Sonnenplätzli (design of the grounds), 1995*

Canton of Solothurn
Deitingen
— Zweien School, multipurpose hall and sports complex, 1992*

Erlinsbach
— Mühlefeld Retirement Home (design of the grounds), 1988–89*

Canton of St. Gall
Bad Ragaz
— Bad Ragaz Cemetery (extension), 1987–89

Gommiswald
— G. Garden, 1978

Mörschwil
— P. Garden (redesign), 1993*

Rapperswil-Jona
— Bühl Retirement and Nursing Home (design of the grounds), 1989–90
— Interkantonales Technikum Rapperswil (shrubbery), 1982
— Jonaportstrasse (design of street area), 1990
— Villa Grünfels (design of the grounds), 1989

St. Gallen
— Achslengut (design of the park), 1997*–2002++
— Bernhardswies (design of the grounds), 1994*
— Historisches Museum (design of the display of a collection of glacial erratics), 1985–88
— Municipal Park (study for the Upper Brühl and Lower Brühl), 1985–86
— Municipal Park (new circular bed of roses and new rose garden), 1988–93
— New driveway to Brühltor parking garage (greenery), 1990*
— Scheitlinspärkli near the Cantonal School (design of the park), 1986–88
— Vonwilpark housing development, Lachen neighborhood (design of the grounds), 1997*
— Wartensteinstrasse 6 housing development (design of the grounds), 1997*

Trübbach
— Schollberg quarry (landscape design), 1994*

Valens
— Spa and Bath Clinic (preliminary project for the design of the grounds), 1986*

Canton of Tessin
Gambarogno-Contone
— Möbel-Pfister AG (design of the grounds), 1989

Ronco Sopra Ascona
— M. Garden (redesign), 1995

Canton of Thurgau
Amriswil
— G. & H. Garden, 1983*–84*, 1989–91

Egnach
— Sp. Garden (redesign), 1991–93

Weinfelden
— Special school (design of the grounds), 1976

Canton of Waadt
Champvent
— Château de Champvent (renovation of historical gardens), 1994–98

Nyon
— Centre Commercial La Combe roof garden (vegetation proposal), 1980

Canton of Wallis
Brig
— Stockalper Palace (renovation of castle garden), 1996 (comp.) –98

Canton of Zug
Hünenberg
— Business Park (design of the grounds), 1995°

Zug
— Landis + Gyr AG (master plan), 1994°

Canton of Zurich

Adliswil
— Allmend (commons) (redesign), 1994–95
— C. Garden, 1994
— Sihlhof housing development (design of the grounds), 1993*°
— Sihlhof Kindergarten (design of the grounds), 1995
— Swiss Re AG (until 1999: Schweizerische Rückversicherungs-Gesellschaft), sports complex (renovation), 1996–98
— Swiss Re AG, Tüfi Administration Building (design of the grounds), 1995–98

Dielsdorf
— Dielsdorf Cemetery (project, extension), 1990*

Dübendorf
— Eidgenössische Materialprüfungs- und Forschungsanstalt (Swiss Federal Laboratories for Materials Science and Technology, Empa) (master plan for the grounds), 1996–2003++
— Zion Retirement Home (design of the grounds), 1985–88

Dürnten
— Felsenburgsenke semi-natural green space (project plan), 1981

Erlenbach
— M. Garden, 1997–98
— M.-M. Garden (redesign), 1987–89

Gockhausen
— B. Garden (redesign), 1992–95

Herrliberg
— B. Garden (proposal), 1994*
— Hechlenberg housing development (design of the grounds), 1996

Horgen
— G. Garden, 1973

Kilchberg
— Lindt & Sprüngli AG, new auditorium (design of the grounds), 1989

Kloten
— Zurich Airport
—— "Airport 2000" (fifth phase of construction) (commissioned study for the design of the grounds), 1996*
—— Zurich-Kloten Airport (interior green space for the operations center), 1988, 1990–91

Küsnacht
— J. Garden, 1995–96

— R. Garden, 1997*–2001++
— Sch. Garden (redesign), 1991–92
— W. Garden (redesign to add a pool), 1989

Maur
— K.-U. Garden (redesign), 1992–95

Nänikon
— Denz & Co. AG, office building (roof garden), 1992–93

Niederhasli
— Lindenstrasse/Huebwiesen housing development (design of the grounds), 1972–75

Oetwil am See
— N. Garden (redesign), 1983, 1991–92, 1994

Opfikon
— Dorfstrasse 55 (design of the grounds), 1987–88
— Obere Wallisellerstrasse (design of streetscape), 1989

Regensberg
— L. Garden (project plans for redesign), 1987*
— Primary school (preliminary project for the design of the grounds), 1972

Rümlang
— C. Garden (redesign), 1994*–95

Rüschlikon
— G. Garden (redesign), 1992°
— Schweizerische Rückversicherungs-Gesellschaft (since 1999: Swiss Re AG), Villa Bodmer Seminar and Training Center (design of the grounds), 1994–97

Schlieren
— Im Hübler housing development (project plans for the design of the grounds), 1987*
— Senior and neighborhood center (proposal for the design of the grounds, not realized), 1994

Stäfa
— Phonak AG (open space design and greenery for interior), 1986–89

Thalwil
— "Bahn 2000," Zurich–Thalwil double-track railway (vegetation concept), 1990*

Uitikon-Waldegg
— E. Garden (redesign), 1989, 1993, 1997

Urdorf
— Limmat Valley Cantonal School (design of the grounds), 1985

Uster
— Zellweger Luwa AG, administration building (design of the grounds), 1997*

Volketswil
— Chimlibach housing development (design of the grounds), 1984

Wädenswil-Au
— Stiftung Bruder Klaus (foundation), Alte Landstrasse development, 1997*

Weiningen
— Fahrweid neighborhood center (design of the grounds), 1973–74

Wermatswil
— W. Garden (redesign), 1992–93

Wettswil
— Dettenbühl housing development (work on grounds), 1995–97

Winterthur
— Garden for the Sammlung Oskar Reinhart (Oskar Reinhart Collection), 1998*
— Neuwiesen housing development (design of the grounds), 1981

Witikon
— Witikon Cemetery (extension), competition entry, first prize, 1992 (comp.)–94
— B. Garden, 1996

Zollikon
— Zollikon Community House (artists' competition / design of square with Esther Gisler), 1987

Zumikon
— S.-T. Garden (redesign), 1990

Zurich
— Apartment building at Hochstrasse 57 (design of the grounds), 1996*°
— Arnold Böcklin Studio Garden (redesign and maintenance proposal), 1986
— Arterstrasse 24/26 garden courtyard (redesign), 1982–83, 1986
— B. Garden, Spiegelgasse, 1998*
— Brahmsstrasse Daycare Center (design of the grounds), 1990*
— "Chance Oerlikon 2000" (development of former industrial sites), overall concept, 1992*

- Conrad Gessner Park, competition entry, second prize, comp. 1996
- Consulting for the Hard neighborhood, 1987–88
- Denner AG, housing development, Rütistrasse (roof garden), 1979
- Doppler Garden, 1982–83
- Ernst Basler + Partner AG (since 2016: EBP Schweiz AG), Mühlebachstrasse Office Building (front yard and rear courtyard), 1995–96 (extension of the front yard by Vogt Landschaftsarchitekten, 2000–2001 and 2012–13)
- ETH Zurich, Hönggerberg (design of the grounds and landscape master plan), start 1990, 1996–2004++
- Freibad Allenmoos (Allenmoos Outdoor Swimming Pool) ("ground wave" and renovation), 1995–96, 1998–99+
- Helvetia (health insurance company), Stadelhofer-/Theaterstrasse (design of interior courtyard), 1989*
- Hirslanden Clinic (design of the grounds), 1989, 1994, 1995, 1997
- Hirslanden Clinic, Forchstrasse residential buildings (design of the grounds), 1989*
- Hohe Promenade Cemetery, Trümpler tomb and von Schulthess tomb, 1972
- Hornbach (design study), 1986–87
- Hotel Zürichberg (design of the grounds), 1992*
- IBM, Bernerstrasse (open space design and greenery for the interior), 1990°
- K. Garden, Parkring, 1994–95
- The Kienasts' Garden, Tujastrasse, 1974, 1976–78, 1985, 1991
- L. & L. and B. & S. Gardens (redesign), 1993–94
- Mathilde Escher Home (design of the grounds), 1987–89
- MC Publishing House building and apartments, competition entry, first prize, 1991(comp.)
- Muggenbühl Restaurant (preliminary projects for the design of the grounds), 1988–90
- Nebelbach (rehabilitation of stream), 1993–94
- Neuco Retail Store (greenery for the interior), 1988*
- Nordstrasse housing development (design of the grounds), 1980–81
- Paraplegic Center of the Balgrist University Clinic (design of the grounds), 1987*°
- R. Garden, 1996
- Rote Fabrik (roof greening), 1989*
- Rütihof Schoolhouse, Höngg (greenery for the interior), 1994
- Sankt Anton Church Center (design of the grounds), 1981
- Schanzengraben and SIA (Swiss Society of Engineers and Architects) Building, competition entry, first prize, 1986 (comp.)–88
- Schule für Körper- und Mehrfachbehinderte (School for the Physically and Multiply Handicapped) (design of recess area and humid biotope), 1987–88
- Schweizerische Bankgesellschaft (since 1998: UBS AG), Altstetten (entrance square), 1984
- Selnau Retirement Home, 1997*
- Shell (Switzerland) AG, administrative building (greenery for the interior), 1991*
- Sihlfeld Schoolhouse (design of the grounds), 1987*–89
- Sihlraum outdoor space design (development study), 1981–82
- Sihltal Zürich Uetiliberg Bahn SZU AG (railway company) (renaturation of the riverbed and banks), 1984–85
- Stiftung Puureheimet Brotchorb (foundation) (design of the grounds), 1988
- Storz AG, multifamily apartment building, Malojaweg (project for the design of the grounds), 1993
- Südstrasse development (design of the grounds), 1993*
- Swiss Re AG (until 1999: Schweizerische Rückversicherungs-Gesellschaft), Gotthardstrasse Office Building (now: Bank Vontobel) (garden courtyard), 1994–95
- Swiss Re AG, Mythenquai Main Administration Building (design of the grounds), 1997–2002++
- Swiss Re AG, Swiss Insurance Training Center (SITC), Seestrasse (garden), 1997–2000++
- Tramline 10 (phytosociological vegetation concept), 1979, 1982–88
- University of Zurich, Irchel Campus (design of a square), 1986°
- Villa Brandt Garden (redesign), 1989–90
- Villa Susenberg Garden, 1997–98
- Villa Wehrli (design of the grounds), 1988*
- Zoo Zürich, Masoala Rain Forest Hall (landscape design), 1994–2003++
- Zurich Stock Exchange, new building (proposal for the design of the grounds), 1987*

Austria
Feldkirch
- Waldbad Gisingen (forest swimming pool), 1996*

Graz-Unterpremstätten
- Internationale Gartenschau (IGS) 2000 (International Horticultural Show 2000), competition entry, first prize, 1996/97 (comp.)–99+
—— Arable Garden (not preserved), 1997–99+
—— Flower Garden (not preserved), 1997–99+
—— Mountain Garden (preserved), 1997–99+
—— Pheasant Garden (preserved), 1997–99+

Lustenau
- Church of Saints Peter and Paul (forecourt), 1996*
- New building for kindergarten (proposal for the design of the grounds), 1993
- Rathausplatz (design of square after extension), 1994*

Wolfurt
- Events hall (design of the grounds), 1994*°

Finland
Helsinki
- Töölönlahti Park, competition entry, no prize, 1997

France
Brunstatt-Didenheim
- Ricola Europe Packing and Distribution Building (design of the grounds), 1994

Chaumont-sur-Loire
- Fifth Festival International des Jardins ("La nature n'existe pas" garden), 1996

Mouans-Sartoux
- Museum park (commissioned study), 1998*

Paris
- Compagnie Suisse de Réassurances S. A. (since 1999: Swiss Re AG), new building for headquarters (garden courtyards), 1996–97

Germany
Bad Münder am Deister
- Bad Münder Spa Gardens, 1993 (comp.)–97

Bad Wiessee
— St. Garden, 1997*

Barnim
— Barnim park landscape (consultant report), 1994*

Berlin
— Berlin-Tempelhof Airport (commissioned study for reuse), 1998–2006++
— Buchholz-Nord (consultant report on urban planning), 1994*
— Embassy of the Czech Republic, residence (garden), 1996–97
— Embassy of Switzerland (garden), 1996–98
— Federal Foreign Office (design of the grounds), 1996
— Internationale Bauausstellung (International Buildinge Exhibition, IBA) 1999, Buchholz-Ost (open space design, unrealized), 1997–98+
— Königsstadt-Terrassen (design of the grounds and roof greening), 1995–96
— Moabiter Werder, promenade and parks, competition entry, first prize, partially realized, comp. 1990–91, 1992–2002++
— Moabiter Werder, promenade, daycare center and school (design of the grounds), 1994*
— Park am Springpfuhl, Helene-Weigel-Platz, 1995*
— Southern bank of the Spree and Spreebogen (Spree Curve) (design study), 1991*
— Urban planning implementation competition for Biesdorf-Süd and Mahrzahn (revision), 1994*
— Wuhlepark, competition entry, first prize, 1996*

Chemnitz
— "Strasse der Nationen" (Street of Nations), competition entry, third prize, partially realized, 1993*

Dreieich
— Baierhansenwiesen, "Dreihundertdreiundreissig Eichen für Dreieich" (Three Hundred and Thirty Oaks for Dreieich), competition entry, first prize, unrealized, 1996 (comp.)–98*

Erfurt
— Bundesarbeitsgericht (Federal Labor Court) (public park, inner courtyard, and roof garden), competition entry, first prize, 1995 (comp.)–99+

Frankfurt am Main
— Frankfurt Green Belt, Hafenbruch and Mainpromenade (consultant report), 1990
— Günthersburgpark (resedign and extension), competition entry, first prize, unrealized, 1991–92 (comp.), 1992–94

Fürstenfeldbruck
— Sparkasse Fürstenfeldbruck (design of the grounds), competition entry, first prize, 1997 (comp.)–99+

Hanover
— Bünteweg housing development (planning for the design of the grounds), 1995*
— Expo 2000, Hanover Trade Fair (open spaces), competition entry, first prize, 1992 (comp.)–2000
—— Boulevard of the United Trees (preserved), 1995–2000
—— Earth Garden (preserved), 1995–2000
—— Expo Lake and Expo Roof (preserved), 1995–2000
—— Park Wave (preserved), 1995–2000
— Kronsberg Landscape Design (master plan), 1994–95, realized in parts through 1999

Karlsruhe
— Oberreut Municipal Cemetery, competition entry, 1994
— ZKM | Center for Art and Media Karlsruhe (green strip on Brauerstrasse and design of square), competition entry, first prize, 1994* (comp.)◊

Kassel
— Bachmann Garden, 1976
— Gesamthochschule Kassel (Integrated University of Kassel, since 2003: University of Kassel), grounds of buildings K 9 and K 10, Organisationseinheit Architektur, Stadt- und Landschaftsplanung (Organizational Unit of Architectural and Urban and Landscape Planning) (ash tree planting action, implementation plan), 1977
— Sommermann Garden, 1975

Koblenz
— Michel-Nusser Garden, 1973

Laufenburg
— Codman'sche Anlagen (grounds)/ Rhine bank, competition entry, 1980

Leipzig
— Augustusplatz, competition entry, 1994

Magdeburg
— Neustädter Feld (proposal for urban planning renewal), 1994

Mechtenberg
— Internationale Bauausstellung Emscher Park, "Dornröschen am Mechtenberg" (Sleeping Beauty at Mechten Hill) (design seminar and revisions), 1993–95

Meiningen
— Dreissigacker-Süd, competition entry, first prize, 1993 (comp.)◊

Oldenburg
— Petschallies Garden, 1975–76

Ulm
— M. Garden, 1995

Luxembourg
Betzdorf
— SES S. A. (Société Européenne des Satellites) (master plan and design of the grounds), 1998–2001++

United Kingdom
London
— Tate Modern (exterior grounds, first phase), 1996–98

USA
New York
— Swiss Reinsurance Company Ltd. (since 1999: Swiss Re AG), headquarters (design of the grounds), 1996*

Bibliography

I. Writings by Dieter Kienast (single-author texts listed chronologically), exhibition catalogs, monographs, films, and documentations about Kienast's teachings

1. Unpublished sources

Kienast, Dieter. No title [description of the housing development Lindenstrasse/Huebwiesen in Niederhasli]. Manuscript, n.d. [ca. 1972–75]. gta Archives (NLS Archive) / ETH Zurich (Dieter Kienast Bequest).
———. "Gestalt öffentlicher Räume." Lecture for the chair of Franz Oswald. Typoscript, n.d. [after 1980]. gta Archives (NLS Archive) / ETH Zurich (Dieter Kienast Bequest).
———. "Der Beitrag des Gartens zur Verbesserung der Wohnumwelt." *SIA Mitteilungsblatt* 2 (1980): 6–10. gta Archives (NLS Archive) / ETH-Zürich (Dieter Kienast Bequest).
———. "Notizen zum Konzept des Stadtparks Brühlwiese in Wettingen." Typoscript, n.d. [ca. 1982]. Archiv SKK Landschaftsarchitekten, Wettingen.
———. "Botanischer Garten Südteil: Naturnahe Biotope." Manuscript, January 4, 1980. Archiv SKK Landschaftsarchitekten, Wettingen.
———. "Arkadien: Stadtraum zwischen Wunsch und Wirklichkeit." Report on the symposium "Pflanze und Stadt: Die Stadt als Garten" on the occasion of the "plantec," Frankfurt am Main, September 30, 1994. Edited by Ständige Konferenz der Gartenbauamtsleiter beim Deutschen Städtetag et al. Archiv Fachgebiet Landschaftsarchitektur, Karlsruhe Institute of Technology (KIT).
———, and Toni Raymann. "Grün 80; oder, Warum lieb' ich alles was so grün ist? Alternatives Konzept zur Durchführung der Gartenschau 1980 in Basel," idea competition for Grün 80. Typoscript, 1976. Archiv SKK Landschaftsarchitekten, Wettingen.
———, and Thom Roelly. "Freiraumplanung Kassel-Nordstadt unter besonderer Berücksich-tigung der Vegetation und ihrer Verwertbarkeit für Analyse und Planung" (Planning of Open Spaces in the Nordstadt District of Kassel, Paying Particular Attention to the Vegetation and Its Utility for Analysis and Planning). Diploma thesis, Gesamthochschule Kassel 1975. gta Archives (NLS Archive) / ETH Zurich (Dieter Kienast Bequest). Published under the title *Standortökologische Untersuchungen in Stadtquartieren – insbesondere zur Vegetation – unter dem Aspekt der freiraumplanerischen Verwertbarkeit. Gesamthochschule Kassel.* Schriftenreihe OE Architektur, Stadt- und Landschaftsplanung 3, no. 2. Kassel: *Gesamthochschule Kassel,* 1978.
———, and Büro Peter Paul Stöckli. "Botanischer Garten – Südteil/ Sektor Schöne Gärten: Trockenbiotop-Pflanzensoziologischer Garten; pflanzen-soziologische Bestandsaufnahme." Typoscript, Wettingen, August 25, 1978. Archiv SKK Landschaftsarchitekten, Wettingen.
Projects of students and staff of the Institut für Landschaft und Garten, Prof. Dr. Ing. Dieter Kienast (Sommersemester 1992–Wintersemester 1996–97). "Bericht des Instituts für Landschaft und Garten der Universität Karlsruhe." Karlsruhe, 1997. Archiv Fachgebiet Landschaftsarchitektur, Karlsruhe Institute of Technology (KIT).
Rothe, Ulrike, editorial staff, Universität Karlsruhe (TH), Institut für Landschaft und Garten, Prof. Dr. Ing. Dieter Kienast. "Dokumentation von Lehre und Forschung Sommersemester 1992 bis Wintersemester 1996/97." Karlsruhe, 1997. Archiv Fachgebiet Landschaftsarchitektur, Karlsruhe Institute of Technology (KIT).

2. Publications

Dieter Kienast. Photographs by Christian Vogt. Translated by Felicity Gloth, Bruce Almberg, and Katja Steiner. Basel: Birkhäuser, 2004.
Dieter Kienast: In Praise of the Sensousness. With a DVD by Marc Schwarz. Exhibition catalog. Zurich: gta Verlag, 1999. German edition: *Dieter Kienast: Lob der Sinnlichkeit.* Zurich: gta Verlag, 1999.
Dieter Kienast: Zwischen Arkadien und Restfläche. 1992. In collaboration with Günther Vogt, with photographs by Christian Vogt. Exhibition catalog. Lucerne: Architekturgalerie Luzern, 1992.
Hülbusch, Karl Heinrich, Heidbert Bäuerle, Frank Hesse, and Dieter Kienast. *Freiraum- und landschaftsplanerische Analyse des Stadtgebietes von Schleswig.* Urbs et Regio: Kasseler Schriften zur Geografie und Planung 11. Kassel: Gesamthochschulbibliothek, 1979.
Kienast, Dieter. "Die Ruderalvegetation der Stadt Kassel: Beiträge zur Vegetationskunde Nordhessens." In *50 Jahre Floristisch-soziologische Arbeitsgemeinschaft, 1927–1977.* Edited by Reinhold Tüxen and Hartmut Dierschke, 83–101. Todenmann-Göttingen: Floristisch-soziologische Arbeitsgemeinschaft, 1977.
———. "Naturschutz: Unkraut in der Henschelei!" *Der Monolith: Studentische Zeitschrift der Organisationseinheit Architektur, Stadt- und Landschaftsplanung, Gesamthochschule Kassel* 12 (December 1977): 2.
———. *Die spontane Vegetation der Stadt Kassel in Abhängigkeit von bau- und stadtstrukturellen Quartierstypen.* Urbs et Regio: Kasseler Schriften zur Geografie und Planung 10, PhD diss., Gesamthochschule Kassel. Kassel: Gesamthochschul-Bibliothek, 1978.
———. "Kartierung der realen Vegetation des Siedlungsgebietes der Stadt Schleswig mit Hilfe von Sigma-Gesellschaften." In *Assoziationskomplexe (Sigmeten) und ihre praktische Anwendung.* Berichte der internationalen Symposien der Internationalen Vereinigung für Vegetationskunde 21, 329–62. Vaduz: Cramer, 1978.
———. "Pflanzengesellschaften des alten Fabrikgeländes Henschel in Kassel." *Philippia* 3, no. 5 (1978): 408–22.
———. "Bemerkungen zum wohnungsnahen Freiraum" (Remarks on Free Spaces Close to Habitations). *anthos* 18, no. 4 (1979): 2–9.
———. "Vom naturnahen Garten; oder, Von der Nutzbarkeit der Vegetation." *Der Gartenbau* 100, no. 25 (1979): 1117–22.
———. "Wohngrün zu Mehrfamilienhäusern in Wollishofen, Zürich"

(Private Verdure of Apartment Blocks at Wollishofen, Zurich). *anthos* 18, no. 4 (1979): 10–13.

———. "Botanischer Garten Südteil: Naturnahe Biotope" (Botanic Garden Southern Section: Biotopes Close to Nature). *anthos* 19, no. 1 (1980): 56–65.

———. "Wilde Gärten." In *Handbuch für Quartier-Verbesserer.* Edited by Martin Weiss and Peter Lanz, n.p. Zurich: Ex Libris, 1980.

———. "'Ein Spiel ist der Weg der Kinder zur Erkenntnis der Welt, in der sie leben und die zu verändern sie berufen sind' (Maxim Gorki)." In *Öffentliche Baumappe der Ostschweiz,* 30–33. Goldach: Fachpresse Goldach, 1981. Republished in Professur für Landschaftsarchitektur ETH Zürich, *Dieter Kienast,* 47–52.

———. "Vom Gestaltungsdiktat zum Naturdiktat – oder: Gärten gegen Menschen?" *Landschaft + Stadt* 13, no. 3 (1981): 120–28.

———. "Zum Ausbau und Unterhalt des Botanischen Gartens Basel." *Brüglinger Mosaik* 5, no. 25 (1982): 15–23.

———. "Gedanken zum Projekt im ersten Rang, Stöckli & Kienast, Wettingen" (Project with 1st prize, Stöckli & Kienast, Wettingen), 26–27, in "Ideenwettbewerb Kurpark Bad Zurzach AG" (Competition of Ideas for Zurzach Bath Spa Parc). *anthos* 23, no. 4 (1984): 24–35.

———. "Kein Raum für Zürich" (No Room for Zurich). *anthos* 24, no. 1 (1985): 33–36.

———. "Bemerkungen zum Wettbewerb Kasernenareal Zürich – gesehen aus landschaftsarchitektonischer Sicht." *anthos* 25, no. 3 (1986): 42–45.

———. "Die Gestalt des öffentlichen Raumes." In *Stadt, Stadtraum, Raumqualität.* Edited by Schweizerischer Werkbund, Ortsgruppe Innerschweiz, 40–46. Lucerne: Schweizerischer Werkbund, 1986.

———. "Kultur oder Natur? Von Rasenmähern und Ökofreaks." *Aktuelles Bauen / Plan* 21, no. 3 (1986): 24–30.

———. "Ohne Leitbild." *Garten + Landschaft* 96, no. 11 (1986): 34–38.

———. "Interiorscaping; or, Pleasure in Verdure Planning Indoors." *anthos* 28, no. 1 (1989): 28–31.

———. "Zur Bedeutung von Freiräumen." *Kommunalmagazin* 6 (1989): 1–7.

———. "Ökologie gegen Gestalt? – oder Natürlichkeit und Künstlichkeit als Programm." *SRL Schriftenreihe* 25 (1990): 102–107.

———. "Über den Umgang mit dem Friedhof" (On Dealing with Cemeteries). *anthos* 29, no. 4 (1990): 10–14.

———. "Aussenräume der Ecole Cantonale de langue francaise, Bern" (The Grounds of the Ecole Cantonale de Langue Française, Bern). *Topos* 1 (1992): 76–77.

———. "Die Poesie der Stadtlandschaft." *Garten + Landschaft* 102, no. 3 (1992): 9–13.

———. "Ein neuer Park in Berlin" (A New Park in Berlin). *anthos* 31, no. 3 (1992): 36–40.

———. "Remarques sur la nature de la ville." *Faces* 24 (1992): 14, 16–17.

———. "Vom Kirchhof zum Erholungsraum." *Der Gartenbau* 113, no. 3 (1992): 86–89.

———. "Die Natur der Sache – Stadtlandschaften." In *Stadt-Parks: Urbane Natur in Frankfurt am Main.* Edited by Tom Koenigs, 10–21. Frankfurt am Main: Campus, 1993.

———. "Zwischen Stadtkante und Schloß." *Garten + Landschaft* 103, no. 12 (1993): 26–28.

———. "Begrünungslust im Innenraum." *Der Gartenbau* 115, no. 16 (1994): 34–35.

———. "Von der Notwendigkeit künstlerischer Innovation und ihrem Verhältnis zum Massengeschmack der Landschaftsarchitektur." Lecture in November 1991. In *Choreographie des öffentlichen Raumes.* Edited by Jürgen Wenzel, 52–64. Berlin: self-published, 1994.

———. "Zur Dichte der Stadt" (Reflections on Urban Density). *Topos* 7 (1994): 103–9.

———. "Zwischen Poesie und Geschwätzigkeit." *Garten + Landschaft* 104, no. 1 (1994): 13–17.

———. "Der Garten als geistige Landschaft" (The Garden as Spiritual Landscape). *Topos* 11 (1995): 68–79.

———. "Garten E. am Uetliberg." In *Gute Gärten,* 31.

———. "Garten Kienast, Zürich." In *Gute Gärten,* 37.

———. "Safari – oder wo ist Madagaskar?" (Safari—or: Where is Madagascar?). *Topos* 13 (1995): 63–72.

———. "Stadtlandschaft." *Zolltexte* 5, no. 2 (1995): 43–46.

———. "Ün decalogo / A Set of Rules." *Lotus International* 87 (1995): 63–81.

———. "Cultivating Discontinuity." Interview with Udo Weilacher. In *Between Landscape Architecture and Land Art.* Edited by Udo Weilacher. Tranlslated by Felicity Gloth 137–56. Basel: Birkhäuser, 1996.

———. "La nuova città paesaggio / The New City Landscape." *Lotus International* 90 (1996): 116–17.

———. "Funktion, Form und Aussage" (Function, Form and Statement). Interview by Robert Schäfer. *Topos* 18 (1997): 6–12.

———. "Madagaskar in Zürich." *anthos* 36, no. 4 (1997): 31–85.

———. "Outlooks and Insights." In *Kienast: Gärten / Gardens,* 96–103.

———. "Retorno a las raicep: Jardin en St. Gallen y entorno del Inselspital en Berna." *Arquitectura Viva* 53 (1997): 50–53.

———. "Stadt und Natur: Gartenkultur im Spiegel der Gesellschaft." *archithese* 27, no. 4 (1997): 4–11. Reprinted in *Passagen / Passages* 24 (1998): 33–40.

———. "Where Is Arcadia?" In *Kienast: Gärten / Gardens,* 82–95.

———. "Die Natur der Stadt: Städtische Dichte und authentische Natur." In *Kulturlandschaft Stadt: Architektur Städtebau Denkmalschutz; Texte für Ursula Koch, Stadträtin von Zürich von April 1986 bis März 1998,* 66–74. Zurich: Hochparterre, 1998.

———. "Naturwandel." *anthos* 37, no. 1 (1998): 10–15.

———. "10 Theses on Landscape Architecture." In *Dieter Kienast: In Praise of the Sensousness,* 84–94.

———. "Zehn Thesen zur Landschaftsarchitektur." *DISP* 138 (1999): 4–6. Reprinted in *Der Gartenbau* 120, no. 13 (1999): 6–9.

———. "Courtyard Design: Offices of Swiss Re, Zurich." In *Kienast Vogt: Aussenräume / Open Spaces,* 184–93.

———. "Expo 2000 Hanover and Fair Grounds." In *Kienast Vogt: Parks und Friedhöfe / Parks and Cemeteries,* 80–97.

———. "Landscape Design for Kronsberg in Hanover." In *Kienast Vogt: Parks und Friedhöfe / Parks and Cemeteries,* 126–31.

———. "Vom Kirchhof zum Erholungsraum" (1992). In Professur für Landschaftsarchitektur ETH Zürich, *Dieter Kienast,* 117–23.

———. "Zehn Thesen zur Landschaftsarchitektur." Version from November 1998. In Professur für Landschaftsarchitektur ETH Zürich, *Dieter Kienast,* 207–10.

———. "Fürstenwald Cemetry in Chur." In *Dieter Kienast,* 116–19.

———. "Garden of the Waldhaus Psychiatric Clinic, Chur." In *Dieter Kienast*, 104–15.
———. "Longing for Paradise." In *Dieter Kienast*, 84–90.
———, and Guido Hager. *Gartenarchitektur – Freiraumgestaltung, Nachdiplomstudiengang 89/90.* Schriftenreihe Abteilung Grünplanung, Landschafts- und Gartenarchitektur ITR – Ingenieurschule Interkantonales Technikum Rapperswil 1. Rapperswil: Abteilung Grünplanung, Garten- und Landschaftsarchitektur, 1992.
———, Hans Loidl, and Holger Haag. "Leidenschaft läßt sich nicht lehren." Interview with Robert Schäfer and K. H. Ludwig. *Garten + Landschaft* 100, no. 2 (1990): 43–48.
———, and Thom Roelly. *Standortökologische Untersuchungen in Stadtquartieren—insbesondere zur Vegetation—unter dem Aspekt der freiraumplanerischen Verwertbarkeit, Gesamthochschule Kassel.* Schriftenreihe OE Architektur, Stadt- und Landschaftsplanung 3, no. 2. Kassel: *Gesamthochschule Kassel*, 1978.
———, and Günther Vogt. "Die Form, der Inhalt und die Zeit" (Form, Content and Time). *Topos* 2 (1993): 6–17.
Kienast: Gärten / Gardens. Photographs by Christian Vogt. Translated by Bruce Almberg and Katja Steiner. Basel: Birkhäuser, 1997.
Kienast Vogt: Aussenräume / Open Spaces. Photographs by Christian Vogt. Translated by Felicity Gloth. Basel: Birkhäuser, 2000.
Kienast Vogt: Parks und Friedhöfe / Parks and Cemeteries. Photographs by Christian Vogt. Translated by Felicity Gloth. Basel: Birkhäuser, 2002.
Lehre des Instituts für Landschaft und Garten, Lehrstuhl für Landschaftsarchitektur und Entwerfen. "Dieter Kienast." In *Ein-Blick, Aus-Blick: Universität Karlsruhe, Fakultät für Architektur; Forschung und Lehre.* Edited by Fakultät für Architektur, Universität Karlsruhe, 120–25. Tübingen: Wasmuth, 1999.
Professur für Landschaftsarchitektur ETH Zürich, ed. *Dieter Kienast: Die Poetik des Gartens; Über Ordnung und Chaos in der Landschaftsarchitektur.* Basel: Birkhäuser, 2002.
Professur für Landschaftsarchitektur, ETH Zürich, Professor Christophe Girot, ed. *Textbuch Landschaftsarchitektur: Texte Dieter Kienast.* Zurich: Institut für Landschaftsarchitektur, ETH Zürich 2000.
Professur für Landschaftsarchitektur, Professor Dieter Kienast, ETH Zürich, ed. *Textbuch Landschaftsarchitektur.* Zurich: self-published, 1998.
Schwarz, Marc. *Dieter Kienast: In Praise of the Sensousness.* DVD. In *Dieter Kienast: In Praise of the Sensousness.*
———, and Annemarie Bucher. *D. K.: Eine Spurensuche.* DVD. Zurich: schwarzpictures, 2008.
Stöckli, Peter Paul, and Dieter Kienast. "Grünplanung Aarau: Ein Versuch, Freiraumplanung zu betreiben" (A Verdure Plan for Aarau: An Attempt to Implement Open-Space Planning). *anthos* 21, no. 3 (1982): 29–38.

II. Further literature and media

1. Unpublished sources

Blumenthal, Silvan. "Das Lehrcanapé und sein Architektenbild: Lucius Burckhardt an der ETH Zürich, 1970–1973." Diploma thesis on an elective subject, ETH Zurich, Professor Laurent Stalder, 2008.
Brockelmann, H. "Das Kasseler Modell der integrierten Gesamthochschule im Studiengang Architektur, Stadt- und Landschaftsplanung." Seminar paper, Architecture and Urban, and Landscape Planning, Gesamthochschule Kassel, 1992. Archiv Grauer Raum, docu:lab, University of Kassel, Department of Architecture, Urban and Regional Planning, Landscape Architecture and Landscape Planning.
Fahrni, Sacha, and Christian Gut. "Gartenräume von Dieter Kienast: Der Gartenraum im Verhältnis zur Landschaft." Diploma thesis on an elective subject, ETH Zurich, Chair of Landscape Architecture, Professor Christophe Girot (supervision: Anette Freytag), 2009.
"Grün 80—Schlussbericht der Direktion: 2. Schweizerische Ausstellung für Garten- und Landschaftsbau 12.04–12.10.1980." Typoscript, Basel, 1980. Archiv SKK Landschaftsarchitekten, Wettingen.
Hoffmann, Heide. "Das Studium der Stadtplanung: Studiengänge im Vergleich; Kassel, Dortmund, Berlin, Kaiserslautern, Hamburg-Harburg." Seminar paper, Architecture and Urban and Landscape Planning, Gesamthochschule Kassel, winter semester 1996–97. Archiv Grauer Raum, docu:lab, University of Kassel, Department of Architecture, Urban and Regional Planning, Landscape Architecture and Landscape Planning.
Rüegg, Arthur. Speech at the memorial for Dieter Kienast at ETH Zurich. Typoscript, Zurich, January 23, 1999. Archiv Arthur Rüegg.

2. Secundary literature, online sources, radio broadcasts

Aberle, Waltraud. "Landschaft der Gärten beim Swisscom-Neubau." *Der Gartenbau* 120, no. 15 (1999): 6–9.
Adams, William Howard. *The French Garden 1500–1800.* New York: Braziller, 1979.
Adina, Michele. *The Oppenheim Fountain.* swissinfo.ch, December 17, 2012, https://www.swissinfo.ch/eng/the-oppenheim-fountain_a-bridge-between-nature-and-the-city/34137508.
Adorno, Theodor W. *Aesthetic Theory.* Edited and translated by Robert Hullot-Kentor. Minneapolis: University of Minnesota Press, 1997.
———. "Functionalism Today." Translated by Jane O. Newman and John H. Smith. *Oppositions* 17 (1979): 31–41.
———. *Ohne Leitbild: Parva aesthetica.* 5th ed. Frankfurt am Main: Suhrkamp, 1973.
Aeberhard, Beat. "Der Garten der Sinne: Dieter Kienast—Lob der Sinnlichkeit an der ETH Zürich." *archithese*, 30, no. 1 (2000): 90.
"Albertslund-Syd." *anthos* 7, no. 3 (1968): 30–31.
Almqvist, Paula. "Solitär am See." *Architektur & Wohnen* 2 (2002): 128–30.
"Alters-Wohn- und Siedlungsheim, Dübendorf/ZH" (Old Age and Residential Development in Dübendorf/Zurich). *anthos* 7, no. 1 (1968): 11–13.
Amstutz, Marcel. "Das kleine Paradies." *Basler Magazin* (April 27, 1985): 1–5.
Andersson, Thorbjörn. "The Functionalism in Gardening Art." In *20th-Century Architecture.* Vol. 4, *Sweden.* Edited by Claes Caldenby, Jöran Lindvall, and Wilfried Wang, 226–41. Munich: Prestel, 1998.
Andritzky, Michael, and Klaus Spitzer, eds. *Grün in der Stadt: Von oben, von selbst, für alle, von allen.* Reinbek bei Hamburg: Rowohlt, 1981.
Atteslander, Peter, and Bernd Hamm, eds. *Materialien zur Siedlungs-*

soziologie. Neue wissenschaftliche Bibliothek 69, Soziologie. Cologne: Kiepenheuer & Witsch, 1974.

"Atrium-Siedlung in Reinach/Basel" (Atrium Development in Reinach/Basle). *anthos* 7, no. 3 (1968): 37–40.

Auböck, Maria. "Wie sieht denn ein gebauter Frühlinstag aus?" In *Dieter Kienast: Zwischen Arkadien und Restfläche,* 7–8.

Bachmann, Heini, Max Bosshard, Rut Fön, Hartmut Frank, Alex Gérad, Hans-Jürgen Herzog, and Hans-Ruedi Müller. *"Göhnerswil": Wohnungsbau im Kapitalismus; Eine Untersuchung über die Bedingungen und Auswirkungen der privatwirtschaftlichen Wohnungsproduktion am Beispiel der Vorstadtsiedlung "Sunnebüel" in Volketswil bei Zürich und dem Generalunternehmung Ernst Göhner AG.* Zurich: Verlagsgenossenschaft, 1972.

Bachmann, Ingeborg. *Frankfurter Vorlesungen: Probleme zeitgenössischer Dichtung.* Munich: Piper, 1980.

Bartetzko, Dieter. "Die Natur der Großstädter: Frankfurts künftiger neu-alter Günthersburgpark." *Stadtbauwelt,* 84, no. 117 (1993): 594–97.

Barzilay, Marianne, Catherine Hayward, and Lucette Lombard-Valentino. *L'Invention du parc: Parc de la Villette, Paris; Concours international / International Competition, 1982–1983.* Paris: Graphite, 1984.

Bauman, Zygmunt. *Intimations of Postmodernity.* London: Routledge, 1992.

Beck, Ulrich. *Risk Society: Towards a New Modernity.* Translated by Mark Ritter. London: Sage Publications, 1992.

Berger, John, and Jean Mohr, with the help of Nicolas Philibert. *Another Way of Telling.* New York: Pantheon, 1982.

Berger, Maurice. *Robert Morris: Minimalism and 1960s.* New York: Harper & Row, 1989.

Berger, Peter L., and Hansfried Kellner. *Sociology Reinterpreted: An Essay on Method and Vocation.* Garden City, N. Y.: Anchor Press/Doubleday, 1981.

Berndt, Heide. *Die Natur der Stadt.* Frankfurt am Main: Neue Kritik, 1978.

Beuys, Joseph. "Not Just a Few Are Called, But Everyone." Interview by George Jappe (1972). Translated by John Wheelwright. In *Art in Theory 1900–2000: An Anthology of Changing Ideas.* Edited by Charles Harrison and Paul Wood, 903–6. New ed. Malden, MA: Blackwell, 2003.

Blair MacDougall, Elisabeth, and F. Hamilton Hazlehurst, eds. *The French Formal Garden.* Dumbarton Oaks Colloquium on the History of Landscape Architecture 3. Washington D. C.: Dumbarton Oaks, 1974.

Blattner, Martin, Markus Ritter, and Klaus C. Ewald. *Basler Natur-Atlas.* Edited by Basler Naturschutz, Sektion des Schweizerischen Bundes für Naturschutz. Basel: Basler Naturschutz, 1985.

Blumenthal, Silvan. "Das Lehrcanapé." *archithese* 39, no. 2 (2009): 96–99.

———. *Das Lehrcanapé: Lucius Burckhardt und das Architektenbild an der ETH Zürich, 1970–1973.* Standpunkte Dokumente 2. Basel: Standpunkte, 2010.

Böhme, Gernot. *Für eine ökologische Naturästhetik.* Frankfurt am Main: Suhrkamp, 1989.

Böhme, Hartmut. *Natur und Subjekt.* Frankfurt am Main: Suhrkamp, 1988.

Bohrer, Karl Heinz. "Nach der Natur: Ansichten einer Moderne jenseits der Utopie." In *Über Politik und Ästhetik,* 209–29. Munich: Hanser, 1988.

Böse, Helmut. *Die Aneignung von städtischen Freiräumen: Beiträge zur Theorie und sozialen Praxis des Freiraums.* Gesamthochschule Kassel: Arbeitsbericht des Fachbereichs Stadtplanung und Landschaftsplanung 22. Kassel: Gesamthochschule Kassel, 1981.

———. "Vorbilder statt Leitbilder." *Garten + Landschaft* 96, no. 11 (1986): 28–33.

Böse-Vetter, Helmut, ed. *Freiraum und Vegetation: Festschrift zum 60. Geburtstag von Karl Heinrich Hülbusch am 21. Mai 1996.* Notizbuch der Kasseler Schule 40, no. 2. Kassel: Arbeitsgemeinschaft Freiraum und Vegetation, 1996.

———. "Zu den Notizbüchern und zur Kasseler Schule." In Böse-Vetter, *Freiraum und Vegetation,* 2.

Bois, Yve-Alain. "A Picturesque Stroll around Clara-Clara." *October* 29 (Spring 1984): 32–64.

Bourdieu, Pierre. *Zur Soziologie der symbolischen Formen.* 5th ed. Frankfurt am Main: Suhrkamp, 1994.

Brecht, Bertolt. "Mr. K and Nature." Translated by Martin Chalmers. In *Stories of Mr. Keuner,* 17. San Francisco: City Lights, 2001.

Boesers, Knut. "Die Folter, das ist die Vernunft: Ein Gespräch Knut Boesers mit Michel Foucault." In *Die Sprache des Grossen Bruders: Gibt es ein ost-westliches Kartell der Unterdrückung?* Literaturmagazin 8. Edited by Nicolas Born and Jürgen Manthey, 60–68. Reinbek bei Hamburg: Rowohlt, 1977.

Bredekamp, Horst. 1995. *Vicino Orsini und der heilige Wald von Bomarzo: Ein Fürst als Künstler und Anarchist.* 2 vols. Worms: Werner, 1995.

Brenken, Heike, Antje Brink, Anka Förster, Christina von Haaren, Kaspar Klaffke, Michael Rode, and Wulf Tessin. *Naturschutz, Naherholung und Landwirtschaft am Stadtrand: Integrierte Landnutzung als Konzept suburbaner Landschaftsentwicklung am Beispiel Hannover-Kronsberg; Abschlussbericht zum E+E-Vorhaben "Naturschutzorientierte Entwicklung im suburbanen Bereich am Beispiel Hannover-Kronsberg" und Zwischenbericht der wissenschaftlichen Begleituntersuchungen.* Angewandte Landschaftsökologie 57. Bonn-Bad Godesberg: Landwirtschaftsverlag, 2003.

Brömer, Jens, ed. *Portrait einer Hochschule: Universität Gesamthochschule Kassel.* 2nd ed. Kassel: Gesamthochschule Kassel, 1994.

Bucher, Annemarie. "Vom Landschaftsgarten zur Gartenlandschaft: Schweizerische Gartengestaltung auf dem Weg in die Gegenwart." In *Vom Landschaftsgarten zur Gartenlandschaft: Gartenkunst zwischen 1880 und 1980 im Archiv für Schweizer Gartenarchitektur und Landschaftsplanung.* Edited by Stiftung Archiv für Schweizer Gartenarchitektur und Landschaftsplanung, 35–86. Zurich: vdf Hochschulverlag, 1998.

———. "Landschaft zwischen Bild und Begriff." *trans* 11 (November 2003): *transScape:* 12–19.

———. "Naturen ausstellen: Schweizerische Gartenbauausstellungen zwischen Kunst und Ökologie." DSc diss., ETH Zurich, 2008.

———, and Martine Jaquet. *"Von der Blumenschau zum Künstlergarten": Schweizerische Gartenbau-Ausstellungen.* Lausanne: 2000.

Bühler, Karl-Dietrich. "Sven Ingvar Anderssons Grünes Reich" (Sven-Ingvar Andersson's Verdant Realm). *anthos* 16, no. 1 (1977): 23–28.

Burckhardt, Lucius. "Das Ende der polytechnischen Lösbarkeit" [1989]. In *Wer plant die Planung,* 119–28. Reprinted as "The End of

Polytechnic Solvability (1989)" in *Who Plans the Planning?*, 110–18.

———. *Das Zebra streifen.* Schriftenreihe des Fachbereichs Stadtplanung und Landschaftsplanung 20. Edited by Helmut Aebischer, with texts by Gerhard Lang et al., 1. Kassel: Infosystem Planung, GhK, 1994.

———. "Der Architekt in der Gesellschaft von morgen." *Werk* 52, no. 11 (1965): 243*–44*.

———. "Der bedeutungsvolle Garten des Ian Hamilton Finlay" (Ian Hamilton Finlay's Expressive Garden). *anthos*, 23, no. 4 (1984): 2–8.

———. "Der kleinstmögliche Eingriff." In *Die Kinder fressen ihre Revolution*, 241–47.

———. *Design ist unsichtbar.* Edited by Hans Höher. Ostfildern: Cantz, 1995.

———. *Die Kinder fressen ihre Revolution: Wohnen—Planen—Bauen—Grünen.* Edited by Bazon Brock. Cologne: DuMont, 1985.

———. "Gärtnern: Kunst und Notwendigkeit." *Basler Magazin* 21 (1977): 1–5. Reprinted as "Gardening—An Art and A Necessity (1977)" in *Writings*, 123–32.

———. "Kritik der sechziger Jahre." *Werk* 60, no. 12 (1973): 1588–90.

———. "A Matter of Looking and Recognizing: In Conversation with Thomas Fuchs (1993)." In *Why Is Landscape Beautiful?*, 282–87.

———. "No Man's Land (1980)." In *Why Is Landscape Beautiful?*, 126–27.

———. "Strollogy – A New Science (1998)." In *Why Is Landscape Beautiful?*, 288–94.

———. "Vom Entwurfsakademismus zur Behandlung bösartiger Probleme." *Canapé News* 29 (1973): 34–37. Reprinted as "From Design Academicism to the Treatment of Wicked Problems (1973)" in *Writings*, 77–84.

———. *Who Plans the Planning? Architecture, Politics, and Mankind.* Edited by Jezko Fezer and Martin Schmitz. Basel: Birkhäuser, 2019. German original, *Wer plant die Planung? Architektur, Politik und Mensch.* Edited by Jezko Fezer and Martin Schmitz. N.p. [Berlin]: Schmitz, 2004.

———. *Why Is Landscape Beautiful? The Science of Strollology.* Edited by Markus Ritter and Martin Schmitz. Translated by Jill Denton. Basel: Birkhäuser, 2015. German original, *Warum ist Landschaft schön? Die Spaziergangswissenschaft.* Textsammlung Lucius Burckhardt 2. Edited by Markus Ritter and Martin Schmitz. Berlin: Schmitz, 2006.

———. *Writings: Rethinking Man-Made Environments; Politics, Landscape & Design.* Edited by Jesko Fezer and Martin Schmitz. Vienna: Springer, 2012.

———, and Walter Förderer. *Bauen ein Prozess,* Teufen: Niggli, 1972.

———, Max Frisch, and Markus Kutter. *achtung: die Schweiz; Ein Gespräch über unsere Lage und ein Vorschlag zur Tat.* Basler politische Schriften 2. Basel: Handschin, 1955.

———, Vittorio Gregotti, and Pierluigi Nicolin. *Die Stadt: Bild und Wirkung; Eine Veranstaltung des Internationalen Design Zentrum Berlin.* Edited by Internationales Design Zentrum Berlin. Berlin: Internationales Design Zentrum, 1974.

———, and Internationales Design Zentrum Berlin. *Design der Zukunft: Architektur, Design, Technik, Ökologie.* Cologne: DuMont, 1987.

———, and Markus Kutter. *Wir selber bauen unsere Stadt: Ein Hinweis auf die Möglichkeiten staatlicher Baupolitik.* Foreword by Max Frisch. Basel: Handschin, 1953.

———, and Vladimir Nikolic. *Sichtbar machen: Documenta Urbana Kassel.* Kassel: Gesamthochschule/Fachbereich 13/Stadt- und Landschaftsplanung, 1981.

Buttlar, Adrian von. *Der Landschaftsgarten: Gartenkunst des Klassizismus und der Romantik.* Cologne: DuMont, 1989.

Canapé News 29 (1973): *Experiment Canapé (Schlussbericht).*

Christian Vogt: Fotografische Arbeiten im Architekturmuseum in Basel / Photographic works in the Architecture Museum in Basel. Exhibition catalog. Basel: Architekturmuseum, 1990.

Clément, Gilles. *Le Jardin en mouvement: De la vallée au parc André-Citroën.* Paris: Sens & Tonka, 1994.

Conan, Michel, ed. *Environmentalism in Landscape Architecture.* Dumbarton Oaks Colloquium on the History of Landscape Architecture 21. Washington D.C.: Dumbarton Oaks, 1999.

———. *Perspectives on Garden Histories.* Dumbarton Oaks Colloquium on the History of Landscape Architecture 22. Washington D.C.: Dumbarton Oaks, 2000.

Conrads, Ulrich, ed. *Die Bauhaus-Debatte 1953: Dokumente einer verdrängten Kontroverse.* Braunschweig: Vieweg, 1994.

Corboz, André. "The Land as Palimpsest." *Diogenes* 31, no. 121 (1983): 12–34.

Danzer, Robert, and Helmut Oberrauner. "Wettbewerb: Internationale Gartenschau – Steiermark 2000." *Wettbewerbe: Architekturjournal* 21, no. 159–160 (1997): 76–85.

De Jong, Erik A. "Paradise Is Just Where You Are Right Now." In *Kienast Vogt: Aussenräume / Open Spaces,* 10–15.

Dean, Martin R. "Places against Oblivion of the Self." In *Dieter Kienast*, 6–9.

De Certeau, Michel. *The Practice of Everyday Life.* Translated by Steven Rendall. Berkeley: University of California Press, 1984.

Derrida, Jacques. "Différance" [1972]. Translated by Alan Bass. In *Margins of Philosophy*, 3–27. Chicago: University of Chicago Press, 1982.

———. *Of Grammatology.* Translated by Gayatri Chakravorty Spivak. Baltimore: Johns Hopkins Univerity Press, 1976. French original published 1967.

———. "Signature Event Context" [1972]. Translated by Alan Bass. In *Margins of Philosophy,* 307–30. Chicago: University of Chicago Press, 1982.

Descombes, Georges. "Superposition." Lecture at the conference "Thinking the Contemporary Landscape: Positions and Oppositions," Hanover-Herrenhausen, June 20, 2013. http://www.video.ethz.ch/conferences/2013/ila/05_thursday/2ea2d207-bec0-44de-89bc-c503917d1ca5.html.

———. *Voie suisse: L'itinéraire genevois; De Morschach à Brunnen.* Geneva: Republic and Canton of Geneva, 1991.

Detzelhofer, Anna. "Zauber der Gärten: Die IGS 2000 in Graz in der Steiermark." *Landschaftsarchitektur* 29, no. 12 (1999): 26–28.

Deutscher Werkbund, Arbeitsgruppe Kassel, ed. *Durch Pflege zerstört: Die Kasseler Karlsaue vor der Bundesgartenschau.* Kassel: n.p., 1980.

Dewey, John. *Art as Experience.* New York: Perigee Books, 2005. 1st ed. New York: Minton, Balch, 1934.

Diedrich, Lisa. "Dieter Kienast, 1945–1998." *Bauwelt* 90, no. 4 (1999): 150.

Documenta 6. Kassel. Exhibition catalog. 3 vols. Kassel: Dierichs, 1977.

Documenta 7. Kassel. Exhibition catalog. 2 vols. Kassel: Dierichs, 1982.

Dünne, Jörg, and Stephan Günzel, ed. *Raumtheorie: Grundlagentexte aus Philosophie und Kulturwissenschaft.* Frankfurt am Main: Suhrkamp, 2006.

Duvigneaud, Paul. "L'écosystème 'urbs': Études écologiques de l'écosystème urbain Bruxellois 1." *Mémoires de la Société Royale de Botanique de Belgique* 6 (1974): 5–35.

Eggenberger, Christian, and Lars Müller, eds. *Photo Suisse.* Translated by Ishbel Flett and Catherine Schelbert. Baden: Lars Müller, 2004.

"Ein Wohnhaus aus der Werkstatt." Conversation with Bruno Reichlin, Marcel Meili, and Markus Peter. *Werk, Bauen + Wohnen* 80, no. 11 (1993): 16–27.

Elias, Norbert. "On Nature." In Norbert Elias. *On the Sociology of Knowledge and the Sciences.* Collected Works 14, Essays I, 53–65. Edited by Richard Kilminster and Stephen Mennell. Dublin: University College Dublin Press, 2009.

Ellenberg, Heinz. *Vegetation Mitteleuropas mit den Alpen in ökologischer Sicht.* 2nd ed. Stuttgart: Ulmer, 1978.

Engelmann, Peter. *Postmoderne und Dekonstruktion: Texte französischer Philosophen der Gegenwart.* Stuttgart: Reclam, 1990.

"Ergebnisse des Wettbewerbs für die Grün 80: Zweite Schweizerische Ausstellung für Garten- und Landschaftsbau in Basel" (Results of the Competition for "Grün 80": 2nd Swiss Exhibition of Garden and Landscape Design 1980 in Basle). *anthos* 16, no. 3 (1977): 32–37.

Fezer, Jesko. "Politics—Environment—Mankind." In Burckhardt, *Who Plans the Planning?,* 12–16.

———, and Martin Schmitz. "The Work of Lucius Burckhardt." In Burckhardt, *Writings,* 7–26.

Foucault, Michel. "Andere Räume." In *Textbuch Landschaftsarchitektur.* Edited by Professur für Landschaftsarchitektur, ETH Zürich, 90–99. Zurich: self-published, 1998.

———. "Les hétéropies." In *Die Heterotopien / Les hétéropies—Der utopische Körper / Le corps utopique: Zwei Radiovorträge,* 37–52. Frankfurt am Main: Suhrkamp, 2005.

Foxley, Alice. *Distance and Engagement: Walking, Thinking and Making Landscape; Vogt Landscape Architects,* Baden: Lars Müller, 2010.

Franzen, Brigitte. *Die vierte Natur: Gärten in der zeitgenössischen Kunst.* Cologne: König, 2000.

Fred Eicher Landschaftsarchitekt: Schulthess Gartenpreis 2004. Zurich: Schweizer Heimatschutz, 2004.

Freytag, Anette. "Back to Form: Landscape Architecture and Representation in Europe after the Sixties." In *Composite Landscapes: Photomontage and Landscape Architecture.* Exhibition catalog. Edited by Charles Waldheim and Andrea Hansen, 92–115. Ostfildern: Hatje Cantz, 2014.

———. "Bildkritik. Ästhetisierungsstrategien in der zeitgenössischen Landschaftsarchitektur." *Die Gartenkunst* 15, no. 2 (2003): 204–24.

———. "Der Garten: Streben nach Glück und Erkenntnis." *anthos* 49, no. 1 (2010): 4–9.

———. "Dieter Kienast and the Topological and Phenomenological Dimensions of Landscape Architecture." *Thinking the Contemporary Landscape.* Edited by Christophe Girot and Dora Imhof, 229–245. New York: Princeton Architectural Press, 2017.

———. "Garden K in Zurich by Kienast Vogt Partner." *Best Private Plots / Die besten Gärten: Internationale Beispiele zur Gartenarchitektur 2006,* 20–22. St. Pölten: Land Niederösterreich, Abteilung Umweltwirtschaft und Raumordnungsförderung, 2006.

———. "Global Dangers—Private Oases? Gardens: Setting the Trend for a New Era of Caring for Ourselves." In *System Landschaft: Zeitgenössische deutsche Landschaftsarchitektur / Landscape as a System: Contemporary German Landscape Architecture.* Edited by Bund Deutscher Landschaftsarchitekten, 78–83. Basel: Birkhäuser, 2009.

———. "Grau und Grün: Dieter Kienasts (1945–1998) Beitrag zur Landschaftsarchitektur." *Gartenhistorisches Forschungskolloquium: Zusammenstellung der Tagungsbeiträge.* Graue Reihe des Instituts für Stadt- und Regionalplanung 17. Edited by Sylvia Butenschön, 47–52. Berlin: Universitätsverlag der TU, 2008.

———. "Siedlung Lindenstrasse/Huebwiesen in Niederhasli (1972–75) von Dieter Kienast und Peter Paul Stöckli." Lecture at the Rapperswiler Tag, theme "Wohnen!" (Housing!), Hochschule für Technik Rapperswil, May 16, 2008. www.rapperswilertag.ch/r08/downloads08/05_RT08_Freytag.pdf (link no longer active).

———. "Topology and Phenomenology in Landscape Architecture." In *Landschaftsarchitekturtheorie: Aktuelle Zugänge, Perspektiven und Positionen.* Edited by Karsten Berr, 195–225. Wiesbaden: Springer, 2018.

———. "Zeitgenössisches aus dem Privatgarten." *Die Brücke: Kärnten, Kunst, Kultur* 71 (October 2006): 28–30.

———, and Wolfgang Kos. "Neue Parkideen in Europa: Zwischen Arkadien und Restfläche." Radio broadcast. *Diagonal: Radio für Zeitgenossen,* Österreichischer Rundfunk (ORF), Österreich 1 (Ö 1), first broadcast on October 10, 1998.

Furter, Fabian, and Patrick Schöck, eds. *Göhner Wohnen: Wachstumseuphorie und Plattenbau.* Exhibition catalog. Baden: Hier und Jetzt, 2013.

Garten der Kunst: Österreichischer Skulpturenpark / Art Garden: Scuplture Park Austria. Edited by Österreichische Skulpturenpark Privatstiftung. Ostfildern: Hatje Cantz, 2006.

Gesamthochschule Kassel, Stadt und Landschaftsplanung, ed. *Bibliografie Stadtvegetation.* Kassel: Gesamthochschule Kassel, Infosystem Planung, GhK, 1987.

Ginzburg, Carlo. "Clues: Roots of an Evidential Paradigm" [1979]. Translated by John and Anne C. Tedeschi. In *Clues, Myths, and the Historical Method,* 87–113. Baltimore: Johns Hopkins University Press, 1989.

Girot, Christophe. "Ansätze zu einer allgemeinen Landschaftstheorie." *Topos* 28 (1999): 33–41. Reprinted as "Towards a General Theory of Landscape" in *Texte zur Landschaft: Essays über Entwurf, Stil, Zeit und Raum / About Landscape: Essays on Design, Style, Time, and Space,* 83–92. Munich: Callwey, 2002.

———. "Graue Landschaften." In Professur für Landschaftsarchitektur ETH Zürich, *Dieter Kienast,* 7–10.

———. "Naturerfahrung und Symbolik im Stadtgrün / The Shifting Meaning of Nature in the City." *Stadtgrün: Europäische Landschaftsarchitektur im 21. Jahrhundert / Urban Green: European Landscape Design for the 21st Century.* Edited by Annette Becker and Peter Cachola Schmal, 218–25. Basel: Birkhäuser, 2010.

———. "Raus aus dem Garten, rein in die Landschaft." In *Unterwegs in Zürich und Winterthur: Landschaftsarchitektur und Stadträume 2000–2009.* Edited by Claudia

Moll and Roderick Hönig, 154–60. Zurich: Edition Hochparterre, 2009.

———. "Towards a Landscape Society." In Thies Schröder. *Changes in Scenery: Contemporary Landscape Architecture in Europe,* 6–9. Basel: Birkhäuser, 2001.

———. "Urbane Landschaften der Zukunft." *anthos* 42, no. 2 (2003): 36–40.

———. "Vom Volkspark zum Themenpark: Ein Plädoyer für Zeit-Räume. In *Mögliche Räume: Stadt schafft Landschaft.* Edited by Diethild Kornhardt, Gabriele Pütz, and Thies Schröder, 162–66. Hamburg: Junius, 2002.

Göbl-Gross, Thomas. "Antwort auf virtuelle Welten." In *Garten + Landschaft* 108, no. 7 (1998): 10–13.

———. "Die Freiflächen des Zentrums für Kunst und Medientechnologie, Karlsruhe" (The Grounds of the Center for Art and Media, Karlsruhe). In *Kienast Vogt: Aussenräume / Open Spaces,* 152–63.

———. "Dieter Kienast gestorben." *Neue Luzerner Zeitung,* May 5, 1999. Also printed in *Neue Nidwaldner Zeitung, Neue Obwaldner Zeitung, Neue Schwyzer Zeitung, Neue Urner Zeitung, Neue Zuger Zeitung.*

Gmür, Otti. "Zwischen Arkadien und Restfläche." In *Dieter Kienast: Zwischen Arkadien und Restfläche,* 5–6.

Greiner, Ulrich. "Falsches Präsens: Warum unterwerfen sich so viele Autoren dem Diktat der Gegenwart?" *DIE ZEIT,* no. 19, April 30, 2009, feuilleton.

Groener, Fernando, and Rose-Marie Kandeler, ed. *7000 Eichen.* Cologne: König, 1987.

Groh, Ruth, and Dieter Groh. "Natur als Maßstab: Eine Kopfgeburt." *Merkur* 47, no. 536 (1993): 965–79.

Grosse-Bächle, Lucia. *Eine Pflanze ist kein Stein: Strategien für die Gestaltung mit der Dynamik von Pflanzen; Untersuchung an Beispielen zeitgenössischer Landschaftsarchitektur.* Beiträge zur räumlichen Planung 71. PhD diss., Universität Hannover, 2003.

Groupement Superpositions: Schulthess Gartenpreis 2012 / Prix Schulthess des Jardins 2012. Zurich: Schweizer Heimatschutz, 2012.

"Grün 80 Basel, Wettbewerbsprojekt, 1977." In Jansen et al., *Architektur lehren,* 258.

———. "Freizeitlandschaft der Olympiade 1972." Interview. *Garten + Landschaft* 79, no. 7 (1969): 215–17.

Grzimek, Günther. "Leitbild für das Studium am Lehrstuhl für Landschaftskultur an der Hochschule für bildende Künste, Kassel." In *Hochschule für Bildende Künste Kassel, Lehrstuhl für Landschaftskultur,* 101–2. Kassel: Lehrstuhl für Landschaftskultur, 1968.

———. "Spiellandschaft der Olympiade 1972." Interview. *Garten + Landschaft* 80, no. 9 (1970): 301–6.

Gute Gärten: Gestaltete Freiräume in der Region Zürich. Zurich: Bund Schweizer Landschaftsarchitekten und Landschaftsarchitektinnen (BSLA), Regionalgruppe Zürich, 1995.

Haarman, Anke. "Artistic Research / Künstlerische Forschung." Lecture at the Universität der Künste Berlin, 2008. www.aha-projekte.de/HaarmannArtisticResearch.pdf.

Hagen-Hodgson, Petra. "Natur entwerfen: Tagung zur Natürlichkeit und Künstlichkeit im Werk von Dieter Kienast (1945–1998)." *werk, bauen + wohnen* 96, no. 1–2 (2009): 64–65.

Hager, Guido. "École cantonale de la langue française de Berne." *Faces* 24 (1992): 15.

———. "Fleeting Moments—an Encounter in Time." In *Dieter Kienast: In Praise of the Sensousness,* 9–11.

———. "Pausenplatz, École cantonale de la langue française." In *Dieter Kienast: Zwischen Arkadien und Restfläche,* 29–33.

Hahne, Heinz. Review of *Die spontane Vegetation der Stadt Kassel* by Dieter Kienast. *Das Gartenamt* 31, no. 11 (1982): 692–93.

Hallmann, Heinz. "Die Entwicklung der Landschaftsarchitekten und ihre Ausbildung in Deutschland." Part 2. *Das Gartenamt: Stadt und Grün* 41, no. 3 (1992): 165–70.

Handke, Peter. *Die Geschichte des Bleistifts.* Frankfurt am Main: Suhrkamp, 1985.

———. "The Lesson of Mont Sainte-Victoire" [1980]. In *Slow Homecoming.* Translated by Ralph Manheim, 139–212. New York: Farrar, Straus, and Giroux, 1985.

———. "The Long Way Around" [1979]. In *Slow Homecoming.* Translated by Ralph Manheim, 3–138. New York: Farrar, Straus, and Giroux, 1985.

Hard, Gerhard. 1996. "Schwierigkeiten beim Spurenlesen." In Böse-Vetter, *Freiraum und Vegetation,* 39–51.

Harvey, David. *The Condition of Postmodernity: An Enquiry into the Origins of Cultural Change.* Cambridge, MA: Blackwell, 1990.

Heintz, Bettina, and Jörg Huber, eds. *Mit dem Auge denken: Strategien der Sichtbarmachung in wissenschaftlichen und virtuellen Welten.* Zurich: Voldemeer, 2001.

Helbich, Ilse. "Von unbetretbaren Gärten." In Freytag and Kos, "Neue Parkideen in Europa."

Henningsen, Jens. "Stadtteilpark und Wuhlegrünzug." *Landschaftsarchitekten BDLA* 1 (1999): 25.

Herzog, Jacques. "Mit allen Sinnen spüren." Interview by Sabine Kraft and Christian Kühn. *ARCH+* 31, no. 142 (1998): 32–40.

Herzog, Jacques, Pierre de Meuron, and Dieter Kienast. "Aufwertung einer gebauten Leere durch Neuordnung." *werk, bauen + wohnen* 87, no. 4 (2000): 10–14.

Hiesmayr, Ernst. "Garten der Siedlung Nussdorf, Wien." *Fred Eicher Landschaftsarchitekt,* 14–15.

Hill, Penelope. *Contemporary History of Garden Design: European Gardens between Art and Architecture.* Basel: Birkhäuser, 2004.

Hoesli, Bernhard. "Addendum: Transparent Form-Organization as an Instrument of Design" [1982]. Translated by Jori Walker. In Rowe and Slutzky, *Transparency,* 84–99.

———. "Commentary" [1968]. Translated by Jori Walker. In Rowe and Slutzky, *Transparency,* 57–83.

———. "Die Abteilung für Architektur." In *Eidgenössische Technische Hochschule Zürich 1855–1980: Festschrift zum 125jährigen Bestehen.* Edited by the Rector of ETH Zurich, 253–90. Zurich: Neue Zürcher Zeitung, 1980.

Hofer, Polo. "Kein grüner Salat an der Grün 80." *Basler-Zeitung,* August 23, 1980: 10.

Hohler, Franz. "The Recapture" [1982]. Translated by Diane Dicks. In *At Home: A Selection of Stories,* 62–77. Basel: Bergli, 2009.

Huber, Silvia. Umgebungsgestaltung Sparkasse D – Fürstenfeldbruck: Mosaikartige Eingriffe. *architektur + technik* 22, no. 4 (1999): 14–18.

Hülbusch, Inge Meta. *Innenhaus und Aussenhaus: Umbauter und sozialer Raum.* 2nd ed. Schriftenreihe OE Architektur, Stadt- und Landschaftsplanung 1, no. 33. Kassel: Gesamthochschule Kassel, 1981.

Hülbusch, Karl Heinrich. "Anmerkungen zu diesem Notizbuch: Krautern mit Unkraut." In *Krautern mit Unkraut; oder, Gärtnerische*

Erfahrung mit der spontanen Vegetation. Notizbuch der Kasseler Schule 2, 1–15. Kassel: Arbeitsgemeinschaft Freiraum und Vegetation, 1986.

———. "Notizbuch der Kasseler Schule: Programmatische Anmerkungen." In *Krautern mit Unkraut; oder, Gärtnerische Erfahrung mit der spontanen Vegetation.* Notizbuch der Kasseler Schule 2, 158–63. Kassel: Arbeitsgemeinschaft Freiraum und Vegetation, 1986.

———. "Vegetationskundige Spaziergänge." In Böse-Vetter, *Freiraum und Vegetation,* 417–20.

Hunt, John Dixon. *L'Art du jardin et son histoire.* Paris: Odile Jacob, 1996.

———, and Peter Wills, eds. *The Genius of the Place: The English Landscape Garden, 1620–1820.* Cambridge, MA: MIT Press, 1993.

"Ideenwettbewerb Kurpark Bad Zurzach AG." *anthos* 23 (1984), no. 4: 24–35.

Jahn, Gisela. No title. In *Assoziationskomplexe (Sigmeten) und ihre praktische Anwendung.* Berichte der Internationalen Symposien der Internationalen Vereinigung für Vegetationskunde 21, 529–32. Vaduz: Cramer, 1978.

Jansen, Jürgen, Hansueli Jörg, Luca Maraini, and Hanspeter Stöckli. *Architektur lehren: Bernhard Hoesli an der Architekturabteilung der ETH Zürich / Teaching Architecture: Bernhard Hoesli at the Department of Architecture at the ETH Zurich.* Zurich: gta Verlag, 1989.

Jenni, Bruno, and Irma Noseda. "Höfe im Hauptsitz der Swisscom, Worblaufen." *werk, bauen + wohnen* 87, no. 9 (2000): 61.

Kappeler, Susanne. "Gärten als Kunstwerke: Neue Arbeiten des Landschaftsarchitekten Dieter Kienast." *Neue Zürcher Zeitung,* January 31, 1998.

———. "Gartenkunst im Dialog mit der Architektur: Landschaftsarchitekt Dieter Kienast gestorben." *Neue Zürcher Zeitung,* December 28, 1998.

Keil, Roger, and Klaus Ronneberger. "Arkadien postmodern: Stadtlandschaft zwischen Streuobst und Gewerbepark." In *Vision offener Grünräume: Grüngürtel Frankfurt,* 196–208. Edited by Tom Koenigs. Frankfurt am Main: Campus, 1991.

Kienast-Lüder, Erika. "Garten Kienast." *anthos* 44, no. 4 (2005): 19–23.

———. "Monsieur Hulot in the Garden." In *Kienast: Gärten / Gardens,* 142–43.

Kittelmann, Marion. "Heils-Wahn und Zeit-Schwindel im Werk von Anna & Bernhard Blume." PhD diss., Universität Wuppertal, 2002. http://nbn-resolving.de/urn:nbn:de:hbz:468-20020428.

Knapp, Rüdiger. *Einführung in die Pflanzensoziologie: Pflanzengesellschaften, Vegetationskunde, Vegetationskartierung und deren Annwendung in Land- und Forstwirtschaft, Landschaftspflege, Natur- und Umweltschutz, Unterricht und anderen Gebieten.* 3rd ed. Stuttgart: Ulmer, 1971.

König, Kathrin. "Ein Garten für zwei Botaniker." *anthos* 35, no. 4 (1996): 4–7.

Krämer, Sybille, Werner Kogge, and Gernot Grube, eds. *Spur: Spurenlesen als Orientierungstechnik und Wissenschaft.* Frankfurt am Main: Suhrkamp, 2007.

Krebs, Stefanie, *Zur Lesbarkeit zeitgenössischer Landschaftsarchitektur: Verbindungen zur Philosophie der Dekonstruktion.* Beiträge zur räumlichen Planung 63. Hanover: Institut für Grünplanung, 2002.

Krippendorf, Jost. *Die Landschaftsfresser: Tourismus und Erholungslandschaft—Verderben oder Segen?,* Bern: Hallwag, 1975.

Kunst bleibt Kunst: Projekt '74; Aspekte internationaler Kunst am Anfang der 70er Jahre. Exhibition catalog. Cologne: Josef-Haubrich-Kunsthalle, 1974.

Kunst und Architektur: Eine Selbstdarstellung der Fachbereiche Kunst, Visuelle Kommunikation, Produkt-Design und Architectur. Kassel: Gesamthochschule: 1990.

Kutter, Markus. "Die Folgen, die eine Gartenbau-Ausstellung hätte haben können." *Tages-Anzeiger Magazin* 3 (1978): 14–18.

———. "Queerbeetein durch die Grün 80." *Tages-Anzeiger Magazin* 32 (1980): 13–21.

———. *Vorwärts zur Natur: was war damit gemeint? Ein Ausstellungsprojekt und seine Hintergründe,* Niederteufen: Niggli, 1977.

Lassus, Bernard. *The Landscape Approach.* Translated by Stephen Bann, Paul Buck, and Catherine Petit. Philadelphia: University of Pennsylvania Press, 1998.

Lauterbach, Iris. *Der französische Garten am Ende des Ancien Régime: "Schöne Ordnung" und "geschmackvolles Ebenmass".* Worms: Werner, 1987.

Lefebvre, Henri. *La Production de l'espace.* 4th ed. Paris: Anthropos, 2000. First published 1974.

Leonardi, Cesare, and Franca Stagi. *The Architecture of the Trees.* Translated by Natalia Danford. New York: Princeton Architectural Press, 2019. Italian original published in Milan: Mazotta, 1982.

Le Roy, Louis G. *Natuur uitschakelen, natuur inschakelen.* Deventer: Ankh-Hermes, 1973.

Lévi-Strauss, Claude. *The Savage Mind.* Chicago: University of Chicago Press, 1966. French original published in Paris: Plon, 1962.

Löbbecke, Anja. "Über Naturgärten: Eine Ideengeschichte und kritische Retrospektive sowie zu ihrer Bedeutung für die heutige Landschaftsarchitektur." PhD diss., Technical University of Munich, 2013. https://mediatum.ub.tum.de/node?id=1111136.

Lüdtke, Hartmut. *Freizeit in der Industriegesellschaft: Emanzipation oder Anpassung?* Opladen: Leske, 1972.

Lührs, Helmut. "Einsatz und Anwendung der spontanen Vegetation in der Freiraumplanung." In *Krautern mit Unkraut; oder: Gärtnerische Erfahrung mit der spontanen Vegetation.* Notizbuch der Kasseler Schule 2, 130–48. Kassel: Arbeitsgemeinschaft Freiraum und Vegetation, 1986.

Mackert, Gabriele. *Skandal und Mythos: Eine Befragung Harald Szeemanns zur Documenta 5 (1972).* Exhibition catalog. Vienna: Kunsthalle, 2002.

Madura, Jonathan. "Sol LeWitt's Structural Methods." In *Sol LeWitt: Structures,* 24–43. New Haven: Yale University Press, 2011.

Magnago Lampugnani, Vittorio. *Die Modernität des Dauerhaften: Essays zu Stadt, Architektur und Design.* Berlin: Wagenbach, 1995.

Manz, Rudolf, *Vito de Onto: Raumerfinder und -erforscher . . . seine Freunde nennen ihn Video.* Zurich 1996.

Marchart, Oliver. *Post-Foundational Political Thought: Political Difference in Nancy, Lefort, Badiou, Laclau; Taking on the Political.* Edinburgh: Edinburgh University Press, 2007.

Marquart, Christian. "Das Wallmeisterhaus." *Bauwelt* 87, no. 16 (1996): 953–57.

Medici-Mall, Katharina. "Die Gärten des Berner Lory-Spital." In *Dieter*

Kienast: Zwischen Arkadien und Restfläche, 34–43.

———. "Lass den Anfang mit dem Ende sich in Eins zusammenziehen! Zum Wettbewerbsprojekt für den Frankfurter Günthersburgpark." In Erik de Jong, Erika Schmidt, and Brigitt Sigel, eds. *Der Garten—ein Ort des Wandels: Perspektiven für die Denkmalpflege,* 161–70. Zurich: vdf Hochschulverlag, 2006.

Mehli, Reto. "Das Lei(d)tbild 'Landschaft': Zur Kritik ästhetischer Leitbilder in der Gartenarchitektur." In *Reise oder Tour?* Notizbuch der Kasseler Schule 26, 128–56. Kassel: Arbeitsgemeinschaft Freiraum und Vegetation, 1992.

———. "Die mit den Förmchen spielen: Über die 'Bühnenbildnerei' in der Gartenarchitektur." In Böse-Vetter, *Freiraum und Vegetation,* 77–88.

Metken, Günter. "Nikolaus Lang: Für die Geschwister Götte." In *Kunst bleibt Kunst,* 238–41.

———. *Spurensicherung: Archäologie und Erinnerung.* Exhibition catalog. Hamburg: Kunstverein in Hamburg, 1974.

Migge, Leberecht. *Die Gartenkultur des 20. Jahrhunderts.* Jena: Diederichs, 1913.

Milchert, Jürgen. Review of *Die Aneignung von städtischen Freiräumen* by Helmut Böse. *Das Gartenamt* 32, no. 2 (1983): 116–17.

Mitscherlich, Alexander. *Die Unwirtlichkeit unserer Städte: Anstiftung zum Unfrieden.* Enl. ed. Frankfurt am Main: Suhrkamp, 2008. First published 1965.

Moll, Claudia, and Axel Simon. *Eduard Neuenschwander: Architekt und Umweltgestalter.* Zurich: gta Verlag, 2009.

Mouffe, Chantal. *The Democratic Paradox.* New York: Verso, 2000.

Müller, Jörg. *Alle Jahre wieder saust der Presslufthammer nieder; oder, Die Veränderung der Landschaft.* Aarau: Sauerländer, 1973. First English ed. *The Changing Countryside.* New York: Atheneum, 1977.

Nachlese Freiraumplanung. Notizbuch der Kassler Schule 10. Kassel: Arbeitsgemeinschaft Freiraum und Vegetation, 1989.

Nadolny, Sten. *Das Erzählen und die guten Absichten: Münchener Poetik Vorlesungen.* 3rd ed. Munich: Piper 1997. First published 1990.

New Topographics: Robert Adams, Lewis Baltz, Bernd and Hilla Becher, Joe Deal, Frank Gohlke, Nicholas Nixon, John Schott, Stephen Shore, Henry Wessel Jr. Exhibition catalog. Göttingen: Steidl, 2009.

Neuner, Stefan, Jörg Huber, and Geza Ziemer, eds. "*Paradoxien der Partizipation.*" Special issue, *31: Das Magazin des Instituts für Theorie der Gestaltung und Kunst der Züricher Hochschule der Künste* 10/11 (December 2007).

New Topographics: Photographs of Man-Altered Landscape. Exhibition catalog. Rochester, NY: International Museum of Photography at George Eastman House, 1975.

Oechslin, Werner. "'In Praise of Garden Art' for Dieter Kienast—Instead of a Superfluous Foreword." In *Dieter Kienast: In Praise of the Sensousness,* 6–7.

Ohff, Heinz. *Der grüne Fürst: Das abenteuerliche Leben des Hermann Pückler-Muskau.* Munich: Piper, 1991.

"Partizpiation: Grün in die Nordstadt." *Der Monolith: Studentische Zeitschrift der Organisationsenheit Architektur, Stadt- und Landschaftsplanung, Gesamthochschule Kassel* 12 (December 1977): front page.

Prigge, Walter, and Kai Vöckler. "Voids." In *Kienast Vogt: Parks und Friedhöfe / Parks and Cemetries,* 9–14.

Proksch, Thomas. "Dieter Kienast und seine Gärten." In *Zauber der Gärten: Ideen zum Nachgestalten,* 15–17. Leopoldsdorf: Agrarverlag, 2000.

Pückler-Muskau, Hermann von. *Hints on Landscape Gardening: English Edition with the Hand-Colored Illustrations of the Atlas of 1834.* Edited by John Hargraves. Basel: De Gruyter, 2014.

Radziewsky, Elke von. "Der Künstler ist immer der Gärtner." *Architektur & Wohnen* 3 (1994): 58–68.

———. "Et in Arcadia Ego." In *Schweiz: Zürich, Basel, Genf; Alpenglühen und Avantgarde,* 104–110. Hamburg: Jahreszeiten-Verlag, 1999.

Raussmüller, Urs, and Christel Sauer, eds. *Carl Andre: Cuts.* Basel: Raussmüller Collection, 2011.

———. "Ich mag Kunst, die nicht lügt." Interview. *Du: Das Kulturmagazin* 796 (2009): 72–79.

———, eds. *Mario Merz: Senza titolo.* Basel: Raussmüller Collection, 2011.

Raymann, Toni. "Privat nutzbarer Freiraum im Geschosswohnungsbau" (Open Spaces for Private Use in Apartment Housing Developments). *anthos* 22, no. 2 (1983): 2–26.

"Results of the Competition for 'Grün 80'—2nd Swiss Exhibition of Garden and Landscape Design 1980 in Basle." *anthos* 16, no. 3 (1977): 32–37.

Reuß, Jürgen von. "Lassus in Kassel: Ein Blick auf die Gartenkunst." In Andrea Koenecke, Udo Weilacher, and Joachim Wolschke-Bulmahn, eds. *Die Kunst, Landschaft neu zu erfinden: Werk und Wirken von Bernard Lassus.* CGL Studies 8, 85–115. Munich: Meidenbauer, 2010.

Rheinberger, Hans-Jörg. *On Historicizing Epistemology: An Essay.* Translated by David Fernbach. Stanford, CA: Stanford University Press, 2010.

Rittel, Horst. *Planen, Entwerfen, Design: Ausgewählte Schriften zu Theorie und Methodik.* Stuttgart: Kohlhammer, 1992.

———. "Systematik des Planens." *Werk* 54, no. 8 (1967): 505–8.

Ritter, Joachim. "Landschaft: Zur Funktion des Ästhetischen in der modernen Gesellschaft." In *Subjektivität: Sechs Aufsätze,* 141–90. Frankfurt am Main: Suhrkamp, 1989. First published 1963.

Ritz, German. "Entering the Autobiographical Garden." In *Kienast: Gärten / Gardens,* 18–33.

Rotzler, Stefan, "Grün in die Nordstadtfen. Symposium zur Aktualität des Werks von Dieter Kienast (1945–1998)." in: *Der Gartenbau* 130, no. 4 (2009): 20–21; no. 5 (2009): 2–3.

Rowe, Colin, and Fred Koetter. *Collage City.* Cambridge, MA: MIT Press, 1978.

Rowe, Colin, and Robert Slutzky. *Transparency.* Basel: Birkhäuser, 1997.

———. "Transparency: Literal and Phenomenal." *Perspecta* 8 (1963), 45–54.

Rossi, Aldo. *The Architecture of the City.* Translateed by Dian Ghirardo and Joan Ockman. Cambridge, MA: MIT Press, 1982. Italian original published 1966.

Rüegg, Arthur. "Die Gärten von Kienast Vogt Partner." In Esther Maria Jungo and Hans Peter von Ah. *Der Hauptsitz der Swisscom in Worblaufen bei Bern,* 13–19. Bern: Gesellschaft für Schweizerische Kunstgeschichte, 1999.

———. "Dieter Kienast: Paradise is . . . ." In *Dieter Kienast: In Praise of the Sensousness,* 12–14.

———. "Komplexe Gärten aus der Gussform." *Garten + Landschaft* 109, no. 5 (1999): 18–21.

———. "Mit Natur Architektur gebaut." *Tages-Anzeiger*, December 30, 1998: 50.

———. "Swisscom Administrative Building, Worblaufen Bern." In *Kienast Vogt: Aussenräume / Open Spaces*, 74–95.

———. "Vielfalt und Dichte: Zum Garten Medici in Erlenbach." In *Dieter Kienast: Zwischen Arkadien und Restfläche*, 44–55.

Salm, Karin. "Dieter Kienast und sein Einsatz für eine progressive Landschaftsarchitektur." Conversation with Günther Vogt and Stefan Rotzler. Radio broadcast. *Reflexe*, Schweizer Radio DRS, DRS 2, first broadcast October 28, 2005.

———. "Grün kann sogar gefährlich sein: Der Landschaftsarchitekt Dieter Kienast." Conversation with Udo Weilacher and Anette Freytag. Radio broadcast. *Reflexe*, Schweizer Radio DRS, DRS 2, first broadcast January 20, 2009, https://www.srf.ch/sendungen/reflexe/gruen-kann-sogar-gefaehrlich-sein-der-landschaftsarchitekt-dieter-kienast.

Sassen, Saskia. "The Global Street: Where the Powerless Get to Make History." Lecture at the conference "Thinking the Contemporary Landscape: Positions and Oppositions." Hanover-Herrenhausen, June 21, 2013, https://video.ethz.ch/conferences/2013/ila/06_friday/c10211a6-3d35-4393-a809-d4223782ae07.html.

Sauer, Christel. "Carl Andre: Cuts, 1967." In Raussmüller and Sauer, *Carl Andre*, n.p.

———. "Ein Einstieg in die Kunst des 20. Jahrhunderts." In Urs Raussmüller and idem, eds. *Main Stations: Newman, Pollock, Beuys, Broodthaers, Klein, Warhol, LeWitt, Johns, Stella, Ryman, Kounellis, Nauman, Weiner*, 7–19. Schaffhausen: Neue Kunst Bücher, 1996.

———. "Joseph Beuys." In Urs Raussmüller and idem, eds. *Main Stations: Newman, Pollock, Beuys, Broodthaers, Klein, Warhol, LeWitt, Johns, Stella, Ryman, Kounellis, Nauman, Weiner*, 21–25. Schaffhausen: Neue Kunst Bücher, 1996.

———, ed. *Mario Merz: Architettura fondata dal tempo, architettura sfondata dal tempo,* Frauenfeld: Raussmüller Collection, 2009.

Schäfer, Robert. "Dieter Kienast: Quality Is in Being Definite." *Topos* 26 (1999): 131.

———. "Zeichen setzen." *Garten + Landschaft* 109, no. 5 (1999): 1.

Schama, Simon. *Landscape and Memory.* New York: Knopf, 1995.

Schenker, Christoph, and Michael Hiltbrunner. *Kunst und Öffentlichkeit: Kritische Praxis im Stadtraum Zürich.* Zurich: JRP Ringier, 2007.

Schenker, Rudolf. "Grün 80-Trockenbiotop: Die Natur hat Vortritt." *Panda Nachrichten: Mitteilungsorgan der Stiftung WWF Schweiz* 13, no. 2 (1980): 15.

Schleich, Elisabeth. "Die Gartenarchitektur-Szene: Gedanken zur aktuellen Garten- und Landschaftsarchitektur." *Der Gartenbau* 106, no. 21 (1985): 46.

Schmid, André. "Internationale Gartenschau Steiermark 2000." *anthos* 39, no. 1 (2000): 34–37.

Schmid, Christian. *Stadt, Raum und Gesellschaft: Henri Lefebvre und die Theorie der Produktion des Raumes.* Stuttgart: Steiner, 2005.

Schmidt, Eike. "Der Naturgarten, ein neuer Weg." *Garten + Landschaft* 91, no. 11 (1981): 877–84.

Schmitz, Martin. "From Critical Urban Studies to the Science of Walking." In Burckhardt, *Who Plans the Planning?*, 7–11.

Schneckenburger, Manfred, ed. *Documenta: Idee und Institution; Tendenzen, Konzepte, Materialien.* Exhibition catalog. Munich: Bruckmann, 1983.

Schultz, Uwe, ed. *Umwelt aus Beton; oder, Unsere unmenschlichen Städte,* Reinbek bei Hamburg: Rowohlt, 1971.

Schürmeyer, Bernd, and Christine Anna Vetter. *Die Naturgarten-Bewegung: Die "Ideen", professionelle Elemente, Professionalisierungstendenzen.* Gesamthochschule Kassel. Arbeitsbericht des Fachbereichs Stadt- und Landschaftsplanung 42. Kassel: Infosystem Planung, GhK, 1982.

Schwarz, Urs. *Der Naturgarten: Mehr Platz für einheimische Pflanzen und Tiere.* Frankfurt am Main: Krüger, 1980.

Seel, Martin. *Eine Ästhetik der Natur.* Frankfurt am Main: Suhrkamp. 1991.

Seleger, Ursula. "C. Th. Sörensens Herausforderung" (C. Th. Sörensen's Challenge). *anthos* 12, no. 4 (1973): 39–40.

Seligman, Werner. "The Texas Years and the Beginning at ETH Zurich, 1956–1961." Jansen et al., *Architektur lehren*, 7–16.

Sennett, Richard, ed. *Classic Essays on the Culture of Cities,* New York: Appleton-Centutry-Crofts, 1969.

———. *The Fall of Public Man.* New York: Knopf, 1977.

———. *The Uses of Disorder: Personal Identity and City Life,* London: Vintage, 1971.

Shklovsky, Viktor. "Art as Device (1917–1919)." *Viktor Shklovsky: A Reader,* 73–96. Edited and translated by Alexandra Berlina. New York: Bloomsbury Academic, 2017.

"Siedlung Nussdorf, Wien" (Nussdorf Vienna Development). *anthos* 7, no. 3 (1968): 33–36.

Simmel, Georg. The Philosophy of Landscape" [1913]. Translated by Josef Bleichner. *Theory, Culture & Society* 24, nos. 7–8 (December 2007): 20–29.

Skulptur im 20. Jahrhundert. Exhibition catalog. Basel: n.p., 1984.

Smithson, Robert. *The Collected Writings.* Edited by Jack Flam. Berkeley: University of California Press, 1996.

———. *The Writings of Robert Smithson: Essays with Illustrations.* Edited by Nancy Holt. New York: New York University Press, 1979.

"Sol LeWitt by Saul Ostrow." Interview. *BOMB* 85 (Fall 2003): 25–29. http://bombmagazine.org/article/2583/sol-lewitt.

Sol LeWitt: Structures, 1965–2006. Exhibition catalog. New Haven: Yale University Press, 2011.

Staeck, Klaus, ed. *Befragung der Documenta; oder, Die Kunst soll schön bleiben.* Göttingen: Steidl, 1972.

Stahel, Urs, and Martin Heller. *Wichtige Bilder: Fotografie in der Schweiz.* Exhibition catalog. Zurich: Der Alltag, 1990.

Steinmann, Martin. "The Presence of Things: Comments on Recent Architecture in Northern Switzerland." In Mark Gilbert and Kevin Alter, eds. *Construction, Intention, Detail: Five Projects from Five Swiss Architects / Fünf Projekte von fünf Schweizer Architekten.* 2nd ed., 8–25. Zurich: Artemis, 1994.

Stern, Christian. "Bedeutung und Entwicklung der Freiraumgestaltung im industriellen Wohnungsbau" (The Significance and Evolution of the Design of Open Spaces in Industrial Dwelling Construction). *anthos* 7, no. 3 (1968): 16–22.

Stippl, Hannah. "Nur wo der Mensch die Natur gestört hat, wird die Landschaft wirklich schön: Die landschaftstheoretischen Aquarelle von Lucius Burckhardt." PhD diss., University of Applied Arts Vienna, 2011.

Stöckli, Peter. "Am Ende der Strasse – ein Nachruf auf Dieter Kienast." *anthos* 38, no. 1 (1999): 58–59.

Stoffler, Johannes. "Gegen die Vereinfachung der Gartenbotschaft." *Stadt und Grün* 53, no. 1 (2004): 16–20.

———. *Gustav Amman: Landschaften der Moderne in der Schweiz.* Zurich: gta Verlag, 2008.

Sukopp, Herbert, *Rückeroberung: Natur im Großstadtbereich.* Wiener Vorlesungen 102. Vienna: Pincus, 2003.

———, and Rüdiger Wittig, ed. *Stadtökologie: Ein Fachbuch für Studium und Praxis.* 2nd ed. Stuttgart: Fischer 1998.

Szeemann, Harald, ed. *Besser sehen durch documenta 5: Befragung der Realität; Bildwelten heute,* Kassel: Bertelsmann, 1972.

Talk show der documenta 6: Stichwortgeber: Georg Jappe. Kunstforum International 21. Mainz: Kunstforum International, 1977.

Tessin, Wulf. *Ästhetik des Angenehmen: Städtische Freiräume zwischen professioneller Ästhetik und Laiengeschmack.* Wiesbaden: Verlag für Sozialwissenschaften, 2008.

Tüxen, Reinhold. Foreword to *Die spontane Vegetation der Stadt Kassel,* by Dieter Kienast, n.p.

Treib, Marc. "The Hedge and the Void: The Landscapes of Dieter Kienast and an Overview of His Career." *Landscape Architecture* 93, no. 1 (2003): 79–107.

Tremp, Andreas. "Wege sind wie Löcher." *Der Architekt* 2 (1995): 103–5.

———. "Die Allee der Vereinigten Bäume." *Garten + Landschaft* 109, no. 5 (1999): 22–25.

Tschan, Richard. "Grusswort des Schweizer Gartenbau-Präsidenten." *Deutscher Gartenbau* 34, no. 15 (1980): 666.

Ulbricht, Annette, ed. *Von der Henschelei zur Hochschule: Der Campus der Universität Kassel am Holländischen Platz und seine Geschichte.* Kasseler Semesterbücher Studia Cassellana 15. Kassel: Kassel University Press, 2004.

Ursprung, Philip. *Die Kunst der Gegenwart: 1960 bis heute.* Munich: Beck, 2010.

———, ed. *Herzog & de Meuron: Natural History.* Exhibition catalog. Baden: Lars Müller, 2002.

Valda, Andreas. "Mit Gärten gegen die Grünen." *Tages-Anzeiger,* December 23, 1997: 49.

Vewijnen, Jan. "Politische Radikalität und poetische Präzision." In "Viele Mythen, ein Maestro," 39–41.

"Viele Mythen, ein Maestro: Kommentare zur Zürcher Lehrtätigkeit von Aldo Rossi." *Werk, Bauen + Wohnen* 84, no. 12 (1997): 37–44.

"Viele Mythen, ein Maestro. Kommentare zur Zürcher Lehrtätigkeit von Aldo Rossi. Teil II." *Werk, Bauen + Wohnen* 85, no. 1–2 (1998): 37–44.

Vogt, Günther. "Die Gleichzeitigkeit des Anderen: Friedhofserweiterung Baden-Rütihof." In *Dieter Kienast: Zwischen Arkadien und Restfläche,* 19–28.

———. *Landscape as a Cabinet of Curiosities: In Search of a Position.* Edited by Rebecca Bornhauser and Thomas Kissling. Translated by Jennifer Taylor and Roderick O'Donovan. Zurich: Lars Müller, 2015.

———, ed. *Miniatur und Panorama: Vogt Landschaftsarchitekten; Arbeiten 2000–2006.* Baden: Lars Müller, 2006.

———, ed. *Miniature and Panorama: Vogt Landscape Architects; Projects 2000–12.* Translated by Laura Bruce and Steven Lindberg. Zurich: Lars Müller, 2012.

Vogt Landschaftsarchitekten: Lupe und Fernglas—Miniatur und Panorama / Magnifiying Glass and Binoculars—Miniature and Panorama. Exhibition catalog. Edited by Kirstin Feiress and Hans-Jürgen Commerell. Berlin: Aedes, 2007.

Vogt Landschaftsarchitekten: Radikal, präzis und poetisch; Schulthess-Gartenpreis 2010. Zurich: Schweizer Heimatschutz, 2010.

Vom Landschaftsgarten zur Gartenlandschaft: Gartenkunst zwischen 1880 und 1980 im Archiv für Schweizer Gartenarchitektur und Landschaftsplanung. Edited by Stiftung Archiv für Schweizer Gartenarchitektur und Landschaftsplanung. Zurich: vdf Hochschulverlag, 1998.

Von Büchern und Bäumen: Vogt Landschaftsarchitekten; Eine Ausstellung im Architekturmuseum Basel vom 27. November 2004 bis 30. Januar 2005 / About Books and Trees: Vogt Landscape Architects; An Exhibition in the Architekturmuseum Basel from November 27, 2004 to January 30, 2005. Exhibition catalog. Basel: Architekturmuseum Basel, 2004.

Weber, Rolf. *Die Besiedlung des Trümmerschutts und der Müllplätze durch die Pflanzenwelt.* Museumsreihe 21. Plauen: Vogtländisches Kreismuseum, 1960.

Weilacher, Udo. "Architektonische Gärten jenseits ökologischer Klischees." *Gartenpraxis* 25, no. 3 (1999): 38–43.

———. "Der Friedhof Fürstenwald bei Chur von Kienast, Vogt & Partner. Abstrakter Totenbezirk und Gegenwärtigkeit von Landschaft." *werk, bauen + wohnen* 87, no. 10 (2000): 40–42.

———. "Die Sehnsucht nach dem Paradies." *Bayerische Akademie der Schönen Künste: Jahrbuch* 19 (2005): 429–43.

———. "Die Sinnlichkeit architektonischer Strenge: Zum Tod des Landschaftsarchitekten Dieter Kienast." *archithese* 29, no. 1 (1999): 74–75.

———. "Dieter Kienast." In Candice A. Shoemaker, ed. *Encyclopedia of Gardens: History and Design.* Vol. 2, 701–703. Chicago: Fitzroy Dearborn, 2001.

———. "Friedhof Fürstenwald bei Chur." *Garten + Landschaft* 109, no. 5 (1999): 14–17.

———. "The Garden as the Last Luxury Today: Thought Provoking Garden Projects by Dieter Kienast (1945–1998)." In Michel Conan, ed. *Contemporary Garden Aesthetics, Creations and Interpretations.* Dumbarton Oaks Colloquium on the History of Landscape Architecture 29, 81–96. Washington D.C.: Dumbarton Oaks, 2007.

———. *In Gardens: Profiles of Contemporary European Landscape Architecture.* Basel: Birkhäuser, 2005.

———. "Minimalistische Kompositionen von poetischer Sinnlichkeit: Zum Tod des Landschaftsarchitekten Dieter Kienast." *Basler Magazin* 3 (1999): 11.

———. "Moderne, unwandelbare Gärten?" In Erik de Jong, Erika Schmidt, and Brigitt Sigel, eds. *Der Garten—ein Ort des Wandels: Perspektiven für die Denkmalpflege,* 263–75. Zurich: vdf Hochschulverlag, 2006.

———. "Natur und Architektur im Dialog." *NZZ-Folio* 3 (March 1999): 64–65.

———. "Scanning Tracks: Effects of the principal creative work of Dieter Kienast." In *Dieter Kienast,* 284–95.

———. "Schweiz: Vom Heimatstil zur guten Form." *Topos* 33 (2000): 81–89.

———. *Syntax of Landscape: The Landscape Architecture of Peter Latz and Partners.* Basel: Birkhäuser, 2008.

———. *Visionary Gardens: Modern Landscapa by Ernst Cramer.* Translated by Michael Robinson. Basel: Birkhäuser, 2001.

———, and Peter Wullschleger. *Landschaftsarchitekturführer Schweiz.* Edited by Bund Schweizer Landschaftsarchitekten and Landschaftsarchitektinnen BSLA. Basel: Birkhäuser, 2002.

Weiss, Evelyn. "Die wiedergefundene Zeit." In *Kunst bleibt Kunst,* 14–25.

Welsch, Wolfgang. *Ästhetisches Denken.* Stuttgart: Reclam, 1990.

———. *Grenzgänge der Ästhetik.* Stuttgart: Reclam, 1996.

———. *Undoing Aesthetics.* Translated by Andrew Inkpin. Thousand Oaks, CA: Sage, 1997.

———. *Unsere postmoderne Moderne.* Weinheim: VCH, 1991.

———, ed. *Wege aus der Moderne: Schlüsseltexte der Postmoderne-Diskussion.* Acta humaniora. 2nd rev. ed. Berlin: Akademie-Verlag, 1994. First published Weinheim: VCH, 1988.

Werkner, Patrick. *Kunst seit 1940: Von Jackson Pollock bis Joseph Beuys.* Vienna: Böhlau, 2007.

"Wettbewerb Internationale Gartenschau – Steiermark 2000." *Wettbewerbe: Architekturjournal* 21, no. 159–160 (1997): 76–85.

Whittaker, Robert H., ed. *Classification of Plant Communities.* The Hague: Junk, 1978.

Wilmanns, Otti. Review of *Die spontane Vegetation der Stadt Kassel* by Dieter Kienast and *Freiraum- und landschaftsplanerische Analyse des Stadtgebietes von Schleswig* by Karl Heinrich Hülbusch et al. *Phytocoenologia* 8, no. 1 (1980): 151–52.

Wolf, Christa. *Voraussetzungen einer Erzählung: Kassandra; Frankfurter Poetikvorlesungen.* Munich: Luchterhand, 1993.

Wormbs, Brigitte. "Aussichten in Zwischenräume." In *Dieter Kienast: Zwischen Arkadien und Restfläche,* 9–10.

———. "Der Güntherburgpark in Frankfurt a. Main." In *Dieter Kienast: Zwischen Arkadien und Restfläche,* 10–13.

———. "Diction and Contradiction: Ways of Reading History and Tales." In *Kienast: Gärten / Gardens,* 8–16.

———. "Die Parkanlage Moabiter Werder Berlin." In *Dieter Kienast: Zwischen Arkadien und Restfläche,* 14–18.

———. "Satz und Gegensatz." *Basler Zeitung,* April 5, 1997: 12–13.

Wrede, Stuart, and William Howard Adams., eds. *Denatured Visions: Landscape and Culture in the Twentieth Century.* Exhibition catalog. New York: Museum of Modern Art, 1991.

Wullschleger, Peter. "Ein letzter Garten: Zum Tod des Landschaftsarchitekten Dieter Kienast." *Der Gartenbau* 120, no. 2 (1999): 13. Reprinted in *Werk, Bauen + Wohnen* 86, no. 4 (1999): 62–63.

Zinsli, Urs, and Konrad Erhard. "Friedhofsanlage bei Chur." *Baumeister* 96, no. 6 (1999): 32.

Zweite, Armin, ed. *Joseph Beuys: Natur, Materie, Form.* Exhibition catalog. Munich: Schirmer/Mosel, 1991.

Illustration Credits

All images were provided by the gta Archives (NSL Archive) / ETH Zurich (Dieter Kienast Bequest) except:

Adams, *French Garden 1500–1800,* 51 fig. 51; 33 fig. 26
    Figs. 74, 105
Georg Aerni
    Figs. 3, 4, 152, 153, 160; figs. pp. 68–74, 92–109, 116–119, 122–124, 126–129, 140–145, 172–77, 198–209, 262–264, 270–281, 328–338
anthos 7, no. 1 (1968): 13, 19–20
    Figs. 59, 118
anthos 7, no. 3 (1968): 36
    Fig. 77
anthos 16, no. 1 (1977): 23, 25
    Figs. 127, 128
anthos 16, no. 3 (1977): 36
    Fig. 140
anthos 19, no. 1 (1980): 60
    Fig. 27
anthos 23, no. 4 (1984): 28, 31
    Figs. 108, 109
archithese 27, no. 4 (1997): 4–5
    Fig. 187
Archiv Institut Entwerfen von Stadt und Landschaft, Karlsruhe Institute of Technology
    Figs. 196–99
Archiv Fabienne Kienast Weber
    Figs. pp. 15, 16; figs. 164, 167; figs. pp. 401, 402
Archiv Arthur Rüegg
    Figs. 79, 81, 141
Archiv Marc Schwarz
    Fig. 207
Archiv für Schweizer Landschaftsarchitektur (ASLA), Rapperswil, Switzerland
    Figs. 76, 80
Archiv SKK Landschaftsarchitekten, Wettingen
    Figs. 23, 25, 26, 48–54, 66, 67, 86, 88–99, 142
Archiv Vogt Landschaftsarchitekten, Zurich:
    Fig. 155
Böse-Vetter, *Freiraum und Vegetation,* 409, 411
    Figs. 12, 35
Burckhardt, *Design ist unsichtbar,* 225
    Fig. 9

Der Gartenbau 100, no. 25 (1979): 1122
    Fig. 1
Deutscher Werkbund, Arbeitsgruppe Kassel, *Durch Pflege zerstört,* cover
    Fig. 11
Dieter Kienast, 206
    Fig. 46
Fabio Fabbrini © Raussmüller Collection
    Figs. 145, 149
Fred Eicher Landschaftsarchitekt, 16, 19
    Figs. 133, 134, 137
Anette Freytag
    Figs. 40, 135, 136, 177
Garten + Landschaft 80, no. 7 (1969): 217
    Fig. 138
Garten + Landschaft 96, no. 11 (1986): 29, 32
    Figs. 5, 6
"Grün 80—Schlussbericht der Direktion," Archiv SKK Landschaftsarchitekten, Wettingen
    Fig. 14
gta Archives / ETH Zurich (Heinrich Helfenstein Bequest)
    Fig. 61
gta Archives (NSL Archive) / ETH Zurich (Hans Marti Bequest)
    Fig. 87
gta Exhibitions, ETH Zurich
    Figs. 208–11
Franz Hohler, *Die Rückeroberung,* illustrations by Karin Widmer (Bern: Zytglogge, 1991), 32–33
    Fig. 34. Courtesy of Karin Widmer
© Timothy Hursley
    Fig. 151
Kienast, *Die spontane Vegetation der Stadt Kassel,* table 7 and attached maps
    Figs. 15, 17, 18
Kienast: Gärten / Gardens, 143, cover
    Figs. 163, 200
Kienast Vogt: Aussenräume / Open Spaces, 75, 40, cover
    Figs. 175, 180, 201
Kienast Vogt: Parks und Friedhöfe / Parks and Cemeteries, 41, 85, 86, 91
    Figs. 82, 171, 172, 173
Kulturstiftung des Hauses Hessen, Museum Schloss Fasanerie, Eichenzell near Fulda
    Fig. 31
Kunst bleibt Kunst, 238–39, 212–13, 166–67
    Figs. 19, 20, 22
Landschaft + Stadt 3 (1981): 121, 124
    Fig. 186
Lauterbach, *Der französische Garten am Ende des Ancien Régime,* 73
    Fig. 75
L'Invention du Parc, 47, 34
    Figs. 106, 107
Manz, *Vito de Onto,* 13
    Fig. 206

Martin Schmitz Verlag, Berlin:
    Figs. 8, 10, 13
Migge, *Die Gartenkultur des 20. Jahrhunderts,* 75
    Fig. 104
Müller, *Alle Jahre wieder saust der Presslufthammer nieder,* n.p.
    Figs. 7, 32
Professur für Landschaftsarchitektur ETH Zürich, *Dieter Kienast,* 192–93
    Fig. 202
Raussmüller Collection, Basel | Fabio Fabbrini © Raussmüller Collection
    Figs. 144, 146–147
Rowe and Slutzky, *Transparency,* 114–15
    Fig. 85
Marc Schwarz
    Fig. 212
Schweizer Illustrierte, September 22, 1980, 17
    Fig. 24
Skulptur im 20. Jahrhundert, 169, 215
    Figs. 150, 158
Smithson, *Collected Writings,* 205
    Fig. 21
Sol LeWitt: Structures, 30
    Fig. 149
Johannes Stoffler
    Fig. 64
University of Kassel, docu:lab, Archiv Grauer Raum, Heidrun Hubenthal
    Figs. 33, 131
Christian Vogt
    Figs. 2, 37, 38, 39, 43, 44, 47, 63, 72, 73, 101–103, 113–115, 126, 154, 157, 159, 181, 183–185, 188–193, 204, 205
Weilacher, *Visionäre Gärten,* 213, 257
    Figs. 60, 78

Further legal representatives:
© The Andy Warhol Foundation for the Visual Arts, Inc. / 2021, ProLitteris, Zurich
© Holt/Smithson Foundation / 2021, ProLitteris, Zurich
© The LeWitt Estate / 2021, ProLitteris, Zurich
© 2021, ProLitteris, Zurich: Carl Andre, Joseph Beuys, Claudio Costa, Gerhard Lang, Mario Merz, Jörg Müller, Christian Vogt

Index

Aachen 319
Aadorf 179
Adlikon 168, 180
Adorno, Theodor W. 146, 216
Aebischer, Helmut 36
Aeschbacher, Kurt 62
Altherr, Jürg 56, 185, 359, 407
Altwegg, Pit 396 n. 146
Amsler, Arnold 219, 299
Amstutz, Marcel 77
Andersson, Sven-Ingvar 255
Anderson, Laurie 345, 403
Andre, Carl 306–7, 311, 345, 403
ARCOOP 115, 300

Bach, Johann Sebastian 345, 403
Baden 22, 161, 182, 186–88
——— Rütihof Cemetery 372
Bad Münder, Spa Gardens 185
Bad Zurzach, Spa Gardens 12, 90, 219–20, 229, 234–35, 254, 255, 299, 344, 346, 376
Barcelona 234
Basel 40, 55, 180, 218, 220, 260, 299, 316, 319, 320, 359, 371, 372, 407
——— Dry Grassland Biotope, Merian Gärten (Merian Gardens) 9, 20, 56, 62–78, 185, 313
Bava, Henri 360
Berlin 28, 45, 216, 317, 413
——— Moabiter Werder 220, 235, 247, 255, 308, 344, 380
Bern 159, 314, 390, 407, 408
——— École cantonale de langue française 85–111, 112, 219, 301
Berndt, Heide 147 n. 2
Beuys, Joseph 9, 11, 25, 38, 306, 308–11
Beyeler, Ernst 313, 407
Biel/Bienne 219, 220
Blanc, Patrick 316
Bloch, Ernst 346, 403
Boesch, Georges 162
Böse, Helmut 26–30
Bosshard, David 85
Botta, Mario 357
Branitz. See Cottbus
Braun-Blanquet, Josias 41, 49
Brecht, Bertolt 260–61, 404–5
Bregenz 319, 359
Brunstatt-Didenheim, Ricola Europe Packing and Distribution Building 115, 314

Brus, Günter 25
Bucher, Annemarie 182, 216
Burckhardt, Annemarie 81
Burckhardt, Lucius 23–25, 29–37, 38–41, 46, 49, 56, 62, 83, 181, 345, 407
Burkard, Meyer, Steiger + Partner 186, 188
Burkhalter und Sumi Architekten 300

Cézanne, Paul 379
Chaplin, Charlie 344
Chur
——— Fürstenwald Cemetery 19, 20, 265–84, 360, 386, 390
——— Waldhaus Psychiatric Clinic 361, 364, 383
Christo 345, 403
Cook, James 34
Costa, Claudio 58
Cottbus 215
Cramer, Ernst 167, 180, 215

Dean, James 344
De Jong, Erik A. 380
De Maria, Walter 345, 403
Descombes, Georges 317, 319
Diedrich, Lisa 10
Diener & Diener Architekten 300
Dübendorf 180, 412

Egnach, Sp. Garden 361, 383
Eicher, Fred 22, 30, 115, 167, 189, 215, 299, 397 n. 163, 407
Elíasson, Ólafur 359
Enzensberger, Hans Magnus 380
Erhard, Franz 265–69
Erlenbach, M.-M. Garden 115, 300, 380
Erni, Peter 56, 359, 407
Ernst, Max 344

Fällanden 168
Fahrni, Sacha 325
Fellini, Federico 344
Finlay, Ian Hamilton 30, 320
Fischli/Weiss 36
Flavin, Dan 306, 311
Folkerts, Thilo 384
Fontainebleau 211, 242
Forster, Georg 34
Foucault, Michel 56, 389
Foxley, Alice 359
Frankfurt am Main, Günthersburgpark 220, 235, 245, 300, 344, 353, 361, 380
Freising-Weihenstephan 22, 407
Frisch, Max 62

Ganz, Daniel 246
Gasser, Christoph 246
Gasser, Martin 371
Gaudí, Antoni 234–35
Geneva 319
Girot, Christophe 372
Glass, Philip 345, 403
Gockhausen, B. Garden 115, 345, 361

Göbel-Groß, Thomas 383
Gorky, Maxim 181
Gosam in the Wachau 301
Graz-Unterpremstätten, Internationale Gartenschau (International Horticultural Show) 2000 185, 218, 255, 346–53, 360
Greenaway, Peter 346. 404
Greifensee 180
Grzimek, Günther 22–24, 30–31, 33, 220, 299, 407
Gut, Christian 325
Gutmann, Rolf 33

Haerle, Christoph 399 n. 230
Hager, Guido 399 n. 230, n. 234
Hallmann, Heinz 26
Hamburg-Fuhlsbüttel 234
Handke, Peter 306, 316–19, 341, 346, 379–80, 389, 403
Hanover
——— Expo 2000 185, 300, 349–56, 357
——— Kronsberg Landscape Design 360–63
Hard, Gerhard 43
Harrison, Newton 59
Heer, René 85
Heizer, Michael 9
Hejduk, John 220
Helbich, Ilse 160, 235
Heller, Martin 371
Herteren 345
Herzog & de Meuron Architekten 115, 300, 306, 310, 313–16, 319, 353
Hiesmayr, Ernst 167, 299
Hirsche, Lee 220
Hitchcock, Alfred 344
Hoesli, Bernhard 219–29, 244–45, 302
Hofer, Polo 63
Hogarth, William 79
Hohler, Franz 81–83
Holzer, Jenny 320
Hülbusch, Inge Meta 26, 260
Hülbusch, Karl Heinrich 10, 23–24, 26, 29–30, 38–46, 49, 54–56, 63, 81–83, 181, 295, 407
Hüsler, Peter 265
Hunziker, Wolf 62

Jacobs, Jane 29
Jarzombek, Mark 227–29
Jureczek, Ruth 36

Kaprow, Allan 58
Kassel 7, 9–13, 18–61, 79, 81–83, 110–11, 115, 121, 132, 138, 139, 159, 161, 219, 220, 244, 245, 260–61, 295 passim
Kienast, Nicole 22, 407
Kienast-Lüder, Erika 22, 220, 235, 246, 254, 265, 380, 407, 408
Kienast Weber, Fabienne 383, 407
König Urmi, Katharina 345
Koeppel, Hans-Dietmar 344, 407, 408
Koetter, Fred 220, 235

Konstanz 319
Kounellis, Jannis 306, 312–14
Kundera, Milan 306
Kutter, Markus 62–63
Kyoto 346

Lang, Gerhard 36
Lang, Nikolaus 57
Lassus, Bernard 30, 34, 235, 244
Latz, Peter 29, 407
Laufen, Ricola Marketing Building 316
Le Corbusier 227
Le Nôtre, André 211
Leonardi, Cesare 346
Le Rouge, George Louis 211
Le Roy, Louis 84, 216
Lévi-Strauss, Claude 27, 56, 58
LeWitt, Sol 283, 306, 312–13, 341, 345, 403
Leymen, Haus Rudin 314
Linnaeu, Carl von 244
Locke, John 79
London, Tate Modern 319
Long, Richard 306, 319
Louafi, Kamel 380
Lyotard, Jean-François 306

Manz, Rudolf 384–86, 389
Marbach, Ueli 300
Marie-Antoinette 211
Martinsson, Gunnar 385
Masina, Giulietta 344
Matta-Clark, Gordon 313, 314
Mattern, Hermann 22
Maur, K.-U. Garden 393 n. 45, 412
Mayer Harrison, Helen 59
Mehli, Reto 10
Meili Peter 300
Merian, Christoph 66
Merz, Mario 306, 311, 312–14
Metken, Günter 57–58
Meyer, Adrian 188
Migge, Leberecht 29, 189, 234
Milchert, Jürgen 27
Mitscherlich, Alexander 29, 84
Moll, Claudia 180
Muchow, Martha 29
Müller, Jörg 31, 79
Müller, Wolfram 385
Mülligen 345, 409

Nauman, Bruce 306
Neuenschwander, Eduard 180
Neuhaus, Max 319
Neukom, Willi 180, 220, 299
Newmark, Nicole 326
New York 57, 313
Niederhasli, Lindenstrasse / Huebwiesen housing development 12, 48, 159, 161–82, 188, 213, 218, 301
Nitsch, Hermann 25
Nouvel, Jean 306
Nüesch, Ernst 162

Oppenheim, Meret 314

Paris 37, 235, 317
Perrin, Carmen 319
Petschek, Peter 396 n. 134
Pfastatt 314
Plato 112
Poussin, Nicolas 340
Prigge, Walter 380
Proksch, Thomas 360
Pückler-Muskau, Hermann von 215

Raetz, Markus 36
Rapperswil 56, 182, 213, 246, 313, 345, 359, 389, 407, 408
Raussmüller, Urs 306, 308–313
Raymann, Toni 62, 63, 168, 260
Reichlin, Bruno 300
Reuß, Jürgen H. von 23, 29
Rheinberger, Hans-Jörg 61
Riehen, L. Garden 185
Rittel, Horst W. J. 33
Ritter, Joachim 41
Ritter, Markus 55, 63
Ritz, German 380
Roelly, Thom 22, 24, 46–49, 78–79, 83–84, 112, 115, 298, 407
Romero & Schaefle 120, 133
Rotzler, Stefan 161
Rowe, Colin 220, 227, 235
Rüegg, Arthur 219–20, 299, 300, 357, 380
Ryhiner, Hans-Peter 62
Ryman, Robert 306

Sassen, Saskia 84
St. Gallen 358, 411, 413
Sauer, Christel 306, 308–13
Schäfer, Robert 380
Schäfer, Wolfgang 57–58
Schaffhausen 306, 308–11
Schmidt, Eike 26–27
Schmithüsen, Josef 49
Schoenberg, Arnold 345
Schwarz, Marc 12, 182, 211, 386, 389
Schwarz, Urs 84, 182, 216
Schwarzkogler, Rudolf 25
Sennett, Richard 346, 403
Siebrecht, Ruedi 219, 299
Signer, Roman 36
Simmel, Georg 41
Simon, Axel 180
Shklovsky, Viktor 19, 22, 307, 317, 340
Slutzky, Robert 220, 227
Smithson, Robert 60, 313–14
Sørensen, Carl Theodor 181
Somazzi Häfliger Grunder 85
Stagi, Franca 346
Stahel, Urs 371
Steinmann, Martin 300, 316
Stella, Frank 312
Stern, Christian 180
Stifter, Adalbert 346, 403
Stöckli, Peter Paul 10, 13, 22, 56, 62, 161–62, 168, 170, 185–86, 188, 213, 218, 299, 345, 407, 409

Stücheli Architekten 133
Sukopp, Herbert 45
Szeemann, Harald 25

Tati, Jacques 345–46, 404
Tischbein, Johann Heinrich 79
Treib, Marc 10
Tschan, Richard 62
Tschumi, Bernard 235, 244
Tüxen, Reinhold 41, 49, 63, 367

Uitikon-Waldegg, E. Garden 320, 324, 326–27
Ulm, M. Garden 324

Vetsch, Walter 185, 358
Vöckler, Kai 380
Vogt, Christian 11–12, 297–98, 320, 324, 367, 371–76, 380, 383, 389
Vogt, Günther 85, 121, 133, 220, 235, 246, 254, 265, 297, 316–17, 319, 344, 357–59, 371, 372, 384, 407, 408
Volketswil 159, 180

Weilacher, Udo 10, 180, 215, 383, 384
Welsch, Wolfgang 112, 146, 306
Wettingen 22, 407, 408
Wettingen, Brühlwiese Municipal Park 10, 12, 182–218, 219–20, 227, 233, 334, 254, 386
Wien-Nussdorf 167, 215
Wilmanns, Otti 54
Wirz, Heinz 376
Worblaufen, Swisscom Administration Building 115, 357, 361, 366
Wormbs, Brigitte 320, 380
Wullschleger, Peter 10

Yountville, Napa Valley 310, 314

Zaugg, Rémy 314
Zinsli, Urs 265, 268–69
Zurich 12, 33–34, 56, 62, 77, 81, 83, 161, 162, 167, 180, 213, 215, 219, 220, 227, 244, 299, 306, 308, 313, 346, 359, 383, 384–89, 407, 408
——— Ernst Basler + Partner, Mühlebachstrasse Office Building 120–33, 138
——— Freibad Allenmoos 185, 390
——— K. Garden 393 n. 45
——— The Kienasts' Garden 159, 245–64, 342–44, 390
——— L. & L. Garden 115–20, 185
——— Swiss Re, Gotthardstrasse Office Building 20, 22, 133–46, 361, 413
——— Zoo, Masoala-Halle (Masoala Rain Forest Hall) 358, 359, 385
Zulauf, Albert 22, 161, 185, 186, 299, 407
Zulauf, Rainer 182
Zumthor, Peter 359

Acknowledgments

In 2016, *Dieter Kienast: Stadt und Landschaft lesbar machen* was published, the first comprehensive critical study of the Swiss landscape architect's work. Now the book is being released in a slightly revised English translation, including a new translation of Kienast's influential "Zehn Thesen zur Landschaftsarchitektur" (Ten Theses on Landscape Architecture).

My gratitude goes to the translator Linden Stevenson and the copyeditor Dawn Michelle d'Atri for their patience and perseverance, especially under the difficult circumstances of 2020, for their research, and critical comments. Thomas Skelton-Robinson gave the text a final polish through his meticulous proof corrections. I am grateful to all three of them for the intense and fruitful exchange.

Kienast's work could never have been assimilated and these two books completed without the generous and enduring support of numerous parties:

I am most grateful for Erika Kienast-Lüder and Christophe Girot making possible the research project that was indispensable for an analysis of the oeuvre. Erika Kienast-Lüder offered her holdings of Dieter Kienast's published papers as a permanent loan to the gta Archives (NSL Archive) of ETH Zurich for many years before donating them in 2013. The plans and illustrations in this volume are largely taken from that collection; the three monographs on Dieter Kienast's oeuvre compiled by her were an important basis for my own study of his work. The hours I was able to spend in the Kienasts' Garden provided valuable inspiration for this book. Christophe Girot, Professor of Landscape Architecture at ETH Zurich, initiated the research project, supervised it as a mentor, and also made the infrastructure and personnel resources of his chair available for the realization of the English version. Much of the work for the English edition was completed in the Bibliothek Dieter Kienast at ETH Zurich, which Erika Kienast-Lüder donated to the Institut für Landschaftsarchitektur in 2016.

Dieter Kienast's firm partners—Peter Paul Stöckli, Hans-Dietmar Koeppel, and Günther Vogt—all granted me unlimited access to the relevant materials in their archives and supplemented my research with personal conversations. The firms SKK Landschaftsarchitekten and Vogt Landschaftsarchitekten also provided logistical assistance, and SKK provided additional materials for illustrations.

The research project that was the foundation for the German edition was fortunate to have inestimable intellectual guidance from Arthur Rüegg and Philip Ursprung. Thanks to the generous support of the Lucius und Annemarie Burckhardt Stiftung, ETH Zurich and its Department of Architecture, Thomas and Marita Klinger-Lohr, the Schweizerische Kulturstiftung Pro Helvetia, and the Bund Schweizer Landschaftsarchitekten und Landschaftsarchitektinnen, the results of that research could be made available to a broad readership in a beautifully designed book with new photographs by Georg Aerni. For lively scholarly exchange and professional support during two fellowships along the way, I wish to thank Charles Waldheim, Iris Lauterbach, and Rosa Weis.

Among Dieter Kienast's numerous contemporaries and colleagues, Annemarie Burckhardt, Karl Heinrich Hülbusch, Thom Roelly, Peter Latz, Heidrun Hubenthal, and Peter Sparla provided detailed information about his years in Kassel; Jürg Altherr and Katharina König Urmi provided insight into his teaching methods at the Interkantonales Technikum Rapperswil and readily discussed with me the interplay of ecology and design. Wolfram Müller and Ulrike Rothe explained his teaching methods at the Institut für Landschaft und Garten of the Universität Karlsruhe (TH). I am grateful to Marc Schwarz for crucial information about Kienast's planned teaching program at ETH Zurich and the exhibition shown there in 1999: *Dieter Kienast: Lob der Sinnlichkeit* (In Praise of Sensuousness). In addition to the firm partners already mentioned, Fabienne Kienast Weber, Rita Illien, and Tony Raymann reported on the ordinary workday in various firms.

Conversations with Norbert Kühn, Markus Ritter, and Johannes Stoffler, and with Steven Handel and Frank Gallagher for the English edition, enriched my reflections on the role that spontaneous vegetation in the city played in the evolution of landscape architecture. I had detailed exchanges with Claudia Moll on the significance and influence of the *Naturgarten* (natural garden) movement. Annemarie Bucher, Paolo Bürgi, Georges Descombes, Marc Treib, Stefan Rotzler, Guido Hager, and Brigitt Sigel made essential contributions to a deeper understanding of how Swiss landscape architecture evolved in the 1980s and 1990s. Carmen Eschler and Kaspar Klaffke described their collaboration with Dieter Kienast from their perspective as clients, and the photographer Christian Vogt from the perspective of their artistic cooperation of many years. In all of these discussions, their personal contact with Dieter Kienast as a person became evident.

My research also received crucial inspiration from intense discussions of various aspects of Kienast´s work with Tarek Münch, Friedrich Teja Bach, Alfons Dworsky, Albert Kirchengast, Tibor Lamoth, Stefan Neuner, Gisela Steinlechner, and Nina Zschocke.

Neither my research project nor the publication would have been conceivable without the gta Archives (NSL Archive). A big thank-you goes to the former director of the NSL Archive, Marta Knieza, who supported my sorting and recording of the unpublished papers, as well as to its current director, Alex Winiger, and his co-worker Marco Cascianelli. Bruno Maurer, as director of the gta Archives, was always well disposed to my project and offered sound advice.

I thank the gta Verlag near the end only because of its position in the book-production process. As with the German edition previously, this publication project was supported with ideas and advice long before the home stretch. Ulrike Steiner, Ulla Bein, and Moritz Gleich once again granted it great attentiveness, patience, and circumspection. The intelligent and appealing visual realization of the content and structure of the book, which was the work of the crew of Büro 146, namely, Maike Hamacher, Madeleine Stahel, and Christa Lanz, was likewise adapted for the English edition. Once again, one could not wish for better interplay of graphic designers, publishers, and author. I owe them my most sincere gratitude.

Last but not least, I wish to thank my friends and family for their great encouragement and support over all of these years, especially Patrick Furrer and Flora, who in the crucial project phase provided the necessary latitude for a favorable conclusion of this project.

Project management
  Ulrike Steiner
Translations
  Linden Stevenson
Copyediting
  Dawn Michelle d'Atri
Proofreading
  Thomas Skelton-Robinson
Graphic concept and design
  Büro 146. Maike Hamacher,
  Valentin Hindermann,
  Madeleine Stahel,
  Zurich, with Christa Lanz
Prepress and printing
  Offsetdruckerei Karl Grammlich
  GmbH, Pliezhausen
Binding
  Spinner, Ottersweier
Cover and front and back endpaper illustrations
  Patterns representing plants and materials from the sample books of the firms of Stöckli, Kienast & Koeppel and Kienast Vogt Partner, Archiv Fabienne Kienast Weber

© 2021
gta Verlag, ETH Zurich
Institute for the History and Theory
of Architecture
Department of Architecture
8093 Zurich, Switzerland
www.verlag.gta.arch.ethz.ch
© Texts: by the author
© Illustrations: by the image authors
and their legal successors;
for copyrights, see the credits

Every reasonable attempt has been made by the authors, editors and publishers to identify owners of copyrights. Should any errors or omissions have occurred, please notify us.

The entire contents of this work, insofar as they do not affect the rights of third parties, are protected by copyright. All rights are reserved. No part of this publication may be reproduced, stored in a retrieval system, or transmitted, in any form or by any means, electronic, mechanical, photocopying, recording, or otherwise, without the written permission of the publisher.

Bibliographic information published by the Deutsche Nationalbibliothek
  The Deutsche Nationalbibliothek lists this publication in the Deutsche Nationalbibliografie; detailed bibliographic data are available on the Internet under http://dnb.dnb.de.

ISBN 978-3-85676-387-9

gta Verlag
ETH zürich